Optical Bistability 2

Optical Bistability 2

Edited by

Charles M. Bowden

U.S. Army Missile Laboratory
U.S. Army Missile Command
Redstone Arsenal, Alabama

Hyatt M. Gibbs

University of Arizona
Tucson, Arizona

and

Samuel L. McCall

AT & T Bell Laboratories
Murray Hill, New Jersey

Plenum Press • New York and London

Library of Congress Cataloging in Publication Data

International Topical Meeting on Optical Bistability (1983: University of Rochester)
  Optical bistability 2.

  "Proceedings of an International Topical Meeting on Optical Bistability, held June 15–17, 1983, at the University of Rochester, Rochester, New York"—Verso t.p.
  Includes bibliographical references and index.
  1. Optical bistability—Congresses. I. Bowden, Charles M. II. Gibbs, Hyatt M. III. McCall, Samuel L. IV. Title.
  QC446.15.I65   1983   535'.2   83-24438
  ISBN-13: 978-1-4684-4720-0   e-ISBN-13: 978-1-4684-4718-7
  DOI: 10.1007/978-1-4684-4718-7

The views, opinions, and/or findings contained in this book are those of the author(s) and should not be construed as an official Department of the Army position or decision. unless so designated by other documentation.

Proceedings of an international topical meeting on Optical Bistability, held June 15–17, 1983, at the University of Rochester, Rochester, New York

©1984 Plenum Press, New York
Softcover reprint of the hardcover 1st edition 1984
A Division of Plenum Publishing Corporation
233 Spring Street, New York, N.Y. 10013
All rights reserved

PREFACE

    This volume is a collection of experimental and theoretical
papers presented at the international "Topical Meeting on Optical
Bistability", held at the University of Rochester, June 15-17,
1983, sponsored jointly by the Air Force Office of Scientific Re-
search; the Army Research Office; and the Optical Society of
America. The Conference, which had 150 attendees, overlapped (on
June 15) with the Fifth Rochester Conference on Coherence and
Quantum Optics with two joint sessions. Some of the topics cover-
ed in this volume are also treated in the Proceedings of that
Conference.

    Since the last international conference on Optical Bistability,
held in Asheville, North Carolina, June 3-5, 1980, there have been
new and important fundamental advances in the field. This is borne
out in papers in this volume dealing with optical chaos and period
doubling bifurcations leading to chaos as well as the report of
results of an experiment using a very simple system exhibiting absorptive optical bistability in a ring cavity using optically pumped sodium atoms, which was successfully analyzed quantitatively
by a simple theory. Other advances discussed here include the observation of optical bistability due to the effect of radiation
pressure on one mirror of a Fabry-Perot cavity, and the prediction
of mirrorless intrinsic opitcal bistability due to the local field
correction incorporated into the Maxwell-Bloch formulation.

    Advances in optical bistability in semiconductors relate closer
to actual device applications. Evidence was reported favoring
optical bistability using the biexciton resonance in $CuC\ell$, where
picosecond switching times are expected. Other work reported milliwatt optical bistability in CdS using bound excitons and room temperature bistability in both InSb as well as GaAs multiple quantum
well structures. Investigations of strong optical nonlinearities
useful for optical bistability were reported for InAs, CdHgTe and
GaAs quantum well structures.

    The Conference concluded with an open forum discussion on
material presented during the Conference and future directions of

the field. These discussions were recorded and edited and are included at the end of the volume. It is hoped that these discussions
will convey some of the flavor and fervor of the meeting, but we
feel that remarks and insights are recorded which should aid anyone
attempting to use this volume to familiarize themselves with the
current state-of-the-art and possible future directions. From the
forum discussions, the reader may speculate upon what strident advances might be eviden in a future international meeting on optical
bistability. One aspect appears clear; the field will continue
to develop even more toward the device- and applications-oriented
investigations. Perhaps in a future Conference, researchers will
be able to report that the elusive practical optical transistor
has been achieved. Will the fields of computer architecture,
information-processing and information transmission then be pre-
pared for the anticipated products of this extremely active and
productive field of Optical Bistability?

C. M. BOWDEN
H. M. GIBBS
S. L. McCALL

12 September 1983

CONTENTS

CHAOS

FLUCTUATIONS (cont'd)

SEMICONDUCTORS

SEMICONDUCTORS (cont'd)

LASERS, LIQUID CRYSTALS AND DYES

TRANSVERSE EFFECTS

THEORIES

THEORIES (cont'd)

PANEL DISCUSSION

OPTICAL BISTABILITY WITH TWO-LEVEL ATOMS

H. J. Kimble and A. T. Rosenberger

Department of Physics
University of Texas
Austin, Texas 78712

P. D. Drummond

Department of Physics and Astronomy
University of Rochester
Rochester, NY 14627

INTRODUCTION

Since the first observation of optical bistability in 1976 (1),
a large number of investigations of this phenomenon have appeared in
the literature (2). Of particular theoretical interest has been the
bistable system composed of "two-level" atoms within an optical
resonator. While the theory in this case is well-advanced, only
recently have experimental results been obtained (3-6). The work
that we report is for absorptive bistability with an intracavity
medium that approximates a collection of two-level atoms. We
discuss our measurements of both the evolution of steady-state
hysteresis and the slowing down of switching times using an
inhomogeneously broadened medium in a standing-wave cavity. We also
present results of the first observation of absorptive bistability
in a predominantly homogeneously broadened medium of two-level atoms
in a ring cavity.

For our intracavity medium we have followed the procedure used
in studies of resonance fluorescence to produce a two-level atom by
optical pumping of sodium (7). A schematic of our apparatus is
shown in Figure 1. The essential elements of the apparatus are the
multiple atomic beams of sodium and the high-finesse interferometer
through which they pass. The atomic beams are collimated by several
sets of .5 mm x .5 mm apertures. These beams pass through a region

1

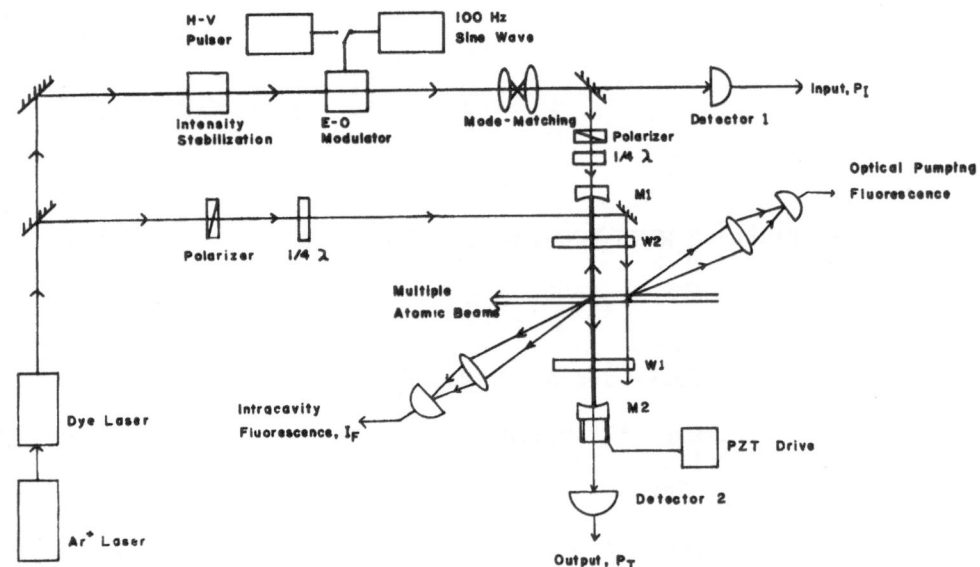

Figure 1.   Experimental configuration. Multiple beams of atomic
             sodium pass through a high finesse optical resonator
             after being optically prepumped.

of uniform magnetic field of 0.5-1.0 Gauss parallel to the interferometer axis. Optical prepumping of the atoms before they enter the mode volume of the interferometer limits absorption and emission therein to transitions between the $3^2S_{1/2}$, F=2, m_F=2 and $3^2P_{3/2}$, F=3, m_F=3 states. Thus two-level atoms are effectively produced. Verification of this scheme is provided by the enhancement in absorption cross-section measured in the transmission of a weak probe.

The optical cavity is operated near the confocal spacing and is excited by a mode-matched beam. However, it is difficult to estimate the efficiency with which the fundamental TEM_{00} mode is excited, since at the confocal spacing some higher order transverse modes are degenerate in frequency with each longitudinal mode of the cavity. The important properties of the cavity are its length L, mode waist w_0, finesse F (2π divided by round-trip loss), transmission T_0 (ratio of cavity output to input on resonance), enhancement ε (ratio of intracavity to input intensities), and the transmittance T_2 of the exit mirror. Experimental details will be given in the following sections as they pertain to each experiment.

BISTABILITY IN A STANDING WAVE CAVITY

The details of the experimental arrangement used in this case have been described previously (5,8). Five primary atomic beams and two weaker secondary sets of four beams each, offset by a small angle to either side of the primary beams, result in an inhomogeneously broadened medium of 30 MHz linewidth, three times the natural width of 10 MHz. The measured resonant absorption $\alpha_m \ell$ ranges from 0 to 1.5. The cavity mirrors are external to the vacuum system, and the cavity properties are as follows: length L=25 cm, waist w_o=150 μm, finesse F=210, transmission T_o=0.018, and intracavity enhancement ε=18.

In order to observe hysteresis in the transmission characteristics of the cavity, the intensity incident upon the cavity is modulated on a time scale long compared with either the cavity decay time or the atomic lifetime. The evolution of the switching points of the hysteresis cycle is investigated as the absorption $\alpha_m \ell$, or equivalently the effective cooperativity C_e, is varied. The experimental results are shown in Figure 2.

The curve shown in Figure 2 results from the single transverse-mode theory of Drummond (9), which takes into account inhomogeneous broadening and standing waves. The theory has as well been extended to include atomic motion longitudinally through the spatially periodic intracavity intensity. This results in a state equation of the form

$$Y = X[1 + 2C \int P(\Delta) \chi(\Delta,X) \, d\Delta]^2, \qquad [1]$$

where $\chi(\Delta,X)$ is a radial average of the nonlinear susceptibility (10) and $P(\Delta)$ is the inhomogeneous distribution function. The cooperativity C and the effective cooperativity C_e are given by

$$C = \frac{\alpha_o \ell}{2\pi} F, \qquad C_e = \frac{\alpha_m \ell}{2\pi} F = C \int \frac{P(\Delta)d\Delta}{1 + \Delta^2}, \qquad [2]$$

where $\alpha_o \ell$ is the line-center homogeneous absorption and $\alpha_m \ell$ is the measured absorption. The dimensionless intensities Y and X are given by (9):

$$Y = f_1 \frac{P_i}{\pi w_o^2 I_s} \varepsilon, \qquad X = f_2 \frac{P_t}{\pi w_o^2 I_s T_2}, \qquad [3]$$

where f_1=3/2, f_2=3, P_i and P_t are the incident and transmitted powers, and I_s is the saturation intensity. As there is a ±30% experimental uncertainty in Y, the theory has been scaled up by a factor of 1.26 resulting in the relatively good fit evident in Figure 2. Since all of the quantities in Equations [2] and [3] have

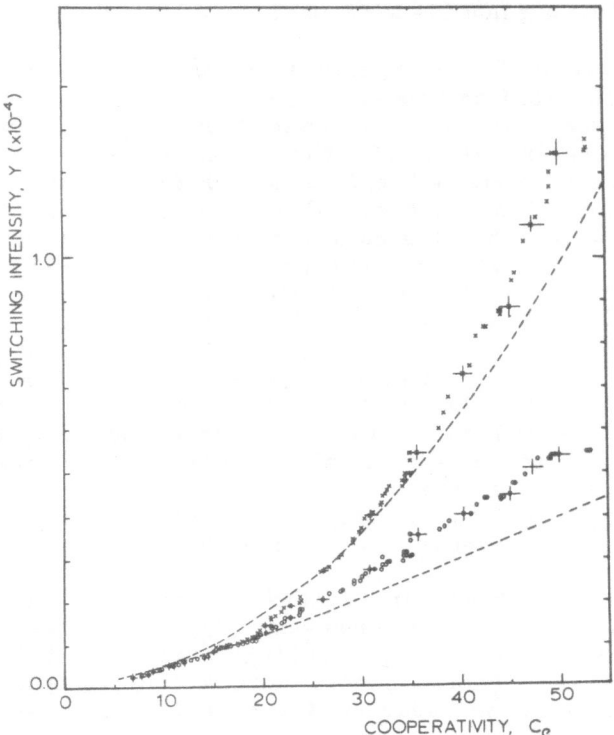

Figure 2.   Experimental switching intensities plotted versus effective cooperativity. The curves represent the theory of
Eq. [1] scaled up by a factor of 1.26 to give a best fit
within the overall experimental uncertainty in Y.

been measured, Figure 2 represents an absolute comparison between
theory and experiment; the scaling done is within the experimental
uncertainty.

TRANSIENT RESPONSE

The experimental arrangement used for the study of transient
response is essentially the same as that described above (6).   In
this case, however, the input intensity is a step function of power
P_0, greater than the power P_2 needed to cause switching to the upper
branch.   The temporal behavior of the approach of the bistable
system to steady state is studied as the switching increment
$\Delta P = (P_0 - P_2)/P_2$ is varied.   For $\Delta P \gg 1$ the behavior is essentially
identical to that of the empty cavity, but as ΔP approaches zero, a
delay in excess of the empty cavity filling time is observed.   This
delay increases as ΔP decreases, and becomes as large as twelve

times the empty cavity filling time for small ΔP. The data are in qualitative agreement with calculations based on the plane-wave mean-field model (11) and with another recent experiment (12). A quantitative comparison with theory has not yet been possible because of experimental sensitivity to fluctuations in laser power and in cavity length.

BISTABILITY IN A RING CAVITY

For this experiment (13), the apparatus is modified in two important ways. First, the number of primary atomic beams is increased to ten, and the collimation is improved such that the secondary beams are eliminated. The absorption linewidth is thus reduced to 13 MHz. The broadening beyond the natural linewidth is predominantly homogeneous and is due to the finite transit time of the atoms through the cavity mode. The resonant absorption $\alpha_m \ell$ can now be varied from 0 to 3.5. The second modification involves a new interferometer, completely enclosed within the vacuum system and operated as a ring cavity. The cavity configuration is shown in Figure 3; by positioning the cavity mirrors at the confocal spacing, a ring cavity is produced using only two mirrors. The empty ring cavity properties are: length L=5 cm, waist w_o=69μm, finesse F=188, transmission T_o=3.8x10^{-3}, transmittance T_2=4.8x10^{-4}, and enhancement ε=7.2.

The state equation that we consider is due to Drummond (9):

$$Y = X[1 + \frac{C}{X} \int P(\Delta) \ln (1 + \frac{2X}{1 + \Delta^2})d\Delta]^2, \qquad [4]$$

where C and C_e are defined as in Equation [2], but with denominators

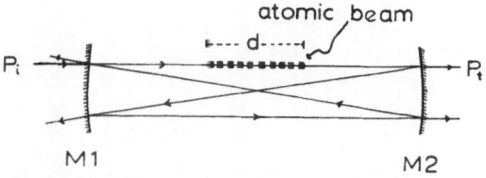

Figure 3.   Schematic illustrating the ring cavity with ten intracavity atomic beams directed out of the plane of the
figure. The distance d=14 mm.

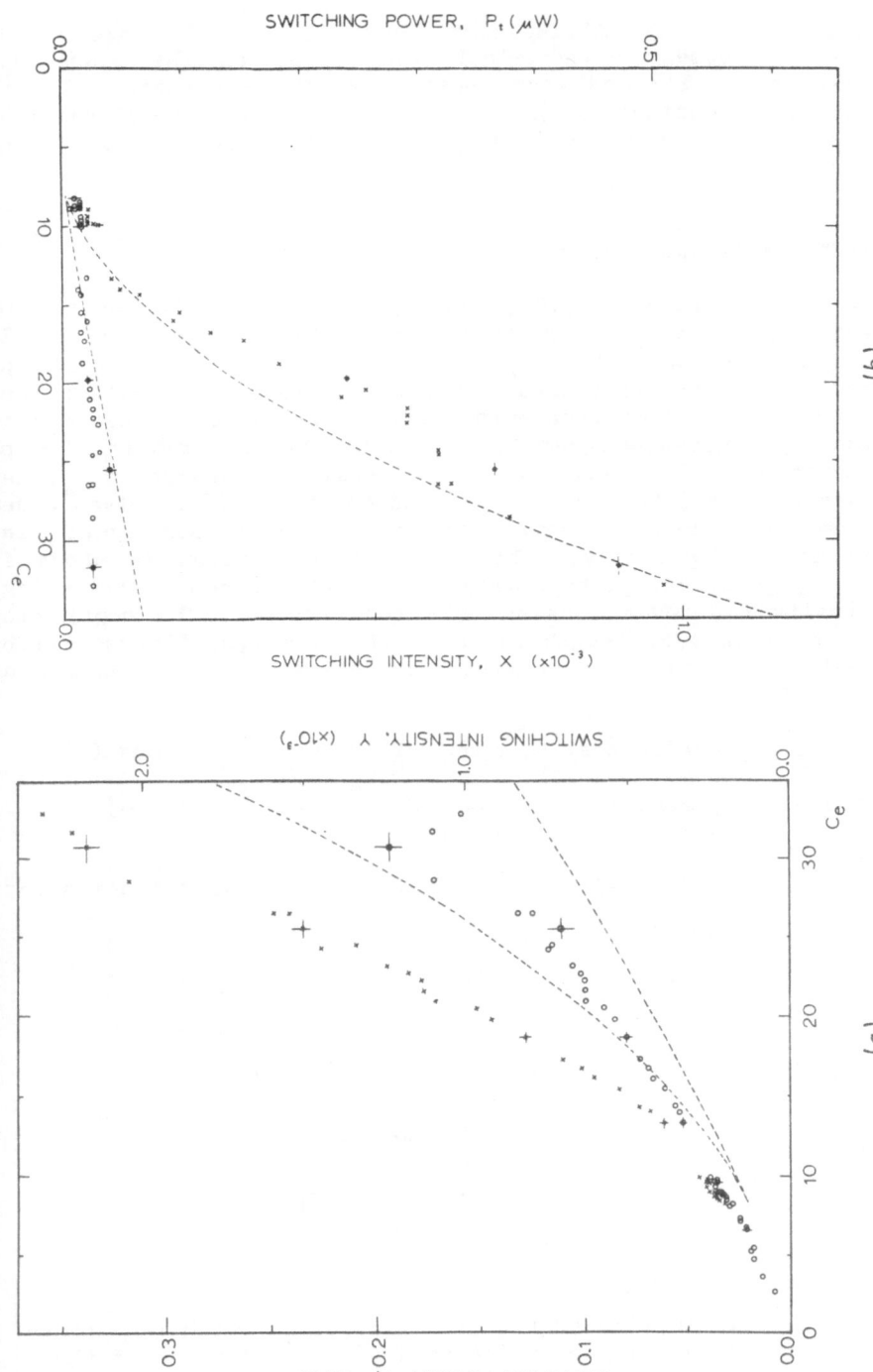

Figure 4. (a) Incident switching powers P_i and intensities y versus effective cooperativity C_e.
(b) Transmitted switching powers P_t and intensities X versus C_e. The curves represent the theory of Eq. [4]. No scaling has been done.

of 4π rather than 2π. $P(\Delta)$ is the narrow inhomogeneous distribution resulting from the improved but imperfect collimation, and Y and X are given by Equation [3] with $f_1=f_2=1$, taking I_s to be the saturation intensity under the conditions of transit-time broadening.

Figure 4 shows switching powers P_1 and P_t and switching intensities Y and X and displays the evolution of bistability as C_e is varied. The X are points on the upper branch, one being the point to which the transmission switches up, the other being the point from which the transmission switches down. In Figure 4 the measured switching powers are given on the left vertical axes and the corresponding intensities Y or X derived from Equation [3] are given on the right vertical axes. The curves plotted represent the theory as described above. Again, the comparison is absolute in the sense that all experimental quantities have been measured and there are no free parameters. There is a ±15% uncertainty in the determination of C_e, ±15% in Y, and ±30% in X. In Figure 5 the ratio of switching powers Y_2/Y_1 is plotted vs. C_e. Note that the critical onset of bistability is much more accurately descried by the Gaussian-beam theory than by the plane wave theory.

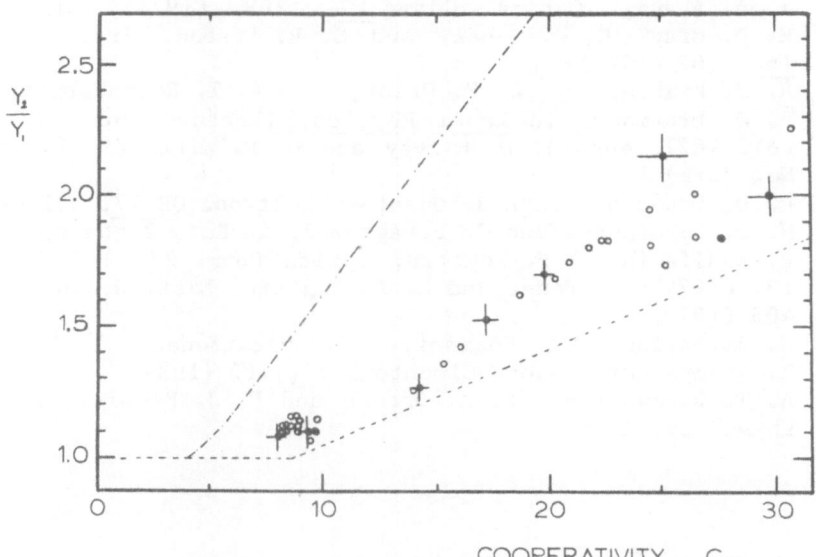

Figure 5. Ratio of switching intensities Y_2/Y_1 versus effective cooperativity. The curve to the left is the prediction of
a plane-wave theory; the curve to the right is the theory
of Eq. [4].

In summary, the agreement of our steady-state measurements with theory is reasonably good. The observed discrepancies, which could be due to breakdown of the two-level approximation at high intracavity intensities, excitation of higher-order transverse modes of the confocal cavity, or departures from mean-field behavior at large $\alpha_m \ell$, are being investigated. Work is continuing on these experiments and is being extended to investigate dispersive bistability and the anomalous behavior that we have observed on the upper branch under conditions of large C_e and nonzero detunings.

Support has been provided by the National Science Foundation under grant number PHY-8211194, and by the Joint Services Electronics Program.

REFERENCES

1. H. M. Gibbs,   S. L. McCall,   and   T. N. C. Venkatesan, Phys. Rev. Lett. 36, 1135 (1976).
2. Optical Bistability, eds. C. M. Bowden,   M. Ciftan,   and H. R. Robl (Plenum Press, New York), 1981.
3. K. G. Weyer, H. W. Wiedenmann, M. Rateike, W. R. MacGillivray, P. Meystre, and H. Walther, Optics Comm. 37, 426 (1981).
4. W. J. Sandle and A. Gallagher, Phys. Rev. A 24, 2017 (1981).
5. D. E. Grant and H. J. Kimble, Optics Lett. 7, 353 (1982).
6. D. E. Grant and H. J. Kimble, Optics Comm. 44, 415 (1983).
7. J. A. Abate,   Optics   Comm. 10,   269   (1974);   M. L. Citron, H. R. Gray, C. W. Gabel, and C. R. Stroud, Jr., Phys. Rev. A 16, 1507 (1977).
8. H. J. Kimble,   D. E. Grant,   A. T. Rosenberger,   and P. D. Drummond, in Laser Physics, (Lecture Notes in Physics, vol. 182), eds. J. D. Harvey and D. F. Walls (Springer-Verlag, New York) 1983.
9. P. D. Drummond, IEEE J. Quantum Electron. QE-17, 301 (1981).
10. H. J. Carmichael and G. P. Agrawal, in Ref. 2 above, p. 237.
11. R. Bonifacio nd P. Meystre, Optics Comm. 27, 147 (1978); 29, 131 (1979); V. Benza and L. A. Lugiato, Lett. Nuovo Cimento 26, 405 (1979).
12. S. Barbarino,   A. Gozzini,   F. Maccarrone,   I. Longo,   and R. Stampacchia, Nuovo Cimento B 71, 183 (1982).
13. A. T. Rosenberger, L. A. Orozco and H. J. Kimble, submitted to Phys. Rev. A.

OBSERVATION OF IKEDA INSTABILITIES AND OPTICAL BISTABILITY

IN AN ALL-OPTICAL RESONATOR CONTAINING NH$_3$ GAS

R.G. Harrison, C.A. Emshary, I.A. Al-Saidi and
W.J. Firth
Department of Physics
Heriot-Watt University
Edinburgh, U.K.

INTRODUCTION

Substantial effort over the last few years on optically bistable systems has more recently been extended to considerations of period-doubling cascades to chaotic behaviour[1,2] in such systems. Passive all-optical systems are particularly interesting here, as basically simple arrangements capable of exhibiting oscillation[3,4] and turbulence, but also because they can be fully quantised. Ikeda[5] showed in 1979 that an optically-bistable ring resonator containing a two-level system can show a period-doubling cascade, a sufficiently strong c.w. input beam yielding an output oscillating at <u>twice</u> the resonator round trip time t_R. On further increasing the input field the output period doubles to chaos. Since then, observations of these phenomena have been made in various optical systems, such as a hybrid bistable device[6] and lasers[7,8], but the nearest approach to Ikeda's system has been a recent demonstration[9] in fibre-optic resonator, using mode-locked excitation to avoid stimulated scattering. None of these systems are particularly simple, nor do they lend themselves to quantisation.

We believe that molecular gases, excited close to resonance by a CO$_2$ laser have unique advantages in this field, and here we report observations of $2t_R$ oscillation (with some indications of $4t_R$) in an all-optical system very similar to Ikeda's original proposal. A passive ring resonator was pumped by a TEA CO$_2$ laser pulse (10R(14) transition, λ = 10.3 μm). This line lies 1.23 GHz above the aR(11)[10] transition of the NH$_3$ gas contained in a 1 m intra-cavity cell at pressures of 9-15 torr, where it acts as a <u>homogeneously-broadened</u> two-level system. Our results, which also include observation of optical bistability are in excellent accord with the theory of Ikeda which is generalised here to include reservoir kinetics; associated with population transfer to and from the manifolds of states not in resonance with the radiation field.

EXPERIMENTAL

The arrangement is illustrated in Fig. 1. The CO_2 hybrid TEA laser/amplifier system yields smooth single transverse and longitudinal mode pulses of FWHM $\sim$ 100 ns (Fig. 2a) and peak power $\sim$ 1 MW. The laser pulses are coupled, using a single-surface Ge flat, R = 36%, into a 3.5 m three-element ring cavity, closed by 100% gold mirrors, containing the gas cell. The input and cavity signals were sampled by KBr beam splitters, and monitored by photon-drag detectors and a Tektronix 7104 oscilloscope: total response time $\lesssim$ 1 ns. For NH_3 pressures $\sim$9-15 torr, significant self-focusing was observed in single-pass experiments, confirming a nonlinear refractive index contribution substantial enough for dispersive optical bistability and Ikeda instability. Closing the ring caused a huge distortion of the pulse shapes (sampled after the NH_3 cell). In particular, a considerable proportion of these showed modulation at the 23.4 ns period expected for Ikeda oscillation in our system.

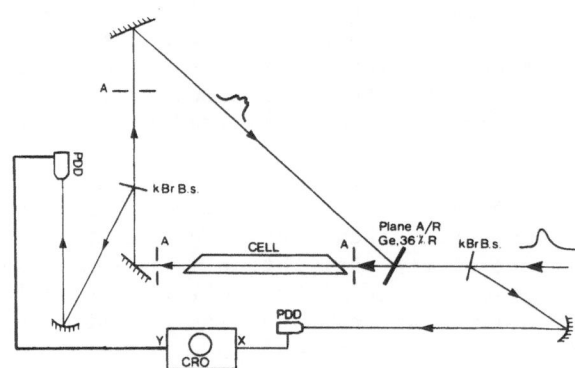

Fig. 1   Schematic of ring cavity system

Fig. 2 shows representative examples of these modulated pulses.
To confirm the period, we have digitised and Fourier transformed the
traces; the resulting spectra show pronounced peaks at ∿ 45 MHz,
confirming our observation of Ikeda instability: we even have, in
two of the cases, a subsidiary peak at $(4t_R)^{-1}$, possibly indicating
a further bifurcation.

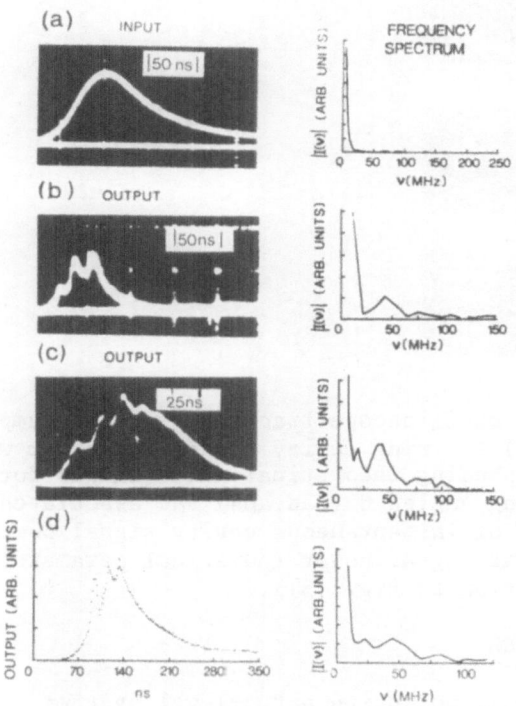

Fig. 2   Sample Oscilloscope traces of (a) the pump signal
          and (b)-(d) the ring cavity signal, together with
          their frequency spectra; obtained from digitized
          traces as shown in (d).

In contrast, the input pulse has an essentially featureless
spectrum (Fig. 2a).

    The oscilloscope traces in Fig. 3 are other representative
examples of the cavity signal showing strong pulse distortion

indicative of optical bistability. We note that because our cavity was free standing, the detuning angle θ inevitably drifted from shot to shot, enabling us to sample the full range of possible pulse shapes.

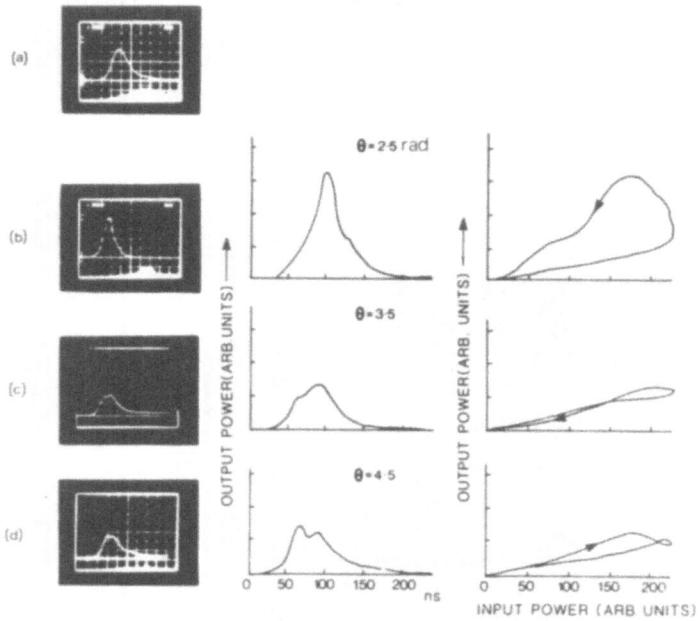

Fig. 3.   Sample oscilloscope traces of (a) the pump signal
           (b)-(d) the ring cavity signal, together with
           corresponding theoretical pulse shapes for various
           detuning angles θ (radians) and associated hysteresis
           curves of instantaneous cavity signal power against
           incident signal power (numerical parameters are given
           in caption to fig. (5)).

THEORETICAL MODEL

Our system is not quite a two-level system: allowance must be made for population transfer within the rotational manifolds, as described in detail elsewhere[11]: here we give a simplified version which demonstrates how the full system may be handled in the context of a passive ring resonator, thereby obtaining a considerable generalisation of Ikeda's model[5].

The level scheme is diagrammed in Fig. 4 and leads to the set of rate equations

$$\dot{n}_1 = -W(n_1 - \frac{g_1}{g_2} n_2) + KN_1 - kn_1 \qquad (1)$$

$$\dot{N}_1 = -KN_1 + kn_1 + k_v N_2 \tag{2}$$

with similar equations for n_2 and N_2. For simplicity, we assume equal level-manifold rate constants in the two levels; g_1/g_2 is the degeneracy factor (here 5/3) and k_v the V-T rate, which is negligibly small for our conditions.

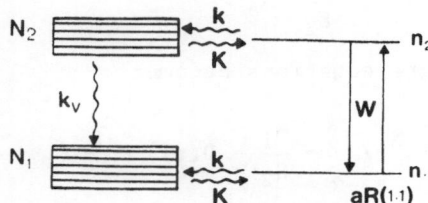

Fig. 4   Kinetic model for NH_3 absorption. Radiative coupling is represented by continuous lines and collisional relaxation (both rotational and vibrational) by wavy lines.

These equations conserve total population, and $(n_1 + n_2) = n_e$ and $(N_1 + N_2) = N_e$ separately ("e" relating to thermal equilibrium); generalisation is straightforward. Detailed balance requires that $k\, n_e = KN_e$; in the infinite-reservoir limit K thus goes to zero and the system has effectively just two levels. In NH_3 however, $n_e/N_e \sim 2\%$ and $K \sim 2$ μs^{-1} torr^{-1} [12] so $kt_R < 1$, as required for Ikeda instability, but $Kt_R > 1$. The major effect of finite K is a leaching of population on a time scale K^{-1}, which eventually kills the instability: note that absence of oscillation on the falling edge is a feature of our data (Fig. 2).

To convert the rate equations from local equations to take account of propagation and feedback effects, we perform a retarded-time integration[5], to obtain

$$\varepsilon(t) = \varepsilon_i + R\varepsilon(t-t_R)e^{-i\theta} e^{-(\alpha L/2)(1-i\Delta)D(t-t_R)} \tag{3}$$

where $\varepsilon(t)$ is the intra-cavity field and ε_i the transmitted incident field, just before the gas cell, both normalised to the saturation field[13], R and θ are the empty-cavity amplitude loss and phase-shift per round trip, α the small signal absorption coefficient, Δ the molecular detuning, $= (\omega_{CO_2} - \omega_{NH_3})/\gamma_1$ and (assuming $n_{2e} = N_{2e} = 0$)

$$D(t) = \frac{1}{n_e L} \int_0^L dz [n_1(z,t') - \frac{g_1}{g_2} n_2(z,t')]_{t'=t-(L-z)/c} \qquad (4)$$

$$\equiv < n_1 - \frac{g_1}{g_2} n_2 > / n_e$$

The population rate equations become:

$$\dot{D} = k[(1 + \frac{g_1}{g_2}) N_1/N_e - \frac{g_1}{g_2} - D(t)] - \frac{k}{2} (1 + \frac{g_1}{g_2}) |\varepsilon(t)|^2 \frac{(1 - e^{-\alpha L D(t)})}{\alpha L}$$

$$(5)$$

Eqns. (1,2) are essentially unchanged, but the populations are now assumed averaged as in (4).

We have numerically integrated these equations, using the pump pulse of Fig. 2a as input, and Fig. 5 shows the predicted intra-cavity pulses as a function of cavity tuning for representative parameter values. Oscillation at $2t_R$ is manifest in three of these traces, while strong pulse distortion indicative of optical bistability occurs at the opposite tuning, as expected[13,14]; further examples being shown in fig. 3 together with oscilloscope traces of the cavity signal for comparison. In view of the fact that our present model neglects self-focusing this range of behaviours matches extremely well, with the pulse shapes we observe.

For comparison of our model with a pure 2-level system we show two traces fig. (g,h) with K set equal to zero. The Ikeda oscillation (compare (h) with (f)) is essentially unaffected, while the bistable pulse shape (compare (g) with (c)) is considerably changed - this is not unexpected, since the latter is a first-order transition, whereas Ikeda instability is second order.

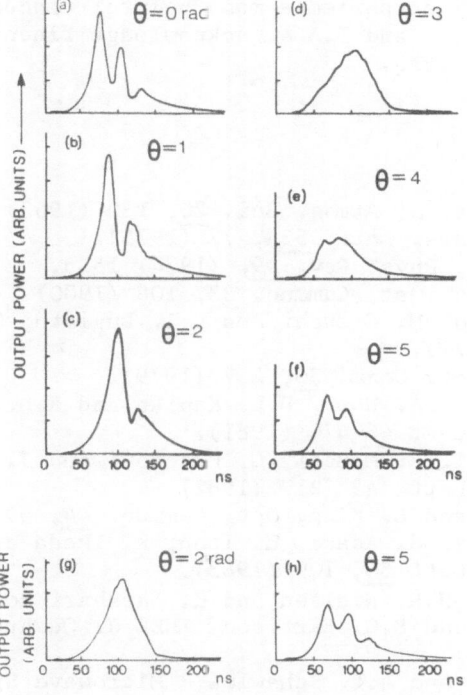

Fig. 5 Numerically determined ring cavity signals for various cavity tuning using fig. 2(a) as the pump signal. ($\alpha L = 3$, $kt_R = 5$, $Kt_R = 0.1$, $|\varepsilon|^2_{max} = 0.9$).

CONCLUSION

These results establish molecular gases pumped by CO_2 lasers as extremely promising media for the demonstration and investigation of optical bistability and chaos in all-optical systems.   As well as a vast range of laser-molecule coincidences, there is an additional flexibility in that pressure, of both absorber and buffer, can be used to control response times.   This flexibility should lead to operation with pulses long compared to t_R - even cw operation - which would enable experimental verification of the predictions of Moloney et al[15] regarding the routes to chaos in systems of this type.

We gratefully acknowledge many helpful discussions with E. Abraham.   C.A.E. and I.A.A. acknowledge financial support from the government of Iraq.

REFERENCES

1.   E.N. Lorentz, J. Atmos. Sci. $\underline{20}$, 130 (1963).
2.   H. Haken, Phys. Lett. $\underline{53A}$, 77 (1975).
3.   S.L. McCall, Phys. Rev. $\underline{A9}$, (1974) 1515.
4.   L.A. Lugiato, Opt. Commun. $\underline{33}$, 108 (1980)
      R. Bonifacio, M. Gronchi and L.A. Lugiato, Opt. Commun. $\underline{30}$, 129 (1979).
5.   K. Ikeda, Opt. Comm. $\underline{30}$, 257 (1979).
6.   H.M. Gibbs, F.A. Hopf, D.L. Kaplan and R.L. Shoemaker Phys. Rev. Lett $\underline{46}$ 474 (1981).
7.   F.T. Arecchi, R. Meucci, G. Puccioni and J. Tredicce Phys. Rev. Lett. $\underline{49}$ 1217 (1982).
8.   C.O. Weiss and H. King, Opt. Commun. $\underline{44}$, 59 (1982).
9.   H. Nakatsuka, S. Asaka, M. Itoh, K. Ikeda and M. Matsuoka, Phys. Rev. Lett $\underline{50}$, 109 (1983).
10.   J.S. Garing,H.H. Nielsen and K. Narahari Row, J. Mol.Spect.3, 496 (1
11.   P.K. Gupta and R.G. Harrison, IEEE J. Quantum Electron $\underline{QE-17}$, 2238 (1981).
12.   C.H. Townes and A.L. Schawlow, Microwave Spectro scopy, New York: McGraw-Hill, 1955, p74.
13.   H.M. Carmichael, R.R. Snapp and W.C. Schieve, Phys. Rev. $\underline{A26}$, 3408 (1982).
14.   W.J. Firth, Opt. Commun. $\underline{39}$, 343 (1981).
15.   J.V. Moloney, F.A. Hopf and H.M. Gibbs, Phys. Rev. $\underline{A25}$, 3442 (1982).

SWITCHING BEHAVIOR OF BISTABLE RESONATORS FILLED WITH TWO-LEVEL ATOMS

G. Cooperman, M. Dagenais, and H.G. Winful

GTE Laboratories Incorporated
40 Sylvan Road
Waltham, Massachusetts 02254

INTRODUCTION

A comprehensive study of the switching dynamics of bistable optical devices is reported. The intracavity medium is modelled as an ensemble of two-level atoms uniformly distributed in a plane parallel Fabry-Perot. By allowing for arbitrary atomic detunings and cavity mistunings, both the absorptive and dispersive contributions to optical bistability are included. Variations in polarization and population over wavelength distances are treated by means of expansions in spatial Fourier series, having as fundamental a half optical wavelength.[1] The Fourier series are truncated after the first harmonic. We mostly concentrate on the dynamics of the turn-off since it is believed to be the limiting factor in the operation of an optical bistable device at high repetition rates. The turn-on of an optical bistable device can be made arbitrarily fast by increasing the intensity of the incident field. We emphasize that switching an optical bistable device on or off from steady state involves going from a certain intracavity distribution of the polarization, inversion, and electric field, to three completely different distributions. All three distributions need to be changed and each of them generally evolves in time at a different rate. For instance, monitoring the time it takes for the output intensity to decay after a sudden turn-off of the input intensity does not necessarily provide any information on how long one has to wait before being able to excite the system again. Generally, this time will be much longer than the time it takes for the output field to decay and will be of the order of T_1, the energy relaxation time, in the bad cavity limit (low reflectivity). We believe that these considerations are of much more general applicability and are not only restricted to an

17

ensemble of two-level atoms in a Fabry-Perot. Experimentally measured turn-off, based only on monitoring the time dependence of the output intensity, might need revision. In our analysis, we have also considered the possibility of obtaining adiabatic following[2] of the polarization and inversion in order to accelerate the time response of an optical bistable device. A steady state analysis yields the very important result that even in the dispersive regime, when one is far detuned from the atomic resonance, for physically realistic parameters, it is only possible to observe optical bistability if the intracavity intensity is of the order of the saturation intensity or way above it. The implication is that even when operating in the dispersive mode, the excited population will not adiabatically follow the switching laser pulse, and thus switch-off cannot occur faster than T_1, the population relaxation time. In the good cavity limit (high reflectivity), this time can even be much longer than T_1. Thus one does not gain in speed by operating under non-resonant conditions. This puts stringent restrictions for the operation of such a device as a memory element.

SWITCHING DYNAMICS

The following dimensionless Maxwell-Bloch equations describe the time evolution of the forward (E_F) and backward (E_B) intracavity electric fields, the forward (P_o^+) and backward (P_o^-) polarizations, the average inversion density (n) and the population grating density (n_1):

$$\frac{\partial E_F}{\partial z} + \frac{\partial E_F}{\partial t} = \frac{\alpha L}{2} P_o^+ \quad ,$$

$$\frac{\partial E_B}{\partial z} - \frac{\partial E_B}{\partial t} = -\frac{\alpha L}{2} P_o^- \quad ,$$

$$\frac{\partial P_o^+}{\partial t} = \frac{1}{T_2} \left\{ -(1+i\Delta) P_o^+ + \frac{1}{2} (\bar{n}E_F + n_1 E_B) \right\} ,$$

$$\frac{\partial P_o^-}{\partial t} = \frac{1}{T_2} \left\{ -(1+i\Delta) P_o^- + \frac{1}{2} (\bar{n}E_B + n_1{}^*E_F) \right\} ,$$

$$\frac{\partial \bar{n}}{\partial t} = \frac{1}{T_1} \left\{ (n^o - \bar{n}) - [P_o^+{}^*E_F + P_o^-{}^* E_B + P_o^+ E_F^* + P_o^- E_B^*] \right\} ,$$

$$\frac{\partial n_1}{\partial t} = -\frac{1}{T_1} \left\{ n_1 + [P_o^-{}^*E_F + P_o^+ E_B^*] \right\} .$$

In obtaining these equations, we have made the following transformations:

$$E = \frac{\hbar}{2\mu\sqrt{T_1 T_2}} E' \quad , \quad P_o = -i\mu N\sqrt{\frac{T_2}{T_1}} P_o' \quad , \quad n=n'N \quad ,$$

$$t = \frac{\eta L}{c} t' \quad , \quad T_1 = \frac{\eta L}{c} T_1' \quad , \quad T_2 = \frac{\eta L}{c} T_2' \quad ,$$

$$z = Lz' \quad , \quad \omega = \frac{\omega' c}{\eta L T_2}' \quad , \quad \alpha = \frac{8\pi\omega\mu^2 T_2 N}{\hbar\eta c} \quad .$$

The primed variables are the new dimensionless variables which have been used in Equation 1. The total electric field E, the total polarization P and the inversion n are defined in the following way:

$$E = E_F e^{i(kz-\omega t)} + E_B e^{-i(kz+\omega t)} + c.c.,$$
$$P = P_o^+ e^{i(kz-\omega t)} + P_o^- e^{-i(kz+\omega t)} + c.c.$$
$$n=\bar{n} + (n_1 e^{2ikz} + c.c.) \quad .$$

T_1 and T_2 are the energy relaxation time and the dipole dephasing time, respectively. N is the density of atoms, $\Delta = \omega_o - \omega$ is the laser detuning from the resonance at ω_o, η is the background index of refraction and n^o is the initial inversion. The boundary conditions appropriate to a Fabry-Perot cavity of length L are:

$$E_T = \sqrt{T} \ E_F(L,t)e^{ikL} \quad ,$$
$$E_F(0,t) = \sqrt{T} \ E_I(0,t) + \sqrt{R} \ E_B(0,t) \quad ,$$
$$E_B(L,t) = \sqrt{R} \ E_F(L,t) \ e^{i2kL} \quad ,$$

where E_I and E_T are the incident and transmitted electric field, respectively and R and T are the reflectivity and transmissivity of the mirror.

As an initial check of the accuracy of our computer code, we have first reproduced the published on-resonance results.[3] We now present our results describing the turn-off of a bistable device from steady state. This time is believed to be the limiting factor in the fast operation of an optical bistable device. We are interested in answering the following question: How fast can we cycle an optical bistable device? In order to answer this question, we note that the turn-off of an optical bistable device from the upper branch to the lower branch of an optical bistability loop involves changing completely the distribution of the polarization, inversion, and intracavity electric field characteristic of the upper branch to the distributions characteristic of the lower branch. Each of these distributions evolves with a different

characteristic time and the turn-off time must be defined by the
longest time if, of course, the cavity lifetime is made sufficiently
short. Because of our goal of determining the ultimate speed of
operation of an optical bistable device, we are not interested in
the good cavity limit (high finesse) since the material properties
do not come into play and since the switching time becomes much
longer than the material lifetime. Figure 1 shows the steady state
hysteresis loop on which the time-dependent switching studies are
based. In Figure 2 we show the turn-off of the input intensity from
steady state when the incident intensity is suddenly turned-off.
The cavity lifetime was chosen to be much shorter than T_1. In
Figure 2a, the dipole dephasing is due to spontaneous emission
only ($T_2=2T_1$) and, in Figure 2b, T_2 was chosen to be of the order
of the cavity lifetime. As can be seen, the output intensity decays
in a time given by the cavity lifetime, a time much shorter than T_1.
When T_2 is made very short, the Rabi oscillations are averaged out.
One also notices a long tail of much lower intensity in the decay
of the output intensity. This tail decays in a time T_1. Figure 3
shows the time dependence of the spatially averaged inversion.
This inversion is proportional to the absorption that would be seen
by a weak laser probing the optical bistable device. This shows
that even though the output intensity can decay rapidly, the inversion of the medium takes a time of the order of T_1 to reach
steady state.   This implies in general that one cannot measure
the response time of an optical bistable device by measuring only
the time it takes for the output to decay, as has customarily been
done. This time has nothing to do with the response and cycling
time of an optical bistable device. We can give a physical argument on why the output intensity can decay much faster than T_1.

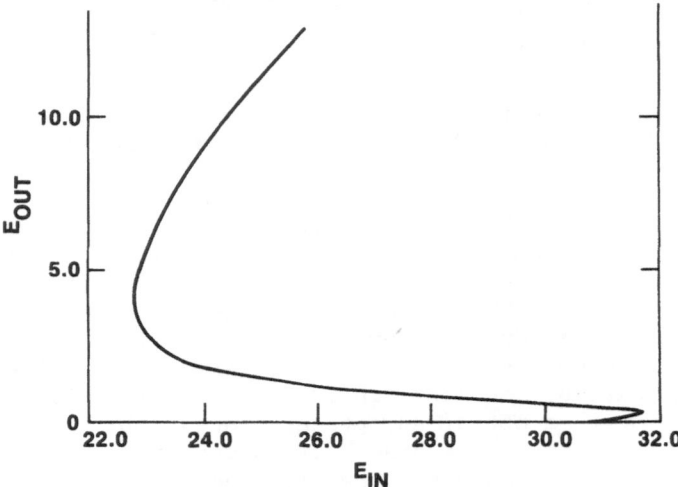

Fig. 1.   Steady state bistability curve used in switching studies.
          The parameters are R-0.5, $\alpha L=1000$, and zero atomic and
          cavity detuning.

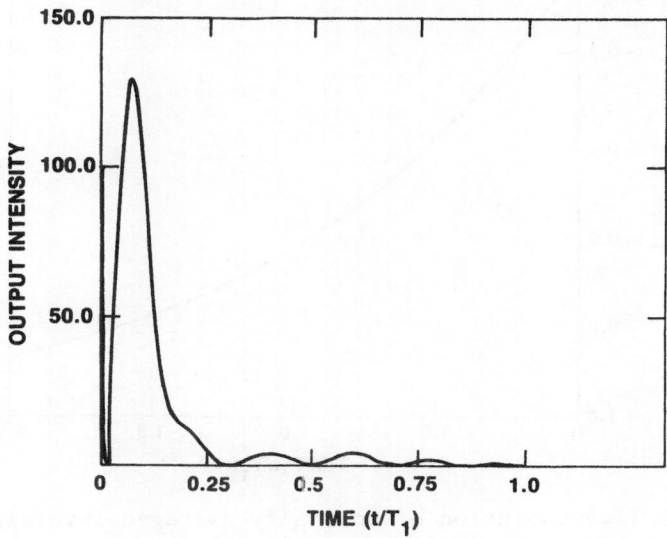

Fig. 2a.   Decay of output intensity for a sudden turn-off of input
intensity.   Here $T_2=2T_1$.

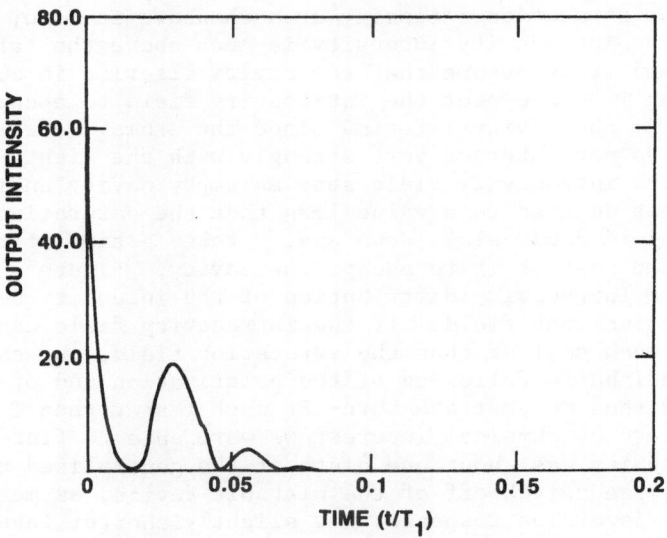

Fig. 2b.   Turn-off dynamics for $T_2 \ll T_1$.

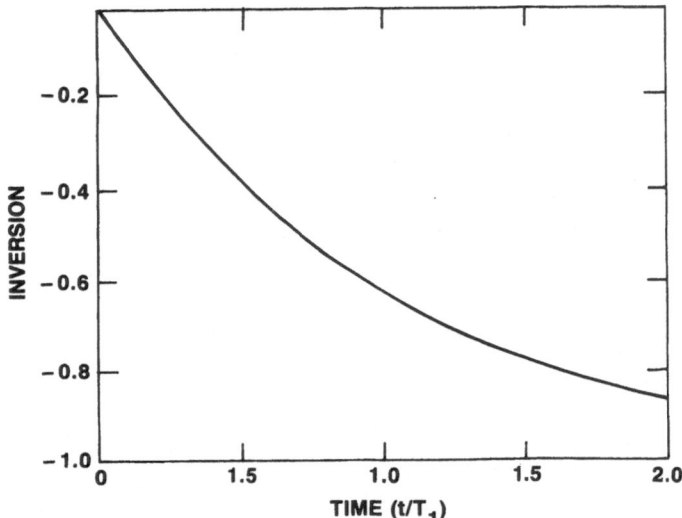

Fig. 3. Time evolution of spatially averaged inversion after
         sudden turn-off.

A steady state analysis has shown that, in most cases of physical
interest (for reasonable values of finesse and absorption coefficients), it is not possible to observe optical bistability
without at the same time, having an intracavity intensity of the
order of the saturation intensity or much above it. So, if
typically the intracavity intensity is much above the saturation
intensity, and if we assume that the cavity lifetime is much
shorter than T_1, we expect the intracavity field to escape in a
time given by the cavity lifetime since the atoms are all saturated and do not interact very strongly with the light. For all
purposes, the intracavity field sees an empty cavity until its
amplitude has decayed to a value less than the saturation field.
Then, the field decay slows down and it takes a time of the order
of T_1 for the rest of it to escape the cavity. Figure 4 shows the
steady state intracavity distribution of the intensity before turning off the incident field. If the intracavity field cannot be
made to be much smaller than the saturation field, one cannot hope
to obtain adiabatic following of the polarization and of the inversion and thus demonstrate turn-off much faster than T_1. In
the best case of physical interest, we were able to find the intracavity intensity was about one ninth of the generalized saturation
intensity. The switch-off of the bistable device, as measured by
the average inversion response, was slightly shorter than T_1. No
cases were found where the turn-off could be accomplished in a
time much shorter than T_1.

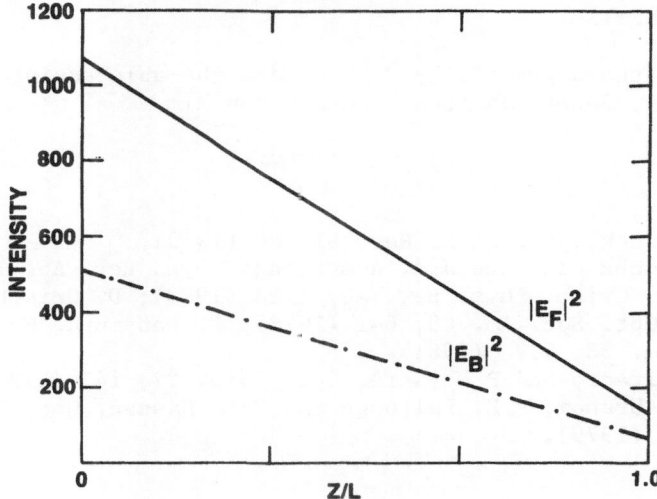

Fig. 4. Spatial distribution of forward and backward cavity fields.

CONCLUSION

In conclusion, we have shown that in cases where the intracavity field is much larger than the saturation field, the output
field can decay much faster than T_1 if the input intensity is
suddenly turned-off. In these cases, it is found that the speed
of operation of an optical bistable device is still limited by T_1,
the energy relaxation time. In general, inferring the response
time of an optical bistable device by only monitoring the output
field following a turn-off of the incident field, will not always
give the right answer. Only if the intracavity field is of the
order of the saturation field can the output field decay in a time
given by T_1. A more reliable technique for measuring the response
time of an optical bistable device involves monitoring the transmission of a weak beam following the turn-off of the incident field.
In this case, the response time of the transmission is, in fact, the
response time of the optical bistable device, irrespective of the
size of the intracavity field. If one uses reasonable finesses and
absorption coefficients, we have found no on or off-resonance cases,
where the intracavity field can be much smaller than the generalized
(including detuning) saturation field. No adiabatic following of
the population or polarization is observed. This is a reflection
of the fact that population cannot decay much faster than T_1, given
that the intracavity field is of the same order or much higher than
the saturation field. T_1 places a fundamental limit on signal
processing speed.

ACKNOWLEDGMENTS

    The authors gratefully acknowledge the able assistance of
Enson Chang, Janet Johnston, and Dae Eun Kim.

REFERENCES

1. J.A. Fleck, Jr., Phys. Rev. B1, 84 (1970).
2. D. Grischkowsky and J.A. Armstrong, Phys. Rev. A6, 1566 (1972);
        M.D. Crisp, Phys. Rev. A8, 2128 (1973); D. Grischkowsky,
        J. Opt. Soc. Am. 68, 641 (1978); E. Hanamura, Solid State
        Comm. 38, 939 (1981).
3. R. Bonifacio and P Meystre, Opt. Comm. 27, 147 (1978);
        E. Abraham, R.K. Bullough and S.S. Hassan, Opt. Comm. 29,
        109 (1979).

ZEEMAN COHERENCE EFFECTS IN ABSORPTIVE POLARIZATION BISTABILITY

Govind P. Agrawal

Bell Laboratories
Murray Hill, N.J.   07974

INTRODUCTION

The phenomenon of optical bistability (OB) is usually treated within the framework of a single optical mode interacting with a two-level system.[1]   Recently, attention has been paid to OB in three-level systems.[2-10]   Here, the physical process of transverse optical pumping,[11] related to coherent population trapping, plays an important role. Of particular interest is the case of $J = 1 \rightarrow J = 0$ transition of an atomic system wherein orthogonally circular-polarized components of a single incident beam interact with $m = \pm 1$ ground state Zeeman sublevels (Λ-configuration). Depending on polarization characteristics of the incident beam, OB is observed in the beam intensity, the beam polarization, or both. Two-photon-induced Zeeman coherence is expected to play a significant role in such systems. In most of previous work,[2-10] however, Zeeman coherence effects have been either ignored or incorporated only within the framework of a closed three-level system.

The purpose of the present paper is to develop a general model of OB in Λ-type three-level systems. In particular, no restrictions are imposed on various longitudinal and transverse relaxation rates so that collisional effects are readily included. We consider an open system allowing for the ground-state population decay due to collisions with foreign perturbers or due to finite atomic interaction time. Considerable simplification is achieved by assuming homogeneous broadening and exact one-photon resonance.   Dispersive effects due to magnetic-field-induced Zeeman splitting are however included. Several specific cases of OB are studied. Particular attention is paid to level-crossing OB

that occurs when the external magnetic field is varied at a fixed
incident beam intensity.

BISTABILITY STATE EQUATIONS

For a simplified discussion of OB, the governing state
equations are obtained by assuming that a weakly absorbing
nonlinear medium is placed inside a high-Q ring cavity. Following
a standard procedure,[12] we obtain

$$Y_n = X_n \left[(1 + Cf_n'')^2 + (\phi - Cf_n')^2 \right], \tag{1}$$

where n equals 1 and 2 for right (σ_+) and left (σ_-) circularly
polarized components, respectively. The incident and the
transmitted fields are normalized using

$$Y_n = E_n^{in}/(I_sT)^{1/2}, \qquad X_n = E_n^{tr}/(I_sT)^{1/2}, \tag{2}$$

where T is the mirror transmittivity and I_s is the saturation
intensity that is usually much lower than that of a two-level
system.[6] The bistability parameter $C = \alpha_0 L/2T$, α_0 is the
absorption coefficient, ϕ is the detuning parameter and
$f_n = f_n' + if_n'' = (\omega/\alpha_0 c) \chi_n$, ω being the frequency of the incident
light. The nonlinear susceptibility χ_n, obtained using a set of
nine density-matrix equations,[13,14] governs the mutual coupling
of the σ_+ and σ_- polarization components. Further details can be
found in Ref. 14. The final expression for f_n is

$$f_n = \frac{1}{D} \frac{\delta+i}{1+\delta^2} \left[1 + \frac{(p+i\delta r)I_{3-n}'}{(1+p+\bar{I}) + i\delta(1+r+\bar{I})} \right], \tag{3}$$

where the saturation denominator D is given by

$$D(I_1,I_2) = 1 + (1 + q)(I_1' + I_2') + (1 + 2q)QI_1'I_2', \tag{4}$$

$$Q = \frac{p(1+p+\bar{I}) + \delta^2(p+p\bar{I}+r^2)}{(1+p+\bar{I})^2 + \delta^2(1+r+\bar{I})^2}. \tag{5}$$

The mode intensities $I_n = X_n^2$, $I_n' = I_n/(1+\delta^2)$ and $\bar{I} = (I_1'+I_2')/2$.
The parameter $\delta = \Omega_L/\gamma_{01}$ is a measure of the Zeeman splitting
and represents the Larmor frequency in units of the homogeneous
line width γ_{01}. Various relaxation rates enter in (3) through
three dimensionless parameters:

$$p = \frac{\gamma_{12}^{ph}}{\gamma_1}, \qquad q = \frac{\gamma_1}{\gamma_0}, \qquad r = \frac{1}{q} + \frac{\gamma_{01}^{ph}}{\gamma_1}, \tag{6}$$

where γ_i is the population (longitudinal) decay rate, $\gamma_{ij} = (\gamma_i + \gamma_j)/2 + \gamma_{ij}^{ph}$ is the coherence (transverse) decay rate (i, j = 0, 1 and 2). The collisional decay of Zeeman coherence is governed by p. The parameter q is the ratio of the lower to upper-level relaxation rates and generally $q \ll 1$ for effective transverse optical pumping. In the following we assume $r \simeq q^{-1}$ and $\phi = 0$.

Equations (1)-(6) provide a general description of OB in three-level systems. Different qualitative features in OB are obtained depending on whether Y_1 and Y_2 are varied simultaneously or one of them is kept fixed. We consider each situation separately.

BISTABILITY OF POLARIZATION ELLIPSE

In this case Y_2 is kept fixed and X_1 and X_2 are obtained as a function of Y_1. Experimentally it amounts to change the input beam polarization and observe OB in the polarization ellipse of the transmitted beam. To illustrate main features, consider the specific case of degenerate Zeeman sublevels ($\delta = 0$) with $q \ll 1$. From Eq. (3), f_n is imaginary and is given by

$$f_n = i(1 + \tilde{p}I_{3-n}) \, / \, (1 + I_1 + I_2 + \tilde{p}I_1 I_2), \tag{7}$$

where $\tilde{p} = 2p/(2 + I_1 + I_2)$. For p = 0 the system exhibits both self- and cross- saturation. However, when the collisional decay of Zeeman coherence dominates ($p \to \infty$), $f_n = i(1 + I_n)^{-1}$ and the two polarization components are uncoupled. This clearly indicates the role of Zeeman coherence in providing mutual coupling. For p = 0 the state equation (1) is identical to that considered in Ref. 12 and the corresponding results can be directly used.

POLARIZATION TRISTABILITY

Kitano et al.[3] first suggested the use of Zeeman coupling to observe optical tristability in the output beam polarization. Their analysis is performed in the rate-equation approximation for a dispersion-dominated system. We have investigated tristability domains for a purely absorptive system ($\delta = 0$) by using Eqs. (1) and (7) with the boundary condition $Y_1 = Y_2$ that corresponds to the linear polarization of the incident beam. Of particular interest is the dependence of tristability on the collisional decay of Zeeman coherence and the results are shown in Fig. 1. A somewhat surprising result is that that a stable asymmetric solution ($X_1 \neq X_2$) does not exist for $p \leqslant 5$. Significant

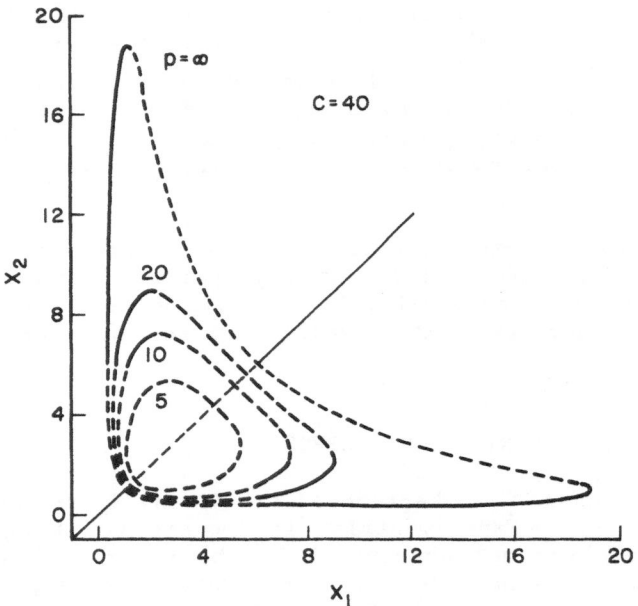

Fig. 1. Symmetric ($X_1 = X_2$) and asymmetric ($X_1 \neq X_2$) solutions of
an absorptive Zeeman system obtained by the parametric
variation of the incident field. Dashed portion of each
curve corresponds to an unstable state. Tristability
disappears for $p \leqslant 5$.

collisional decay of Zeeman coherence appears to be a prerequisite
for optical tristability in the present absorptive system.

LEVEL-CROSSING OPTICAL BISTABILITY

In this case OB is observed as a function of the Zeeman-
splitting parameter δ (varied through an applied magnetic field)
at a given intensity of the linearly polarized ($Y_1 = Y_2 \equiv Y$)
incident field. Only the symmetric solution $X_1 = X_2 = X$ is
considered so that the transmitted field is also linearly
polarized. Figure 2 shows the effect of the collisional decay of
Zeeman coherence on OB for $Y = 20$, $C = 40$, and $q = 0.01$. Although
bistability occurs for all values of p, its detailed features are
substantially affected. For $p = 0$ (no collisional decay of Zeeman
coherence), a narrow central peak in the cavity transmission
occurs due to transverse optical pumping.[11] As p increases, this
peak broadens until it completely disappears for sufficiently
large values of p.

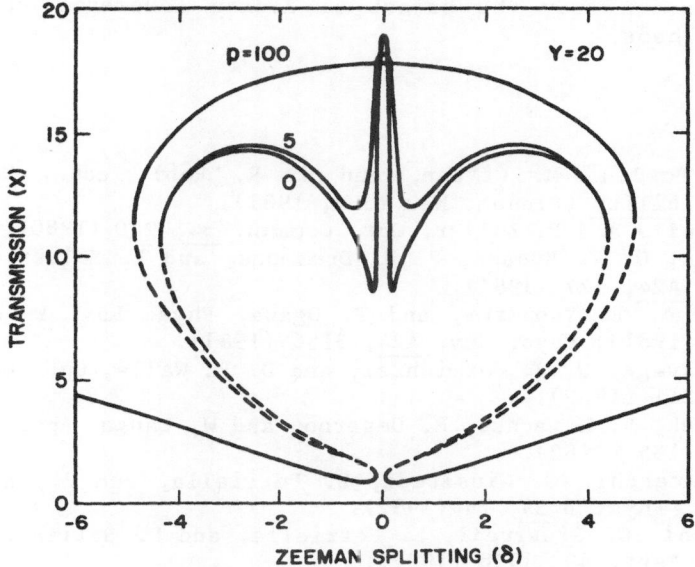

Fig. 2. Transmitted field versus δ for several values of p at a
         fixed value of the incident field-amplitude.   Narrow
         central peak due to transverse optical pumping disappears
         for sufficiently large values of p.

Further calculations were performed to study the dependence
of level-crossing OB on the relaxation-rate ratio $q = \gamma_1/\gamma_0$. As q
increases for a fixed value of p, the central transmission peak
broadens and the hysteresis width decreases.   For the specific
case of equal relaxation rates (q = 1), bistability disappears
altogether.   This clearly indicates that transverse optical
pumping[11] is the main physical mechanism that leads to OB in
Zeeman systems.

CONCLUSIONS

A theoretical analysis of Zeeman-coherence effects in
absorptive optical bistabilty is presented after modeling the
nonlinear medium as a homogeneously broadened Λ-type three-level
system. It is found that the OB features are strongly dependent
on the relative level-relaxation rates and the collisional decay
of Zeeman coherence. The predicted behavior should be observable
using J = 1 → J = 0 transition of an atomic beam (Na or Sm, for
example).   Recently, transient effects such as critical slowing
down and magnetically induced self-pulsing have been observed[9] in
Zeeman systems using Na vapor. It would be of interest to extend

present work to study the effect of Zeeman coherence on self-pulsing and chaos.[8]

REFERENCES

1. C. M. Bowden, M. Ciftan, and H. R. Robl, eds., Optical Bistability (Plenum, New York, 1981).
2. D. F. Walls and P. Zoller, Opt. Commun. 34, 260 (1980); D. F. Walls, C. V. Kunasz, P. D. Drummond, and P. Zoller, Phys. Rev. A24, 627 (1981).
3. M. Kitano, T. Yabuzaki, and T. Ogawa, Phys. Rev. Lett. 46, 926 (1981); Phys. Rev. A24, 3156 (1981).
4. C. M. Savage, H. J. Carmichael, and D. F. Walls, Opt. Commun. 42, 211 (1982).
5. J. Mlynek, F. Mitschke, R. Deserno, and W. Lange, Appl. Phys. B28, 135 (1982).
6. F. T. Arecchi, G. Giusfredi, E. Petriella, and P. Salieri, Appl. Phys. B 29, 79 (1982).
7. S. Cecchi, G. Giusfredi, E. Petriella, and P. Salieri, Phys. Rev. Lett. 49, 1928 (1982).
8. H. J. Carmichael, C. M. Savage, and D. F. Walls, Phys. Rev. Lett. 50, 163 (1983).
9. J. Mlynek, F. Mitschke, and W. Lange, these proceedings and Phys. Rev. Lett. 50, 1660 (1983).
10. J. Mlynek, F. Mitschke, R. Deserno, and W. Lange, Phys. Rev. A (to be published).
11. G. Orriols, Nuovo Cimento 53B, 1 (1979).
12. G. P. Agrawal, Appl. Phys. Lett. 38, 505 (1981).
13. J. Heppner, C. O. Weiss, U. Hubner, and G. Schinn, IEEE J Quantum Electron. QE-16, 392 (1980).
14. G. P. Agrawal, Phys. Rev. A (to be published).

POLARIZATION SWITCHING WITH J=½ TO J=½ ATOMS IN A RING CAVITY

W.J.Sandle, M.W.Hamilton and R.J.Ballagh

Physics Department
University of Otago
Dunedin, New Zealand

INTRODUCTION

The aim of this paper is to summarize theoretical results for the mean-field, steady-state response of a plane-wave ring cavity containing atoms with a homogeneously broadened J=½ to J=½ transition. The driving laser radiation can have arbitrary polarization and detuning. The phenomena encompassed by the results presented here include conventional optical bistability, both with linearly polarized light (LOB)[1] and with circularly polarized light (COB)[2,3] as well as the more complex phenomenon we term "polarization switching" (PS).[4-9]

The latter occurs when a linearly polarized input (equal superposition of left and right circular polarizations) produces an output where one of the circular polarizations dominates. Kitano et al[4] were the first to investigate aspects of this behaviour - which they named "optical tristability" - and they employed a simple three-state system to model the behaviour in the non-saturation regime.

In the following we summarize extension of our own previous treatment (HBS)[5] - polarization switching in a J=½ to J=½ transition - to include atom-laser and cavity-laser detunings as well as a steady applied magnetic field.

STEADY-STATE MEAN-FIELD STATE EQUATIONS (LONGITUDINAL MAGNETIC FIELD)

The electric field experienced by the atoms can be written in units of the saturation field (as defined by Eq.(8) in HBS):

$$\vec{E}(z,t) = \{X_+(z)\hat{e}_{-1}^* + X_-(z)\hat{e}_1^*\}\exp\{i\omega(z/c-t)\} + c.c., \qquad (1)$$

where $X_\pm$ are complex slowly varying amplitudes corresponding to $\sigma^\pm$ radiation.

The evolution of the radiation field in the atomic medium is determined by Maxwell's equations in the form

$$\frac{\partial}{\partial z}\, \underset{\sim}{X}_\pm(z) = -\frac{\alpha}{2}\, \underset{\sim}{\eta}_\pm \underset{\sim}{X}_\pm(z) \ . \tag{2}$$

The complex susceptibilities $(\frac{ic\alpha}{\omega}\, \eta_+$, where α is the weak-field resonant absorption coefficient) are given via

$$\eta_\pm = (1-i\Delta_\pm)\left\{1+\Delta_\pm^2+8X_\pm^2+(2\beta-4)\ \frac{X_\pm^2(1+\Delta_\mp^2)-X_\mp^2(1+\Delta_\pm^2)}{1+\Delta_\mp^2+4\beta X_\mp^2}\right\}^{-1} \tag{3}$$

where $X_\pm$ are the (real) magnitudes of $X_\pm$, and $\Delta_\pm$ are the detunings of the incident field from the $\sigma^\pm$ transitions (see Fig.1), in units of $\Gamma_1(\ell u)$, the dipole optical coherence decay rate. The collision parameter β is defined (see HBS) in terms of the upper-state radiative decay rate γ and the upper (lower) level orientation decay rates[10] $\Gamma_1(u)\,(\Gamma_1(\ell))$:

$$\beta = \frac{\gamma}{\Gamma_1(u)\Gamma_1(\ell)}\ \{\Gamma_1(\ell)+\Gamma_1(u) - \frac{\gamma}{3}\} \ . \tag{4}$$

Expression (3) is obtained by solving the atomic density matrix equations, and includes the effects of saturation, optical pumping and a steady magnetic field applied along the propagation (z) direction, and expressed through the detunings $\Delta_\pm$.

In steady state, and within the mean-field approximation, we obtain the following state equations for a ring cavity

$$\underset{\sim}{Y}_\pm = \underset{\sim}{X}_\pm(1+2C\eta_\pm+i\phi) \ , \tag{5}$$

where $Y_\pm$ are the $\sigma^\pm$ components of the field incident upon the cavity (scaled by the mirror transmittance), C is the usual cooperativity parameter, and ϕ is the empty cavity-field detuning in units of the cavity linewidth.

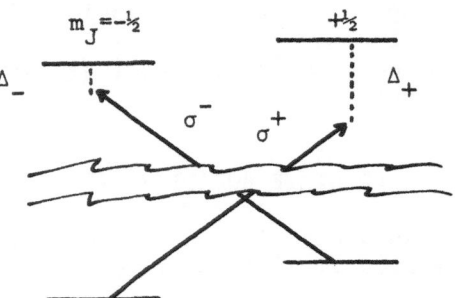

Fig.1. Atom-laser detunings $\Delta_\pm$

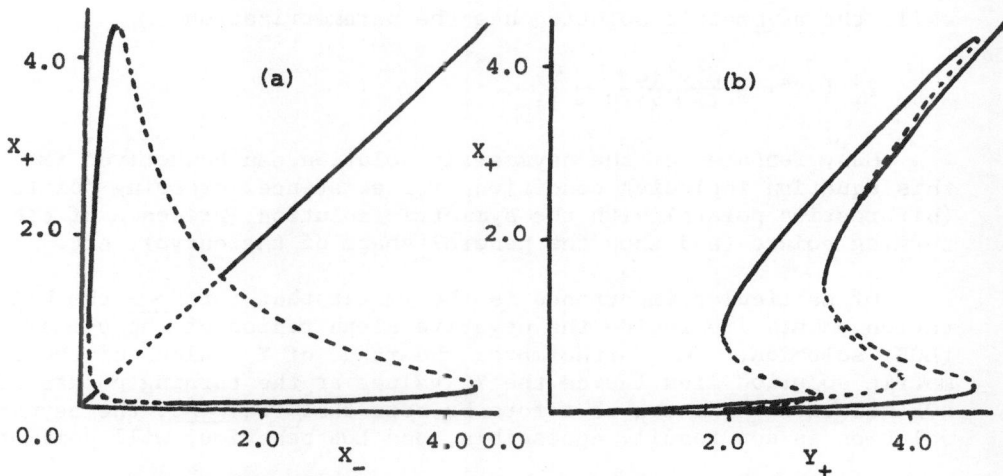

Fig.2.   Projections of the solution for linearly polarized input.
Parameter values are given in the text.

RESONANT CASE ($\Delta_+ = \Delta_- = \phi = 0$)

Many interesting features of the solution of Eq.(5) are contained
in the exactly resonant case with zero applied magnetic field.   Then
$\eta_\pm$ and $X_\pm$ are real quantities and, for fixed input polarization X_+/Y_-,
Eq.(5) describes a trajectory in (X_+,X_-,Y_+) space.   This trajectory
can be portrayed as projections onto the (X_-,X_+) or (X_+,Y_+) planes.
Figs.2(a) and 2(b) illustrate these projections for the case of input
linear polarization with C=10 and β=8.8, and at exact resonance.

It can be seen from Fig.2(a) that "asymmetric" (in the sense that
$X_+ \neq X_-$) as well as "symmetric" ($X_+ = X_-$) solutions exist.   The features of these solutions can be derived most readily from an analytic
treatment (see HBS for details) which is summarized below.

Analytic Treatment

For linearly polarized resonant excitation, the difference between the two equations (5) factors into symmetric $(X_+ - X_-) = 0$ and
asymmetric solutions.   The latter takes the simple form
$1 + (2 + \beta)u + 2\beta v^2 + 2C(1-v) = 0$ with the parametrization

$$u = 2(X_+^2 + X_-^2) , \qquad v = 4X_+X_- \quad . \tag{6}$$

The symmetric solution obeys the conventional OB state equation,

$$Y_+ = X_+(1 + \frac{2C}{1 + 8X_+^2})$$

$$(Y_+ = Y_-, \quad X_+ = X_-) \tag{7}$$

while the <u>asymmetric</u> solution has the parametrization

$$Y_+^2 \; (=Y_-^2) \; = \; \frac{\beta^2 v^2 (C + \tfrac{1}{2} - v)}{(\beta + 2)(\beta v - 1)} \; . \tag{8}$$

Many features of the asymmetric solution can be derived from this equation including conditions for existence, crossing points (bifurcation points) with the symmetric solution, presence of other turning points (and thus the general shape of the curve), etc.

Of particular importance is the result that for $\beta < 2$ the bifurcation points lie inside the negative slope region of the symmetric (LOB) solution, and furthermore, the range of Y_+ values of the asymmetric solution lies inside the Y_+ values at the turning points of the LOB solution. It can therefore be seen that for $\beta < 2$, the asymmetric solution is not readily accessible, and LOB behaviour will dominate.

A numerical analysis shows that the symmetric solution is always unstable between the bifurcation points. This has significance for the $\beta > 2$ case, where the bifurcation points lie outside the negative slope region of the symmetric solution. Between the bifurcation points the system must follow the asymmetric solution; the output then becomes elliptically polarized, and PS behaviour will be observed.

PHYSICAL INTERPRETATION

A steady-state output field that is asymmetric will occur if the atoms preferentially absorb one of the two (equal) input circular polarizations. This will be the case when the atoms become optically pumped in the ground state. We focus on the situation $\beta \gg 2$, because then $\gamma \gg \Gamma_1(\ell)$, giving the possibility for significant optical pumping to occur at power levels below those for saturation; e.g. $\beta X_+^2 > 1 >$ βX_-^2, $1 > X_+^2 > X_-^2$. In the following we distinguish three power regimes.

In the <u>low-power regime</u> ($\beta X_\pm^2 \ll 1$), neither saturation nor optical pumping occur to any significant extent. The absorption is strong, and equal for both polarizations ($\eta_+ = \eta_- \simeq 1$. The cavity transmits linearly polarized light (for linear polarization input) but with weak intensity

$$X_\pm^2 \simeq Y_\pm^2 / (1 + 2C)^2 \; . \tag{9}$$

In the <u>optical pumping regime</u> ($\beta X_\pm^2 \gg 1$, $X_\mp^2 \ll 1$), there is sufficient power to allow optical pumping of the atoms, but saturation is not important. A small fluctuation pumps atoms preferentially into one (or other) of the ground substates, causing differential absorption which leads to an enhancement of the imbalance. One component (e.g. X_+) becomes dominant. This favoured circular polarization is transmitted at an intensity near to the empty cavity value;

$$Y_+^2 - X_+^2 \simeq 2C/(\beta + 2) \quad . \tag{10}$$

The other circular polarization is transmitted even more weakly than it would have been given no optical pumping;

$$X_-^2 \simeq Y_-^2/(1 + 4C)^2 < Y_-^2/(1 + 2C)^2 \quad . \tag{11}$$

Finally, in the <u>saturation regime</u> ($X_\pm^2 \gg 1$) strong saturation occurs for both σ^+ and σ^- transitions, and absorption is now again equal for both polarizations, but is weak;

$$\eta_+ = \eta_- \simeq 1/(8X_\pm^2) \quad . $$

The cavity transmits linearly polarized output (again for linearly polarized input) with intensity near to the empty cavity value,

$$Y_\pm^2 - X_\pm^2 \simeq C/2 \quad , \tag{12}$$

but not as close to the empty cavity value as for the dominant polarization in the optical pumping regime. Physically, this is because intensity is dissipated by spontaneous radiation, which is more intense in the saturated regime since there is more population in the upper level.

Fig.3 (fully resonant case) illustrates the points above.

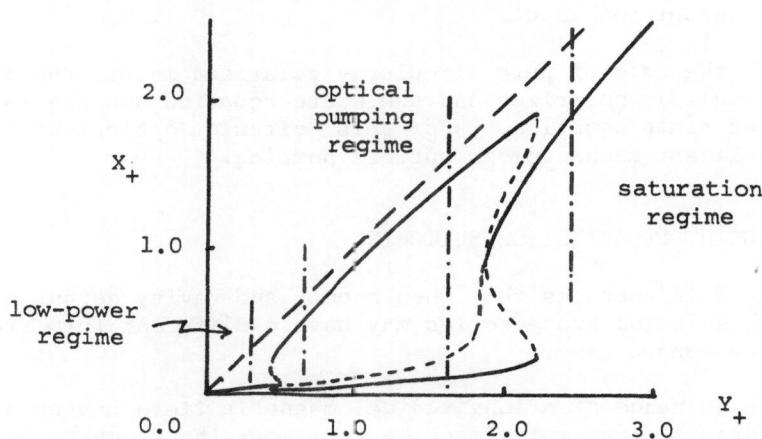

Fig.3. Solution for linearly polarized input illustrating behaviour in the three regimes (C = 4, β = 94).

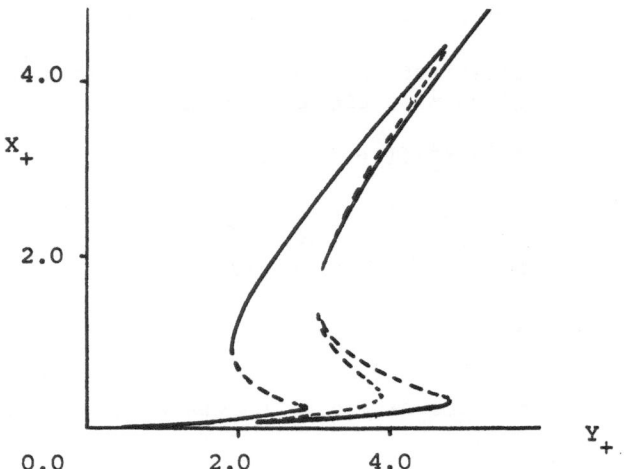

Fig.4.   Solution for Y_-/Y_+ = 0.99 (exact resonance, C=10, β=8.8).

ELLIPTICAL INPUT POLARIZATION

When the initial symmetry is lowered by giving the input circular polarization components different amplitudes, the detailed character of the solution trajectories changes.   Particularly, the bifurcation points disappear, as shown in Fig.4.   The initially larger component remains dominant, although the ellipticity will in general differ between output and input.

For the case of pure circularly polarized input, the output is also circularly polarized and the state equation has the form of the simple OB state equation.   For this "circular optical bistability" the non-linear mechanism is optical pumping.

NON-RESONANT POLARIZATION SWITCHING

Fig.5 illustrates that when atomic and cavity detunings are present, solution trajectories may have a different form from those for the resonant case.

The presence of a longitudinal magnetic field causes Δ_+ and Δ_- to differ;   the general effect is to remove the symmetry between X_+ and X_-, eliminating bifurcation points in a manner similar to that shown in Fig.4.

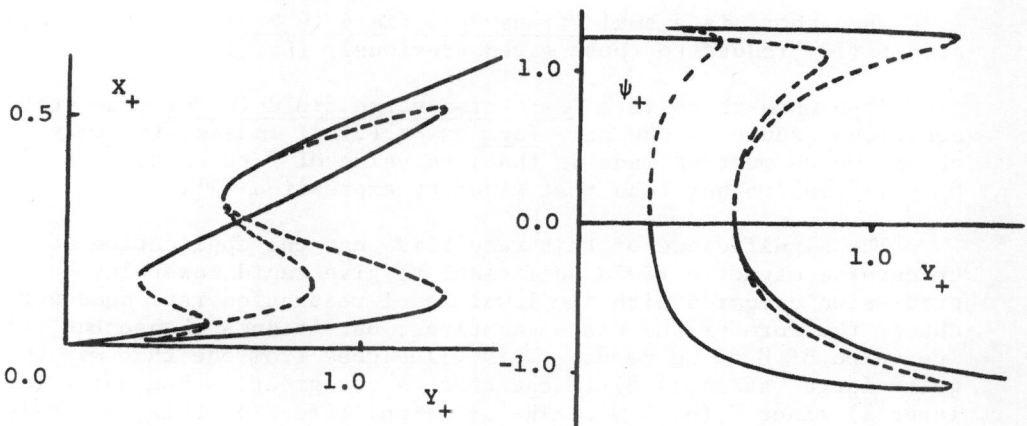

Fig.5.　Solution for linearly polarized input for the detuned case. The magnitude (X_+) and phase (ψ_+) of X_+ are illustrated: $\tilde{X}_+ = X_+ \exp(i\psi_+)$. $C=35$, $\beta=19$, $\Delta_+=\Delta_-=5$ and $\phi=3$.

TRANSVERSE MAGNETIC FIELD CASE

We have solved the full set of atomic density matrix equations in steady state for the case of a transverse magnetic field B. We have employed the assumptions that Larmor evolution in the upper level and of the optical dipole coherence can be ignored compared with Larmor evolution (frequency ω_L) in the lower level.[9] Further, we have restricted the treatment to the fully resonant case with a linearly polarized driving field, and we have taken the cavity field to have large ellipticity (i.e. close to linear polarization) with major principal axis parallel to B.

We then find that

$$\eta_\pm = \frac{\{1+\beta u\}\{1+4\beta X_\mp^2\} \;+\; \Omega_L^2\{1+\beta_B u\}\{1+4\beta_B X_\mp^2\}}{\{1+\beta u\}\{1+(2+\beta)u+8\beta v^2\}+\Omega_L^2\{1+\beta_B u\}\{1+(2+\beta_B)u+8\beta_B v^2\}} .$$

Here, β is given by Eq.(4) and β_B is defined as $\beta_B = \gamma/\Gamma_1(u)$. The quantities u and v are given in Eq.(6). Finally, the Larmor frequency enters through the symbol Ω_L defined as $\Omega_L = \omega_L/\Gamma_1(\ell)$.

The state equations are

$$Y_\pm = X_\pm(1 + 2C\eta_\pm) . \tag{13}$$

In general Eqs.(13) are complicated. However, in two cases of interest simple results follow.

When there is a <u>small transverse field</u> ($\Omega \ll 1$), the state equations (13) reduce to those given previously (Eq.(5)).

Also when there is a <u>large transverse field</u> ($\Omega \gg 1$) the state equations reduce to the same <u>form</u> as Eq.(5).   Indeed, the only change which must be made is that the value of β to be used is $\beta_B = \gamma/\Gamma_1(u)$ rather than that given by expression (4).

The significance of this result is that the application of a transverse magnetic field sufficient to give rapid lower-level precession compared with the lower-level relaxation rate does not change the form of the state equations, but it does change the effective value of β to be used.   This value goes from one that may be quite large (at small B) to one that is not greater than unity (at large B) since $\Gamma_1(u) \geqslant \gamma$.   The principal effect of this change is to modify the character of the switching.   For small B, and given that $\gamma/\Gamma_1(\ell) > 2$, polarization switching as illustrated in, for example, Fig.2 is predicted.   However for large B, the asymmetric part of the solution will move inside the normal linear optical bistability hysteresis loop.   Polarization switching will thus be suppressed and linear optical bistability occurs.

REFERENCES

1.   C. M. Bowden, M. Ciftan and H. R. Robl, "Optical Bistability", Plenum, New York (1981).
2.   R. J. Ballagh, J. Cooper and W. J. Sandle, <u>J. Phys. B: At.Molec. Phys.</u> 14:3881 (1981).
3.   F. T. Arecchi, G. Guisfedi, E. Petriella and P. Salieri, <u>App. Phys. B.</u> 29:79 (1982).
4.   M. Kitano, T. Yabuzaki and T. Ogawa, <u>Phys. Rev. Lett.</u> 46:926 (1981).
5.   M. W. Hamilton, R. J. Ballagh and W. J. Sandle, <u>Z. Physik B.</u> 49:263 (1982).
6.   C. M. Savage, H. J. Carmichael and D. F. Walls, <u>Opt. Comm.</u> 42:211 (1982).
7.   H. J. Carmichael, C. M. Savage and D. F. Walls, <u>Phys. Rev. Lett.</u> 50:163 (1983).
8.   S. Cecchi, G. Guisfredi, E. Petriella and P. Salieri, <u>Phys. Rev. Lett.</u> 49:1928 (1982).
9.   W. J. Sandle and M. W. Hamilton, "Laser Physics Proceedings (Hamilton, New Zealand)" <u>Lecture notes in Physics</u>, Springer-Verlag, Berlin, 182:54 (1983).
10. A. Omont, <u>Prog. Quant. Electron.</u> 5:69 (1977).

POLARIZATION SWITCHING WITH SODIUM VAPOR IN A FABRY PEROT

M. W. Hamilton[*] and W. J. Sandle

Physics Department, University of Otago
Dunedin, New Zealand

It was first predicted by Kitano et al. [1] that an optical
cavity which contained three-state atoms (in the so-called lambda
configuration) would show switching and hysteresis in the polarization state of the output as one varied the input power. Their
model was subsequently generalized to include saturation effects
[2] and a model was developed using a more realistic atomic model,
the J = 1/2 to J = 1/2 transition [3]. Furthermore, it has been
predicted that when the atomic susceptibility is primarily dispersive a self-pulsing of the polarization state, with period doubling bifurcations leading to chaotic behavior, can occur [2,3b]. A
different type of self-pulsing has been predicted in the case of
an applied magnetic field with a direction perpendicular to that
of the radiation field wave vector [4].

If, with a J = 1/2 to J = 1/2 transition, one has a sufficiently slow relaxation rate of the lower level orientation, one
finds that there are two regimes for polarization switching when
one varies the input power to the cavity. The first of these,
occurring at relatively low input power, is ascribed to optical
pumping effects: the second, at high power, to saturation of the
transition. For linearly polarized input radiation, tuned to the
transition, the sequence of polarization switching is as follows.
At low input power the system has linearly polarized output and a
low transmission coefficient due to absorption and/or dispersion

---

[*]Present address: Joint Institute for Laboratory Astrophysics,
University of Colorado and National Bureau of Standards, Boulder,
Colorado 80309 USA.

by the atoms. However, at intermediate powers the system is unstable to small deviations from linearly polarized output because
the differential absorption/dispersion resulting from such a deviation is amplified by a combination of optical pumping and cavity
feedback. The system is then driven into a state of near circularly polarized output. At high input powers, saturation of the
transition restores the stability of linearly polarized output.
When the input power is reduced the polarization switchings occur
at lower values of input power than the corresponding switchings
for increasing input power — one thus sees hysteresis in polarization switching.

The low power (optical pumping based) switching has been observed by Cecchi et al. [5] with a Fabry-Perot containing sodium
vapor. More recently this switching along with magnetic field
induced self-pulsing has been observed by Mitschke et al. [6].
The purpose of this paper is to present what we believe to be
the first observation of the complete steady state polarization
switching sequence including the saturation based switching at
high input power [7].

Our experimental setup is shown in Fig. 1. The dye laser is
tuned to the sodium D1 transition which is homogeneously broadened
by the addition of 50 to 100 Torr of argon buffer gas. Under these
conditions the collisional broadening is comparable to or greater
than both the ground level hyperfine splitting (1.8 GHz) and the
Doppler width (1.5 GHz). One may then use the $J = 1/2$ to $J = 1/2$
transition as an approximate model for the D1 transition [8]. Up
to 500 mW of linearly polarized input power was available to the
Fabry-Perot after transmission through a Faraday isolator, mode
matching lenses and an intensity modulator (electro-optic). The
spherical mirror Fabry-Perot is in a near concentric arrangement
with a free spectral range of 255 MHz and has mirrors of transmissivity $T = 0.02$. In most of the experiments we used a sealed
sodium cell, 10 cm long, with perpendicular windows which each had
a single layer antireflection coating on the outer surfaces. The
cell was connected to a high vacuum and buffer gas handling system and was heated to between 130°C and 150°C. The output of the
Fabry-Perot was split into its σ^+ and σ^- circularly polarized components by a Fresnel rhomb and a suitably oriented calcite prism.

The observation of polarization switching may be made in two
ways; by keeping the laser tuned to an appropriate frequency and
varying the input power to the Fabry-Perot or by maintaining a
constant input power and scanning the laser frequency. We shall
first present results obtained with the former method. Figure 2
shows oscilloscope traces which display the σ^+ and σ^- polarized
components of the output power as functions of the total input
power. The upper trace (σ^-) shows switching with hysteresis
whereas no switching or other major change in transmission is

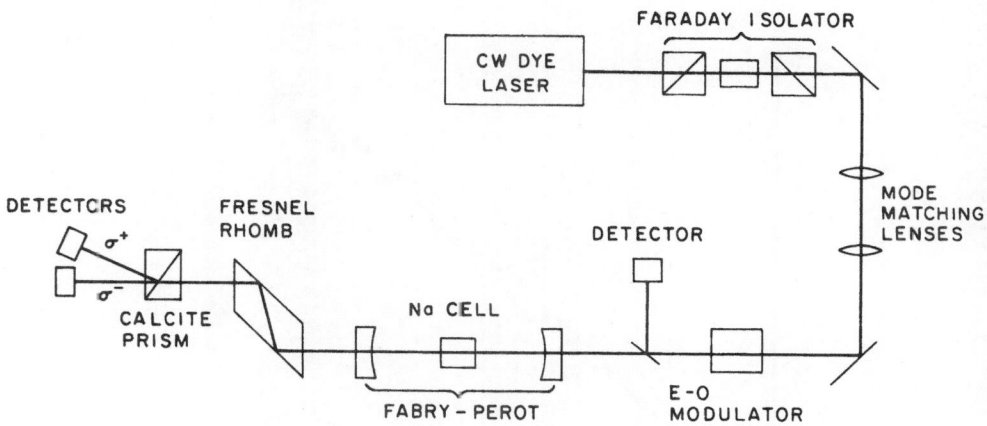

Fig. 1. Schematic of the experiment.

visible in the lower trace (σ^+). Thus this figure shows switching
from linearly polarized output at low input power to nearly circularly polarized output at intermediate input power. This example
is for the laser tuned exactly to resonance with both the atoms
and the Fabry-Perot, i.e. the purely absorptive case. For this
case we were unable to observe the saturation based switching at
higher input powers with the sealed cell arrangement because the
losses caused by the cell windows limited the attainable intracavity power. In later experiments, we have been able to overcome
this problem by using a windowless sodium cell where the cavity
finesse (without sodium) is ≈ 140, compared to ≈ 20 with the sealed
cell. Such a cell is possible because the Fabry-Perot and cell
are enclosed in an evacuable tank (originally to eliminate convec-
tion currents in the air between the Fabry-Perot mirrors) which
is filled with buffer gas to the desired pressure. With this new
cell we have observed the complete polarization switching sequence
in the absorptive case.

It is, however, possible to observe the complete sequence
with the sealed cell arrangement if we detune the laser from exact
resonance with the atoms. Figure 3 shows the entire sequence for
the laser detuned 1.5 GHz on the low frequency side of the transition. The oscilloscope traces are displayed in the same manner as
in Fig. 2. We can again see switching to predominantly σ^- circularly polarized output at intermediate input powers, but now at
high input powers we see the polarization state of the output re-
verting to approximately linear polarization. The regimes of optical pumping based switching and saturation based switching are
indicated and in each, significant hysteresis can be seen. The
laser-cavity detuning is not quoted because, as yet, this parameter is not well controlled, except in the purely absorptive case.

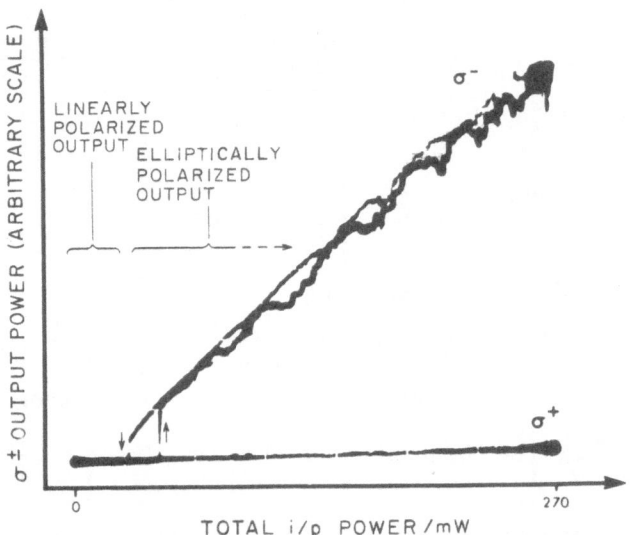

Fig. 2.   Optical pumping based polarization switching in the absorptive case. Sodium density 11×10^{10} cm^{-3}, 50 torr
Ar buffer gas.

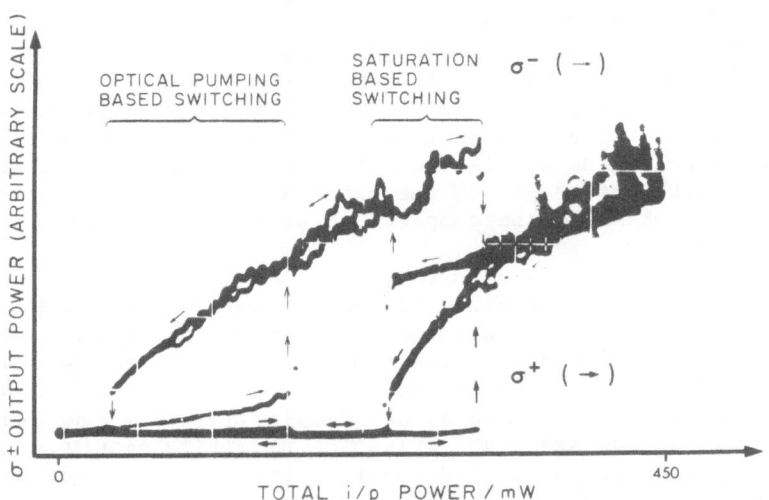

Fig. 3.   The complete steady state polarization switching sequence
for laser frequency 1.5 GHz below line center. Bold(fine)
arrows indicate the behavior of σ^+ (σ^-) polarization.
Sodium density 9×10^{10} cm^{-3}, 50 torr Ar buffer gas.

For purely linearly polarized input one would expect to see
switching from linearly polarized to predominantly σ^+ or σ^- circularly polarized output with equal probability. In the examples
shown, which show repetitive up/down variation of input power,
this is not seen because of residual polarization anisotropies in
the cell window. By adjusting the plane of polarization of the
input we can make the σ^+ polarization switch preferentially to
high transmission (instead of the σ^- polarization). Using the
windowless cell it is possible to observe random switching from
linearly polarized to either mainly σ^+ or mainly σ^- polarized
output.

When we use the second method for observing polarization
switching (scanned frequency, constant power) we scan through
Fabry-Perot resonances and observe that as we pass through a resonance, the transmission for one of the σ^+ or σ^- polarized components increases (not always abruptly) whereas that for the other
decreases. An example of such behavior is the oscilloscope traces
shown in Fig. 4. Here the traces showing the σ^+ (upper) and σ^-
polarizations are separated vertically for clarity. The scan
direction is from low to high frequency (left to right). Abrupt
switching occurs only on the low-frequency side of the Fabry-Perot
resonance, the transition back to linear polarization on the high

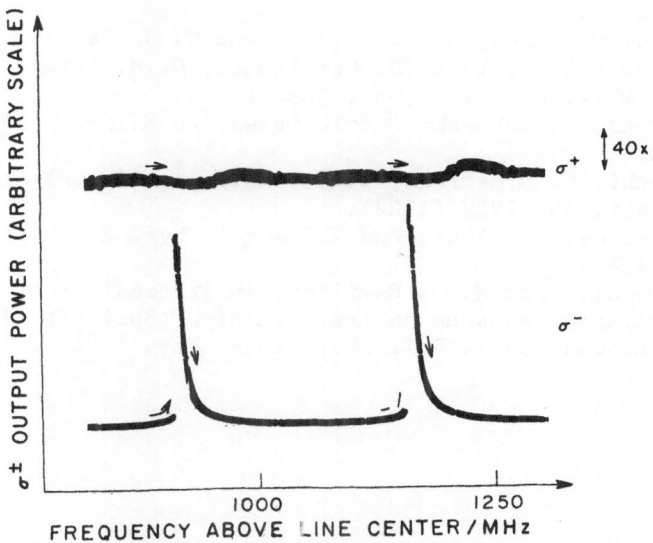

Fig. 4. An example of polarization switching with scanned input
frequency; laser power 5.6 mW, sodium density 2×10^{11}
cm^{-3}, 100 torr Ar buffer gas. The arrows indicate the
direction of the scan.

frequency side being smooth. Note also that the scale for the σ^+ polarized output power is 40 times larger than that for the σ^- polarization. These distorted Fabry-Perot resonances may be understood as being sections, of constant input power, taken through the surfaces that are generated by the transmission characteristics of Fig. 3, for a continuous range of laser-cavity detunings.

The results of the experiment give qualitative support for the theory based on the $J = 1/2$ to $J = 1/2$ transition provided that the detuning from the atomic resonance is not too great (i.e., $\lesssim 3$ GHz). For larger detunings, more complicated behavior than that predicted by the theory has been seen, the description of which is beyond the scope of this paper. It seems likely that in this regime, the validity of our approximating the Dl transition as a $J = 1/2$ to $J = 1/2$ transition breaks down.

It is a pleasure to acknowledge the contributions made to this work by Drs. J. T. Chilwell and R. J. Ballagh and by Mr. J. S. Satchell.

REFERENCES

[1]   M. Kitano, T. Yabuzaki and T. Ogawa, Phys. Rev. Lett. 46, 926 (1981).
[2]   C. M. Savage, H. J. Carmichael and D. F. Walls, Opt. Commun. 42, 211 (1982).
[3]   a) M. W. Hamilton, R. J. Ballagh and W. J. Sandle, Z. Physik B49, 263 (1982); b) H. J. Carmichael, C. M. Savage and D. F. Walls, Phys. Rev. Lett. 50, 163 (1983).
[4]   M. Kitano, T. Yabuzaki and T. Ogawa, Phys. Rev. A 24, 3156 (1981).
[5]   S. Cecchi, G. Giusfredi, E. Petriella and P. Salieri, Phys. Rev. Lett. 49, 1928 (1982).
[6]   F. Mitschke, J. Mlynek and W. Lange, Phys. Rev. Lett. 50, 1660 (1983).
[7]   W. J. Sandle and M. W. Hamilton, in Proceedings of the Third New Zealand Symposium on Laser Physics (Springer, 1983).
[8]   W. J. Sandle and A. Gallagher, Phys. Rev. A 24, 2017 (1981).

CHAOS AND OPTICAL BISTABILITY: BIFURCATION STRUCTURE

K. Ikeda

Department of Physics
Kyoto University
Kyoto 606, Japan

ABSTRACT *

The phase of the output light from a bistable optical cavity
containing a nonlinear dielectric medium obeys the following differential equation with time delay:

$$dx(t)/dt = -x(t) + \pi\mu f(x(t-t_R)).\tag{1}$$

Although the equation of this class is familiar in various areas
such as ecology, neurobiology, acoustics and study of electric circuit, the behavior of its solution has not been investigated in
detail. In this paper we report the results of our recent numerical study of this equation. It is found that, with increase of
parameter μ, which measures the intensity of the incident light,
or delay t_R, the solution of Eq. (1) exhibits transition from a
stationary state to periodic and chaotic states. In the course
of this transition, there appear successive bifurcations of a novel
type, which form, so to call it, a hierarchy of coexisting periodic
solutions: As μ is increased, the stationary solution becomes unstable at the first bifurcation point and breaks into a number of
periodic ones. These periodic solutions form a set of higher-
harmonics which can coexist with each other. With further increase
of μ, each of these solutions further bifurcates into a new set
of coexisting periodic solutions. Such a bifurcation takes place
successively, causing an accelerative accumulation of coexisting
periodic states and making the time evolution of the solution more
and more complicated, until a chaotic state sets in. In the chaotic regime, the coexisting periodic states in turn coalesce

successively into fewer sets and are finally reduced to a single chaotic state with totally complicated time evolution. This type of behavior, which has never been before in any other nonlinear system, appears generic in the class of Eq. (1).

---

*Article in Coherence and Quantum Optics V, edited by L. Mandel and E. Wolf, (Plenum, NY, 1983), p. 875.

OPTICAL BISTABILITY, CHAOS IN THE COHERENT TWO PHOTON PROCESSES

G. S. Agarwal

School of Physics
University of Hyderabad
Hyderabad 500 134, India

and

Surendra Singh

Department of Physics
University of Arkansas
Fayetteville, Arkansas 72701

ABSTRACT*

The cooperative behavior of a system of atoms contained in a ring cavity is investigated under the condition that the fundamental atomic transition is a two photon transition[1]. Maxwell-Bloch equations are used to obtain the input vs. output relation by adiabatically eliminating all the atomic variables. The fundamental equation that results is a complex two dimensional map--which in the mean field limit reduces to the standard bistability equations. The characteristics of the two dimensional map are numerically investigated. Such a map is shown to lead to chaotic behavior following the Feigenbaum[2] scenario. The sequence of events following first regime of chaos is a set of period halving bifurcations. The characteristics of the power spectrum in the region of chaos are presented. The effect of noise on the period doubling bifurcations in the present model, is also investigated and the connection with the theoretical predictions[3] is established. The charges in the dynamical characteristics of the system when the atomic inversion relaxes on the same time scale as the cavity round trip time will also be discussed in detail.

REFERENCES

1. F. T. Arecchi and A. Politi, Nuovo Cim. Lett. 23:65 (1978);
G. S. Agarwal, Opt. Commun. 35:149 (1980); G. P. Agrawal and C.
Flytzanis, Phys. Rev. Lett. 44:1058 (1980).
2. M. J. Geigenbaum, J. Stat. Phys. 19:25 (1978).
3. J. Crutchfield, M. Nauenberg and J. Rudnick, Phys. Rev.
Lett. 46:933 (1981); J. P. Crutchfield and B. A. Huberman, Phys.
Lett. 77A:407 (1980); B. Shraiman, C. E. Wayne and P. C. Martin,
Phys. Rev. Lett. 46:935 (1981).

*Article in Coherence and Quantum Optics V, edited by L. Mandel and
E. Wolf, (Plenum, NY, 1983), p. 885.

SELF-PULSING, BREATHING AND CHAOS IN OPTICAL BISTABILITY AND THE LASER WITH INJECTED SIGNAL

L. A. Lugiato

Instituto Di Fisica Dell'Universita
Milano, Italy

and

L. M. Narducci

Physics Department, Drexel University
Philadelphia, Pennsylvania 19104, USA

ABSTRACT*

The subject of spontaneous pulsations in Optical Bistability has stimulated considerable interest following its prediction by Bonifacio and Lugiato[1] in the framework of the plane-wave, ring cavity model and the realization by McCall[2] of an electro-optical converter of cw coherent light into pulsed radiation. In a subsequent important development, Ikeda[3] showed that, in the dispersive case, the Bonifacio-Lugiato instability leads to chaotic behavior (optical turbulence). This was later observed in a hybrid device by Gibbs, et al.[4].

In this paper, we review some of the most relevant manifestations of self-pulsing in Optical Bistability. First, we consider the absorptive case in which self-pulsing is a multi-mode phenomenon because the resonant mode remains stable while some of the sidebands can develop instability[5]. On the strength of the "dressed mode formalism" of optical bistability[6], which represents an extension of Haken's theory of generalized Ginzburg-Landau equations for phase-transition like phenomena in systems far from equilibrium[7], we have arrived at an essentially analytical understanding of this phenomenon. When only the nearest two sidebands of the resonant mode become unstable, we have identified the entire domain of existence of the self-pulsing state in the plane of the control parameters. This region is divided into a soft- and a hard-excitation domain, where self-pulsing develops with a behavior that is

49

reminiscent of second- or first-order phase transitions, respectively. In the hard-excitation region, one self-pulsing and two steady-state solutions are found to coexist, yielding hysteresis cycles that involve self-pulsing states along one of the branches. On approaching the boundary between the self-pulsing and the precipitation domains, the self-pulsing state becomes unstable. The instability is evidenced by a marked "breathing" behavior of the pulse envelope, whose transient nature has been linked to the unstable character of the limit cycle that bifurcates from the self-pulsing solution at the instability threshold.

When dispersive effects become important, self-pulsing behavior can arise even in the single-mode regime described by the well known mean field model[8,9]. For appropriate values of the control parameters, a large segment of the high transmission branch becomes unstable leading to periodic and irregular (chaotic) self-pulsing. On approaching the chaotic domain from either boundary, one finds a sequence of period doubling bifurcations of the type described by Grossman and Thomae[10] and by Feigenbaum[11] in the framework of the theory of discrete maps. As true also with most other dynamical models, the chaotic domain of this single-mode dispersive bistability contains "windows" of periodicity. We have characterized the behavior of bifurcating solutions and their critical slowing down in the neighborhood of each threshold with asymptotic power laws. These will be reviewed in detail. In the purely dispersive limit, we also show that our model reduces to the one analyzed recently by Ikeda and Akimoto[12]. If the absorbing medium is converted into an amplifier, the optically bistable system becomes a laser with an injected signal. We assume that the laser is above threshold and that the incident field frequency is detuned from the operating frequency of the laser. Obviously, when the injected field has a vanishingly small amplitude, the laser oscillates at its own frequency. On the other hand, when the incident field intensity is large enough, the entire system is expected to oscillate with the incident frequency (this phenomenon is well known as injection locking). We have analyzed the behavior of this system over the entire range of variation of the incident intensity below the locking threshold[13]. For a small incident field, the output power exhibits oscillations with a frequency equal to the mismatch between the incident and the laser frequencies, as originally proposed by Spencer and Lamb[14]. On increasing the incident intensity, the output becomes chaotic. A further increase in the injected signal leads to an inverted sequence of period doubling bifurcations with one- and two-periodic solutions exhibiting undamped breathing over limited domains. Finally, in the vicinity of the injection locking threshold, the system develops a chaotic spiking regime of an entirely different character from the irregular oscillations noted above. From our results, it is reasonable to speculate that the observations of chaos should be accessible with an all-optical system.

REFERENCES

1. R. Bonifacio and L. A. Lugiato, Lett. al Nuovo Cimento
21:510 (1978).
2. S. L. McCall, Appl. Phys. Lett. 32:284 (1978).
3. K. Ikeda, Opt. Comm. 30:257 (1979).
4. H. M. Gibbs, F. A. Hopf, D. L. Kaplan, and R. L. Shoemaker,
Phys. Rev. Lett. 46:474 (1981).
5. M. Gronchi, V. Benza, L. A. Lugiato, P. Meystre, and M.
Sargent III, Phys. Rev. A24:1419 (1981).
6. V. Benza and L. A. Lugiato, Zeit. fur Physik B35:383 (1979);
ibid., 47:79 (1982); L. A. Lugiato, V. Benza, L. M. Narducci and
J. D. Farina, Opt. Comm. 39:405 (1981).
7. H. Haken, Zeit. fur Physik B21:105 (1975); ibid. 22:69
(1975).
8. R. Bonifacio and L. A. Lugiato, Opt. Comm. 19:172 (1976).
9. L. A. Lugiato, L. M. Narducci, D. K. Bandy and C. A.
Pennise, Opt. Comm. 43:287 (1982).
10. S. Grossman and S. Thomae, Zeit. Naturforsch. 32A:1353
(1977).
11. M. Feigenbaum, J. Stat. Phys. 19:25 (1978); ibid. 21:669
(1979).
12. K. Ikeda and O. Akimoto, Phys. Rev. Lett. 48:617 (1982).
13. L. A. Lugiato, L. M. Narducci, D. K. Bandy and C. A. Pennise,
Opt. Comm. (to be published).
14. M. B. Spencer and W. E. Lamb, Jr., Phys. Rev. A5:884 (1972).

*Article in Coherence and Quantum Optics V, edited by L. Mandel and
E. Wolf, (Plenum, NY, 1983), p. 941.

MULTISTABILITY, SELF-OSCILLATION AND CHAOS IN NONLINEAR OPTICS

H.J. Carmichael

Department of Physics, University of Arkansas
Fayetteville, AR 72701

C.M. Savage and D.F. Walls

Physics Department, University of Waikato
Hamilton, New Zealand

INTRODUCTION

It has been an important development in optical bistability to learn that bistable systems may become unstable, and for a cw input produce an oscillating output, either periodic or chaotic. Instabilities leading to periodic self-oscillation were first discussed by McCall[1] and Bonifacio and Lugiato.[2] The rapid growth of interest in optical chaos has followed Ikeda's prediction of chaotic oscillations in a ring cavity with a dispersive nonlinearity.[3] Experimental investigations of Ikeda's proposal have been made[4,5] and further predictions of chaos in bistable and multistable systems have been reported.[6-10]

We have studied chaos in three systems comprising a pair of ring-cavity modes interacting via a nonlinear susceptibility. Each of these systems is described by the coupled oscillator equations

$$\kappa^{-1}\hat{\dot{E}}_{1,2} = \hat{E}_I^{1,2} - \hat{E}_{1,2}[1+i(1-R)^{-1}(\theta_{1,2}-\tfrac{1}{2}k_{1,2}L\chi_{1,2}^{NL})], \qquad (1)$$

where $\hat{E}_1$ and $\hat{E}_2$ are appropriately scaled dimensionless field amplitudes for modes with wavenumbers k_1 and k_2 and phase detunings θ_1 and θ_2 from cavity resonance, $\hat{E}_I^1$ and $\hat{E}_I^2$ are corresponding dimensionless incident field amplitudes, κ^{-1} is the cavity decay time, R is the reflection coefficient at the cavity input and output mirrors, L is the length of the nonlinear medium, and χ_1^{NL} and χ_1^{NL} are nonlinear susceptibilities, functions of both $\hat{E}_1$ and $\hat{E}_2$. Chaotic solutions to these equations have been found for:[1] (1) two circularly polarized modes coupled via a J=½ to J=½ transition,[8] (2) nondegenerate modes coupled via a two-photon transition,[9] and (3) intracavity second harmonic generation.[10] The major part of our paper is devoted to the J=½ to J=½ model for polarization switching. Here we will look carefully at the sequence of bifurcations leading to chaos (the route to chaos) where we find a period-doubling sequence of a new type and the suggestion of a relationship between our equations and the Lorenz equations. We then conclude with a few brief comments on chaos in two-photon bistability and intracavity second harmonic generation.

A J=½ TO J=½ MODEL FOR POLARIZATION SWITCHING

Dispersive polarization switching due to optical pumping between ground state Zeeman sublevels was first proposed by Kitano et al.[11] and has recently been observed in sodium vapor.[12,13] We have found chaos in a model for polarization switching which we originally proposed to include absorptive and saturation effects.[9] We consider a linearly polarized input field with a frequency ω_0 and amplitude $E_I (E_I^1 = E_I^2 = E_I/\sqrt{2})$ and decompose the cavity field into two circularly polarized modes, writing

$$\vec{E}(z,t) = (1/\sqrt{2})[E_1(t)\vec{e}_+ + E_2(t)\vec{e}_-]e^{-i(\omega_0 t - k_0 z)} + c.c., \qquad (2)$$

where $\vec{e}_\pm = (\vec{x} \pm i\vec{y})/\sqrt{2}$ are unit polarization vectors. To calculate χ_1^{NL} and χ_2^{NL} in Eq.(1) we adopt the atomic model illustrated in Fig. 1. Here T_1 and T_1' are relaxation times for the excited state population, T_2 is the relaxation time for the atomic dipole coherences, τ_1' and τ_1 are relaxation times for the population differences between ground and excited state magnetic sublevels, and $\delta\omega = \omega_a - \omega_0$ is a detuning from the atomic resonance. The steady-state solutions to atomic density matrix equations define the susceptibilities[14]

$$\chi_{1,2}^{NL} = i(kL)^{-1}\alpha L(1-i\Delta)(1+|\hat{E}_{2,1}|^2)/S(|\hat{E}_1|^2, |\hat{E}_2|^2) \qquad (3)$$

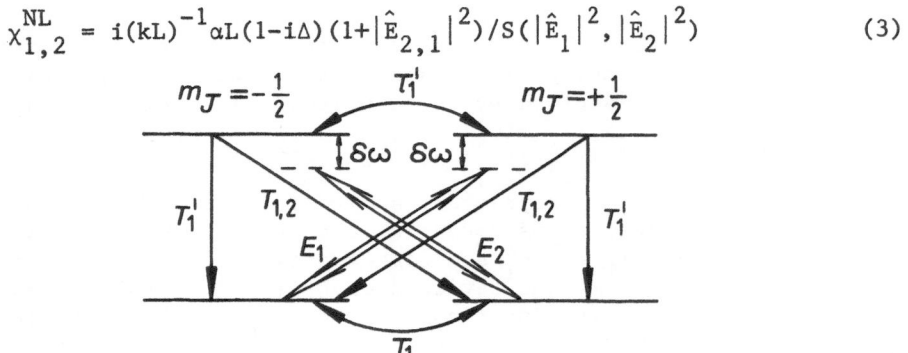

Fig. 1. Energy-level diagram for J=½ to J=½ transition.

with

$$S(|\hat{E}_1|^2, |\hat{E}_2|^2) = 1 + \tfrac{1}{2}(1+\eta)(|\hat{E}_1|^2 + |\hat{E}_2|^2) + |\hat{E}_1|^2|\hat{E}_2|^2, \tag{4}$$

and Eq.(1) becomes (with time measured in units of κ^{-1}),

$$\dot{\hat{E}}_{1,2} = \hat{E}_I - \hat{E}_{1,2}[1 + i\phi + C(1-i\Delta)(1+|\hat{E}_{2,1}|^2)/S(|\hat{E}_1|^2, |\hat{E}_2|^2)], \tag{5}$$

where α is the absorption coefficient (off resonance), $C = \alpha L/2(1-R)$, $\Delta = T_2 \delta\omega$, $\phi = \theta_1/(1-R) = \theta_2/(1-R)$, and

$$\eta = [(\tfrac{1}{T_1} + \tfrac{1}{T_1'})^{-1}/\tfrac{\tau}{2}1](\tfrac{1}{\tau_1} + \tfrac{1}{\tau_1'} + \tfrac{1}{T_1'})(\tfrac{2}{\tau_1'} + \tfrac{1}{T_1} + \tfrac{1}{T_1'})^{-1}, \tag{6}$$

and we have introduced dimensionless field amplitudes

$$\hat{E}_{-,2} = (2\sqrt{2/3}|\mu|/\hbar)[T_2(\tfrac{1}{T_1} + \tfrac{1}{T_1'})^{-1}/(1+\Delta^2)\eta]^{\frac{1}{2}} E_{1,2}/\sqrt{2},$$

$$E_I = \sqrt{T}(1-R)^{-1}e^{i\phi_T}(2\sqrt{2/3}|\mu|/\hbar)[T_2(\tfrac{1}{T_1} + \tfrac{1}{T_1'})^{-1}/(1+\Delta^2)\eta]^{\frac{1}{2}}, \tag{7}$$

where T is the transmission coefficient and ϕ_T the phase change on transmission at the cavity input mirror, and μ is the reduced atomic dipole matrix element.

Steady-state solutions to Eq.(5) are given in Fig. 2 where mode intensities $X_{1,2} = |\hat{E}_{1,2}|^2$ are plotted against the incident intensity $Y = |\hat{E}_I|^2$ for fixed values of C, Δ, ϕ and η. The three-dimensional plots are an aid to understanding the projections $X_{1,2}$ v Y. The special case $\eta = 1$ [Fig. 2(a)] is particularly instructive. Here $S(|\hat{E}_1|^2, |\hat{E}_2|^2)$ factorizes and we obtain two independent state equations

$$Y = X_{1,2}[(1 + \frac{C}{1+X_{1,2}})^2 + (\phi - \frac{C\Delta}{1+X_{1,2}})^2]. \tag{8}$$

Each mode satisfies the familiar cubic bistability equation and the projections $X_{1,2}$ v Y are the familiar S-shaped curves. However, in three dimensions, what is a bistable region for X_1 and X_2 individually is in fact a region of quadrastability. There are two stable symmetric branches (linear polarization) with X_1 and X_2 both in either high transmission or low transmission states, and two stable asymmetric branches (elliptical polarization) with X_1 in high transmission and X_2 in low transmission, or vice versa. In Fig. 2(a) the loop of asymmetric solutions is such that it overlaps the symmetric

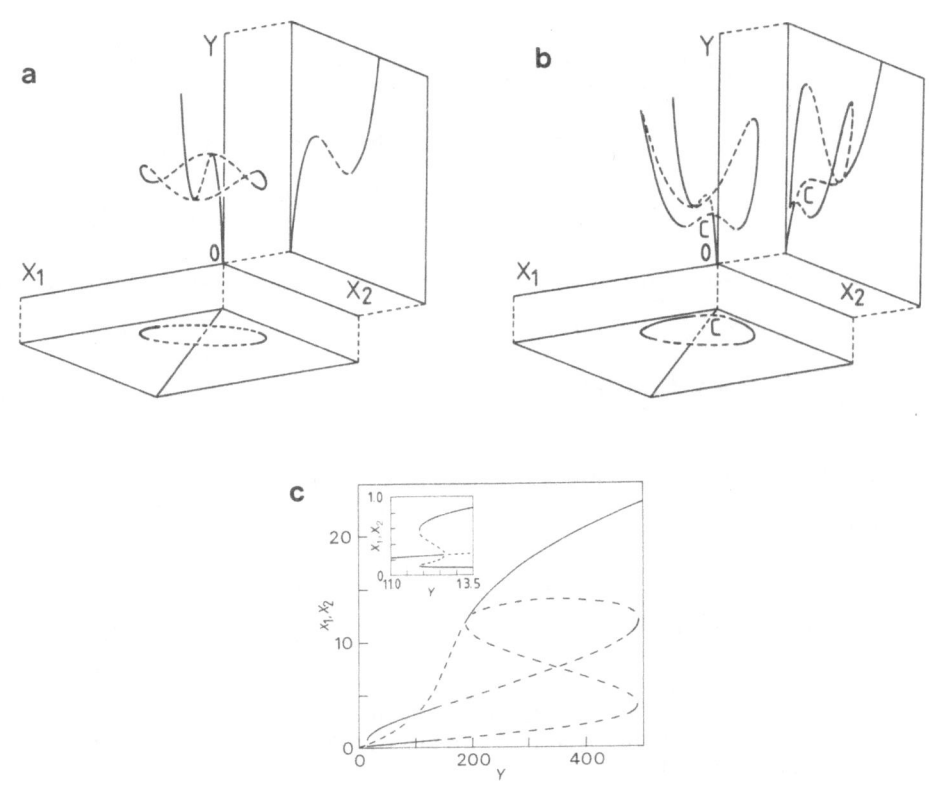

Fig. 2. Steady-state solutions to Eq.(5): (a) η = 1; (b) η = 0.3;
         (c) C = 4, Δ = 5, φ = 15 and η = 0.03. Solid (dashed)curves
         are stable (unstable).

curve in projections on the (X$_{1,2}$,Y) planes. Fig. 2(b) shows how
this loop moves as η is changed.[2] In this figure the point C marks
the bifurcation to optical tristability [the inset in Fig. 2(c)]
predicted by Kitano et al.[11] Our inclusion of saturation in the
present model brings the added structure at higher intensities. With
the inclusion of saturation it is also possible for the asymmetric
branches to become unstable. Fig. 2(c) shows the unstable region as
a function of Y for C = 4, Δ = 5, φ = 15, and η = 0.03. Over a significant range both the symmetric and asymmetric steady states are
unstable and some form of self-oscillation must occur. We will investigate the oscillations which are found by solving Eq.(5) numerically.

    The solutions to Eq.(5) may be traced as trajectories in a four-
dimensional phase space. We plot the solutions found in the long-
time limit as projections on the (X$_1$, X$_2$) plane. A periodic oscillation X$_{1,2}$(t+T) = X$_{1,2}$(t) is represented by a closed curve (limit

cycle), and an aperiodic oscillation by a curve which never retraces
itself.   In the range 145 ≲ Y < 170 periodic oscillations, period-
doubling, and chaos are all observed.   For the upper part of this
range these are illustrated in Figs. 3 and 4(a)-(d).   Before discussing these results we first briefly describe the situation as Y
is increased from Y < 145 where initially the system will be on one
or other of the two stable asymmetric branches [Fig. 2(c)].

At Y ≃ 145 both of the asymmetric branches become unstable via
a Hopf bifurcation.   Beyond the bifurcation point a stable limit cycle
exists around each of the unstable asymmetric steady states.   These
cycles grow with increasing Y, until for Y ≃ 152 they become unstable and the long-time trajectories wind onto a "chaotic figure eight"
encircling both unstable asymmetric steady states.   A chaotic attractor is observed qualitatively similar to that illustrated in Fig.
3(d).   If Y is now decreased the chaotic behavior persists until at

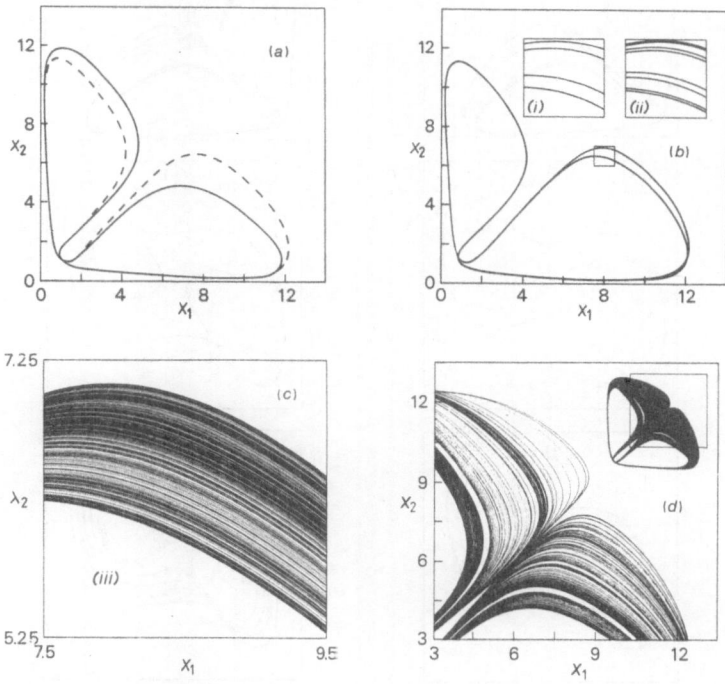

Fig. 3.   Bifurcations to chaos for Y decreasing.   (a) A symmetric
          cycle (solid curve) at Y = 170 has bifurcated to a pair of
          asymmetric cycles (dashed curve and its reflection about
          $X_1 = X_2$) at Y = 166; (b) and (c) each asymmetric cycle
          period doubles to chaos, period two at Y = 165, (i) period
          four at Y = 164.9, (ii) period eight at Y = 164.88, (iii)
          chaos at Y = 164.8; (d) the two asymmetric chaotic attrac-
          tors have merged at Y = 163.7.

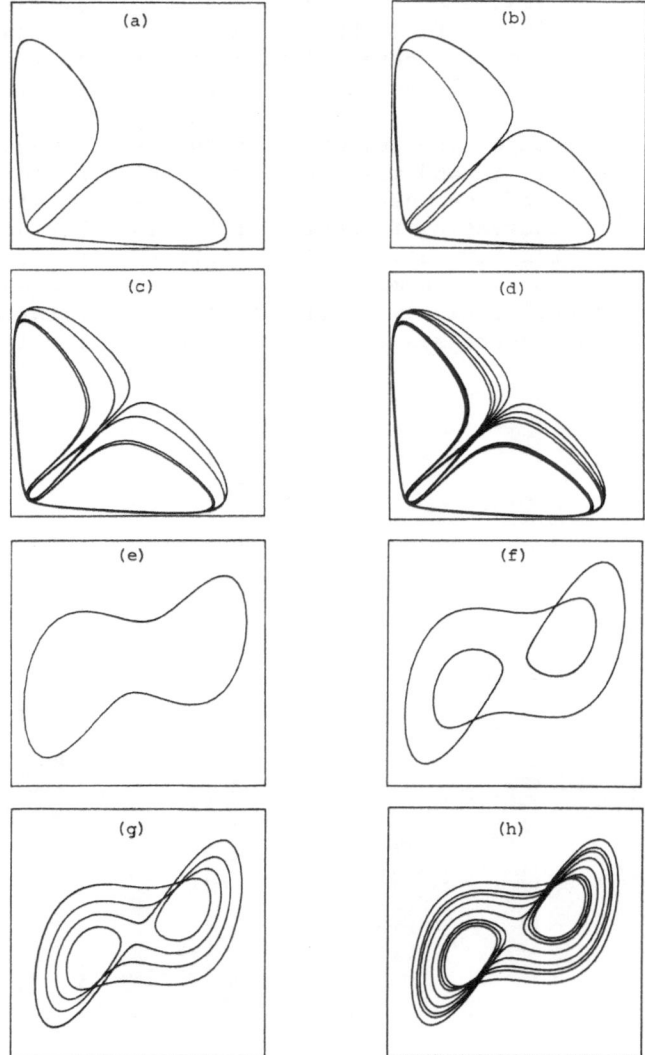

Fig. 4.   Period-doubling of a new type.   (a)-(d) Four symmetric cycles
          from Eq.(5) at Y = 170, 163.6, 161.0, and 160.88; (e)-(h)
          the corresponding four cycles in the Lorenz equations at
          r = 250, 126.1, 105.5, 103.2 (σ = 5, b = 1).

Y ≃ 149 the long-time trajectories return to one or other of the
stable limit cycles. In summary, for 149 ≲ Y ≲ 152 there is a hysteresis involving the coexistence of stable limit cycles about each
of the unstable asymmetric steady states and a chaotic attractor
encircling both. This is reminiscent of a similar hysteresis observed as a function of Rayleigh number in the Lorenz equations.[15]

   Returning now to Figs. 3 and 4, these illustrate the interesting
bifurcation structure which unfolds as Y is decreased from Y = 170.
This initial value of Y is beyond the value over which chaos is observed and a stable limit cycle exists in the shape of a symmetric
figure eight. The bifurcations to chaos from this symmetric cycle
are illustrated in Fig. 3: (1) The symmetric cycle becomes unstable
and is replaced by two asymmetric cycles [Fig. 3(a)]. (2) The asymmetric cycles period-double to chaos to form coexisting asymmetric
chaotic attractors [Fig. 3(b) and (c)]. (3) The two asymmetric
chaotic attractors merge on a single symmetric chaotic attractor
[Fig. 3(d)]. This behavior is also seen in the Lorenz equations as
the Rayleigh number is decreased.[15] It appears, however, that it is
not the last word on this route to chaos. As Y is decreased further
a new stable limit cycle appears, looking very like a superposition
of the two asymmetric cycles previously observed. This new symmetric
cycle bifurcates to chaos as before. Then in a second periodic window, and a third, the same bifurcations from a symmetric limit cycle
to chaos occur. These observations have led us to suggest that this
is the beginning of an infinite sequence of periodic windows based on
a sequence of symmetric cycles which are related via a period-doubling
of a new type.[9] The first four cycles in this sequence are plotted
in Figs. 4(a)-(d). We expect this new period-doubling sequence to
occur in other systems which possess a symmetry like the reflection
symmetry in Eq.(5). The Lorenz equations provide one such example,
and indeed, there the same sequence does occur. The corresponding
cycles are plotted for the Lorenz equations in Fig. 4(e)-(h).

TWO-PHOTON BISTABILITY AND INTRACAVITY SECOND HARMONIC GENERATION

   Two-photon bistability was first proposed by Arecchi and Politi[16]
and has been observed in rubidium vapor.[17] In the early theoretical
work on two-mode two-photon bistability Stark shifts were neglected.
Modeling the dynamics in terms of Eq.(1) Parriger et al.[9] have shown
that with their inclusion instabilities may arise leading to period-
doubling and chaos.

   Some years ago McNeil et al. showed that a driven resonant ring
cavity mode coupled to its second harmonic may produce periodic self-
oscillations.[18] Savage and Walls have recently shown that with the
inclusion of a cavity detuning this system also has chaotic solutions.[10]

ACKNOWLEDGEMENTS

    This research has been supported in part by the United States
Army through its European Research Office.

REFERENCES

1. S. L. McCall, Appl. Phys. Lett. $\underline{32}$, 284 (1978).
2. R. Bonifacio and L. A. Lugiato, Lett. Nuovo Cimento $\underline{21}$, 510
    (1978).
3. K. Ikeda, Optics Commun. $\underline{30}$, 257 (1979); K. Ikeda, H. Daido, and
    O. Akimoto, Phys. Rev. Lett. $\underline{45}$, 709 (1980).
4. H. M. Gibbs, F. A. Hopf, D. L. Kaplan, and R. L. Shoemaker, Phys.
    Rev. Lett. $\underline{46}$, 474 (1981).
5. H. Nakatsuka, S. Asaka, H. Itah, K. Ikeda, and M. Matsuoka, Phys.
    Rev. Lett. $\underline{50}$, 109 (1983).
6. K. Ikeda and O. Akimoto, Phys. Rev. Lett. $\underline{48}$, 617 (1982).
7. L. A. Lugiato, L. M. Narducci, D. K. Brandy, and C. A. Pennise,
    Optics Commun. $\underline{43}$, 281 (1982).
8. C. M. Savage, H. J. Carmichael, and D. F. Walls, Optics Commun.
    $\underline{42}$, 211 (1982); H. J. Carmichael, C. M. Savage, and D. F. Walls,
    Phys. Rev. Lett. $\underline{50}$, 163 (1983).
9. C. Parigger, P. Zoller, and D. F. Walls, Optics Commun. $\underline{44}$, 213
    (1983).
10. C. M. Savage and D. F. Walls, Optica Acta, in press.
11. M. Kitano, T. Yabuzaki, and T. Ogawa, Phys. Rev. Lett. $\underline{46}$, 926
    (1981).
12. S. Cecchi, G. Giusfredi, E. Petriella, and P. Salieri, Phys. Rev.
    Lett. $\underline{49}$, 1928 (1982).
13. W. J. Sandle and M. W. Hamilton, in Proceedings of the Third New
    Zealand Symposium on Laser Physics (Springer) in press.
14. M. W. Hamilton, R. J. Ballagh, and W. Sandle, Z. Physik B, $\underline{49}$,
    263 (1982).
15. C. Sparrow, The Lorenz Equations: Bifurcations, Chaos and Strange
    Attractors, Appl. Math. Sciences $\underline{41}$ (Springer, New York, 1982).
16. F. T. Arecchi and A. Politi, Lett. Nuovo Cimento $\underline{23}$, 65 (1978).
17. A. Giacobino, M. Devaud, F. Biraben, and G. Grynberg, Phys. Rev.
    Lett. $\underline{45}$, 434 (1980).
18. K. J. McNeil, P. D. Drummond, and D. F. Walls, Optics Commun. $\underline{27}$,
    292 (1978).

THE PHYSICAL MECHANISM OF OPTICAL INSTABILITIES

Y. Silberberg and I. Bar-Joseph

Department of Electronics
Weizmann Institute of Science
Rehovot, Israel

INTRODUCTION

Nonlinear optical systems can transform a continuous wave input beam, if it is intense enough, into an oscillating light output. One of the most thoroughly studied systems is an optical ring resonator containing a Kerr medium, first discussed by Ikeda et al[1]. The usual approach of most investigations has been to search, using linear stability analysis, for the first bifurcation. Beyond which the system is studied numerically. Not much has been said about the physical mechanisms leading to instabilities.

In this work we consider the same system, and show that gain, generated by a four-wave mixing process, together with feedback supplied by the resonator, leads to oscillation of sidebands of the input field. Self oscillation is thus interpreted as beating between the input field and the excited sidebands. This approach, beside pointing out the process responsible for instabilities, has an intuitive appeal, and may be used to predict results. Moreover, it enables understanding of the system's evolvement beyond the first bifurcation, and in particular, present a mechanism for period doubling.

It should be mentioned that sideband amplification has been recognized[2] as the process leading to instabilities both in lasers and in bistable systems containing two-level medium[3,4]. Yet, in those cases the nature of the gain process is complicated enough to prevent easy interpretation. The Kerr medium we consider here is characterized by a single time constant τ and thus has a simple tractable behavior. In the following we describe in detail the gain mechanism, and then discuss briefly the results with respect to bistable ring resonators.

Four Wave Mixing Gain Mechanisms

Consider two monochromatic waves, $E_o = A_o \exp[i(\omega_o t - k_o \cdot r)] + c.c.$ and $E_1 = A_1 \exp[i(\omega_1 t - k_1 z)] + c.c.$ interacting in a Kerr medium which is described by a Debye relaxation equation:

$$\tau \dot{n}_{NL} + n_{NL} = n_2 <E^2> \tag{1}$$

Assuming the two waves are not colinear and $|A_1| << |A_o|$, Maxwell-Debye equations yield:

$$\frac{dA_1}{dz} = -i\beta \, A_o A_o^* \, A_1 - i\beta \, \frac{1}{1+i\Delta} \, A_o A_o^* \, A_1 \tag{2}$$

where $\beta = k_1 n_2$ and $\Delta = (\omega_1 - \omega_o)\tau$. This equation can be easily solved to get:

$$I_1(z) = I_1(o) \, \exp(\frac{-2\Delta}{1+\Delta^2} \, \beta \, I_o z) \tag{3}$$

where $I_i = A_i A_i^*$. Obviously, it is amplified for $\omega_1 < \omega_o$, assuming $n_2 > o$. The gain curve is depicted in Figure 1.

The gain peaks for $\Delta = -1$, i.e. $\omega_1 = \omega_o - \tau^{-1}$, vanishes for $\omega_1 = \omega_o$ and turns to loss for $\omega_1 > \omega_o$. This is the simplest mechanism which leads to gain due to the nonlinear interaction between the waves in a sluggish Kerr medium.

Assume now that the two waves E_o and E_1 are colinear. An additional wave E_2 with frequency $\omega_2 = 2\omega_o - \omega_1$ will be formed due to the term $E_o E_o E_1^*$, which is now phase matched. The Maxwell-Debye equations yield:

$$\frac{dA_1}{dz} = -i\beta A_o A_o^* \, A_1 - i\beta \, \frac{1}{1+i\Delta} \, A_o A_o^* \, A_1 - i\beta \, \frac{1}{1+i\Delta} \, A_o A_o \, A_2^*$$

$$\frac{dA_2}{dz} = -i\beta A_o A_o^* \, A_2 - i\beta \, \frac{1}{1-i\Delta} \, A_o A_o^* \, A_2 - i\beta \, \frac{1}{1-i\Delta} \, A_o A_o \, A_1^* \tag{4}$$

assuming $|A_1|$, $|A_2| << |A_o|$. These equations are solved by:

$$A_1(z) = \{A_1(o) - [A_1(o) + A_2^*(o)] \, P_o \, \frac{i}{1+i\Delta} \, z\} \, \exp(-iP_o z)$$

$$A_2(z) = \{A_2(o) - [A_1^*(o) + A_2(o)] \, P_o \, \frac{i}{1-i\Delta} \, z\} \, \exp(-iP_o z) \tag{5}$$

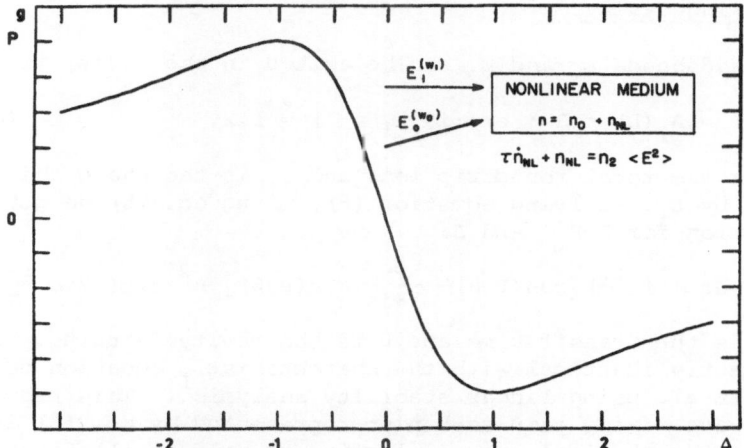

Fig. 1: Exponential gain vs. normalized frequency difference Δ for
         two wave interaction. Insert = Interaction geometry.

where $P_o = \beta I_o$. Note that the gain, which was exponential in the first
case (eq. 3) is now only linear, and is strongly dependent on the
initial conditions. In Figure 2 the intensities $I_1(L)$ and $I_2(L)$ are
depicted for the case $P_o L = 0.1$ as a function of the frequency for two
different initial conditions. By changing the relative phase between
the two sidebands the peak of the gain curve can be shifted between
Δ=o and Δ=-1.

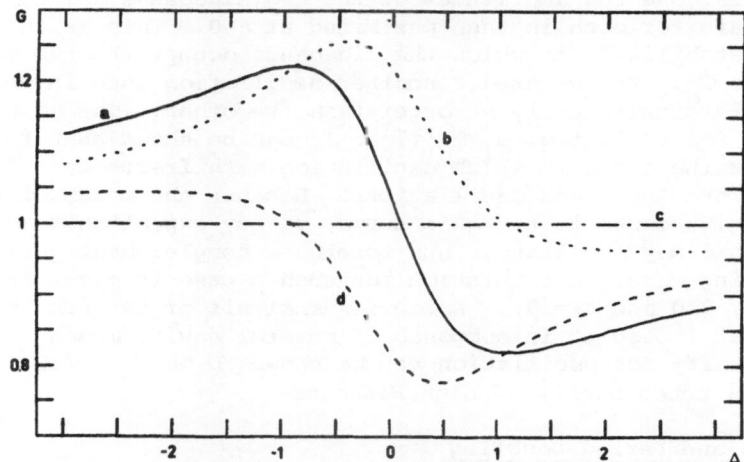

Fig. 2: Four-wave mixing gain vs. normalized frequency difference
         Δ in a medium where $\beta I_o L = 0.1$ for different initial condi-
         tions: a) $A_1(o) = A_2(o)$. b) $A_1(o) = i\, A_2(o)$.
         c) $A_1(o) = -A_2(o)$. d) $A_1(o) = -i\, A_2(o)$.

Bistable Ring Cavity

The sidebands E_1 and E_2 can be exited in the cavity if:

$$A_i(o) = A_i(L) \cdot B \cdot \exp(i\phi_i), \quad i = 1,2. \tag{6}$$

where B is the total roundtrip loss and ϕ_i is the phase shift experienced by E_i. Solving equation (6), using eq. (5) we get a complex equation for $P = P_o L$ and Δ:

$$1 - 2B \exp(-i \frac{tr}{\tau}) [\cos(P-\theta) - \frac{P}{1+i\Delta} \sin(P-\theta)] + B^2 \exp(-2i\Delta \frac{tr}{\tau}) = 0 \tag{7}$$

where tr is the transit time and θ is the cavity detuning. Equation (7) is exactly identical with the characteristic equation derived by Ikeda et al. using linear stability analysis[1]. This proves that the four-wave mixing gain mechanism represented by eq. (5) is fully describing the physical process leading to self-oscillations.

Equation (7) is general, in the sense that it describes the oscillation condition for any value of the parameters τ, tr, θ and B, yet because of its complexity it is not very useful. It has many solutions, and one should better know in advance where to look for those. Understanding of the gain process described in the previous section enables prediction of various modes of oscillation. A detailed discussion of the various self-oscillating regime will be published elsewhere. Here we will describe only one such case, which we believe represents a new type of self-oscillation.

Consider the case $\tau \ll tr$, that is there are many cavity modes within the width of τ^{-1}. We expect two distinct oscillation regimes. In the first one the amplitudes of the two sidebands are comparable, and the gain for both is then maximized at $\Delta=0$. This is the famous "Ikeda Instability", in which the sidebands occupy the two modes nearest to E_o. Yet we expect another oscillation mode in which one sideband is significantly stronger than the other. The gain is then maximized for $\Delta \approx -1$, i.e. $\omega_1 \approx \omega_o - \tau^{-1}$. It can be shown that there is a large detuning range in which oscillation with frequency τ^{-1} will set on before the Ikeda oscillations. Because the modes of the cavity are quite dense in the gain curve, we may expect that more than one sideband may oscillate, thus forming a complex beat pattern in the outgoing wave. A simulation for such a case is given in Fig. 3 for B=0.9, $\theta=0$ and tr=20τ. A Fourier analysis of the same pattern proves that indeed it is composed of several cavity modes. The threshold intensity for oscillation can be shown to be $P_o \simeq 2.9 (1-B)$, assuming a tuned cavity of high Finnesse.

Harmonics and Period Doubling

Probably the most attractive feature of our viewpoint is the ability to comprehend the behavior of the nonlinear system beyond the

first instability point. It is well known that as the power is
increased beyond the first threshold the oscillating output changes
its profile, until a second threshold is reached, where period
doubling occurs. This process repeats itself successively. The
change in the oscillation pattern is explained by the appearance of
higher harmonics of the basic frequency, while period doubling marks
the addition of a new Fourier component at half the basic frequency.

Both these phenomena are well understood in the optical
domain. Harmonics of an oscillating sideband E_1 are generated by
the wave mixing process $E_1E_1E_o{}^*$. This term descibes a source for
oscillations at $\omega_o + 2(\omega_1-\omega_o)$, and therefore harmonics are
always present. On the other hand, the phenomena of period doubling appear only at certain threshold intensity. A field
E_3 at $\omega_3 = (\omega_1+\omega_o)/2$ is amplified through the process $E_oE_1E_3{}^*$.
Note the difference between this gain term (which is proportional
to E_3) and a source term. Oscillations at ω_3 occur only if gain
and resonance conditions are satisfied for E_3. Obviously, once E_3
is present, the beating period is doubled. This process, repeating
itself, is the basic mechanism of successive period doubling in a
nonlinear optical system.

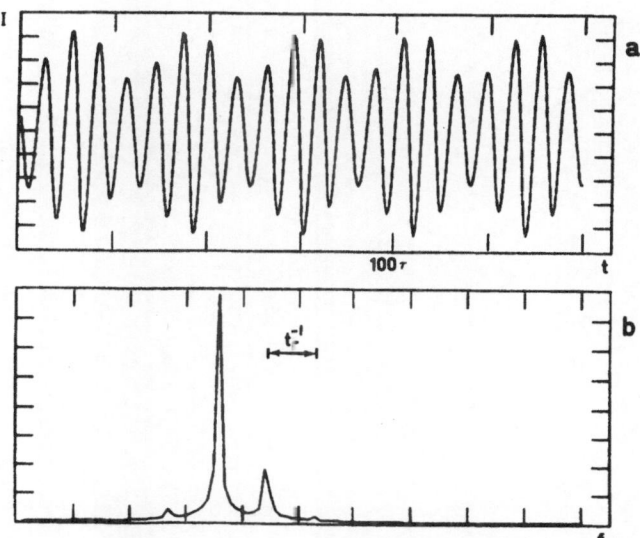

Fig. 3:   a) Light intensity vs. time for a tuned cavity with B=0.9
and tr/τ=20. The oscillation period is ∿ 2πτ.
b) Fourier spectrum of the oscillation, showing that several cavity modes oscillate.

REFERENCES:

1. K. Ikeda, H. Daido and O. Akimoto, Optical Turbulence: Chaotic
    Behavior of Transmitted Light from a Ring Cavity, Phys. Rev.
    Lett. 45:709 (1980).

2. S. L. McCall, Instabilities in Continuous-Wave Light Propaga-
    tion in Absorbing Media, Phys. Rev. A 9:1515 (1974).

3. S. T. Hendow and M. Sargent III, The Role of Populations in
    Single-Mode Laser Instabilities, Opt. Commun. 40:385 (1982).

4. L. W. Hillman, R. W. Boyd and C. R. Stroud Jr., Natural Modes
    for the Analysis of Optical Bistability and Laser Instability,
    Opt. Lett. 7:426 (1982).

CHAOS IN OPTICS

F.A. Hopf, M.W. Derstine, H.M. Gibbs, and M.C. Rushford

Optical Sciences Center

University of Arizona

Tucson, Arizona   85721

I. Introduction

Recently there has been a renewed interest in the random motion of deterministic systems, a subject dating back to Poincare. The original interest lay in trying to explain the fact that many-body problems in physics are well described by theories based on random behavior. Classical physics, however, views such systems as deterministic. It was discovered that most classical motions are, in fact, erratic, but the discovery of quantum mechanics caused the interest in this subject to die down. The revived interest lies in trying to understand turbulent behavior, which may be related to these erratic motions.

The interest in these problems has been spurred by two recent advances in understanding the erratic motions. One is a formal definition of the term chaos, which has helped clarify the problem.[1] To understand the phenomenon, consider the way we understand transitions between ordered and disordered states in a laser.[2] The difference α between the gain g and loss ν/Q, i.e. $\alpha = g - \nu/Q$, is a stability eigenvalue that determines when the disordered state is stable ($\alpha < 0$) vs. unstable ($\alpha > 0$). If the disordered state is unstable, the device lases, and we find it in the ordered single-mode state. This ordered state is stable, but with further increases in the gain it may become unstable, and an ordered multi-mode state may be stable. The stability of each state is determined by a Taylor's expansion about the motion of interest. This expansion gives rise to eigenvalues λ (α for a laser at threshold), and the motion is said to be stable or not depending on whether all λ's are negative or not. A system is said to be chaotic if at least one λ is positive for all possible motions. Then, all conceivable motions of the system are unstable.

67

The chief prediction of chaos is that the power spectrum is a continuum possibly combined with discrete frequencies.[1] The original concept of turbulence, pioneered by Landau,[3] was that the motion was described by a large number of discrete but noncommesurate frequencies giving rise to a complex motion. This concept is not in accord with experiment.[4] Instead, continuous spectra are found which is in accord with the prediction of chaos.[1] Unfortunately, contiuous spectra can arise from many phenomena in physics, e.g. spontaneous emission, which are not related to chaos. Experiments that measure only continuous spectra cannot, in principle, prove that chaos is a correct explanation for the facts. Moreover, the original theories of chaos were vague as to details, so there was little an experimentalist could do to test the theory. This situation was improved substantially by the work of Feigenbaum,[5] who showed that, in certain special cases, there were quantitative features of chaotic systems that were amenable to measurement. These features are discussed below, and apply to the measurements described here.

II. Chaos in optics

The pioneering theoretical work on chaos in optical bistability was done by Ikeda.[6] He emphasized that the dynamics of Fabry-Perot cavities is described by difference-delay equations. Such equations have stability properties that are different from those of the differential equations of mean-field theories that were in vogue at the time.[7] Solutions that were thought to be stable were, according to Ikeda, unstable. Moreover, the temporal motions predicted in the region of instability have the property that their periods are multiples of the cavity round trip time T (specifically 2T, 4T, 8T, etc.). This sequence is the one treated by Feigenbaum (see above), and is called the period doubling sequence[5]. The remarkable feature of these motions is that they cannot be described by an expansion in the modes of the cavity, which always results in motions whose period is an integer fraction of T (e.g. T, T/2, T/3, etc.).[2] Thus the observed motions, which occur in cavities with arbitrarily high Q, cannot be predicted by a theory based on mode expansions. Hence a correlary of Ikeda's prediction is that this standard procedure in laser theory is not self-consistently valid. One can get predictions that are sensible, self-consistent and wrong.

III. Experimental layout

The initial experiments at the University of Arizona were motivated by this peculiar periodicity, rather than by chaos itself.[8] The experiments reported here were done using a hybrid device, but they have since been verified by experiments using cavities.[9] Fig. 1 shows our experimental set-up. A helium-neon

laser beam passes through a Glan prism polarizer, then through a
KDP crystal four times before coming back through the polarizer

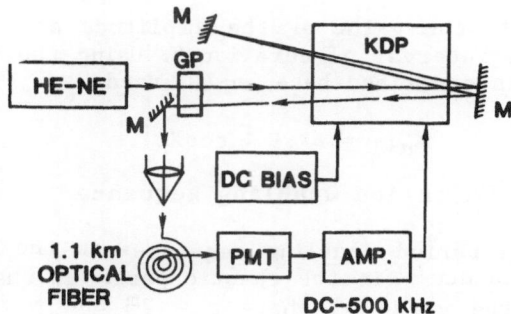

Fig. 1. Experimental layout: He-Ne laser; GP, Glans prism;
KDP, crystal; M. mirrors; PMT, photomultiplier.

where it is coupled into the optical fiber. The light emerging
from the fiber is detected with a photomultiplier, and an amplified
electrical signal is applied to the KDP crystal.

The equation that describes our system is:[8]

$$\tau \dot{X}(t) + X(t) = \mu\pi[1 - \xi\cos(X(t - T) + X_b)] \tag{1}$$

where $X = \pi V/V_h$, V is the voltage applied to the modulator, V_h is
the halfwave voltage of the modulator, and $X_b = \pi V_b/V_h$ is a
variable bias. The quantity μ is proportional to the product of
the input laser intensity and the amplifier gain. The bifurcation
parameter in the experiments is μ, which is measured directly in
the apparatus by procedures discussed below. The ability of the
system to achieve extinction is measured by $\xi = 0.96 \pm 0.01$. We
have explored the ranges of $-\pi < X_b < -\pi/2$. All of the data that
is shown in detail was taken for $X_b = -\pi$, which is representative
of the other cases. This choice of bias maximizes the domains of μ
over which individual waveforms are stable and thus minimizes the
consequences of the 1% drift in the laser power, which is the
major uncertainty in measuring μ. The range of μ that we can test
at this bias lies below the "upper branches" of the bistable
device. The procedure to measure μ is to first break the feedback
loop between the final amplifier and the modulator. The input to
the modulator is then shorted to insure a zero voltage input (the
bias voltage is not changed), and the voltage V at the output of
the final amplifier is measured. This determines μ as $\mu = V/V_h$.

Ikeda showed that for $\tau \ll T$ the representation of the experiment could be simplified to the difference equation[6]

$$X_{n+1} = \mu\pi(1 - \sin(X_n+X_b)), \tag{2}$$

where $X_n = X(nT)$ is the value of the amplitude at one instant in time within a time interval of duration T. Using the bias $X_b = -\pi$ for the experiments described here, eq.(2) reads

$$X_{n+1} = \mu\pi(1 + \cos X_n) \tag{3}$$

IV. **Period doubling sequence**

Eq.(3) gives a period doubling sequence[5] in the periodic part and an inverse sequence[10] in the chaotic part. In these sequences, all periods have the value pT where $p = 2^m$ and m is an integer. We denote such waveforms as P_p, or N_p if chaotic. In Fig. 2 the waveforms that are observed in the device are shown as a function of increasing μ. The stationary waveform found for small μ is not shown. As μ is increased the waveform changes to the form P_2, which has a period twice that of the optical fiber. This waveform is shown in the top row of the figure. The next two waveforms are P_4 and P_8 respectively. The next four waveforms are chaotic. The first three chaotic waveforms are those labeled N_8, N_4 and N_2. These are characterized by periodic waveforms superimposed on a random background. Spectrally these have sharp peaks with a continuous background. The final waveform is one without sharp spectral peaks. For reasons discussed later, it is unclear what this waveform really is (i.e. it may not be the waveform N_1).

The prediction of Feigenbaum has to do with the values of μ at which the transitions between the waveforms take place.[5] Unfortunately, his prediction is only precise in the limit of large p which, because of noise in the device, we cannot reach (the period doubling sequence truncates). Our results are, nonetheless, in qualitative agreement with his predictions. Ikeda has calculated the bifurcation points from Eq. (1), and finds good agreement with experiment.[11] In practice, one gets decent agreement with experiment using eq. (3).

IV. **Character of the transitions**

The term "bifurcation" is used to describe qualitative changes in the character of the waveform (e.g. changes in period). The prediction of Eq. (3) is that the change should look like a second-order phase transition.[5] In the spectra of Fig. 2 one can see that the periodic waveforms have characteristic peaks which uniquely specify the waveforms. For example, a P_8 waveform is characterized by a peak at the frequency 8/T. To determine the character of the transitions, we measure the spectral power in these identifying

μ=

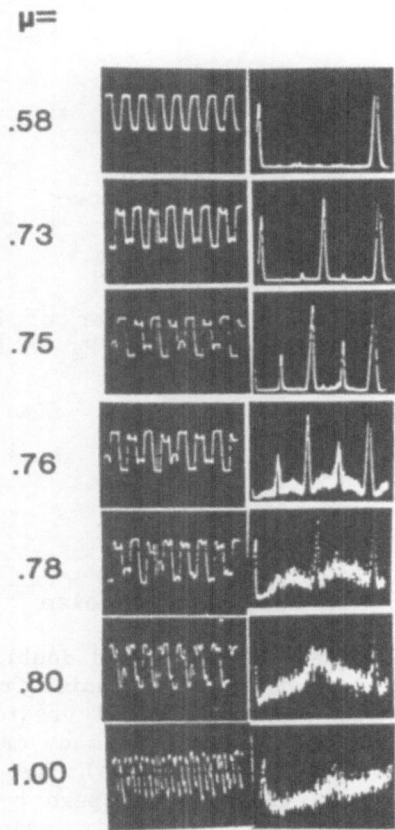

.58

.73

.75

.76

.78

.80

1.00

Fig. 2 Waveforms (right column) and spectra (left column) of the device shown in Fig. 1 as a function of increasing μ.

features and plot the results as a function of μ. The results of these measurements are shown in Fig. 3. The power in the continuous background was used for the transition to chaos. The bifurcations between the periodic waveforms are predicted to be like second order phase transitions, and the experimental curves have the proper shape for such a transition. The transition to chaos is not signifigantly different (to within the 1% uncertainty in μ). Within the error of experiment we can say that the transition to chaos is like a second-order phase transition. We show below cases that are very different from this.

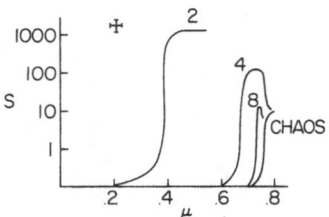

Fig. 3. Measurement of the character or che bifurcations for the transitions from P_1 to P_2, P_2 to P_4, P_4 to P_8 and from P_8 to chaos.

V. Response to noise

One of the difficulties of the period doubling sequence is that it is very sensitive to noise. In a noise-free case, the device should give very large periods (Eq. (3) predicts infinitely large periods, but Eq. (1) has been shown, in many cases,[12-14] to support only finite periods, even without noise). The major attraction of optical devices, which is their great speed compared, for example, to hydrodynamic flows,[4] leads to their major disadvantage. The speed of the device means that we are limited by shot noise.

In Fig. (4) we show the response of our system to varying levels of shot noise. The noise levels are altered by attenuating the light at the output of the fiber, and then increasing the gain in the amplifier to compensate for the loss. The bifurcation point is the value of μ at which the transition takes place. The reported value of μ is determined by the rapidly-rising point of the curves illustrated in Fig. 3. As the noise increases, the higher period solutions drop out. The way the period eight waveforms disappear is mirrored by the way the period four waveforms disappear. This suggests that there is no systematic truncation of the period doubling sequence due to the properties of Eq. (1), which is in agreement with the predictions of Ikeda.[15] Instead, the truncation is due to noise throughout the experimentally acessable domain. Note that, for a given p the periodic waveform disappears at a lower level of noise than the chaotic one. This feature of the experiment is in disgreement with theory.[12]

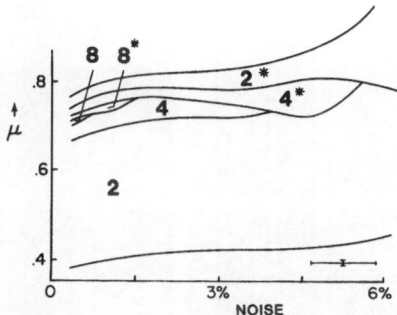

Fig. 4. Bifurcation points in the hybrid device as a function of increasing noise levels.

VI. Other waveforms

At the point that Eq. (3) predicts that there should be a transition between N_2 and N_1, we observe something altogether different.[16] We observe waveforms whose period is either 4T, 8T, or an odd fraction of 2T (e.g. 2T/3, 2T/5, etc.). The waveforms are not like those of Fig. 2; instead they oscillate much more rapidly in time. The strongest spectral features are at frequencies $f = n/2T$, where n is an odd integer. Waveforms of this type are referred to as "frequency-locked" when observed in the transition to chaos, and we denote them as $L_{2/n}^n$, where the subscript denotes the period and the superscript denotes the n that defines the fundamental frequency (i.e. n/2T). Exactly which n's are observed is a sensitive function of the parameters of the device. The waveforms and spectra for the periods 2T/n are shown in Fig. 5.

It is possible to observe transitions between $L_{2/n}^n$ and L_4^n. These appear to be similar to those illustrated in Fig. 3. It is also possible to observe transitions between different n's, in which case, one observes hysteresis loops. There is a range of μ over which waveforms with a given n are stable. At high μ the waveform changes discontinuously to one of higher n; at low μ it changes discontinuously to one of lower n. It is tempting to call these first-order transitions except that they are not of this type at all. The discontinuous transition at high μ is always associated with a transition to chaos (the waveform in the bottom row of Fig. 2 is probably one of these waveforms). The process by which the waveform changes is similar to what has been called a "precipitation".[17] The chaotic waveform lasts for an indeterminate

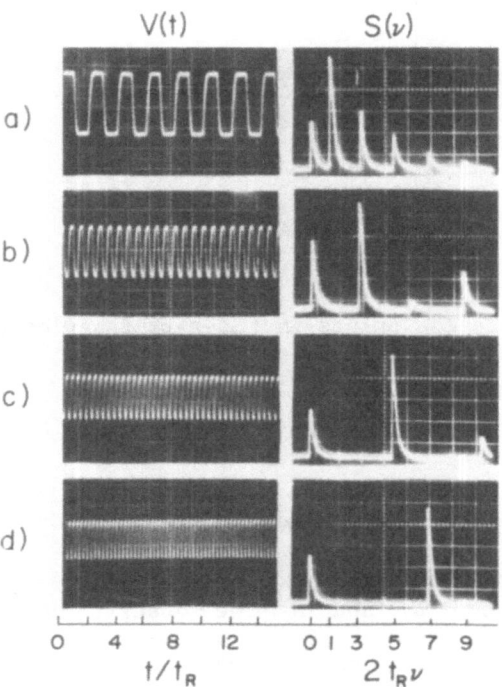

Fig. 5. Waveforms of the type $L^n_{2/n}$ (period $2t_r/n$). The rows show the cases n = 1, 3, 5, and 7, respectively. the left column is the vaveform. The right column is the power spectrum.

amount of time, and then suddenly (within a few T) changes to a frequency-locked form. At the low μ end, the waveform of period 2T/n becomes unstable. It is not, as it is with bistability, that the waveform is no longer a solution to the equations; instead, it is still a solution, but it is unstable.

The reason for this instability is not, at present, understood. Theory says that the frequency-locked waveforms should be stable for much lower μ than is observed experimentally.[18]   Present speculation is that noise destabilizes the waveforms.[19]

The various waveforms in the frequency-locked regime can be shown to be various ways of building solutions of Eq. (1) from solutions of Eq. (3).   These waveforms are thus alternative versions of the period doubling sequence. A waveform with period 2T/n is also periodic on an interval 2T. Hence the transition 2T/n to 4T, which seems anomalous at first, is also a transition from 2T to 4T, i.e. period doubling.

VII. Summary of results

In summary, we have seen that the experiment is, at best, in qualitative agreement with the Feigenbaum theory[5] (which is based on Eq. (3), and the experiment is usually in good agreement with predictions based on Eq. (1)). There are, however, many qualitative disagreements with the period doubling concept. There are observed waveforms that are not anticipated by the theory, but these turn out to be variations on the period-doubling scheme. The experiment is not clean enough to test these other sequences to see whether they conform to Feigenbaum's predictions. In addition, while the abserved truncation by noise is in agreement with theory, the qualitative features are wrong.

Because of the sensitivity of the period doubling sequence to noise, it is likely to be a poor tool for exploring chaos in optical systems. Moreover, theoretical calculations have shown that all-optical systems have alternative paths to chaos, just as other systems do.[20] (There are, at present, no quantitative predictions to guide experiments in cases where the path to chaos is nonunique.)

At present, there are no reliable quantitatve methods, other than for the period doubling sequence, to determine whether a given system is erratic because of chaos or because of some other reason. Techniques that have been developed are dependent on devices being entirely drift free.[21] It is unknown what, if any, artifacts might occur due to inevitable drifts in real systems. For that reason, there is still no general way of being sure that chaos is taking place. One should thus use the term "alleged chaos" for most optical measurements. This qualification is implicit in the following discussion.

VIII. Other measurements and predictions of chaos in optics.

Chaos has been observed in two laser systems, pulsed, multimode CO_2,[22] and in the Casperson instability[23] using He-Xe.[24] It is likely that more examples will be forthcoming in the future. Lasers are poorer candidates for studying chaos than bistable devices, since they are much noisier. There is the all-optical bistability experiment mentioned at the beginning of section III, which uses a pulsed, mode-locked YAG and a fiber-optic cavity configured as a ring. Such pulsed experiments do not permit detailed investigation of the bifurcation structure. We have investigated self-focussing devices of the sort described by Bjorkholm, Smith, Tomlinson and Kaplan.[25] We use a modified version of the their device as illustrated in Fig. 6. It differs from their device only by the location of the pinhole. The pinhole in our device is located before the detector rather than between the mirror and the sodium cell. The changes in the optical

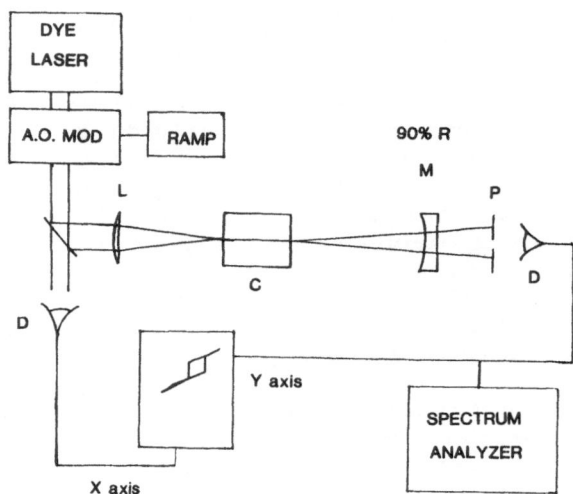

Fig. 6. Schematic of the all-optical bistable device. C,
Sodium cell; L, Lens; M, Mirror; P, Pinhole; D, Detector.

beam are radial in this case, and the fluctuations are in the power
density, rather than in the total power. The pinhole assures
detection of a small portion of the beam.

This experiment differs from the hybrid in one important way.
In the hybrid, the response time of the device is short compared to
the round trip time. The opposite is the case here. Recent
theoretical work by Winful[26] and others[27] has shown that chaos can
occur in this regime of operation. A major difference in the
experimental results is that, in the hybrid, chaos and bistability
are two independent phenomena (all data shown above is taken in
regimes where there are no upper branches), while in the device in
Fig. 6, erratic motions are observed only on the upper branch of of
bistable loops. In Fig. 7, a sequence of waveforms and power
spectra are shown.

Since a time-average upper branch is observable, these
oscillations cannot be due to regenerative oscillations that take
the device between the upper and lower branch. If that were the
case, then only one time-averaged state would be observed.

The progression of waveforms and spectra in Fig. 7 are
qualitatively similar to those in the hybrid, except that there is
no bifurcation structure to the periodic regime. Cases without
bifurcation sequences (with or without noise) are known in
hydrodynamics.[4] We have no way of knowing whether our device has
a sequence that has been truncated by noise, or whether this is a
case in which there is no sequence.

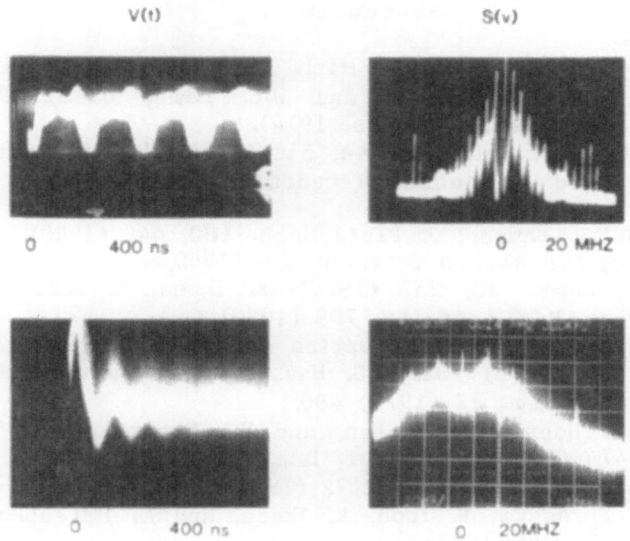

Fig. 7. Waveforms (left column) and power spectra of the
waveform (right column) from the device in Fig. 6.

IX. Conclusion

The experiment described in Figs. 6 and 7 illustrates the
questions raised earlier. On what basis can we conclude that we
are observing chaos? We can see that an instability has occurred,
and that the device makes a transition from a periodic to a
nonperiodic waveform, but is this enough? There is one other
device in optics, namely the inhomogeneously-broadened laser
amplifier[28] (the alleged astronomical maser) that makes a
transition from a single frequency to a continuum, and it is
unclear whether this bears any relationship to chaos. Until more
reliable tests come along, we must regard chaos in optics as
hypothetical. We have made a promising beginning in this problem,
but firm conclusions cannot yet be reached.

Acknowlegment

The authors would like to acknowledge support by the National
Science Foundation under contract No. PHY-8104982 and would like
to thank the Corning Corp. for supplying the optical fiber.

References

1. D. Ruelle and F. Takens, Commun. Math. Phys. **20**, 167-192 (1971).
2. M. Sargent III, M.O. Scully and W.E. Lamb, **Laser Physics** (Addison-Wesley, Reading, Mass. 1974).
3. L. Landau, C.R. Acad. Sci. URSS **44**, 311 (1944); L. Landau and E.M. Lifshitz, **Fluid Mechanics** (Academic, Reading, Mass., 1959), pp. 103-107
4. J.P.Gollub and S.V. Benson, J. Fluid Mech. **100**, 449 (1980).
5. M.J. Feigenbaum, Los Alamos Science **1**, 4 (1980).
6. K. Ikeda, Opt. Comm. **30**, 257 (1979); K. Ikeda, H. Daido, and O. Akimoto, Phys. Rev. Lett. **45**, 709 (1980).
7. R. Bonifacio and L.A. Lugiato, Optics Comm. **19** (1976) 172; Phys. Rev. Lett. **40** (1978) 1023; SS. Hassan, P.D. Drummond and D.F. Walls, Optics Comm. **27** (1978) 480.
8. H.M. Gibbs, F.A. Hopf, D.L. Kaplan, and R.L Shoemaker, Phys. Rev. Lett. **46**, 474 (1981). F.A. Hopf, D.L. Kaplan, H.M. Gibbs, and R.L Shoemaker, Phys. Rev. A **25**, 2172 (1982).
9. H. Nakatsuka, S. Asaka, H. Itoh, K. Ikeda, and M. Matsuoka, Phys. Rev. Lett. **50**, 109 (1983).
10. E.N. Lorenz, in **Nonlinear Dynamics**, edited by R.H.C. Helleman (New York Academy of Sciences, New York, 1980).
11. K. Ikeda (private communication)
12. J.P. Crutchfield and B.A. Huberman, Phy. Lett. **77A**, 407 (1980). J. Crutchfield, M. Nauenberg, and J. Rudnick, Phys. Rev. Lett. **46**, 933 (1981); B. Shraim, L.E. Wayne, and P.C. Martin, Phys. Rev. Lett. **46**, 935 (1981).
13. P.S. Lindsay, Phys. Rev. Lett. **47**, 1349 (1981); W. Lauterborn and E. Cramer, Phys. Rev. Lett. **47**, 1445 (1981).
14. M.W. Derstine, H.M. Gibbs, F.A. Hopf, and D.L. Kaplan, Phys. Rev. A **26**, 3720 (1982).
15. K. Ikeda (private communication)
16. M.W. Derstine, H.M. Gibbs, F.A. Hopf, and D.L. Kaplan, Phys. Rev. A **27**, 3200 (1983).
17. M. Gronchi, V. Benza, L.A. Lugiato, P. Meystre, and M. Sarget III, Phys. Rev. A **24**, 1419 (1981).
18. K. Ikeda, K. Kondo, and O. Akimoto, Phys. Rev. Lett. **49**, 1467 (1982).
19. K. Ikeda and F.A. Hopf (private communication)
20. R.R. Snapp, H.J. Charmichael, and W.C. Shieve, Opt. Comm. **40**, 68 (1981). J. Moloney to be published.
21. R.H. Simoyi, A. Wolf, H.L. Swinney, Phys. Rev. Lett. **49**, 245 (1982); J.D. Farmer (personal communications), discussion of sensitivity of chaotic systems to noise and drift.
22. I.T. Arecchi, R. Meucci, G. Puccioni, and J. Tredicce, Phys. Rev. Lett. **49**, 1217 (1982).
23. L.N. Casperson, A. Yariv, App Phys Lett. **17**, 259 (1970); L.N. Casperson: IEEE JQE-**14**, 756 (1978); Phys. Rev. A, **21**, 911(1980); **23**, 248 (1981).

24. N.B. Abraham, M.D. Coleman, M. Maeda, and J.C. Wesson, App. Phys. B, **28**, 169 (1982); M. Maeda and N.B Abraham, Phys. Rev. A, 20, 3395 (1982).

25. J.E. Bjorkholm, P.W. Smith, W.J. Tomlinson, and A.E. Kaplan, Optics Lett. **6**, (1981).

26. H.G. Winful and G. Cooperman, Appl. Phys. Lett. **40**, 29 (1982).

27. K. Ikeda and O.Akimoto, Phys. Rev. Lett. 48, 617 (1982); Y. Silberberg and L Bar Joseph, Phys. Rev. Lett. **48**, 1541 (1982); J.A. Goldstone and E.A. Garmire, IEEE JQE-**19**, 208 (1983).

28. M.M. Litvak, Phys. Rev. A **2**, 2107 (1970);L.N. Menegozzi, and W.E. Lamb Jr., Phys. Rev. A **17**. (1978). For experiments see e.g. J.H. Parks, D.R. Rao, and A. Javan, Appl. Phys. Lett. **13**, 142, (1968).

TURBULENCE AND 1/f NOISE IN QUANTUM OPTICS

F. T. Arecchi

Istituto Nazionale di Ottica - Firenze

and Università di Firenze - Italy

Generalized multistability and chaotic behavior are observed in several quantum optical systems. Inducing jumps between independent attractors leads to a low frequency power spectrum of 1/f type. Experiments and theory are here presented.

An exciting chapter of physics has been the study of fluctuations and coherence in lasers: how atoms or molecules, rather than radiating e.m. field independently, decide to "cooperate" to a single coherent field; then, for still higher excitation, how and why they organize in a complex pattern of space and time domains, with small correlations with one another (optical turbulence). Here we discuss these new features of quantum optical systems.

It is generally known [1] that $n \geq 3$ degrees of freedom nonlinearly coupled may lead to multiperiodic or chaotic oscillatory behavior (turbulence). Since quantum optics, in the finite-boundary (single mode) plus semiclassical approximations, is ruled by the 5 Maxwell-Bloch equations, one expects similar behavior in quantum optical devices [2]. Often these instabilities are ruled out by time scale considerations. When the atomic variables have fast damping times, at any instant polarization and inversion are in quasi-equilibrium with the rather slow field amplitude, hence the evolution reduces to a one-equation dynamics (adiabatic elimination of atomic variables). That is why a gas laser beyond threshold assumes a smooth coherent behavior. But make a bad cavity, or add an external modulation as done for Q-switching or mode-locking: then one easily gets a three-variable dynamics, sufficient to yield chaos, for particular values of the coupling constants. What was initially considered as a "bad" or "dirty" behavior (self-pulsing, irregular mode-locking) is nowadays studied as a relevant phenomenon. Furthermore, when many domains of attraction coexist (optical multistability) and an external noise allows for jumps among them, a low frequency power spectrum appears with a shape $f^{-\alpha}$ ($\alpha \sim 1$) like the 1/f noise familiar in many systems [3,4].

Equivalent to the three Lorenz first order eqs. [1] is a system of two 1st order eqs. (or one 2nd order eq.) plus an external modulation. An example is the driven Duffing oscillator

$$\ddot{x} + \gamma \dot{x} + \omega_i^2 x - \beta x^3 = A \cos \omega t \qquad (1)$$

which can be experimentally realized with an electronic circuit [5,3]. The potential corresponds to a single minimum. For different control parameters μ (either modulation amplitude A or frequency ω) it may give a sequence of subharmonic bifurcations leading eventually to chaos.

Noise is not essential (deterministic chaos), but if we add it, the number of subharmonic bifurcations before chaos becomes smaller and smaller. This can be put in terms of a scaling law where the variance of the external noise appears somewhat as a modification of the control parameter [5].

Let us now change the sign of the potential, getting two stable valleys. Depending on the initial conditions, we have two independent attractors. Increase μ until they both get strange. Now, addition of a random noise may trigger jumps from one to the other. These jumps give a low frequency divergence in the power spectrum [3]. Here, random noise is essential to couple the two strange attractors otherwise independent. After the first evidence of the jumping phenomenon, a similar effect was observed in a Q-modulated CO_2 laser [4]. It corresponds to a set of 2 coupled rate eqs., with time dependent cavity losses k(t), that is, calling Δ the population inversion and n the photon number, to

$$\dot{\Delta} = R - 2 G n \Delta - \gamma_{\parallel} \Delta$$
$$\dot{n} = \qquad G n \Delta \quad - k(t) n \tag{2}$$

where $k(t) = k_o (1 + m \cos \omega t)$.

Fig. 1a shows generalized bistability, that is, the simultaneous coexistence of two attractors in the phase space $(\dot{n}, n)$. Increasing the modulation depth m, the attractors become strange and the power spectrum displays a low frequency divergence (fig. 1b).

Experiments [3,4] show that nonlinear driven systems yield power spectra with a low frequency divergence $f^{-\alpha}$, α being around 1 whenever the following conditions are fulfilled: i) the system is multistable, that is, it has 2 or more attractors; ii) the attractors are near to be destabilized or they have just become unstable; and iii) the system is "open" to external fluctuations, i.e., the presence of noise is essential to yield jumps between different basins of attraction. The above conditions show that we are in presence of a phenomenon which occurs beyond the usual approach to chaos by either one of the current scenarios. A simple jump between two at-

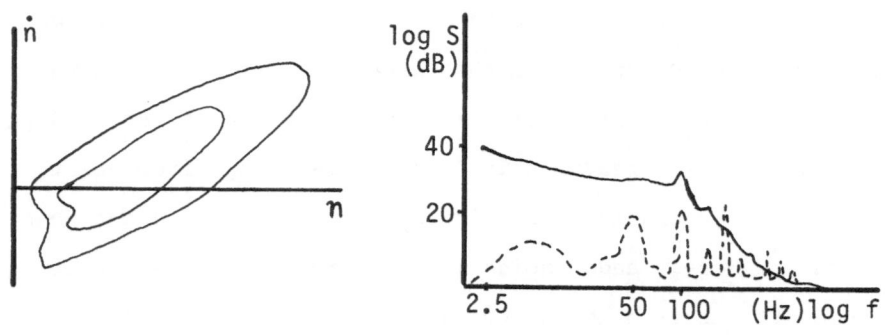

Fig. 1

Bistability and 1/f noise in a CO_2 laser with loss modulation; a) coexistence of two attractors (period 3 and 4 respectively). The two superposed spectra correspond to two starts with different initial conditions; b) comparison between the low frequency cut-off (dashed line) when the two attractors are stable and the low frequency divergence (solid line, slope α =0.6) when the two attractors are strange.

tractors is not sufficient to explain the phenomenology. Indeed, experiments [3,4] show that, when leaving an attractor, the representative point in phase space has a long erratic motion before landing onto another attractor. This "transient" regime is made of motions among repulsive orbits.

A dynamics in terms of a recursive map must allow for at least two independent attractors. The simplest one-dimensional map with two attractors must have two extrema [7]. Hence we study a cubic map in the interval (-1, 1)

$$x_{n+1} = (a-1)x_n - ax_n^3 \ . \tag{3}$$

To account for item iii) the map will be disturbed by white noise with r.m.s. between 10^{-7} and 10^{-5} . Up to a value $a=\bar{a}=3.598\ 076\ \ldots$, the motion is confined either on the interval (-1,0) or (0,1) with qualitative features alike the well known logistic map. For a = $\bar{a}$, we may still have two independent attractors, whose domain, however, are interlaced in complicated ways over the interval (-1,1).

The simplest stable pair of attractors A,B above $\bar{a}$ is a pair of period-3 attractors which are superstable for a_s=3.981 797 ... . These period-3 attractors disappear for a=a=3.982 000 642 ... . For a_s < a < $\tilde{a}$ (fig. 2) the presence of a small amount of noise

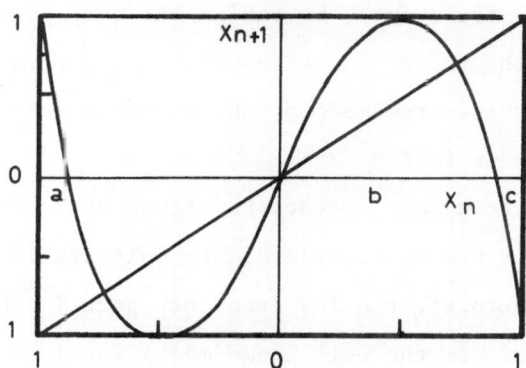

Fig.2 - Cubic map for $\tilde{a}$. a,b,and c show (not in scale) the intervals covered by one of the two period-3 attractors.

makes it easy to leave one attractor and jump toward the other one. Before landing into the other attractor, the representative point wanders on the available space through a long transient because of the complex structure of the two basins of attraction. The corresponding low frequency power spectra, for different noise levels, are given in fig. 3.

These spectra show a power law region extending over about one decade with a slope between 0 and 2. The appear qualitatively in agreement with the experimental spectra of Refs. 3 and 4.

A model explanation of the above spectra was given [7] in terms of jump processes among three regions of phase space (the two at-

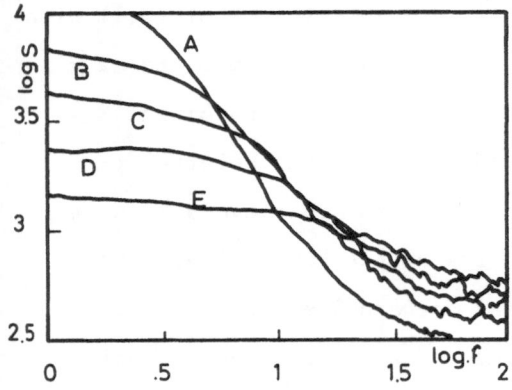

Fig.3 - Power spectra for a=ã, and for increasing noise levels σ ,that is, a) 5×10^{-7} ,b) 10^{-6}, c) 2×10^{-6}, d) 4×10^{-6}, e) 10^{-5}. They can be fitted by $f^{-\alpha}$, with α decreasing from 1.5 for a) to 0.5 for e)

tractors and the intermediate transient), by taking the jump probabilities decorrelated from the internal motions within the three
regions.This leads to a power spectrum made of three Lorentzians
which fit well the spectra shown in fig. 3.

Extrapolating the model to a many attractors situation we formulate the conjecture that for the limit of a highly multistable
system (large number of attractors), the low frequency part $f^{-\alpha}$ has
the exponent α =1. This is akin to the well known model for 1/f noise
in equilibrium systems [8], where the 1/f slope is the sum over a
large number of Lorentzian curves suitably weighted.

REFERENCES
1. E.Lorenz, Jour.Atmos.Sc., 20:130-141 (1963)
2. H.Haken, Phys.Lett.,53A:77 (1975)
3. F.T.Arecchi and F.Lisi,Phys.Rev.Lett., 49:94 (1982) and 50:1330 (1983)
4. F.T.Arecchi, R.Meucci, G.Puccioni and J.Tredicce, Phys.Rev.Lett., 49:1217 (1982)
5. J.P.Crutchfield and B.A.Huberman, Phys.Rev.Lett., 43:1743 (1979)
6. J.P.Crutchfield and B.A.Huberman, Phys.Lett. 77A:407 (1980)
7. F.T.Arecchi,R.Badii and A.Politi, to be published
8. R.F.Voss and J.Clarke, Phys.Rev. B13:556 (1976)

A MECHANISM FOR CHAOS IN A NONLINEAR OPTICAL SYSTEM

J.V. Moloney

Optical Sciences Center
University of Arizona
Tucson, AZ 85721

S. Hammel and C.R.T. Jones

Program in Applied Mathematics
University of Arizona
Tucson, Az 85721

ABSTRACT: We show, using a global phase space analysis, that the chaotic attractor appearing beyond the accumulation point of period doubling bifurcations in a ring bistable cavity is the accumulation of homoclinic orbits associated with unstable fixed points which remain on each successive bifurcation. In contrast to the one dimensional map, periodic cycles of any arbitrary period may, in principle, appear abruptly in the phase space and these in turn can undergo period doubling.

I. INTRODUCTION

Numerous examples of period doubling to chaos in nonlinear optical systems have appeared in the literature since Ikeda's pioneering paper in 1979.[1] Considerable effort has been expended to establish that the period doubling sequences are consistent with Feigenbaum's universality conjecture.[2] In order to determine critical parameter values, these works relied on a linear stability analysis of the appropriate maps or coupled ordinary differential

equations. Period doubling was established by direct numerical
iteration or integration.

We show that the above local stability analysis provides a far
from complete picture of the dynamical structure of the complex
map describing bifurcations in a ring bistable cavity.[3] In
particular, we emphasise that the latter map is two dimensional,
invertible and area contracting having little in common (besides
period doubling) with Feigenbaum's one dimensional logistic map.
The fact that the map is invertible means that we can use the
computer as a powerful analytic tool to construct global phase
space pictures. These pictures show that the "precursor" of the
chaotic attractor appears in the phase space at the early stages
of period doubling. The final chaotic attractor results from the
accumulation of homoclinic orbits in the phase space; the latter
orbits give rise to Cantor set structures which are the signatures
of chaos. Furthermore, homoclinic tangencies induce saddle-node
bifurcations which cause the sudden appearance of new periodic
cycles. These new cycles can undergo period doubling to higher
order cycles. The end result is an incredibly complex (but
understandable) dynamical behavior having little in common with
one dimensional maps.

II. SOME MATHEMATICAL PRELIMINARIES

Although we will seek to minimize technical mathematical
jargon, we need to introduce a few simple mathematical ideas
(through simple pictures) which will be essential for discussion of
the results that appear in later sections. We list (for
convenience) the simplest nonlinear dynamical systems in which
chaos can appear

1) Noninvertible 1-Dimensional Maps (e.g. Feigenbaums logistic
map $X_{n+1} = \mu X_n(1-X_n)$ and variants thereof). Nonuniqueness of the
inverse mapping introduces the complex dynamics but the dimension
is too low for the existence of a strange attractor.

2) Invertible 2-Dimensional Maps (e.g. Henon map[4]). These maps
can be viewed as two dimensional Poincare surfaces of section of a
three dimensional flow (three coupled nonlinear ordinary
differential equations). Unlike the 1-dimensional noninvertible
map they can be iterated forward and backward in direct analogy to
backward and forward integration in time of a system of ordinary
differential equations. The map describing the dynamics of the
intracavity field in a bistable ring cavity belongs to this class.[3]

3) Flows in Three Dimensions (or continuous time solutions of
coupled ordinary differential equations). The uniqueness theorem for
solutions to ordinary differential equations establishes that the

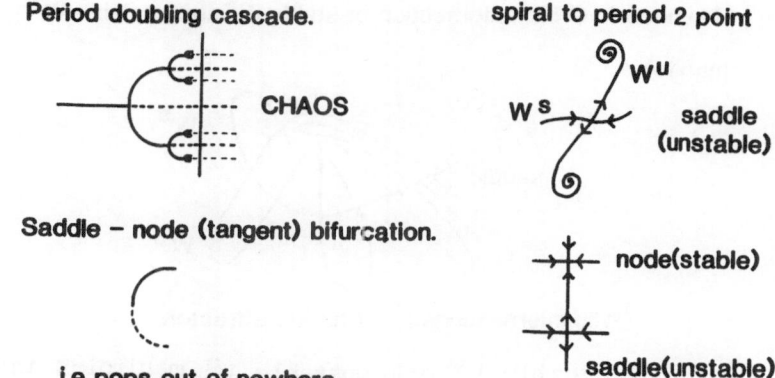

Figure 1. Two types of bifurcation can occur in the complex map for the intracavity field. (a) Period doubling pitchfork bifurcation. Solid lines denote stable fixed points of the map and its higher compositions; dashed lines are the unstable fixed points corresponding to the original stable solutions before bifurcation. The vertical line is the accumulation point immediately beyond which no stable solutions exist. (b) Phase space picture showing unstable saddle and stable period 2 points after bifurcation. (c-d) Saddle-node (tangent bifurcation). A saddle-node pair is simultaneously created in the phase space (see Fig. 1d). The node is stable while the saddle is unstable.

complex dynamical motion associated with chaos can only be observed in three dimensions or greater. Cases 2) and 3) are obviously closely connected.

II.1 PROPERTIES OF 2-D INVERTIBLE MAPS

Immediately beyond the accumulation point of the period doubling cascade, an infinite number of unstable fixed points remain (see Figure 1a). We will show that the chaos observed in the plane wave map is intimately linked to these unstable fixed points. Physically, these are the original linear and nonlinear cavity modes which are now all unstable. Immediately after the stable solution has gone unstable and a stable period two cycle has appeared, the phase space picture is as depicted in Figure 1b. The stable solution which immediately preceding the bifurcation was a stable node, has now changed to an unstable saddle. Linear

Homoclinic Orbit: intersection of stable W^S and unstable W^u

manifolds.

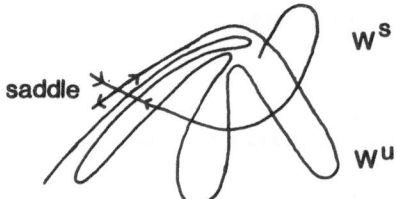

Wild intersections !   Chaotic attractor.

Figure 2.   If the stable W^S and unstable W^u manifolds intersect
once, then they will intersect an infinite number of times. The
stable W^S (unstable W^u) manifold cannot intersect itself, so the
oscillations become progressively wilder as it begins to approach
itself.   The result is a homoclinic orbit.   Before the manifolds
touch one another (homoclinic tangency) stable periodic sinks can
be created within the lobes.[3]

stability analysis at this point shows that one eigenvalue, λ_1 say,
has magnitude greater than unity (it actually passes through -1),
while the other λ_2 has magnitude less than unity.   Motion in the
vicinity of the saddle, therefore, involves expansion along the
unstable manifold W^u (locally the eigenvector associated with λ_1)
and contraction onto W^u, in the direction of the stable manifold W^S
(locally the eigenvector associated with the eigenvalue λ_2).   The
stable and unstable manifolds, however, exist globally in the phase
space (in our case the complex E-plane) and have the important
property that they are invariant curves under the mapping.   Our
picture of motion in the phase plane therefore is that points off
the unstable manifold are forced onto it, are rapidly accelerated
along it and usually end up being attracted to attracting periodic
fixed points (such as the period 2 points in Figure 1b which are
fixed points of the second composition of the map). Besides their
invariance under the mapping, the stable (W^S) and unstable (W^u)
manifolds, can intersect in the phase space unlike solution curves
to differential equations which because of the uniqueness theorem
cannot cross.   If they do intersect once, there is a theorem which
proves that they must intersect an infinite number of times and
the end result is a homoclinic orbit as depicted in Figure 2.

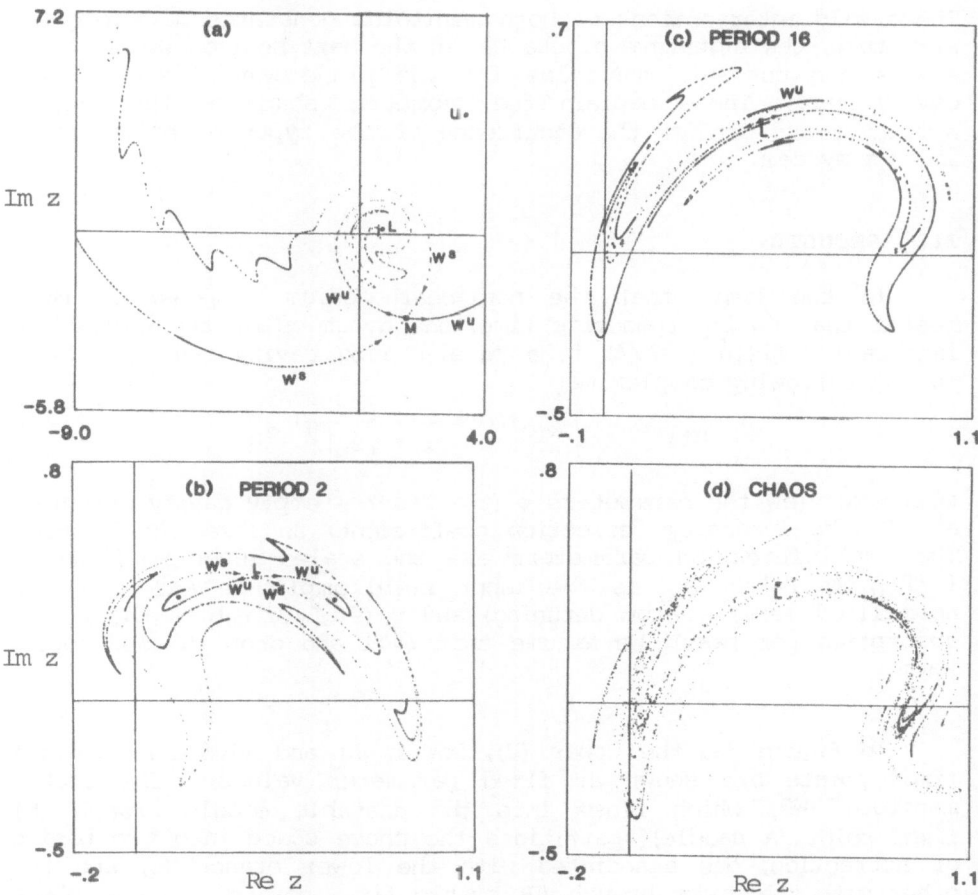

Figure 3.   (a)   Phase space (complex z-plane) pictures showing
upper (U), middle (M) and lower (L) branch fixed points at
parameter values p=7 and γ=.4.   Arrows indicate the direction of
flow away from and towards the middle branch fixed points.   (b)
Shows a small window about the lower branch unstable fixed point
(L) (a saddle) just after it has period doubled (p=7, γ=.7).   Flow
along the unstable manifold W^u of this unstable saddle now spirals
into the stable period 2 points (heavy dots).   The wild
oscillations of the stable manifold W^s are induced by near
interaction with the unstable manifolds of both this and the
original middle branch fixed points (Figure 1a).   (c) Unstable
manifold at period 16 (period 16 points are barely visible).   The
stable manifold at this stage is wildly intersecting W^u generating
a homoclinic orbit and is not shown.   (p=7, γ=.84).   (d) Chaos in
this figure involves erratic bouncing about in the vicinity of the
unstable manifold W^u (p=7, γ=.87).

These wild intersections of both manifolds generates a Cantor set structure, the signature of chaos. In the next section we show the stable and unstable manifolds for the plane wave bistable ring cavity map. These explain the geometric shape of the chaotic attractor and predict the occurrence of new types of bifurcations in this system.

III. RESULTS

In the limit that the nonlinear medium response is much faster than cavity roundtrip time, the dynamics of the normalized intracavity field ($z=E/\Delta$) in a passive ring cavity can be reduced to the following complex map.

$$z_{n+1} = \gamma + R\, e^{i(\phi - \frac{p}{1 + |z_n|^2})}\, z_n$$

where we view the parameters ϕ (the laser - empty cavity detuning) and R (the intensity reflection coefficient) as fixed (R=.9, ϕ=.4). The two bifurcation parameters are the scaled input amplitude γ ($=\sqrt{T}E_{in}/\Delta$; T=1-R, E_{in} is the input field amplitude and Δ is the normalized laser - atom detuning) and p ($=\alpha_0 L/\Delta$; $\alpha_0 L$ is the linear absorption per pass). We assume that $\Delta \gg 1$ and drop the absorption term.

In Figure 3a, the upper (U), lower (L) and middle branch (M) fixed points are shown at fixed parameter values. The stable manifold (W^s) which flows into the unstable middle branch (M) fixed point (a saddle), partitions the phase space into two basins of attraction, one associated with the lower branch (L) and the other with the upper branch (U) stable fixed points. We now place a small window about the lower branch (L) fixed point and follow the development of its phase space structure (once it goes unstable it becomes a saddle with its own stable and unstable manifolds W^s and W^u, respectively) as we go through a period doubling cascade to chaos (varying γ). In Figure 3b we see the unstable lower branch fixed point (L) just after it has gone unstable and period doubled. The period 2 points are indicated by heavy dots; we note here that had we been iterating the map (Eq. (1)) in the conventional manner we would have a rather featureless picture containing just two dots. At this point the stable manifold W^s is showing some wild oscillations but it does not intersect the unstable manifold W^u yet. We emphasize that these manifolds are regular continuously generated curves (although complicated) in the phase space. Before the next period doubling to a four cycle both manifolds touch giving rise to a homoclinic tangency. In the vicinity of this tangency new periodic cycles can be created.[3] At this stage the stable manifold is so wildly oscillatory that we don't plot it.   In view of the earlier discussion we expect all motion to occur in the neighborhood of W^u.   As long as attracting periodic points exist in the phase plane, motion in the vicinity of W^u will be swept onto them. Figure 3c shows the unstable manifold at period 16 (period 16 points are barely visible).   Finally in Figure 3d we show the chaotic attractor obtained by conventional iteration of the map beyond the accumulation point. The chaotic attractor contains the image of the unstable manifold W^u. We note immediately that the chaos involves complicated erratic motion in the vicinity of the unstable manifold.   We remind the reader that we have only followed the original lower branch unstable fixed point.   It is evident from Figure 1a however that an infinity of such unstable fixed points remain on passing the accumulation point.   These in turn are unstable saddles with their own stable and unstable manifolds.   As the map is area contracting (R = 0.9) these unstable saddles occupy progressively smaller regions of the phase space and their homoclinic orbits tend to fill in the details of the chaotic attractor without affecting its overall shape.   Comparing Figure 3d with c, for example, we note that the attractor is dense in the regions where the period 16 points originally existed.

IV.   SUMMARY

By tracking the unstable fixed points which remain after each period doubling bifurcation, we have identified the chaos in a bistable ring cavity with erratic motion in the vicinity of the unstable manifolds of these unstable saddle points.   Using a global phase space analysis we predict that a new type of bifurcation can arise in the vicinity of a homoclinic tangency.   We anticipate that many new nonlinear dynamical phenomena predicted recently in the literature[4] will also occur in the bistable ring cavity.

ACKNOWLEDGMENTS

We acknowledge support from the U.S. Air Force Office of Scientific Research, the U.S. Army Research Office and the National Science Foundation.

REFERENCES

1. K. Ikeda, Opt. Commun. 30, 257 (1979); K. Ikeda, H. Daido and O. Akimoto, Phys. Rev. lett. 45, 709 (1980).
2. R.R. Snapp, H.J. Carmichael and W.C. Schieve, Opt. Commun. 40, 68 (1981)
3. J.V. Moloney, S. Hammell, C.R.T. Jones, to be published.
4. "Nonlinear Dynamics" ed. R.H.G. Helleman, in Annals of the New York Academy of Sciences, 357, (1980)

ROUTES TO OPTICAL TURBULENCE IN A DISPERSIVE RING BISTABLE

CAVITY CONTAINING SATURABLE NONLINEARITIES

J.V. Moloney, H.M. Gibbs and F.A. Hopf

Optical Sciences Center
University of Arizona
Tucson, AZ   85721

ABSTRACT

Using a plane wave analysis we show that co-existent attractors, representing distinct routes to optical turbulence, appear on a single branch of a dispersive ring bistable cavity. Some of these routes appear to arise via a tangent (saddle-node) bifurcation and are not found in the logistic map analyzed by Feigenbaum. When transverse variations in the beam profile are accounted for, a new transition sequence arises which is consistent with recent observations in a fluid dynamical experiment.

I.   INTRODUCTION

The transition to turbulence in fluid dynamical systems involves many different pathways depending on parameters such as the Reynolds number, aspect ratio of the fluid, etc. Remarkable progress has been made over the past 10 years in understanding qualitatively, transitions to low level turbulence in low aspect ratio, low Reynolds number systems by studying mappings of one and two dimensional nonlinear functions.[1] However no quantitative relation appears to exist between fluid parameters (Reynolds #, Prandtl # etc.) and the parameters appearing in the mappings.

95

Numerous papers citing period doubling bifurcations in nonlinear optical systems have appeared since Ikeda's original paper in 1979. With only two exceptions,[2,3] all theoretical studies have been confined to plane wave analyses. These studies have focussed on establishing that Feigenbaum sequences exist for instabilities in these systems. Our contention is that the Feigenbaum period doubling sequences represent a subset of a much more general class of instabilities.[4] When we extend our study to an input beam with a Gaussian spatial profile (we consider a single transverse dimension) we discover strong departures from the plane wave predictions, at least under self-focusing conditions. For parameter values where the plane wave analysis predicts the existence of a simple period 2 cycle on the low transmission branch, a low Fresnel Gaussian beam exhibits radically different bifurcation sequences on both low and high transmission branches.[2] These latter sequences are more typical of the Ruelle-Takens picture[5] of the transition to turbulence. One of our computed sequences on the high transmission branch, is remarkably similar to measured radial velocity profiles in a recent Rayleigh-Benard fluid experiment.[6]

II. PLANE WAVE INSTABILITIES

As an explicit example, we study a unidirectional ring bistable cavity containing a nonlinear saturable medium. In the plane wave approximation and in the limit that the nonlinear medium response is much faster than a cavity roundtrip time $t_R (\equiv \cdot \mathscr{L}/c$, $\mathscr{L}$ = total cavity length), the dynamics of the complex intra-cavity field reduces to the following map[7]

$$E_{n+1} = \sqrt{T}\, E_{in} + B\, \exp\left[\frac{\alpha L |E_n|^2}{1 + \Delta^2 + |E_n|^2}\right] \exp[i\delta(|E_n|^2)]E_n \qquad (1)$$

where

$$\delta(|E_n|^2) = \phi - \alpha L \Delta \left[1 - \frac{|E_n|^2}{1 + \Delta^2 + |E_n|^2}\right]$$

is the intensity dependent total cavity mistuning. Eq. (1) maps the field (E_n) at roundtrip nt_R to the field (E_{n+1}) a cavity roundtrip later $((n + 1)t_R)$; E_{in} is the input amplitude (all amplitudes are scaled to the saturation amplitude), $B = Re^{-\alpha L}$ ($\alpha L = \alpha_0 L/2(1 + \Delta^2)$; $\Delta = (\omega - \omega_{ab})/\gamma_\perp$) and ϕ is the laser-empty cavity detuning.

In previous studies of instabilities in a bistable ring cavity,[8,9] a linear stability analysis of equation (1) yielded parameter values at which a stable solution (on either low or high transmission branches) went unstable. Numerical iteration of

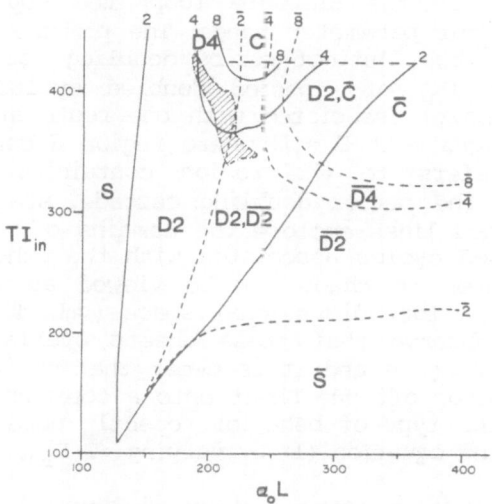

Figure 1.   Bifurcation diagram for Eq. (1) in ($\alpha_0 L$, TI_{in}) parameter space.   Fixed points in S and $\bar{S}$ bifurcate to the respective chaotic attractors C and $\bar{C}$.   The crosshatched domain encloses a stable six cycle which appears to arise by a tangent bifurcation.   All three bifurcating attractors coexist in some regions of parameter space.

equation (1) established that the instabilities involved period doubling cascades to chaos; furthermore the period doubling sequences were found to be consistent with Feigenbaum's universality conjecture.   In Equation (1) the input amplitude ($\sqrt{T}$ E_{in}) was varied, holding the other parameters fixed.   Usually the intracavity field E_n is used as an initial condition for the next increment of $\sqrt{T}$ E_{in}.

It is evident however that equation (1) has a multiparameter dependence and is a two dimensional real map.   Choosing both $\sqrt{T}$ E_{in} and $\alpha_0 L$ as our bifurcation parameters we map out the parameter space picture of the stability domains of both stable solutions and higher period doubled cycles of equation (1). Figure 1 summarizes the dynamical behavior of equation (1) on the low transmission branch of the bistable loop; the upper branch is always stable over this parameter range. The picture is extremely complicated, showing two distinct period doubling routes to chaos. The stable domains (D_n) of period doubled cycles (2^n-cycles, n=1,2,3 explicitly shown) associated with one route are bounded by solid lines; we designate it S → C where region S contains stable solutions and C refers to the region containing the chaotic attractor on which the period doubling cascade, starting from S, terminates. The dashed lines enclose the domains of stability ($\overline{D}_n$) of the period doubled cycles associated with the other route ($\overline{S}$ → $\overline{C}$). These two routes to chaos can be viewed as lying on two connected sheets in a four dimensional space ($\alpha_0 L$, TI_{in}, ReE, ImE). From Figure 1 we observe that these sheets overlap in certain regions of parameter space and it is clear that on varying either parameter, one may drop off one sheet onto a coexisting attractor. This accounts for the type of behavior recently observed by Snapp et al.[9] when iterating equation (1) by varying $\sqrt{T}\, E_{in}$.

One particularly interesting feature of Figure 1 is the abrupt appearance of a period six cycle in the cross-hatched domain. This cycle undergoes it's own period doubling cascade over a finite region of parameter space. Unlike the other period doubling cascades, however, the period 6 cycle does not involve a stable solution going unstable. Instead it appears discontinuously in the complex E-plane via a saddle-node bifurcation. The origin of these bifurcations is discussed elsewhere.[10]

III. GAUSSIAN BEAM INSTABILITIES

When the problem is generalized to include input beams with Gaussian spatial profiles, new routes to optical chaos arise.[2] Instabilities typically involve an initial sequence of Hopf bifurcations from a stable solution to a limit cycle and from the limit cycle to quasiperiodic motion on a two dimensional torus. This initial sequence is consistent with the Landau picture of the transition to turbulence where on increasing a parameter, new incommensurate frequencies appear in the dynamics representing motion on the surface of tori of ever increasing dimension. The turbulent state in this picture consists of an infinity of such incommensurate frequencies. An alternative picture due to Ruelle and Takens[6] argues that motion on tori of dimension greater than three should be unstable and they predict breakdown of the torus followed by a rapid transition to a complicated motion on a strange attractor.

In treating the Gaussian beam problem we must solve the full nonlinear wave equation on each pass around the cavity.[2] In Figures 2 and 3 we summarize our results for a Gaussian input with

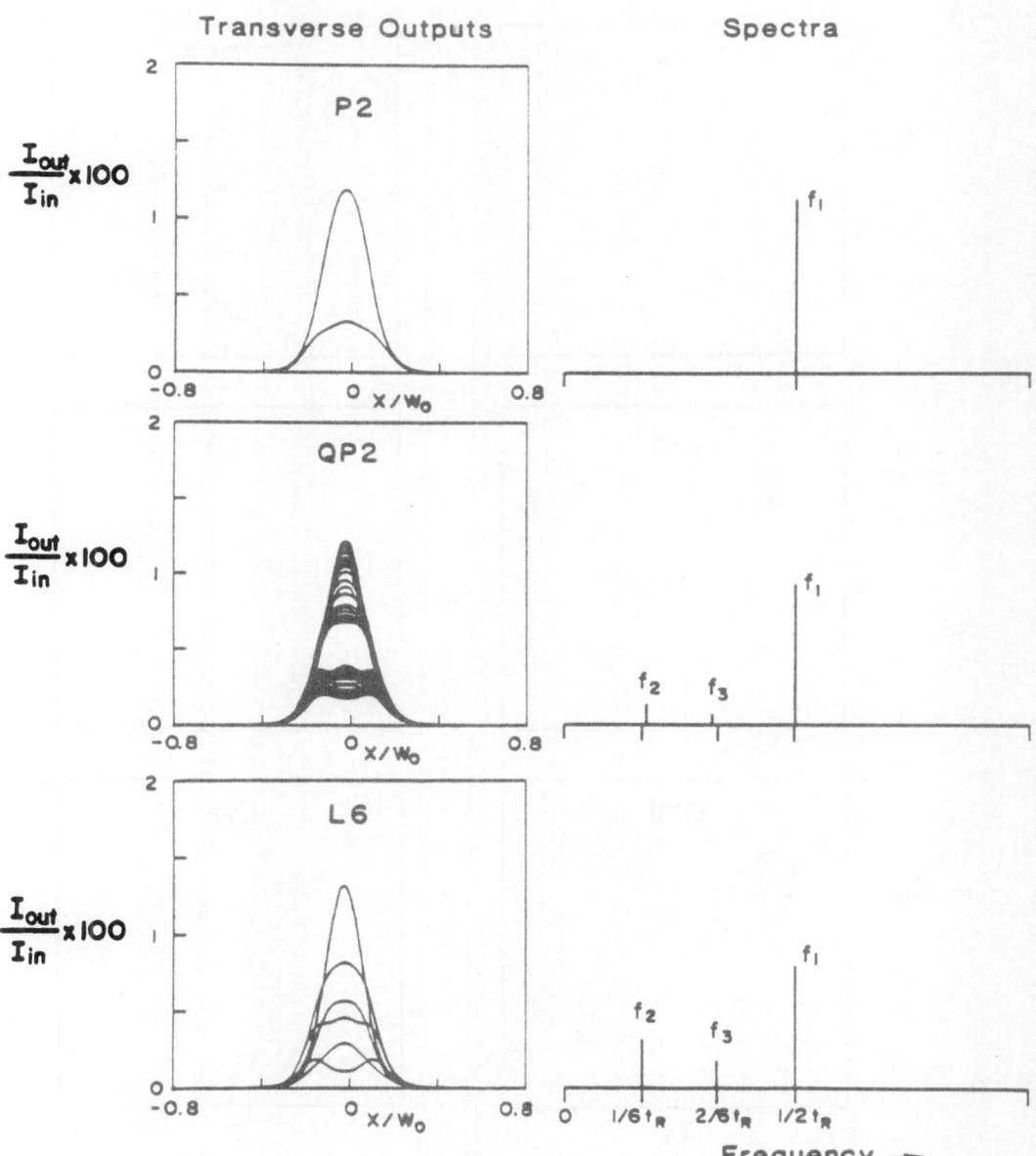

Figure 2.   Transverse spatial profiles (40 successive outputs are plotted) showing period 2 (P2), quasiperiodic (QP2) and frequency locking (L6) outputs on the low transmission branch.   On the right are schematic frequency spectra of on-axis time series showing the period 2 subharmonic $f_1 = 1/2t_R$, the appearance of a new incommensurate frequency f_2 and finally locking of f_2 to f_1 at f_2 = $1/6t_R$. Frequency f_3 is just a linear combination of f_1 and f_2.

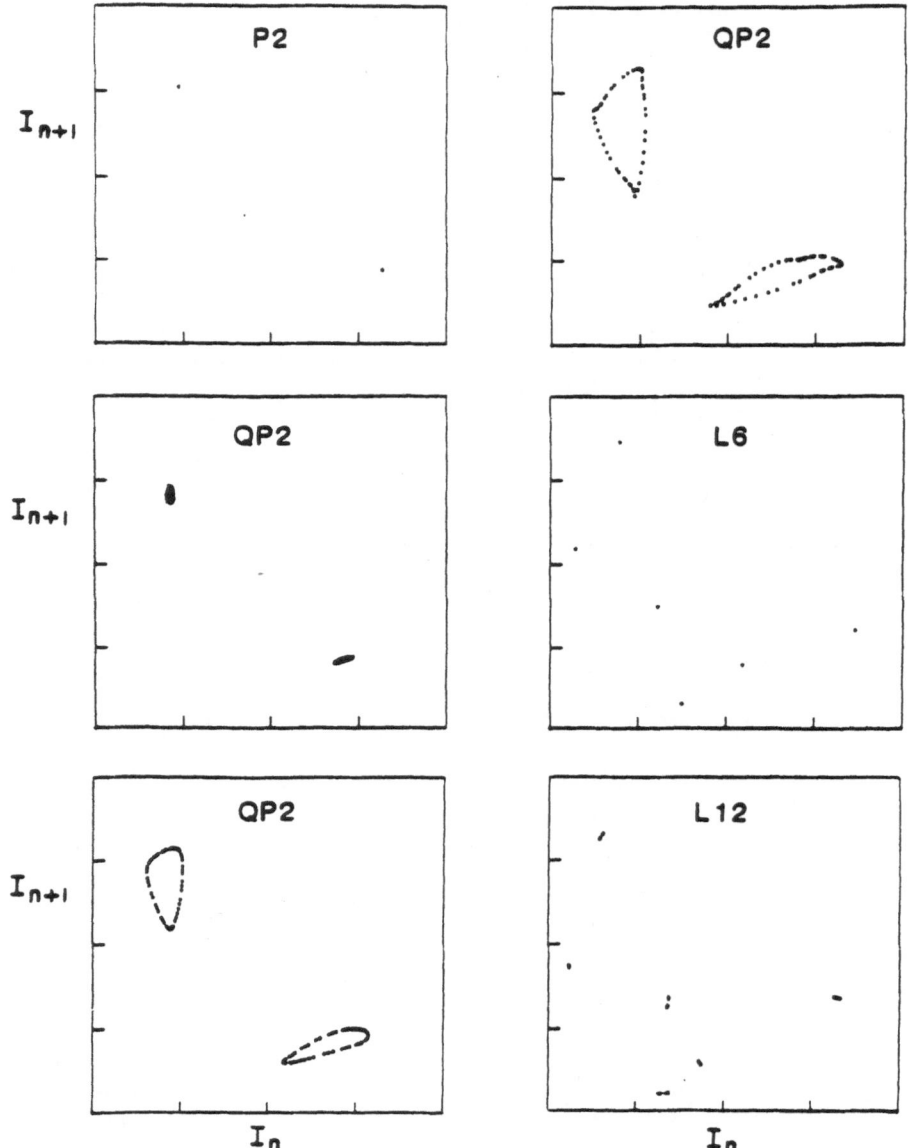

Figure 3.   Poincare surface of section (plot of on-axis output intensity I_{n+1} at roundtrip n+1 versus output intensity I_n at roundtrip n). The surface of section can be viewed as a two dimensional plane through phase space on which crossings of a continuous time orbit are recorded. The sequence shows from top to bottom a limit cycle (two dots on surface of section) undergoing a Hopf bifurcation to motion on a torus, followed by a frequency locking on the torus (L6) and finally a period doubling to L12.

parameters $\alpha L = .25$, $\phi = .4$ and Fresnel number $F = .54$. The corresponding plane wave solution indicates that bifurcation occurs as a function of increasing input amplitude to a period 2 cycle and returns to a stable solution on the low transmission branch; the upper branch is stable for the plane wave map but unstable for the Gaussian input. On the low transmission branch, (see Figure 2) instead of a stable 2-cycle, we observe as a function of increasing peak amplitude a sequence going through period 2 (P2), quasiperiodic (QP2) and frequency locking to a period 6 cycle (L6). The frequency locked output then goes through a period doubling cascade (L12 is shown) to a chaotic attractor. The sequence then returns to a stable solution via a reverse cascade. In Figure 2 we show the transverse profiles for the period 2 (P2), quasiperiodic (QP2) and frequency locked (L6) outputs. Schematic frequency spectra show how the incommensurate frequency f_2 locks to the period 2 subharmonic $f_1 = 1/2t_R$ (t_R - cavity roundtrip time). These sequences are consistent with the Ruelle-Takens picture mentioned above. Figure 3 shows Poincare surfaces of section as we follow the bifurcation sequence up to the first period doubling of the frequency locked output. These sequences show a Hopf bifurcation from a limit cycle to motion on a torus followed by frequency locking and breakdown of motion on the torus.

IV.   SUMMARY

We have established, by numerical iteration of the multiparameter plane wave map describing the intracavity field in a dispersive bistable ring cavity, that coexistent attractors appearing on the low transmission branch arise from bifurcations from stable solutions in different regions of parameter space. Furthermore, new periodic cycles are predicted to appear abruptly via a saddle-node bifurcation. The existence of these latter cycles cannot be a priori predicted using a linear stability analysis but require a global phase space analysis as discussed elsewhere.[10]   Numerical simulations with input Gaussian beams, operating under self-focusing conditions, show strong departures from the plane wave predictions. An open question at the moment is whether or not there are situations where the plane wave predictions can be extrapolated directly to the Gaussian beam case. Clearly, a full quantitative understanding of the dynamics of the plane wave model[10] is necessary before we can hope to properly address this important question.

ACKNOWLEDGMENTS

We acknowledge support from the U.S. Air Force Office of Scientific Research and the National Science Foundation.

REFERENCES

1.   "Nonlinear Dynamics", ed. R.H.G. Helleman, in Annals of the New York Academy of Sciences, 357, (1980).
2.   J.V. Moloney, F.A. Hopf and H.M.Gibbs, Phys. Rev. A 25, 3442 (1982).
3.   W.J. Firth, E. Abraham and E.M. Wright, Appl. Phys. B 28, 170 (1982).
4.   J.V. Moloney, to be published.
5.   D. Ruelle and F. Takens, Commun. Math. Phys. 20 ,167 (1971).
6.   J.P. Gollub and S.V. Benson, J. Fluid Mech. 100, 449 (1980).
7.   The arguments of the exponents in equation (1) should really involve integrals over the nonlinear medium (to account for propagation effects) corresponding to the formal solution to Maxwell's equation.   In the map given by equation (1), inclusion of propagation effects just rescales the argument of the exponent but does not effect the global parameter space picture in Figure 1 beyond a slight shift of the boundaries.
8.   K. Ikeda, Opt. Commun. 30, 257 (1979); K. Ikdea, H. Daido and O. Akimoto, Phys. Rev. Lett. 45, 709 (1980).
9.   R.R. Snapp, H.J. Carmichael and W.C. Schieve, Opt. Commun. 40, 68 (1981).
10. J.V. Moloney, S. Hammell and C.R.T. Jones, to be published.

SELF-PULSING AND CHAOS IN INHOMOGENEOUSLY BROADENED

SINGLE MODE LASERS

R. Graham and Y. Cho [+) *)]

Fachbereich Physik, Universität Essen

D-4300 Essen - W. Germany

ABSTRACT

Self-pulsing and chaos in inhomogeneously broadened single mode lasers is investigated in the framework of two models with 4 and 6 degrees of freedom, respectively. Numerical examples of self-pulsing and chaotic dynamics are presented.

Homogeneously broadened single mode lasers are well-known to exhibit self-pulsing instabilities and chaotic dynamics under the combined conditions of large ratio of gain over losses and low cavity quality [1, 2]. In fact, a homogeneously broadened single mode laser in resonance has essentially 3 degrees of freedom only and is realistically described by the Lorenz model [3, 4], which has served as a prototype model for investigating chaos in continuous dynamical systems [5]. However, the above mentioned conditions for the occurrence of chaos have not yet been realized, experimentally.

Recently, bad-cavity instabilities have also been discussed for inhomogeneously broadened lasers [6, 7], and seem to be more easily accessible, experimentally [8, 9], but models of comparable simplicity as the Lorenz model have not yet been proposed for inhomo-

[+)] Permanent address:
The Institute of Scientific and Industrial Research, Osaka University, Yamadakami, Suita, Osaka 565, Japan,
[*)] supported by the Alexander von Humboldt-Stiftung

geneously broadened lasers. It is the purpose of the
present note to show how such models can be constructed.
We find that an infinite hierarchy of models exist which
increase in accuracy and complexity. We present a 4-di-
mensional model and a 6-dimensional model, which are the
two simplest members of this hierarchy. The results of
a linear stability analysis of the time-independent
states and some numerical solutions are given to show
the various types of dynamical behavior which may occur
in these models. The dynamical behavior is found to be
much richer than in the homogeneously broadened case
and is obtained under physically more realistic conditions.

The equations of motion of an inhomogeneously
broadened laser in single mode action are given by [10]:

$$\dot{b} = -\varkappa b + \sum_\mu g_\mu \alpha_\mu$$

$$\dot{\alpha}_\mu = -(i\nu_\mu + \gamma_\perp)\alpha_\mu + g_\mu \sigma_\mu b \tag{1}$$

$$\dot{\sigma}_\mu = \gamma_\parallel (d_{0\mu} - \sigma_\mu) - 2g_\mu (\alpha_\mu^* b + \alpha_\mu b^*)$$

For the details of the derivation and notation in these
equations see [10]. We merely note that b is the slowly
varying complex mode amplitude at the line center of
the inhomogeneously broadened transition, α_μ, σ_μ are,
respectively, the complex polarisation and population
inversion of the group of atoms with frequency ν_μ off
the line center, g_μ is the dipole coupling constant of
this group of atoms, $\gamma_\parallel d_{0\mu}$ their pumping rate, $\gamma_\perp$, $\gamma_\parallel$
are transverse and longitudinal relaxation rates of the
atomic transition, $2\varkappa$ is the photon decay rate of the
empty cavity. In the following we assume a running mode,
in which case we may take g_μ = g, see [10].

Since the single mode couples only to the combined dipole moments of all groups of atoms, it is useful to introduce this quantity as a new variable. The
equation of motion for this new variable will again contain new variables, which are represented by sums over
the different groups of atoms. Repeating this process
one generates an infinite hierarchy of equations of
motion, which is equivalent to eqs. (1). Approximations
of increasing accuracy can be made by closing this
hierarchy at increasingly higher levels. In this way one
arrives at approximations of (1) by a sequence of finite
dimensional systems. In the following we present the two
simplest approximations obtained in this way and discuss

their dynamical behavior. We introduce the scaled
variables

$$E = 2g b/\gamma_\perp \quad , \quad P = (2g^2/\kappa\gamma_\perp) \sum_\mu \alpha_\mu \quad , \quad D = (g^2/\kappa\gamma_\perp) \sum_\mu \sigma_\mu \qquad (2)$$

and obtain the exact equations

$$dE/d\tau = -(\kappa/\gamma_\perp)(E - P) \quad , \qquad \tau = \gamma_\perp t$$

$$dP/d\tau = -P + DE + (\omega_1^2/\delta_\perp^2) S \qquad (3)$$

$$dD/d\tau = (\gamma_\parallel/\gamma_\perp)(\tau - D) - \tfrac{1}{2}(PE^* + P^*E) \qquad (4)$$

where

$$\tau = (g^2/\kappa\gamma_\perp) \sum_\mu d_{o\mu}$$

is the pumping parameter, and

$$(\omega_1/\delta_\perp)^2 S = (2g^2/\kappa\delta_\perp^2) \sum_\mu (-i \nu_\mu) \alpha_\mu \qquad (5)$$

is a new dynamical variable. We note that by putting
$S = 0$, eqs. (3) are reduced to the equations of motion
of a homogeneously broadened single mode laser. Inhomogeneous broadening is taken into account by writing
down the equation of motion of S obtained from (1)

$$dS/d\tau = -S - iE\Delta - (P - P_0) \qquad (6)$$

Here we introduced the new variables

$$\Delta = (g^2/\kappa\omega_1^2) \sum_\mu \nu_\mu \sigma_\mu \quad , \quad P_0 = \frac{2g^2}{\kappa\gamma_\perp} \sum_\mu \alpha_\mu \left(1 - \frac{\nu_\mu^2}{\omega_1^2}\right) \qquad (7)$$

$$(8)$$

which satisfy the equations of motion

$$d\Delta/d\tau = -(\gamma_\parallel/\gamma_\perp)\, \Delta - i(SE^* - S^*E)/2$$

$$dP_0/d\tau = -P_0 + D_0 E + (\omega_1^2 - \omega_2^2)S/\gamma_\perp^2 + R \qquad (9)$$

$$dD_0/d\tau = -(\gamma_\parallel/\gamma_\perp)(D_0 + \lambda\tau) - (EP_0^* + E^*P_0)/2$$

We have assumed symmetrical pumping $\sum_\mu \nu_\mu d_{o\mu} = 0$
and introduced the parameter

$$\lambda = \frac{\Delta\nu^2}{\omega_1^2} - 1 \qquad (10)$$

and the new variables

$$D_0 = \frac{g^2}{\kappa\gamma_\perp} \sum_\mu \left(1 - \frac{\nu_\mu^2}{\omega_1^2}\right) \sigma_\mu \qquad (11)$$

$$R = \frac{2g^2}{\kappa\gamma_\perp^2} \sum_\mu (\omega_2^2 - \nu_\mu^2)(-i\nu_\mu) \alpha_\mu / \omega_1^2$$

The hierarchy of equations of motion can be continued by
writing an equation of motion for R. It should be noted
that the parameters ω_1, ω_2, are yet arbitrary. We now
consider the two simplest approximations, which close
the hierarchy, but still contain effects due to inhomogeneous broadening. The simplest approximation is obtained by putting $P_0 = 0$. We may choose the free parameter ω_1 in such a way that this approximation becomes
exact at least in the steady state $\alpha_\mu = \alpha_\mu^\circ$, $E = E^\bullet$

$$\omega_1^2 = \left(\sum_\mu \alpha_\mu^\circ \, \gamma_\mu^2 \right) / \sum_\mu \alpha_\mu^\circ \tag{12}$$

Using eqs. (1) we rewrite ω_1 as

$$\omega_1^2 = \left(\sum_\mu W_\mu^{(1)} \, \gamma_\mu^2 \right) / \sum_\mu W_\mu^{(1)} \tag{13}$$

where $\hspace{9cm}$ (14)

$$W_\mu^{(1)} = d_{o\mu} / (\gamma_\perp^2 + \gamma_\mu^2 + \gamma_\perp^3 |E^\circ|^2 / \gamma_\shortparallel)$$

$$|E^\circ|^2 = \left\{ \begin{array}{ll} 0 & t \leq 1 + \omega_1^2 \\ \gamma_\shortparallel (t - 1 - \omega_1^2)/\gamma_\perp & t \geq 1 + \omega_1^2 \end{array} \right.$$

In this approximation, the model consists of eqs. (3),
eq. (6) with $P_0 = 0$, and eq. (7). It still has 8 real
degrees of freedom. However, the solutions in this 8-
dimensional space are attracted to a 4-dimensional subspace in which E, P, S have equal and constant phases
and may be taken as real without loss of generality, and
$\Delta = 0$. Thus we obtain a minimal real 4-dimensional model
of the form

$$\dot{E} = -(\kappa/\gamma_\perp)(E-P) \; , \qquad \dot{P} = -P + DE + \omega_1^2 S/\gamma_\perp^2$$

$$\dot{D} = (\gamma_\shortparallel/\gamma_\perp)(t-D) - PE \; , \qquad \dot{S} = -S - P \tag{15}$$

which contains inhomogeneous broadening via S. The model
contains 4 independent parameters. A more sophisticated
approximation is obtained by closing the infinite
hierarchy via the assumption $R = 0$. We keep the value of
ω_1^2 as given by eq. (12) and choose ω_2^2 in such a way that
$R = 0$ is satisfied at least in the steady state and ob-
tain

$$\omega_2^2 = \frac{\sum_\mu W_\mu^{(2)} \, \gamma_\mu^2}{\sum_\mu W_\mu^{(2)}} \tag{16}$$

where $\quad\quad\quad W_\mu^{(2)} = V_\mu^2\, W_\mu^{(1)}$ $\quad\quad\quad\quad\quad\quad\quad\quad$ (17)

From the requirement that $P_0 = 0$ in the steady state, which is imposed by our choice of ω_1^2, we obtain the condition

$$(\omega_2^2 - \omega_1^2)/\gamma_\perp^2 = \lambda r \quad\quad\quad\quad\quad (18)$$

The resulting model consists of eqs. (3), (6), (9) and has 11 real degrees of freedom. However, the solutions in the 11-dimensional phase-space are attracted to a 6-dimensional subspace, where the phases of E, P, S, P_0 are all constant and equal and may be taken as 0 without loss of generality. In addition $\Delta = 0$ in this subspace. Thus, we arrive at the following 6-dimensional model

$$\dot{E} = -K(E-P)/\gamma_\perp \quad , \quad \dot{P} = -P + DE + \omega_1^2 S/\gamma_\perp^2$$

$$\dot{D} = \gamma_u (r-D)/\gamma_\perp - PE \quad , \quad \dot{S} = -S - P + P_0 \quad\quad (19)$$

$$\dot{P}_0 = -P_0 + ED_0 - \lambda r S \quad , \quad \dot{D}_0 = -\gamma_u (D_0 + \lambda r)/\gamma_\perp - E P_0$$

It contains 5 independent parameters. The minimal 4-dimensional model is reobtained by taking $P_0 = 0$, and omitting in (18) the equation for P_0. It is now clear, how an infinite hierarchy of models of increasing dimensionality may be constructed along these lines.

The 6-dimensional model and the 4-dimensional model reduce to the Lorenz model for the case $\omega_1^2/\gamma_\perp^2 = 0$, which is the special case of homogeneous broadening. For $\lambda = 0$, the 6-dimensional model is reduced to the 4-dimensional model. In this special case the characteristic frequencies of the system, ΔV, ω_1, ω_2 are all equal.

The analysis of the full set of equations (1) and the two models show that there are two time independent solutions, the trivial one, $E = P = S = P_0 = 0$, $D = -D_0/\lambda = r$, and a nontrivial one, $P_0 = 0$, $D_0 = -\lambda r$, $D = 1 + \omega_1^2/\gamma_\perp^2$, $E = P = -S = (\gamma_u (r - 1 - \omega_1^2/\gamma_\perp^2)/\gamma_u)^{1/2}$ which exists only for $r \geqslant 1 + \omega_1^2/\gamma_\perp^2$. A linear stability analysis of the <u>full</u> set of equations (1) shows that the trivial time-independent solution is linearly stable in the whole domain $0 \leqslant r \leqslant 1 + \omega_1^2/\gamma_\perp^2$ and becomes unstable for $r > 1 + \omega_1^2/\gamma_\perp^2$.

A linear stability analysis of the non-trivial time-independent solution of the 4-dimensional model is straightforward but involves tedious algebra, which can be largely avoided by considering the special case $\gamma_\perp = \gamma_n$. The result of the stability analysis for this special case is that the non-trivial time-independent solution is stable, provided $\omega_1^2/\gamma_\perp^2 < (\kappa + \gamma_\perp)/(\kappa - \gamma_\perp)$ and

either $\qquad\qquad\qquad \kappa < 2\gamma_\perp$ $\qquad\qquad\qquad\qquad$ (20)

or

$$\kappa > 2\gamma_\perp , \quad r < r_{th} = \left[\kappa\left((\kappa + 4\gamma_\perp) - \kappa\omega_1^2/\gamma_\perp^2\right)\right]/\gamma_\perp(\kappa - 2\gamma_\perp) \quad (21)$$

Instability occurs in the case of a bad cavity

$$\kappa > 2\gamma_\perp , \quad r > r_{th} \qquad\qquad\qquad\qquad (22)$$

In the limit $\omega_1^2 = 0$, this is the self-pulsing instability of a homogeneously broadened laser for $\kappa > \gamma_\perp + \gamma_n$. Our result shows that a corresponding instability still occurs in an inhomogeneously broadened laser, but at a greatly reduced threshold.

We now present some numerical results which illustrate the dynamical behavior of the 4-dimensional and the 6-dimensional model and allow a comparison with the Lorenz model. In Fig. 1 we present projections of typical non-transient trajectories on the P-D plane in arbitrary units which are scaled with the parameter r in order to preserve the size of the attractors. The fixed parameter values $\gamma_n/\gamma_\perp = 1$, $\omega_1^2/\gamma_\perp^2 = 2$ and $\kappa/\gamma_\perp = 2.8$ were selected in order to satisfy the bad cavity condition and in order to avoid a spurious instability for $\omega_1^2/\gamma_\perp^2 > (\kappa + \gamma_\perp)/(\kappa - \gamma_\perp)$ which exists in the 4-dimensional model but is not present in the exact full equations (1).

Fig. 1 a shows typical trajectories of the 4-dimensional model along the r-axis. The threshold of instability of the time-independent state r = 4.2 according to eq. (21), and much lower than the corresponding threshold r = 23.8 of the Lorenz model, whose trajectories (for $\omega_1^2 = 0$) are shown in Fig. 1 c. Immediately above threshold a limit cycle appears in the 4-dimensional model, which, for somewhat larger values of r, bifurcates to a strange attractor similar to the Lorenz attractor. For still larger values of r there are windows of periodic behavior and period doubling of symmetrical and asymmetrical limit cycles, which are qualitatively similar to corresponding results in the Lorenz model. For r $\gtrsim$ 100 the two models are very similar.

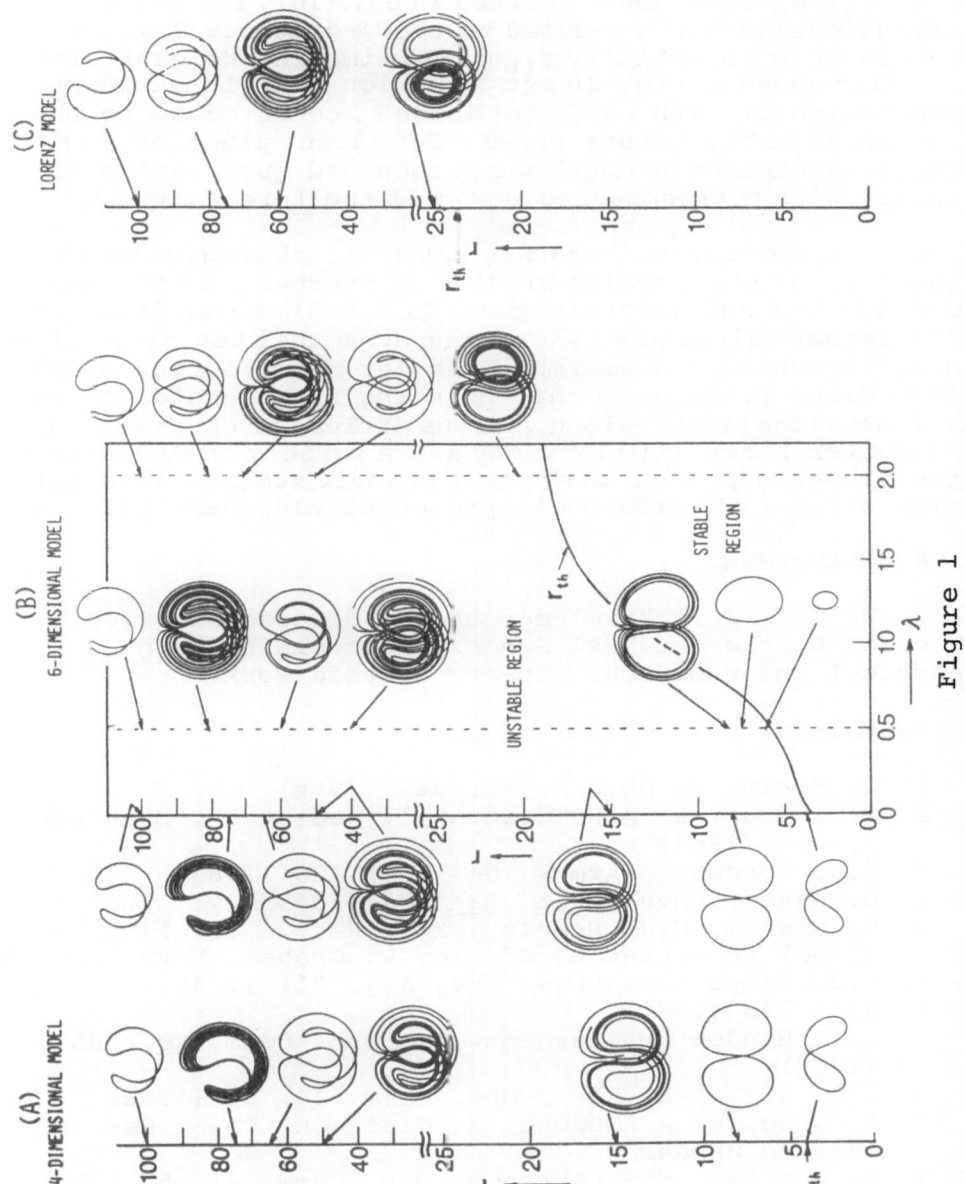

Figure 1

In Fig. 1b we present typical trajectories of the 6-dimensional model for different values of the parameters r (vertical axis) and λ defined in eq. (10). For $\lambda=0$ the 6-dimensional model is reduced to the 4-dimensinal model. The threshold of instability r_{th} of the time-independent state is also shown in Fig. 1b as a function of λ. It increases monotonically with λ and, for large λ, comes close to the threshold of the Lorenz model. For large values of r ($r \gtrsim 60$) the trajectories again show a pronounced qualitative similarity with corresponding states in the Lorenz model.

For smaller values of r, but still above threshold, there exist limit cycles without counterparts in the Lorenz model if λ is sufficiently small ($\lambda \gtrsim 0.5$), which disappear for larger values of λ, where the dynamical behavior, also near threshold, becomes more similar to the Lorenz model. It is clear from Fig. 1 that changing the parameter of the 6-dimensional model along various different curves in the parameter space (λ, r) one may see a variety of routes from the time-independent state to a chaotic state. A more extensive account of this work is presented elsewhere [11].

ACKNOWLEDGEMENT

We wish to acknowledge useful discussions with Michael Dörfle and Axel Schenzle concerning both physical and numerical aspects of this work.

REFERENCES

1   H. Haken, Z. Physik 190, 327 (1966)
2   A.Z. Grasyuk, A.N. Oraevskii, Radiotekh. Elektron. 9, 524 (1964)
3   E.N. Lorenz, J.Atmos. Sci. 20, 130 (1963)
4   H. Haken, Phys. Lett. 53A, 77 (1975)
5   R. Graham, H.J. Scholz, Phys. Rev. A 22, 1198 (1980) to appear M. Dörfle, R. Graham, Phys. Rev. A
6   L.E. Casperson, Phys. Rev. A21, 911 (1980); A23, 248 (1981)
7   S.T. Hendow, M. Sargent III, Opt. Comm. 40, 385 (1982)
8   C.O. Weiss, H. King, Opt. Comm. 44, 59 (1982) C.O. Weiss, A. Godone, A. Olaffson, Phys. Rev. A 1983 to appear
9   J. Bentley, N.B. Abraham, Opt. Comm. 41, 52 (1982), M. Maeda, N.B. Abraham, Phys. Rev. A26, 3395 (1982)
10 H. Haken, Laser Theory, Encyclopedia of Physics 25/2c (1970)
11 R. Graham, Y. Cho, Opt. Comm. (1983) to be published

CONNECTION BETWEEN IKEDA INSTABILITY AND PHASE CONJUGATION

W.J. Firth, E.M. Wright and E.J.D. Cummins

Department of Physics
Heriot-Watt University
Edinburgh, U.K.

Period-doubling to chaos [1] and phase conjugation are two non-linear optical phenomena which have attracted enormous but largely independent interest in recent years. In this paper we establish a close connection between the two phenomena, both being forms of near-degenerate four-wave mixing. We also report a new and rather unexpected feature of the Ikeda problem: the existence of a stable $3t_R$ oscillation <u>below</u> the period-doubling threshold.

First, we show that the threshold condition for a doubly-resonant four-wave parametric oscillator is identical to that for Ikeda instability.

Consider the system shown in Fig. 1. A strong "pump" wave E, frequency ω, interacts in a nonlinear medium with a "signal" wave $\in$ at $(\omega + \Omega)$ and an "idler" μ at $(\omega - \Omega)$. Neglecting in-essential complications, the propagation equations are

$$\frac{\partial E}{\partial z} = i|E|^2 E \tag{1}$$

$$\frac{\partial \in}{\partial z} = 2i|E|^2 \in + i E^2 \mu^* \tag{2}$$

$$\frac{\partial \mu^*}{\partial z} = -2i|E|^2 \mu^* - i E^{*2} \in \tag{3}$$

The right most terms in (2) and (3) clearly describe 4-wave mixing and/or <u>forward</u> phase conjugation, which is not quite perfectly phase-matched, but considerably simpler than "true" phase conjugation with counterpropagating pump beams. This process is rather similar to 3-wave mixing, except that E acts twice instead of once. In the latter case, however, there is no counterpart to the leading terms in these equations: in (1) this is just nonlinear refraction, responsible for dispersive optical bistability, but the terms in (2) and (3) describe a change in refractive-index at $\omega \pm \Omega$ induced by E: in particular they will cause the resonant frequencies of a cavity to tune with pump intensity: we term this process "transphasing". Fig. 2 clearly shows this effect as a lateral movement of the gain spectrum with increasing pump intensity.

Suppose now one feeds back the signal and idler in ring resonators: on each round trip the signal and idler experience an attenuation B and a phase shift ϕ_0. In principle, we could consider separate cavities for signal and idler - which might lead to some interesting phenomena - but here we wish to concentrate on the case where the cavity is shared, i.e. ε and μ are collinear with E. Self oscillation is then possible when

$$e^{-i\Omega t_R} \begin{pmatrix} \varepsilon \\ \mu^* \end{pmatrix} = \begin{pmatrix} Z(1+i|E|^2) & iZE^2 \\ -iZ^*E^{*2} & Z^*(1-i|E|^2) \end{pmatrix} \begin{pmatrix} \varepsilon \\ \mu^* \end{pmatrix} \tag{4}$$

where t_R is the cavity round trip time and $Z = B\exp(i(\phi_0+|E|^2))$. This condition is exactly that governing Ikeda instability[1,2], if we interpret E as the internal field. Fig. 2 shows the gain of an external probe field (i.e. (1) with a source term) versus Ω and $|E|^2$. Clearly the maximum gain occurs whenever transphasing allows both signal and idler to be cavity resonant. If ω is also resonant, this coincides with bistability, but if ε and μ are adjacent cavity modes, then $2\Omega = 2\pi/t_R$ so that four-wave oscillation gives a $2t_R$ modulation to the total field - the Ikeda instability.

It is important to note that transphasing is of the utmost importance in this interpretation. If transphasing were not present, then above a certain critical internal (pump) intensity $(|S|^2)$ the device would always exhibit spontaneous four-wave parametric oscillation, thus giving no 'islands of stability' on the positive-slope-branches above this intensity. The interplay between parametric gain and interference of the signal and idler, as produced by transphasing, leads to these islands of stability, which occupy less and less of the positive-slope-branches as the internal intensity increases[1].

There are an infinity of degenerate solutions $\Omega = (2n+1)\pi/t_R$, corresponding to more widely separated modes: a finite medium response time (or phase mismatch) will obviously impose a finite bandwidth on the gain spectrum in Fig. 2, thereby raising this degeneracy: c.f.[3]

If we now consider a $\underline{folded}$ (Fabry-Perot) resonator, we expect the same general picture, but with differences due to the fact that each of E, ε and μ now comprise two counter-propagating components, giving rise in particular to phase-conjugate reflections.   Fig. 3 shows that the probe spectrum for such a cavity is indeed considerably more complex than Fig. 2.   We draw particular attention to the pair of prominent peaks in the wings with no corresponding central peak.   These peaks can be shown to arise from the phase grating :if washed out by diffusion, these peaks vanish[4]:   they are thus a consequence of nonlinear nonreciprocity.

Given the above physical considerations it is not surprising that these instabilities survive when transverse effects are included: Ref. 4 shows this for a Gaussian pump beam in a Fabry-Perot cavity.   There is a band of t_R-oscillation due to nonlinear nonreciprocity, the first indication of a dynamic effect of this kind.   Physically, the nonlinear medium (a thin central slice in this case) is acting as an intra-cavity phase conjugate mirror.

A close physical connection thus exists between Ikeda instabilities and four-wave mixing, especially phase conjugation.   One important facet of this connection is that a full quantum optical treatment exists for phase conjugation, and predicts that "squeezing" can occur[5]:   adaptation of these results to nonlinear resonators will be of great significance.

Given the above, how are subsequent bifurcations to be interpreted?   We suggest the following picture.   Above the Ikeda threshold, the nonlinear resonator no longer has constant optical properties - the $2t_R$ oscillation forbids this.   The standard cavity mode condition, that the field repeats itself (up to a factor) after a round trip then loses its force, since the medium itself has changed after t_R.   It follows that one can now only insist on repitition after $\underline{two}$ round trips.   Then the effective cavity length is $\underline{doubled}$, and so its free spectral range is $\underline{halved}$.   At the $2t_R$ threshold, these cavity modes are coincident with the pump frequency and the $2t_R$ peaks, but increasing the input will cause the modes, by transphasing, to split off and move in frequency.   We now have a rescaled version of the previous argument: when transphasing brings these "dressed modes" just halfway between the pump and Ikeda ($2t_R$) modes, we again have a double resonance and thus, if the gain is big enough, will get self-oscillation, this time at $4t_R$.   This, in turn, breaks the $2t_R$ symmetry, doubles the mode spectrum, and the "new" modes, through transphasing to double resonance, give $8t_R$ oscillation: and so on.   The period-doubling cascade is thus a progressive spontaneous breaking of the time-translation symmetry of the system, leading in the end to no time symmetry at all, or chaos. It is also reasonable, on this picture, that there should be scaling relations between one bifurcation and the next, and also that the extra input required to go from one bifurcation to the next should decrease as the period increases.

NEW PERIOD-3 SOLUTIONS

We had assumed that all physically relevant solutions of the Ikeda mapping are constant, chaotic, period 2^n or period $j2^n$, the last as "windows" in the chaotic regime. Recently, however, we came across a period-3 solution which is stable in a narrow region below the onset of period-2 (Fig.4). This solution, which we term P3, appears to have rather interesting properties which we have only begun to explore: here we will summarise our present knowledge. The crucial question, of course, is whether P3 is physical. Unlike other non-stationary solutions, P3 does not evolve from the steady solution S, so does not arise in perturbation theory, as Fig.4 shows, and thus cannot be generated quasistatically. A pulsed input beam, however, can couple to P3, and Fig.5 shows strong P3 modulation of a gaussian input pulse, which should thus show up in experiments. Even more striking, Fig.6 shows that the "basin of attraction" of P3 seems to be as large as that of S. In fact we would guess that the boundary between the basins is a Cantor set, and may be of some mathematical interest[6].

The local stability of P3 can be tested by a technique which is best framed in a more general way.

Consider a PN solution of the Ikeda map:

$$E_{n+1} = A + B E_n e^{i|E_n|^2} \tag{5}$$

The first problem is to generate the N complex field values E_n. This can be reduced to the solution of N equations for the real "intensities" $I_n = |E_n|^2$, as follows

$$I_1 = A^2|F_1|^2 = A^2 \frac{|1 + Be^{iI_2} + B^2 e^{i(I_2+I_3)} + \ldots + B^{N-1} e^{i(I_2+..I_N)}|^2}{|1 - B^N e^{i(I_1+I_2+\ldots+I_n)}|^2} \tag{6}$$

with similar equations for the others (we allow B to be complex here, incorporating ϕ_o). The complex fields themselves can then be constructed by using $E_1 = AF_1$ and generating the others from the map. This technique may be used e.g. to generate $P2^n$ solutions within the period-doubling cascade, but we here concentrate on P3.

We found that there are two P3 solutions, which are "born" simultaneously at $A = 1.083$ in Fig. 4. The local stability of these and other solutions is found by finding the eigenvalues of products of the matrix M in (4), but evaluated at the fields E_n i.e. the PN solution is stable if the eigenvalues λ_1, λ_2 of the matrix

$$M(E_N) \quad M(E_{N-1}) \cdots \quad M(E_1)$$

obey $|\lambda_1|$, $|\lambda_2| < 1$. We thus find that the P3 traced by the solid line in Fig. 4 is stable, and the other unstable.

NONLINEAR MEDIUM

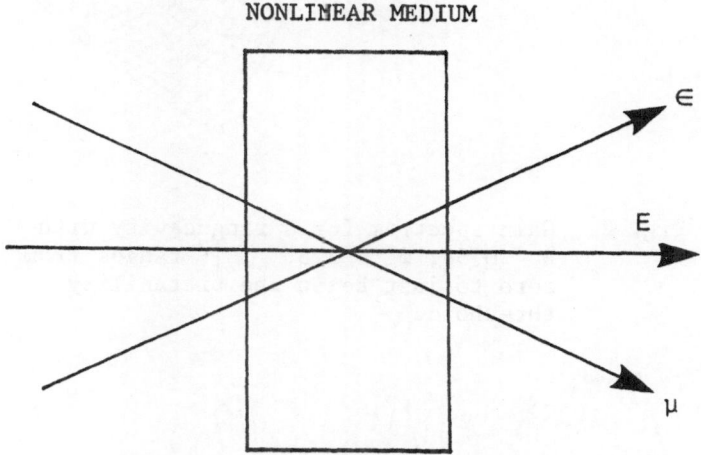

Fig. 1 The "pump" E, the "signal" $\in$ and the "idler" μ interact within the medium which has a third order dispersive nonlinearity.

REFERENCES

1. K. Ikeda, H. Daido and O. Akimoto, Phys. Rev. Lett. <u>45</u>, 709 (1980).
2. W.J. Firth, Opt. Comm. <u>39</u>, 343 (1981).
3. K. Ikeda and O. Akimoto, Phys. Rev. Lett. <u>48</u>, 617 (1982).
4. W.J. Firth and E.M. Wright, Phys. Lett. <u>92A</u>, 211 (1982).
5. H.P. Yuen and J.H. Shapiro, Opt. Lett <u>4</u>, 334 (1979).
6. L. Gribogi, E. Ott and J.A. Yorke, Phys. Rev. Lett. <u>50</u>, 935 (1983).

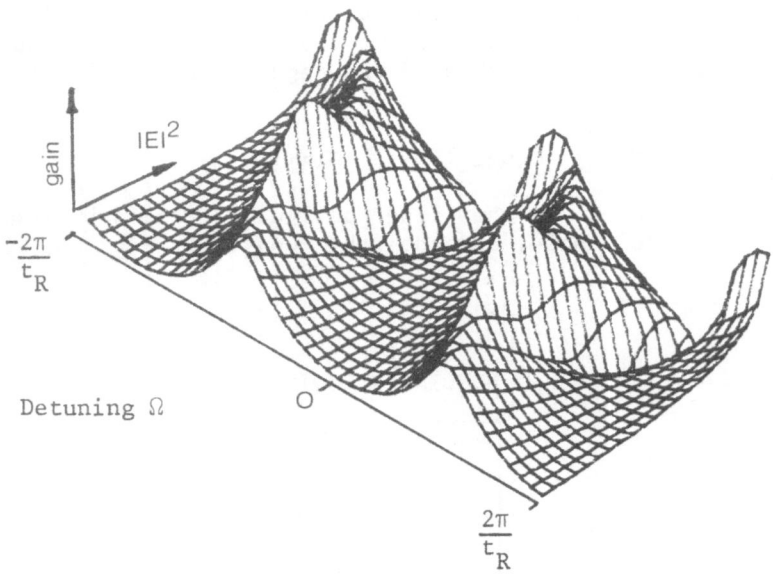

Fig. 2.   Gain spectrum for a ring-cavity with
          B = 0.58, ϕ_0 = 2.3. $|E|^2$ ranges from
          zero to just below the bistability
          threshold.

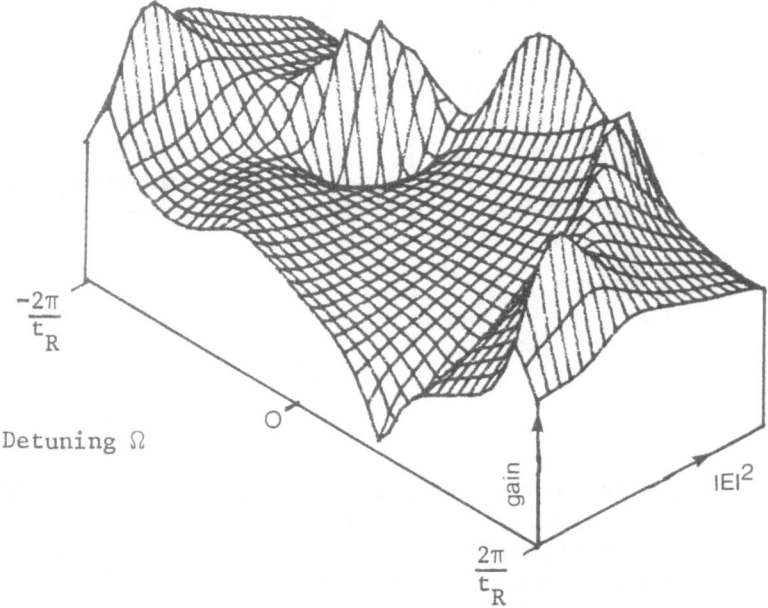

Fig. 3.   Gain spectrum for a folded resonator with
          the phase-grating retained: B = 0.3, ϕ_0 = 3.0.
          The range of $|E|^2$ covers the (first) upper
          branch.

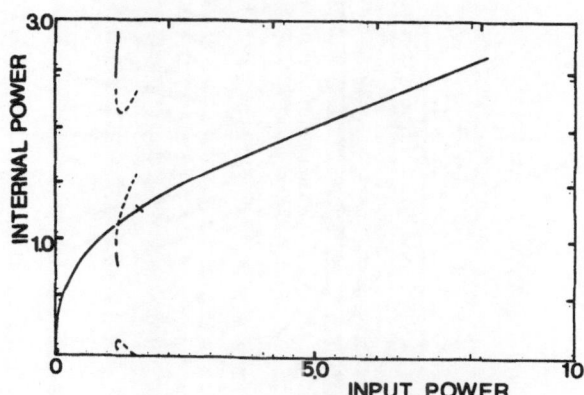

Fig. 4.   Stable and unstable P3 below P2 threshold
           power (x)

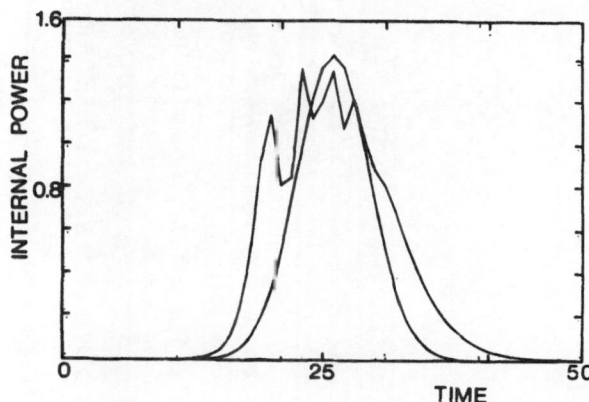

Fig. 5.   Internal pulse and input pulse (FWHM 8.5 t_R)
           for a nonlinear ring resonator (B = 0.8).

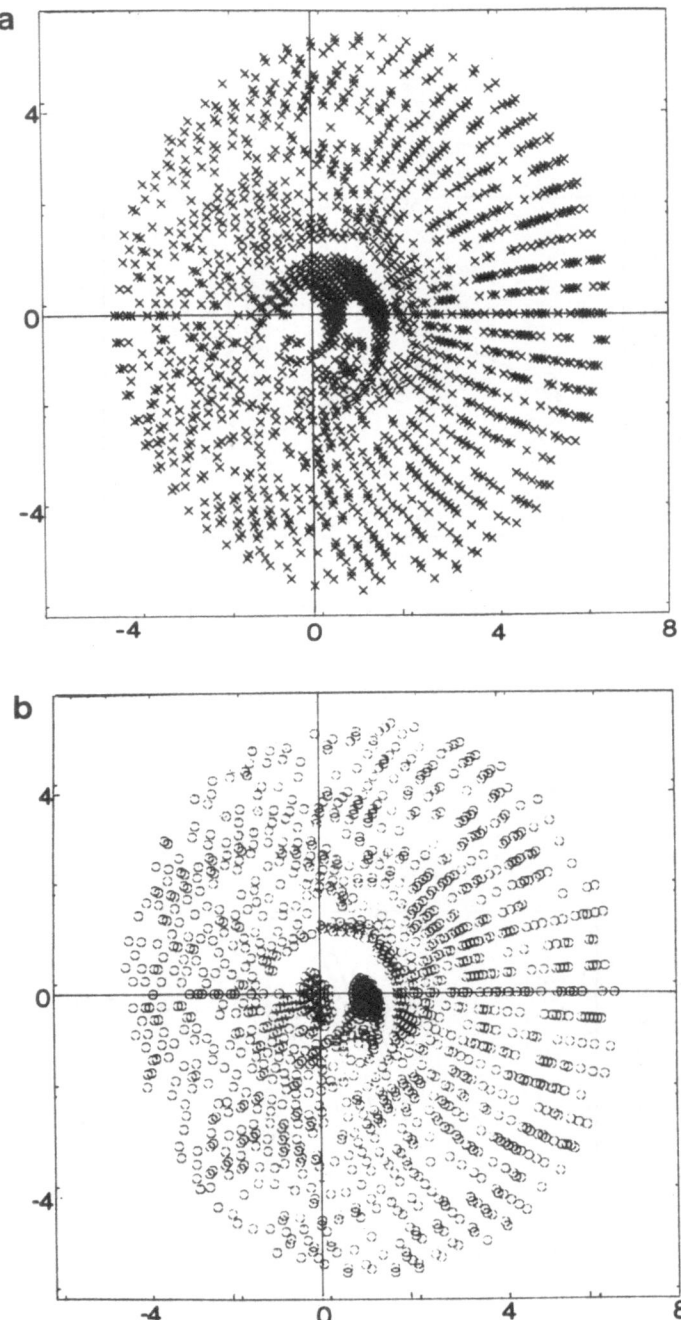

Figure 6.   (a) Basin of attraction for steady state,
            (b) Basin of attraction for P3 ; P_{in} = 1.1 ; B = 0.8.

MULTI-PARAMETER UNIVERSAL ROUTE TO CHAOS IN A FABRY-PEROT

RESONATOR

Eitan Abraham and William J. Firth

Department of Physics
Heriot-Watt University
Edinburgh, U.K.

INTRODUCTION

The ever increasing interest in chaos and related phenomena
in the context of optical bistability[1], has emerged from the predictions by Ikeda and Ikeda et al[2] and subsequent observations by
Gibbs et al[3] in a hybrid system. As Ikeda et al treated systems
with a delayed feedback, e.g. a ring cavity, it seemed an 'open
question ' the existence of chaos in a Fabry-Perot cavity.
Firth[4] was the first to show that a standing-wave resonator containing a cubic non-linearity with relaxation time $\tau = 0$,
exhibited Ikeda-type instabilities; Firth's calculations were
extended by Abraham et al[5] to the case where $\tau \neq 0$ and strong
diffusion (to obliterate standing-wave effects).

In the limit $\tau = 0$, Ikeda's system reduces to a complex
difference equation for the electromagnetic field at the input
mirror: a Feigenbaum cascade of period-doubling bifurcations
leading to chaos occurs as the input field is increased from a
specific threshold value. In this paper we show numerically and
justify analytically that our system[5] also undergoes successive
bifurcations in its route to chaos: a Feigenbaum type of universality emerges as <u>any</u> of the parameters i.e. the cavity detuning
θ_o, the ratio τ/τ_R (t_R: round-trip time) or the input field E_i are
varied.

EQUATIONS OF MOTION

We assume a Kerr medium placed in a Fabry-Perot resonator with
mirror reflectivity R and length L. The Maxwell-Bloch equations
in the limits of high dispersion and intensities well below

saturation, describe the present system.   In addition, if strong diffusion is allowed for then we get[5]

$$\tau \, \partial \chi^{NL}/\partial t \; + \; \chi^{NL} \; = \; |F|^2 \; + \; |B|^2$$

$$\partial F/\partial x \; + \; n_o c^{-1} \, \partial F/\partial t \; = \; i \, \chi^{NL} - \alpha F/2 \qquad\qquad (1)$$

$$-\partial B/\partial x \; + \; n_o c^{-1} \, \partial B/\partial t \; = \; i \, \chi^{NL} - \alpha B/2$$

where $F(B)$ is the forward (backward) field in the cavity, χ^{NL} the non-linear susceptibility, n_o the linear refractive index, c the speed of light in vacuum, and α the linear absorption coefficient. The boundary conditions are

$$F(o,t) \; = \; A \; + \; R^{\frac{1}{2}} \, B(o,t)$$
$$\qquad\qquad (2)$$
$$B(L,t) \; = \; R^{\frac{1}{2}} \, F(L,t) \, \exp(-i\theta_o)$$

where $A = T^{\frac{1}{2}} E_i$ ($T = 1 - R$) and $\theta_o = 4\pi n_o L/\lambda (\mathrm{mod} \; 2\pi)$.   Throughout this paper E_i is constant and $F(t) \equiv F(o,t)$.

Integration of (1) along characteristics yields[5]

$$F(t + t_R) = A + Z \, \exp(i \, \Phi \, [\, |F|^2 \,]) \qquad\qquad (3)$$

where

$$Z = R \, \exp(-\alpha L - i\theta_o)$$

$$t_R = 2n_o L/c$$

$$\Phi[\, |F|^2 \,] = \tau^{-1} \int_{-\infty}^{t} dt' \, \exp[-(t-t')/\tau] \{ |F|^2 \; K \; +$$

$$\int_o^1 dy \, \exp[-\alpha L(1 - y)] [\, |F(t' + yt_R)|^2 + K' |F(t'-t_R+yt_R)|^2] \}$$

K and K' are constants

This pseudo-difference equation is much more complicated than the simple Ikeda map in which the functional $\Phi[\, |F|^2 \,]$ is replaced by $|F|^2$.

The steady state solution of (1) gives a typical multiple-branch transfer curve.   The negative-slope regions are always unstable whereas the instability of the positive-slope ones, by

way of self-oscillations and chaos, depends on the choice of parameters.   The instability edges can be obtained by linearising
(3) about a steady state value: an independent numerical integration of (1) showed that our stability analysis was accurate to
within 1%.

UNIVERSALITY

In the analysis of (3) the major obstacle is that the period
which doubles has no necessary relation to t_R.   Time-domain
analysis is thus difficult and we examine instead the frequency
domain by considering the evolution of the spectrum as a control
parameter λ is varied.

At some value λ_n the period doubles from $2^n T$ to $2 \times 2^n T$,
where T is some base period of order t_R.   Below λ_n, the spectrum
contains all harmonics of $2\pi/2^n T$ while above λ_n a 'new' component
appears half-way between each pair of 'old' components.   It is
thus convenient to represent the Fourier components of $F(t)$, namely

$$F(t) = \sum_m c_m^n \exp (it\, 2\pi m/2^n T)$$

as a vector

$$\underline{f} = (c_1^n,\ c_2^n,\ \dots,\ c_j^n,\ \dots)$$

Eqn.(3) and a broad class of similar relations can then be Fourier-
transformed to take the form (after Taylor expansion of the exponential)

$$E\underline{f} = G(\lambda)\underline{f}\ +\ H(\lambda)\ \underline{ff}\ +\ N(\lambda)\ \underline{fff}\ +\dots \tag{4}$$

where e.g.

$$\underline{ff}\ =\ \sum_{jm} c_j^n\, c_m^n,\ \text{etc}$$

and E and G are diagonal matrices, the mth component of E being
$\exp(-it_R\, 2\pi m/2^n T)$;   H and N are connecting 'tensors'.

Above but close to the bifurcation point where $\lambda - \lambda_n \ll 1$, we
have the new components $\underline{f}_N$ of $\underline{f}$ which are very small compared with
the old ones f_o.   For our analysis we partition $\underline{f}$ as $(\underline{f}_o, \underline{f}_N)$ in
order to linearize in $\underline{f}_N$, and write the relevant part of (4) as

$$E_n\, \underline{f}_N\ =\ M_n\, (\underline{f}_o,\ \lambda)\ \underline{f}_N \tag{5}$$

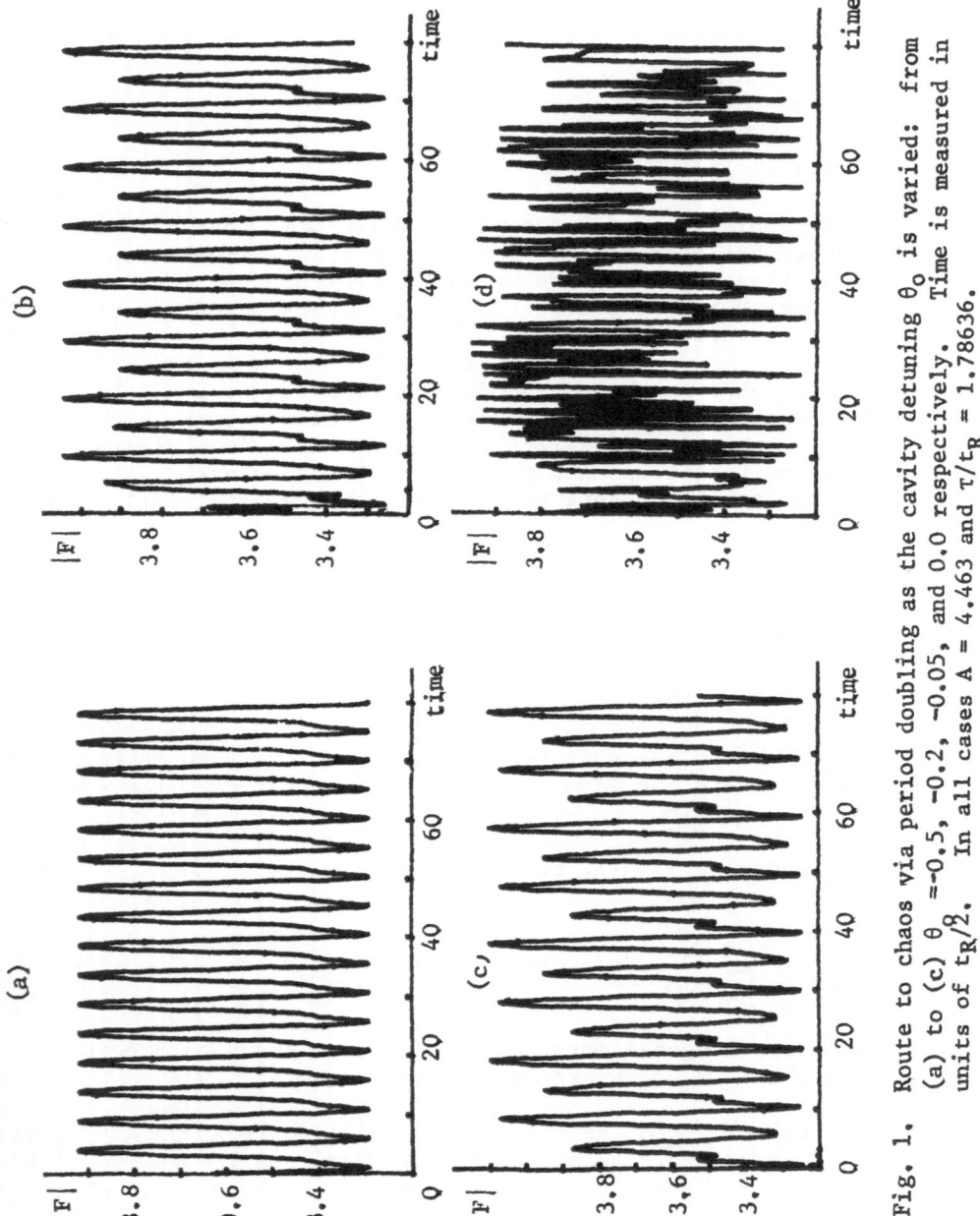

Fig. 1. Route to chaos via period doubling as the cavity detuning θ_0 is varied: from (a) to (c) $\theta_0 = -0.5, -0.2, -0.05,$ and 0.0 respectively. Time is measured in units of $t_R/2$. In all cases $A = 4.463$ and $\tau/t_R = 1.78636$.

Table 1. Universal sequence as the control parameter λ approaches
the limit point λ_∞ ($\delta n = (\lambda_n - \lambda_{n+1})/(\lambda_{n+1} - \lambda_{n+2})$).

Control Parameter λ	λ_8:P4→P8	λ_∞	δ_8	δ_{16}	δ_{32}
A	4.476	4.463	3.902	4.768	4.612
θ_o	.09046	0	4.411	4.601	4.658
τ/t_R	1.826	1.78636	4.506	4.643	4.618

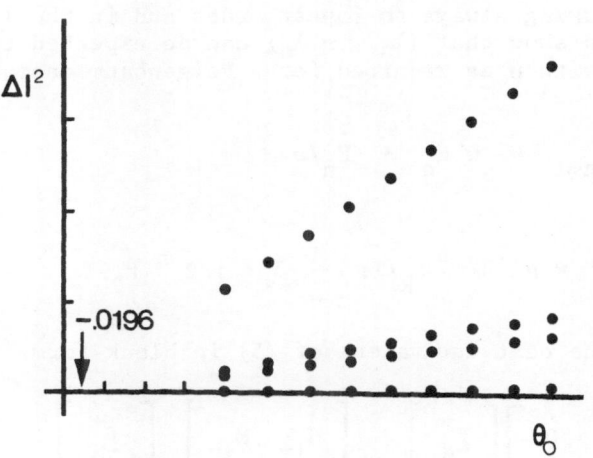

Fig. 2.   Behaviour of ΔI^2 (eqn.(9)) as a function of
θ_o in the neighbourhood of the 4P → 8P threshold.

We can try to solve (5) if we know $\underline{f}_o$: there will be a non-trivial solution $\underline{f}_N^n$ at $\lambda = \lambda_n$.

Just above the bifurcation we take (4) to third order in $\underline{f}_N$ by assuming that $\underline{f}_N \propto \underline{f}_N^n$ where calculating the non-linear effect of $\underline{f}_N$ on $\underline{f}_o$ and itself. We then obtain a linear equation for $\underline{f}_N$ which matches smoothly onto (5) as $\lambda \to \lambda_n$ only if the dependence of M_n on λ is compensated by that due to $\underline{f}_N^n$. As each term in (4) must have the same net frequency, it follows that $\underline{f}_N$ can affect $\underline{f}_o$ only in second order and itself in third order. In both cases, M_n contains $\underline{f}_N^n$ at second order while it contains $(\lambda - \lambda_n)$ in first order. Hence,

$$\underline{f}_N \sim (\lambda - \lambda_n)^{\frac{1}{2}} \underline{f}_N^n \tag{6}$$

This argument provides some justification for the extrapolation procedure explained below since the differences of corresponding maxima are functions of 'new' components only.

At the next bifurcation point, $\lambda = \lambda_{n+1}$, a new component appears either side of each member of $\underline{f}_N$. The latter can be split into two sets $\underline{f}_{N+}$ and $\underline{f}_{N-}$ according to whether the component has frequency above or below that of the corresponding member of $\underline{f}_N$. Thus, working always to lowest order and in the limit of large n, we can show that $(\lambda_{n+1} - \lambda_n)$ can be expected to decrease geometrically with n as required for a Feigenbaum cascade. Noting the identity

$$E_{n+1} = w E_n + E_n/w \tag{7}$$

with

$$w = \exp(i\pi 2^{-n} t_R/T) \underset{n\to\infty}{\to} 1 + i\pi 2^{-n} t_R/T$$

we can write the next equivalent of (5) in block form:

$$\begin{bmatrix} E_n w & 0 \\ 0 & E_n/w \end{bmatrix} \begin{bmatrix} \underline{f}_N^+ \\ \underline{f}_N^- \end{bmatrix} = \begin{bmatrix} M_{++} & M_{+-} \\ M_{-+} & M_{--} \end{bmatrix} \begin{bmatrix} \underline{f}_N^+ \\ \underline{f}_N^- \end{bmatrix} \tag{8}$$

Physically, M provides the difference frequency which enables one new component to scatter and couple to another. For both M_{++} and M_{--}, this difference frequency belongs to $\underline{f}_o$ and, provided $2^n T$ is large compared to all intrinsic response times of the system, both are equal to M_n of (5) (in lowest order). M_{+-} and M_{-+}, on the other hand, couple pairs of frequencies differing by a member of $\underline{f}_N$ so each element of them is proportional to $(\lambda - \lambda_n)^{\frac{1}{2}}$.

Finally, requiring a non-trivial solution of (8) while (5) remains true to lowest order, leads to

$$(w - 1)^2 \sim (\lambda_{n+1} - \lambda_n)$$

and hence using (7)

$$(\lambda_{n+1} - \lambda_n)/(\lambda_{n+2} - \lambda_{n+1}) \simeq (i\pi 2^{-n})^2/(i\pi 2^{n-1})^2 = 4.$$

This demonstrates the required geometrical decrease in $\Delta\lambda$ and we also obtain an approximate value of Feigenbaum's $\delta(= 4.669...)$. On the other hand (4) is extremely general, covering ring as well as Fabry-Perot resonators. A closer examination of (4) should indeed give δ accurately.

TRAPPING BIFURCATION POINTS

For a quantitative explanation of bifurcations in systems of differential equations, a considerable amount of computer time is needed to generate a single time sequence. In these circumstances it is difficult to pin down the most stable control value, while the bifurcation points themselves show poor convergence. We have found a technique, however, that has enabled us to find Feigenbaum's δ to < 1% (see Table I). This is probably the most complex system in which Feigenbaum sequences have been found.

The technique is based on the hypothesis that at each bifurcation the new Fourier components grow as $(\lambda-\lambda_n)^{\frac{1}{2}}$ as λ is increased through its nth bifurcation value λ_n. Consider any peak of $I(t) \equiv F(t)$ which first repeats a time T_n later. If λ is increased from λ_n, $I(t + T_n) - I(t) \neq 0$, and we assert that

$$\Delta I^2 = [I(t + T_n) - I(t)]^2 = const \times (\lambda - \lambda_n) \qquad (9)$$

Thus plotting (9) we get a straight line that intersects the λ-axis at λ_n. As Fig. 2 shows, this rule is very well obeyed and the accuracy in fixing λ_n is enhanced because the number of straight lines doubles at each bifurcation, thus improving the statistics which partially offsets the decline of ΔI^2 for large n.

REFERENCES

1.   For a review with a tutorial approach see E. Abraham and
      S.D. Smith, Rep. Prog. Phys. 45, 815 (1982).

2.   K. Ikeda, Opt. Comm. 30, 257 (1979); K. Ikeda, H. Daido, and
      O. Akimoto, Phys. Rev. Lett. 45, 709 (1980).

3.   H.M. Gibbs, F.A. Hopf, D.L. Kaplan and R.L. Shoemaker, Phys.
      Rev. Lett. 46, 474 (1981).

4.   W.J. Firth, Opt. Comm. 39, 343 (1981).

5.   E. Abraham, W.J. Firth and J. Carr, Phys. Lett 91A, 47
      (1982); E. Abraham and W.J. Firth, Optica Acta (to appear).

6.   H.J. Carmichael, R.R. Snapp and W.C. Schieve, Phys. Rev. A26,
      3408 (1982).

THE NMR-LASER - A NONLINEAR SOLID STATE SYSTEM SHOWING CHAOS

E. Brun, B. Derighetti, R. Holzner and D. Meier

Physik-Institut, University Zürich

CH-8001 Zürich

1. INTRODUCTION

We present experimental observations along the roads to chaos[1] of a tuned low-Q solid state spin-flip NMR laser (raser)[2] which, according to our earlier work[3], should show the universal Feigenbaum scenario under certain conditions. The free running ruby raser is periodically perturbed by modulating one of its physical parameters. For example, one may modulate the pumping magnetization sinusoidally with the angular frequency Ω close to the eigenfrequency of the linearized raser equations, typically $f = \Omega/2\pi \simeq 30$ Hz. If F is the modulation strength, the raser may thus be modelled by[3]

$$dM_v/dt = (-9/2\mu_o\eta Q\gamma M_v + 9\gamma B_1^d)M_z - M_v/T_2 \tag{1}$$

$$dM_z/dt = (1/2\mu_o\eta Q\gamma M_v - \gamma B_1^d)M_v - [M_z - M_e(1-F \sin \Omega t)]/T_e \tag{2}$$

2. SUBHARMONICS AND CHAOS: RASER EXPERIMENTS

With the periodic low-frequency drive we add two control parameters to our system, the strength F of the drive and its natural period $T = 2\pi/\Omega$ which we call period-1. For F-values below a certain threshold F_1 the system is attracted towards a stable limit cycle of period-1. Increasing the control parameter then leads to a series of bifurcations at F_n, $n = 1, 2, 3...$ where attracting limit cycles of period-2^n are born. In Fig. 1 we show the results of an experimental attempt to observe such a sequence of bifurcations which ends in a chaotic state. Extreme care is necessary to produce the depicted asymptotic stable limit-cycle behavior of defined period-2^n. Unidenti-

fied noise sources lead to spurious spectral components which already
show up in the phase-space portraits and Fourier power spectra of
low-order limit cycles. These effects often lead to a truncation of
an expected bifurcation sequence. Any attempt to estimate universal
constants from such data is fortuitous.

In our search for a Feigenbaum scenario we have observed a complex structure of the various basins of attraction. We have found

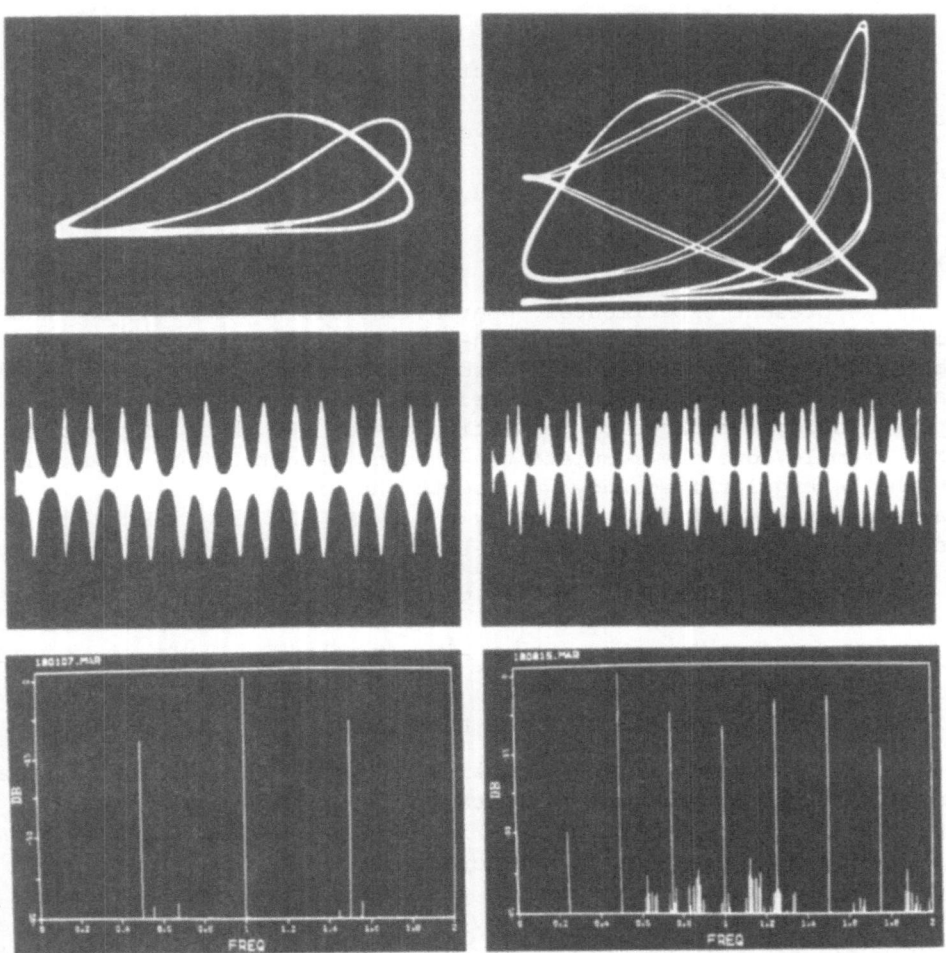

Fig. 1a. Experimental phase-space portraits, time and Fourier
power spectra of a modulated raser in a limit-cycle behavior of
period-2 and 4 with erratic noise. The portraits depict the raser
output vs. modulator signal.

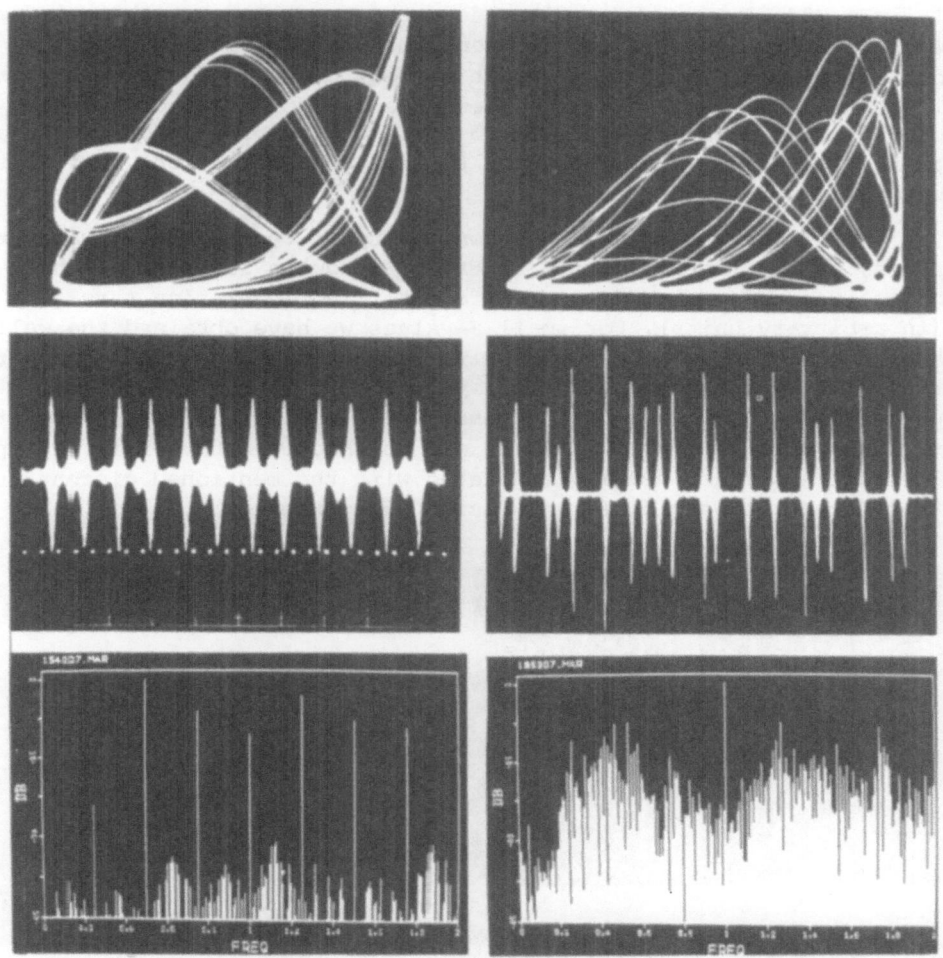

Fig. 1b. Experimental phase-space portraits, time and Fourier
power spectra of a modulated raser in a weakly noisy period-8 limit-
cycle and in a chaotic state.

many discontinuous jumps from one particular limit-cycle to another.
For example, from a limit-cycle of period-3 (Fig. 2) to one of
period-2 and, from limit-cycles of period-N directly to chaos. Transitions of this sort are reminescent of a first order phase transition
and are accompanied by strong hysteresis effects. They are highly
sensitive to noise and slow drifts. We have further found chaotic regions where the spectra vary intermittently in time between phases of
weakly noisy quasi-periodic behavior and phases with broadband noise.

In addition, we have often detected so-called breathing modes where a limit-cycle is low-frequency modulated which leads to an oscillatory behavior. This effect is most commonly found in higher limit-cycles, but it has also been observed for a period-2 cycle where such a breather seems to be extremely stable. We believe that there is a close correnspondence to similar observations as reported by Giglio et al.[4]

To illustrate some of the complexity of the modulated raser response, we show in Fig. 3 the raser amplitude in function of the drive frequency $f = \Omega/2\pi$ for different fixed modulation strengths F (in arbitrary units). For small F-values we have obtained the well known nonlinear response curve with fold-over and switching properties. The system remains for all values of f in a limit-cycle of period-1. For F > 200, the response curve becomes multi-peaked with strong sensitivity to f. This is due to chaotic behavior for example. The high jumps are usually connected with the mentioned discontinuous transitions between different basins of attraction.

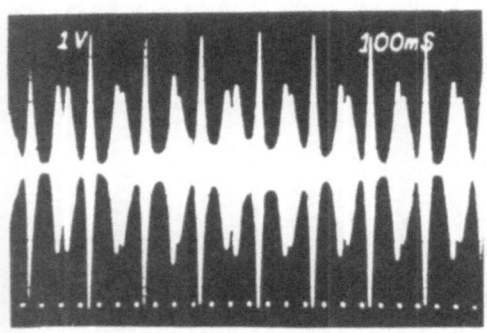

Fig. 2. Predominately a period-3 limit-cycle behavior of the raser. The dots are markers of period-1.

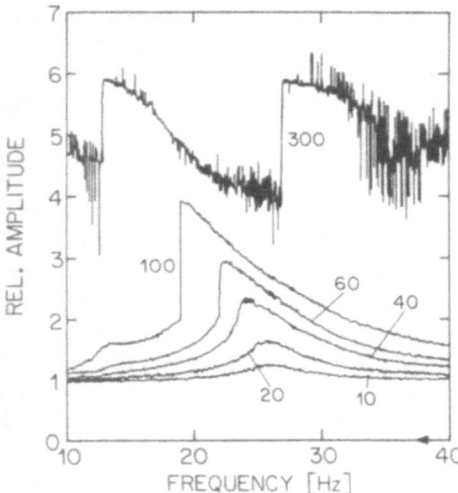

Fig. 3. Experimental low-frequency response curve (peak values) of a modulated raser in function of the drive frequency f for different values of the modulation in arbitrary units.

3. SUBHARMONICS AND CHAOS: COMPUTER EXPERIMENTS

In order to model some of our experiments we have looked for numerical solutions of (1) and (2). We present selected results which we consider as representative. The assumed system parameters are close to the actual physical values, e.g. $Q = 100$, $B_1^q = -10^{-10}T$, $M_e = -1.6$ A/m, $T_2 = 3\times10^{-5}s$, $T_e = 0.1s$, $\Omega = 280s^{-1}$, F has been varied between 0 and 0.3.

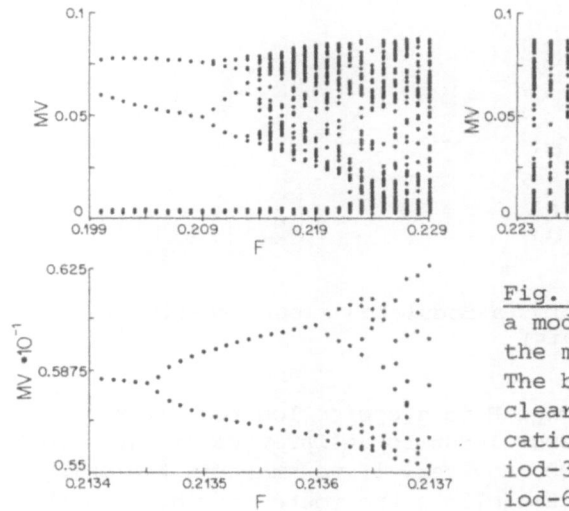

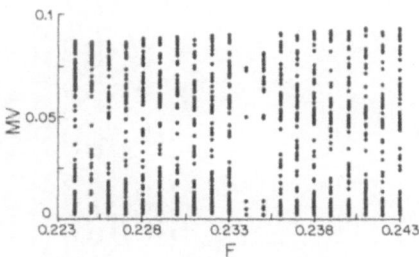

Fig. 4. Bifurcation diagrams for a modulated raser. F refers to the modulation of the DNP pump. The blow up to the left shows clearly the two pitchfork bifurcations from a period-16 to period-32 and period-32 to a period-64 limit-cycle.

Figure 4 shows three bifurcation diagrams for F between 0.2 and 0.245. For each value of F, 128 dots (one dot per period-1) are plotted which represent the asymtotic solution $M_y(t)$ of (1) and (2) at discrete times. We see clearly a sequence of bifurcations with period-doubling which leads to noisy bands (with an inverse cascade of period-doubling bifurcations which are not evident in the plot), a wide region with broadband noise and with a period-5 window near F = 0.234. From blow-ups, as shown in Fig. 4(last graph), one may determine the series F_n of bifurcations with period-doubling by 2^n, n = 1, 2, 3, 4... which converge to F_c. With our moderate resolution in the F-scale we have obtained the values $F_1 = 0.159000$, $F_2 = 0.196830$, $F_3 = 0.209650$, $F_4 = 0.212785$, $F_5 = 0.213445$, $F_6 = 0.213595$, $F_7 = 0.213625$ from which one finds the series of convergence rates $\Delta_1 = 2.95$, $\Delta_2 = 4.09$, $\Delta_3 = 4.75$ $\Delta_4 = 4.40$, $\Delta_5 = 5.00$ which have to be compared with the Feigenbaum limit $\Delta_\infty = 4.6692016...$. Similar results have been found with other system parameters and for sequences in different windows inside the chaotic bands.

The bifurcation diagrams in Fig. 5 illustrate a different asymptotic behavior of the model system within the same range of the control parameter F as in Fig. 4. It is the outcome of the different

choice of initial conditions. With a few percent change in the initial
conditions the system is attracted towards a stable period-3 cycle e.g.
for F = 0.215. Increasing F does not lead to bifurcations. Slightly
below F = 0.2336 a jump to chaotic behavior results. Further increasing F brings the system after a narrow chaotic region into a period-
5 window, followed by a sequence of bifurcations to limit-cycles of
period-5×2^n. The chaotic region and the period-5 window of Fig. 5
are identical to the ones of Fig. 4.

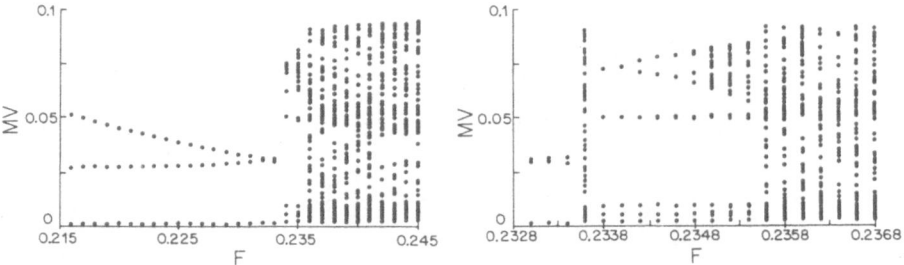

Fig. 5. Bifurcation diagrams for a modulated raser starting in
period-3 limit-cycle (F = 0.215).

Figure 6 shows what happens when F is stepwise lowered. Starting in
the period-3 cycle at F = 0.2005 leads to an irreversible jump to
period-2 cycle near F = 0.1900. However, if we start in the noisy
region at F = 0.225, the system follows the route of Fig. 4 in the
reverse sense. Hence, this route to and from the chaos is reversible.

From (1) and (2) we may obtain the phase-space and time behavior
together with Fourier power spectra. From Fourier spectra of low-order
period-2^n cycles we have found that the power per doubling goes down
by about 19 dB.

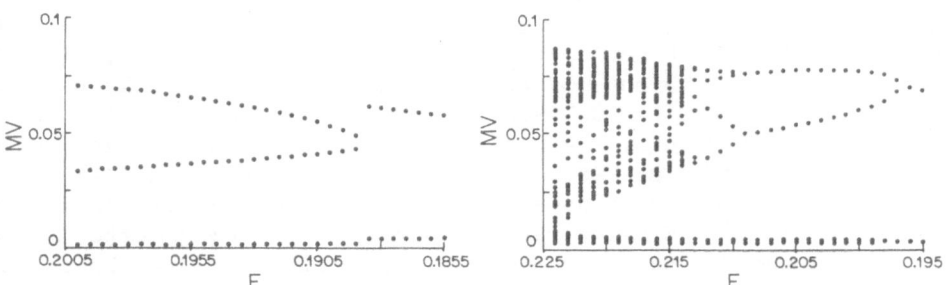

Fig. 6. Bifurcation diagrams of a modulated raser showing an
irreversible transition (3.2) and a reversible path from and
to the chaos.

Figure 7 illustrates the dynamics of the irreversible transition from
the period-3 to the period-2 cycle near F = 0.1886. The given phase-
space portraits are taken at various consecutive times, starting from
a nonequilibrium initial state. The 1. portrait shows a transient

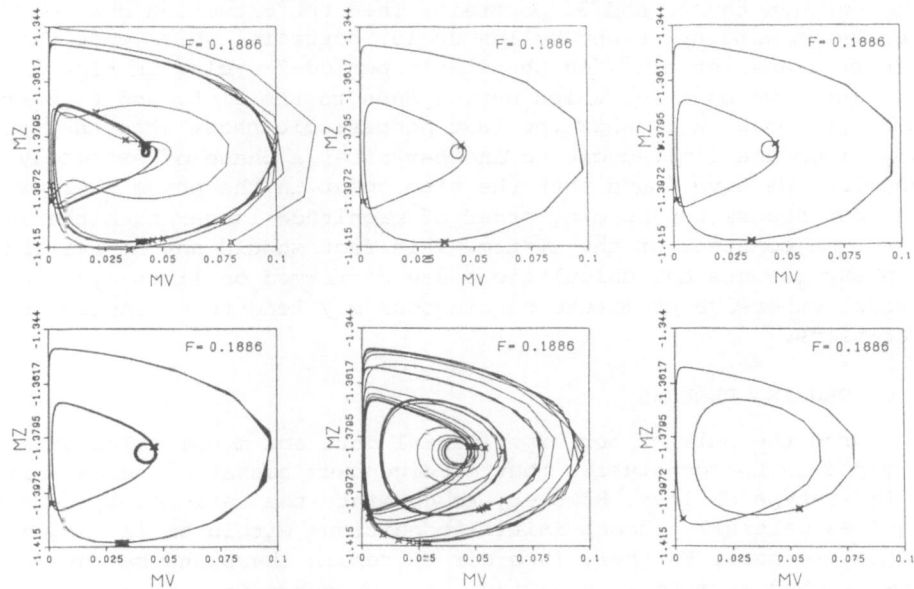

Fig. 7. Sequence of consecutive phase space portraits (32 natural
time-1 units).

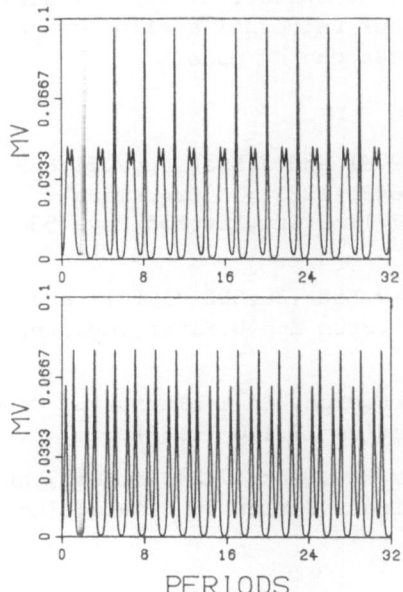

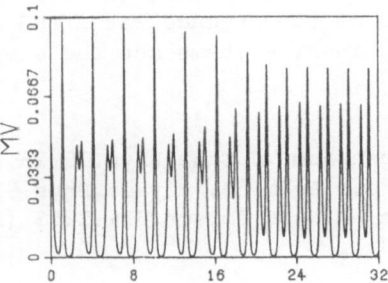

Fig. 8. Time behavior of the modulated raser corresponding to 3., 5.
and 6. portrait of Fig. 11.

which seems to lead to the stable period-3 cycle of the 2. portrait.
However, this is an illusion. As time goes on, slow changes become
manifest which can be recognized as weak nonperiodicities (smeared
out time-1 markers) in the 3. and 4. portraits after a long waiting
time between the 2. and 3. portrait. Then the situation changes drastically. A wild transient builds up (5. portrait) which ends, after
a short transient time, in the stable period-2 cycle. In Fig. 8 we
give the time behavior which corresponds to the 3. 5. and 6. portrait.
These graphs clearly show the fast nonperiodic phase when the system
jumps from one limit-cycle to another after a phase of extremely slow
dynamics. We have found that the time spent in the phase where slowing down occurs can be many order of magnitude longer than the longest time constant of the system. This fact should not be overlooked
when experiments and calculations are performed on limit-cycle behavior, otherwise premature conclusions may lead to erroneous inter-
pretations.

4. CONCLUDING REMARKS

From the bulk of our experimental data and model calculations
we may draw the conclusion that a Feigenbaum scenario indeed exists
in laser-type devices. However, to get into the realm of its attractor
requires carefully chosen initial conditions within small ranges in
parameter space. Further, in order to remain there and be sure to
have reached asymptotic behavior, extreme conditions have to be met,
not only with respect to noise but also to the long time stability,
in particular. It seems that experimenter (including ourselves) have
not reached the goal yet where realiable quantitative tests of limiting values of convergence rates and other universal scaling properties of nonlinear systems but the simplest can be made.

REFERENCES

1. M.J.Feigenbaum, J.Stat.Phys.19:25 (1978) and J.Stat.Phys.21:669
   (1979); J.P.Eckmann, RevMod.Phys.53:643 (1981); S.Grossmann,
   S.Thomae, Z.Naturforsch.32a:1353 (1977); E.Ott,Rev.Mod.Phys.53:
   655 (1981).

2. P.Bösiger, E.Brun and D.Meier,Phys.Rev.Lett.38:602 (1977): and in
   Phys.Rev.A18:671 (1978); P.Bösiger, E.Brun and D.Meier,Phys.Rev.
   A20:1073 (1979).

3. D.Meier, R.Holzner, B.Derighetti and E.Brun, in Evolution of Order
   and Chaos, Springer Series in Synergetics 17:146 (1982).

4. M.Giglio, S.Musazzi and U.Perini,Phys.Rev.Lett.47:243 (1981); and
   in Evolution of Order and Chaos, Springer Series in Synergetics
   17:174 (1982).

SELF-BEATING INSTABILITIES IN BISTABLE DEVICES

J.A. Martín-Pereda and M.A. Muriel

E.T.S. Ingenieros de Telecomunicación U.P.M.
Ciudad Universitaria
Madrid-3, SPAIN

INTRODUCTION

Since the observation of optical bistability by Gibbs et al.[1], optical bistability has been the field where researchers from many fields have found a common place to work. More recently, when Ikeda and co-workers[2-3] discussed the effect of a delayed feedback on instability of a ring cavity containing a non linear dielectric medium, and pointed out that the transmitted light from the ring cavity can be periodic or chaotic in time under a certain condition, optical bistable devices have shown new possibilities to be applied in many different fields. The novel phenomenon has been predicted to be observed in the hybrid optical device[3] and has been confirmed by Gibbs et al.[4]. Moreover, as we have shown[5], a similar effect can be obtained when liquid crystal cells are employed as non linear element.

In this paper sumarize the empirical and theoretical results that have been obtained by us. The electrooptic light intensity modulator has been a nematic liquid crystal cell with twisted configuration. A He-Ne laser beam, 5 mW of power, was incident to the device. Some of the here reported results were presented previously by us. As it will be shown, if certain conditions are achieved, a sustained oscillatory optical output can be obtained. Moreover, a selfbeating phenomena is achieved both, theoretical and empirically.

EXPERIMENTAL

The experimental configuration is the conventional one employed in hybrid optical bistable systems and it is shown, schematically, in Fig.1.

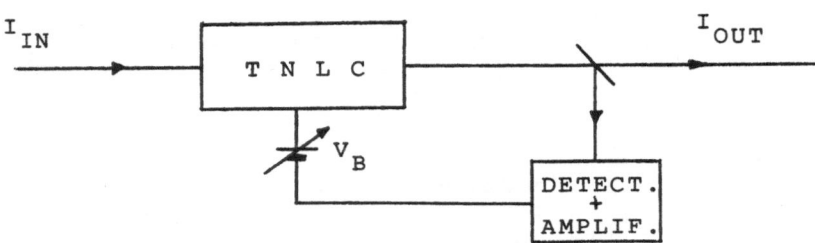

Fig. 1. Hybrid optical bistable system.

The main point concerning this arrangement is the electrooptic light intensity modulator, in our case, a twisted nematic liquid crystal cell. When this cell is orthogonal to the incident laser beam, its transmission curve, as a function of the cell applied voltage, is the one shown in Fig. 2, for 0°. In this case, polarizes are crossed. But when the cell is forming two angles, Fig. 3, with the input beam direction, this transmission no longer verifies.

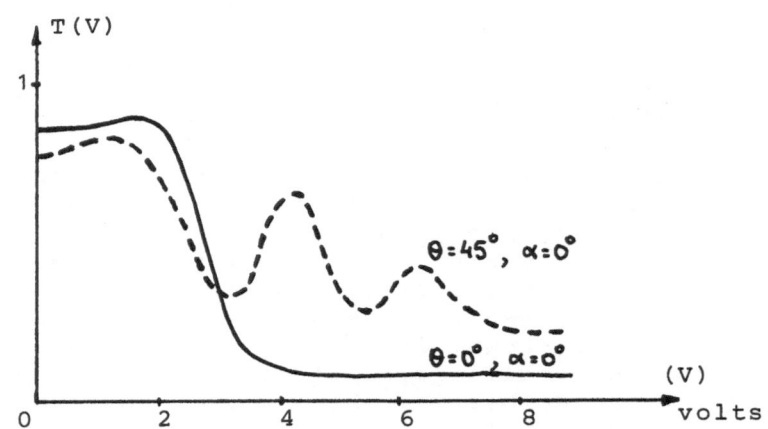

Fig. 2. Optical transmission versus voltage in T.N. cells with crossed polarizers.

The experimental results give the appearance of
several maxima and minima. A theoretical study is being
under progress and it will published elsewhere. For
θ = 45° and   α= 0° the new transmission curve is shown
in Fig. 2. This curve has been obtained for the static
case. When the applied voltage is varying with time, its
shape changes to a more complex form.

With the above considerations, the final set up is
shown in Fig. 4.

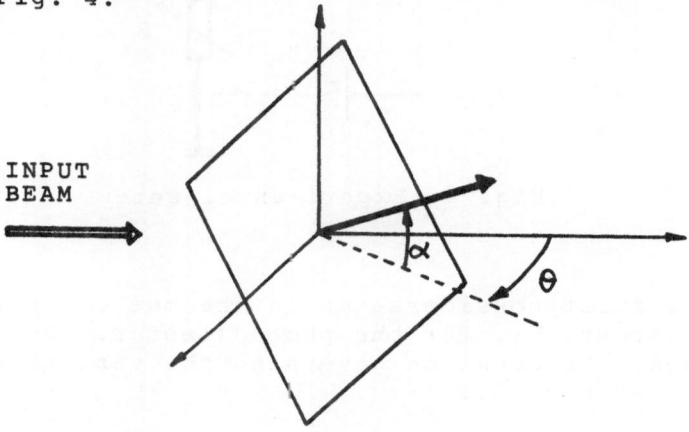

Fig. 3. Orientation of the TNLC cell.

The light, after crossing the liquid crystal cell
and a crossed polarizer, impinges on a phototransistor,
in our case a TIL 78, working as a current source. The
obtained current is afunction of the output intensity
level. Feedback is obtained through the variable resistor R. Its value gives the feedback coefficient. This
electrooptical system is equivalent, from an alectronic
standard point of view, to the circuit shown in Fig.5.a,
where C stands for the liquid crystal cell capacitance.
This circuit, because the phototransistor is operating
as a current source, is equivalent to the one shown in
Fig. 5.b. Moreover, by Norton and Thevenin theorems ,
its electrical equivalent is shown in Fig. 5.c. This
circuit has been the basis of our study.

THEORETICAL MODEL

Several are the considerations involving the circuit
of Fig. 5.c. Everyone of them gives a certain contribution to the total behaviour of the system. Moreover ,
some are responsible for the peculiar results obtained
by us.

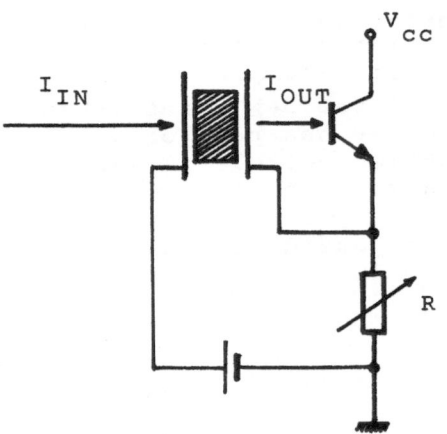

Fig. 4. Experimental setup.

The first consideration is the one concerning the time constant, T_1, for the photodetector. Because its apperance, the equation governing the variation of voltage V_O at Fig. 5.c is

$$V_O + T_1 \frac{dV_O}{dt} = \beta \, I_{OUT} \qquad\qquad (1)$$

where β is the feedback coefficient.

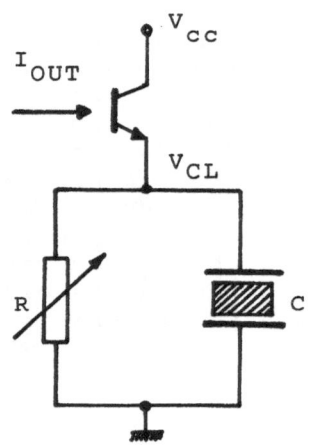

Fig.5.a
Experimental circuit

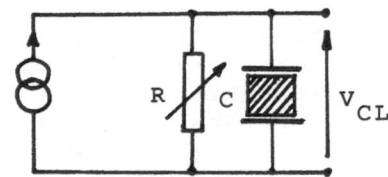

Fig.5.b. Norton circuit

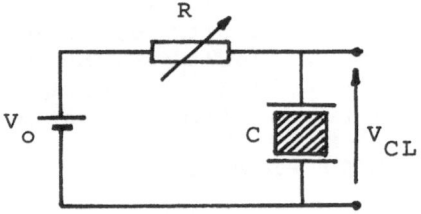

Fig.5.c. Thevenin circuit

Together with the above time constant, is the one associated with R and C components. If we call it, T_2, we will have

$$V_{CL} + T_2 \frac{dV_{CL}}{dt} = V_o \qquad (2)$$

where V_{CL} stands for the effective voltage applied to the liquid crystal cell. C is due to the capacitive effect due to the cell. Moreover, this cell has associated with it capacitive and resistive effects, both giving the value for T_2.

Finally, a third time constant, τ , appears. It is the time needed by the molecules to reorientate according to the electric field inside the cell. This time constant is, certainly, not a constant. Its value depends on the voltage value and on its derivative with time. The first dependence is the strongest and it will be main source for our model. The equation will be

$$V_{ef} + \tau \frac{dV_{ef}}{dt} = V_{CL} \qquad (3)$$

From these three equations we can obtain

$$V + \frac{dV}{dt}\left[\tau + (T_1 + T_2)(1 + \frac{d\tau}{dt}) + T_1 T_2 \frac{d^2\tau}{dt^2} \right] +$$

$$+ \frac{d^2V}{dt^2}\left[\tau(T_1 + T_2) + T_1 T_2 (1 + 2\frac{d\tau}{dt}) \right] +$$

$$+ \frac{d^3V}{dt^3}\left[\tau T_1 T_2 \right] = \beta \, I_{IN} \, T \, (V + V_B) \qquad (4)$$

where $V \equiv V_{ef}$ and $T(V + V_B)$ stands for the transmission function. For $\tau \neq \tau(t)$ this equation simplifies to the one previously reported by us[6]. A further aproach to solve equation (4), is to consider $T_1 \simeq 0$. This aproximation is valid because T_1 is around two orders of magnitude smaller than either T_2 and τ . Equation (4) hence becomes

$$V + (T_2 + \tau + T_2 \frac{d\tau}{dt})\frac{dV}{dt} + T_2\tau \frac{d^2V}{dt^2} =$$

$$= \beta \, I_{IN} \, T \, (V + V_B) \qquad (5)$$

Numerical solutions of this equation, for different values of T_2 and β , are shown in Figs. 6-9. The variation with time of τ , has been taken as

$$\tau \simeq \frac{K'}{(V-V_{th})^2}$$ when voltage is rising, with

V_{th}= V of threshold, and

$\tau \simeq K= 10^{-2}$s. when voltage is going to zero.

The corresponding values for T_2 and $\beta\ I_{IN}$ are given in Figs. 6-9.

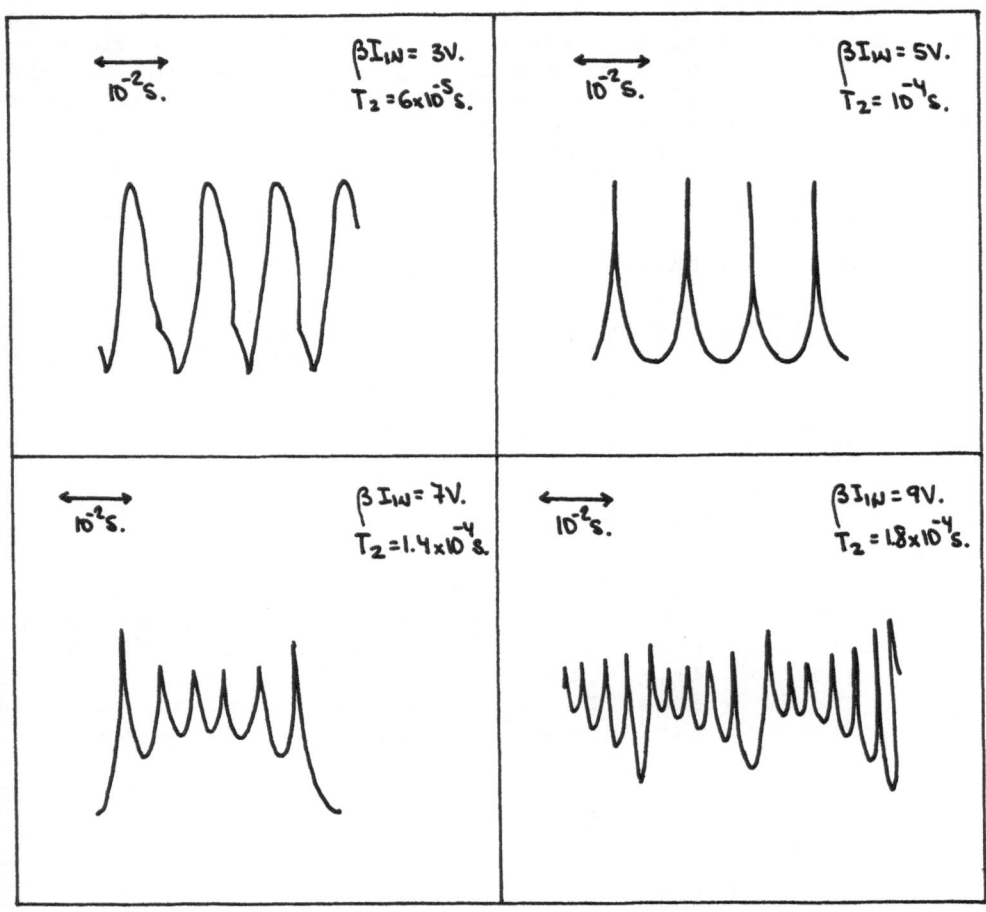

Figs. 6, 7, 8 y 9.   Numerical results.

EXPERIMENTAL RESULTS

The main experimental results are shown in Figs. 10
13.   Their different waveforms correspond as before to
different values for both β I_{IN} and T_2.   As it can be
seen, a good agreement with the empirical results obtained from our theoretical model is obtained for the lowest
values.   This agreement is not so good for higher values.
In particular, for β I_{IN} = 9 volts   the discrepancy is
evident.   The difference is due, according to our calculations, to the simplification had from equation (4)   to
equation (5).   Moreover, in the present model the value
for C, liquid crystal cell capacitance, has been taken
as a constant, but its value depend on V.

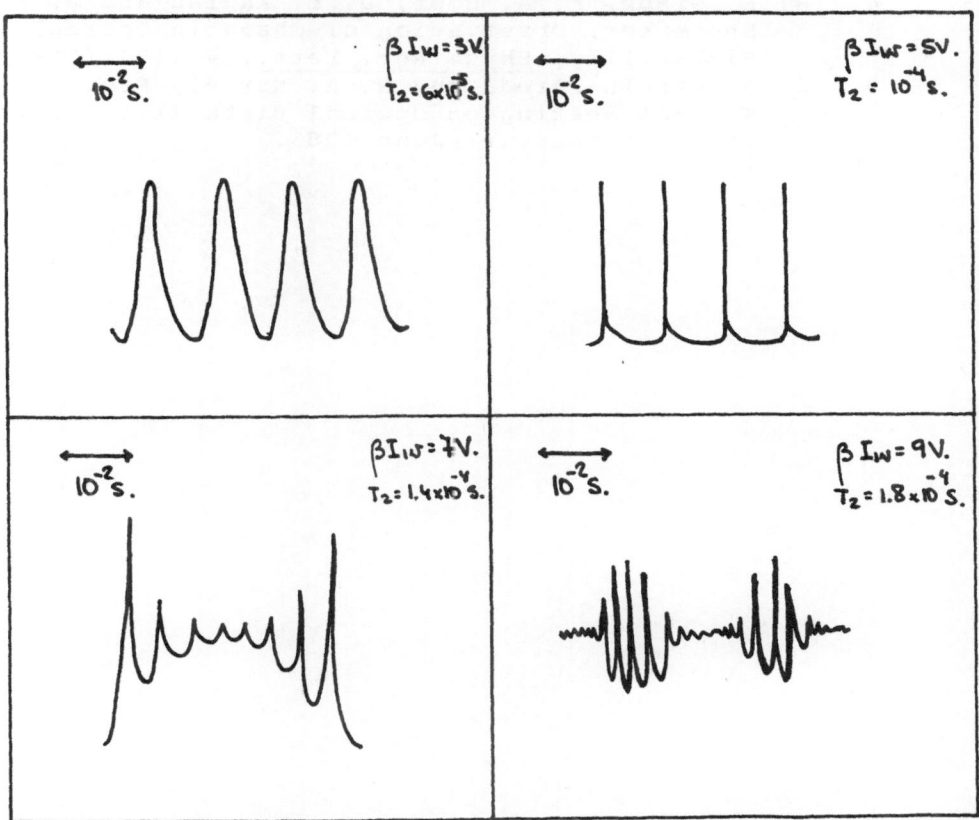

Figs. 10, 11, 12 y 13.   Experimental results.

We thank Dr. Otón for helpful suggestions. Finantial
support from Spanish CAICYT (Grant 3864-79) is appreciated.

REFERENCES

1. H. M. Gibbs, S. L. McCall and T. N. C. Venkatesan
    Differential gain and bistability using a
    sodium-filled Fabry-Perot interferometer,
    Phys. Rev. Lett., 36:1135 (1976).
2. K. Ikeda, Multiple-valued stationary state and
    its instability of the transmitted light by
    a ring cavity system, Opt. Commun., 30:257
    (1979).
3. K. Ikeda, H. Daido and O. Akumoto, Optical tur-
    bulence: Chaotic behavior of transmitted
    light from a ring cavity, Phys. Rev. Lett.,
    45:709 (1980).
4. H. M. Gibbs, F. A. Hopf, D. L. Kaplan and R. L.
    Shoemaker, Observation of Chaos in Optical
    Bistability, Phys. Rev. Lett., 46:474 (1981).
5. J. A. Martín-Pereda and M. A. Muriel, Proc.
    Topical Meeting on Optical Bistability,Poster
    ThB16, Rochester, June 1983.

EMPIRICAL AND ANALYTICAL STUDY OF INSTABILITIES

IN HYBRID OPTICAL BISTABLE SYSTEMS

J.A. Martín-Pereda and M.A. Muriel

E.T.S. Ingenieros de Telecomunicación U.P.M.
Ciudad Universitaria
Madrid-3, SPAIN

INTRODUCTION

As it is well known from the work by Gibbs et al[1] optical turbulence and periodic oscillations are easily seen in hybrid optical bistable devices when a delay is added to the feedback. Such effects, as it was pointed out by Gibbs, may be used to convert cw laser power into a train of light pulses. Furthermore, Okada and Takizawa[2] investigated, theoretically and experimentally, the effect of a delayed feedback on an hybrid electrooptic BOD, with the restriction that the delay time should be less than or comparable to the response time of time of the system. Neyer and Voges[3] demonstrated the neat effect of this feedback delay on the behaviour of electrooptic BOD's, neglecting all the time constants of the system components (photodiode, amplifier, modulator,etc.).

The main aim of this paper is to determine the characteristics needed by the transmission curve of the system to obtain instabilities in hybrid optical bistable devices.

As it will be shown, it is possible to predict optical instabilities merely from geometrical consideration of the transmission plot. An experimental part is given to show the validity of these predictions.

FUNDAMENTAL EQUATIONS

The hybrid optical bistable device employed by us is similar to those reported by Garmire[4], Smith[5] and other workers (Fig. 1).

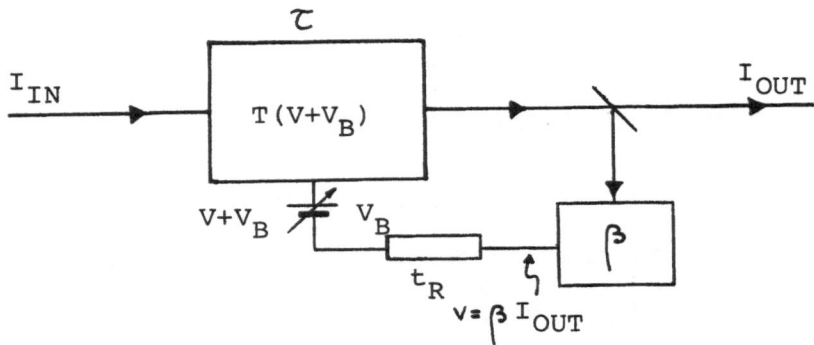

Fig. 1. Hybrid optical bistable system.

Moreover, our theoretical analysis can be applied to any other hybrid system. In any case, it is always verified that

$$I_{OUT}(t) = T(V(t)+V_B)\ I_{IN} \tag{1}$$

where $T(V(t)+V_B)$ stands for the transmission characteristics of the electrooptic device and V_B and $V(t)$ are the bias and feedback voltages, respectively. If a delay, t_R , is added, then

$$V(t)+\tau\ \frac{dV(t)}{dt} = \beta\ I_{OUT}(t-t_R) \tag{2}$$

From these two equations we can obtain

$$V(t)+\tau\ \frac{dV(t)}{dt} = \beta\left[T\big(V(t-t_R)+V_B\big)\right]I_{IN} \tag{3}$$

As it is known, the working point for the case of no-delay can be obtained from the intersection of the transmission curve with the straight line described by the equation

$$V = V_B + \beta\ I_{IN}\ T(V_B + V) \tag{4}$$

whose slope is given by $tg^{-1}(\beta I_{IN})$. This situation is shown in Fig. 2 where A corresponds to a particular transmission curve and B to equation (4). This working point would correspond to a situation where $t_R= 0$ (Fig.2).

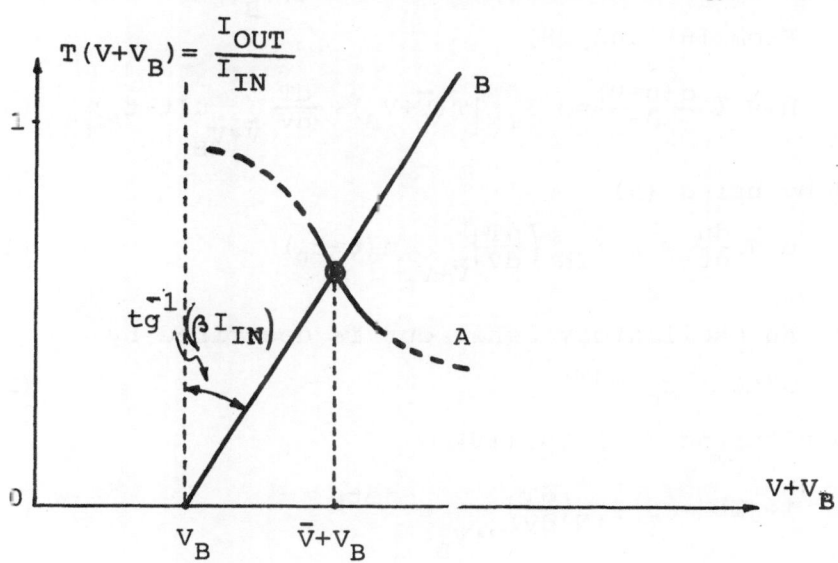

Fig. 2. Equilibrium point.

Let $V(t)$ be $\bar{V}$, the voltage corresponding to the equilibrium point. Hence, $dV(t)/dt = 0$. So,

$$\bar{V} = \beta \ \overline{I_{OUT}} = \beta \ I_{IN} \ T(\bar{V} + V_B) \tag{5}$$

Making $u= V-\bar{V}$

$$u+\bar{V}+ \tau \ \frac{d(u+\bar{V})}{dt} = \beta \ I_{OUT}(t-t_R)= \beta \ I_{IN}T(V(t-t_R)+V_B) \tag{6}$$

A way to solve this equation is by expansion in Taylor series about the working point. This method is valid for a certain number of conditions whose detailed analysis is beyond the scope of this work. The transmission function will be

$$T(V+V_B) = T(\bar{V}+V_B) + \left(\frac{dT}{dV}\right)_{\bar{V}+V_B} (V-\bar{V}) + \ldots \tag{7}$$

and

$$T(V(t-t_R)+V_B) = T(\bar{V}+V_B) + \left(\frac{dT}{dV}\right)_{\bar{V}+V_B} u(t-t_R) + .. \quad (8)$$

From (6) and (8)

$$u+\bar{V}+\tau\frac{d(u+\bar{V})}{dt} = \beta \; I_{IN}\left[T(\bar{V}+V_B) + \frac{dT}{dV}\Big|_{\bar{V}+V_B} u(t-t_R)\right] \quad (9)$$

and by using (5)

$$u+\tau\frac{du}{dt} = \beta \; I_{IN} \left(\frac{dT}{dV}\right)_{\bar{V}+V_B} u(t-t_R) \quad (10)$$

An oscillatory behaviour is described by

$$u(t) = u_O \; e^{j\omega t} \quad (11)$$

substituting (11) in (10)

$$1+j\omega\tau = \beta \; I_{IN}\left(\frac{dT}{dV}\right)_{\bar{V}+V_B} e^{-j\omega t_R} \quad (12)$$

separating the modulus and argument part of each side,

$$\left| \beta \; I_{IN}\left(\frac{dT}{dV}\right)_{\bar{V}+V_B} \right| = \sqrt{1+(\omega\tau)^2} \quad (13)$$

$$tg \; (\omega t_R) = -\omega\tau \quad (14)$$

Hence, taking into acount that $\beta \; I_{IN}\left(\frac{dT}{dV}\right)_{\bar{V}+V_B} < 1$, at

the equilibrium point, we have

$$\beta \; I_{IN}\left(\frac{dT}{dV}\right)_{\bar{V}+V_B} \leqslant -1 \quad (15)$$

and so

$$-\infty \leqslant \left(\frac{dT}{dV}\right)_{\bar{V}+V_B} \leqslant - \frac{1}{\beta \; I_{IN}} \quad (16)$$

This expresion gives the oscillation condition for the $T(V)$ versus V plane. It is shown in Fig.3 for a general case.

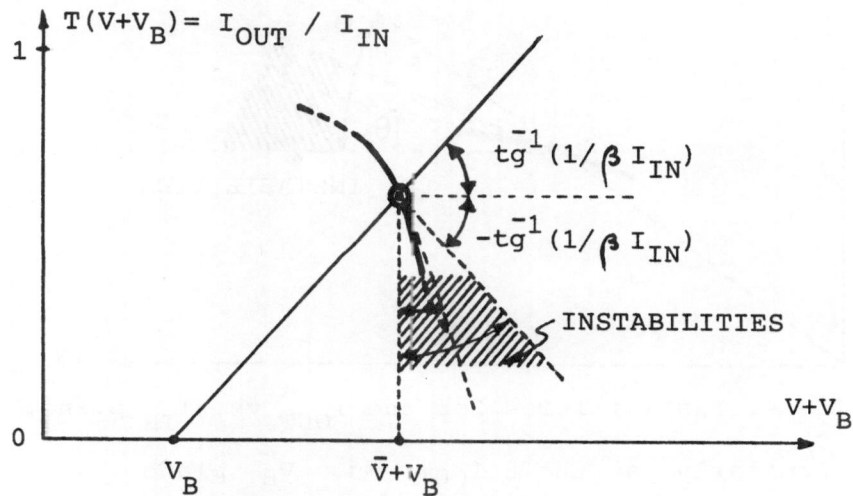

Fig. 3. Geometrical condition for instabilities.

The shadowed region corresponds to the zone where instabilities appear. If the slope of the transmission curve is inside this region, the system will oscillate.

The above condition can be extended for I_{OUT} vs. I_{IN} and $\beta\ I_{OUT}$ vs. V_B planes. In the first case,

$$\frac{dI_{OUT}}{dI_{IN}} = \frac{T}{1-\beta\ I_{IN}\left(\frac{dT}{dV}\right)} \qquad (17)$$

From this equation, expression (16) can be written

$$0 \leqslant \frac{dI_{OUT}}{dI_{IN}} \leqslant \frac{T}{2} \qquad (18)$$

where $T = I_{OUT}/I_{IN}$. This unequality is shown in Fig.4 where instabilities appear for

$$0 \leqslant \mathrm{tg}\ \theta_2 \leqslant \frac{1}{2}\ \mathrm{tg}\ \theta_1 \qquad (19)$$

with $tg\theta_1 = I_{OUT} / I_{IN} = T$ and $tg\theta_2 = d(I_{OUT}) / d(I_{IN})$

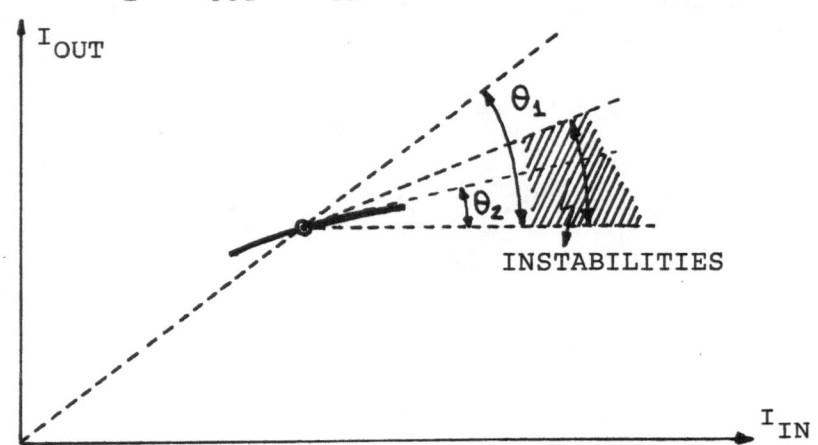

Fig. 4. Instabilities for the I_{OUT} vs. I_{IN} plane.

Similarly, at the βI_{OUT} vs. V_B plane

$$\frac{d\ I_{OUT}}{d\ V_B} = \frac{I_{IN}(\frac{dT}{dV})}{1-\beta\ I_{IN}(\frac{dT}{dV})} \tag{20}$$

and with (16)

$$-1 \leqslant \frac{d(\beta\ I_{OUT})}{dV_B} \leqslant -\frac{1}{2} \tag{21}$$

This situation is depicted in Fig. 5.

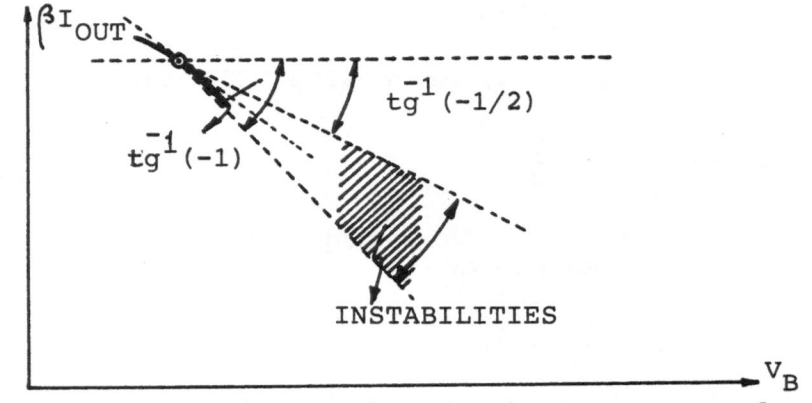

Fig. 5. Instabilities for the βI_{OUT} vs. V_B plane.

EXPERIMENTAL VERIFICATION

The validity of the above model has been tested in
two ways. The first one was to apply our theory to the
empirical results previously reported by Gibbs et al.[1]
and Okada et al.[2]. The regions where their instabilities
appear are in total agreement with our theory.

Moreover, to study unequality (21) the experimental
setup of Fig. 1 was used. A very low frequency (~.1 Hz)
triangular voltage bias was added to the feedback
instead of the usual constant one. A twisted nematic
liquid crystal cell was employed as non-linear element.
Further experimental details are given elsewhere[4]. The
laser beam impinges orthogonally with respect to the
cell.

A periodic oscillation, 20 Hz of frequency, was found
at the regions where instabilities had been predicted by
our theory.

The instabilities region corresponds to the shadowed
zone in Fig. 6, where transmission vs. voltage applied to
the cell is represented.

We can conclude that empirical results are in excellent
agreement with the predictions of our theory.

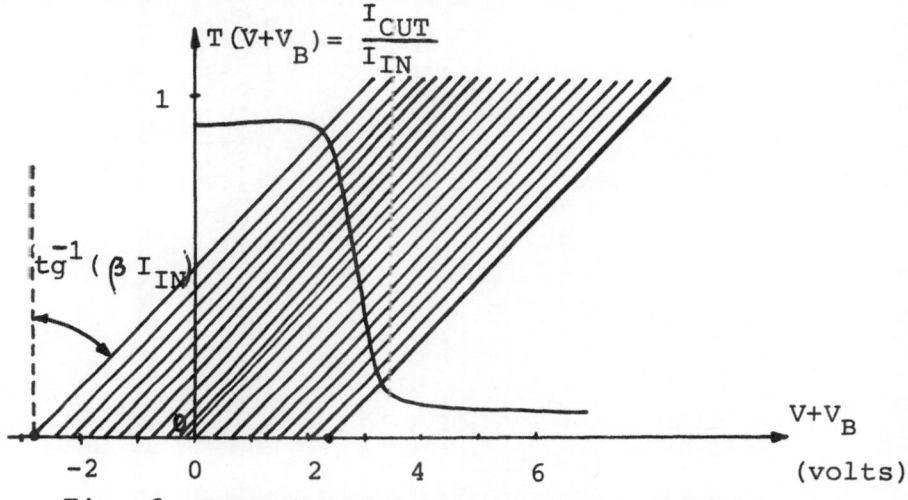

Fig. 6. Experimental region for oscillation.

We thank Dr. Otón for helpful suggestions. Finantial
support from Spanish CAICYT(Grant 3864-79) is appreciated.

REFERENCES

1. H. M. Gibbs, F. A. Hopf, D. L. Kaplan and R. L. Shoemaker, Observation of Chaos in Optical Bistability, Phys. Rev. Lett., 46:474 (1981).
2. M. Okada and K. Takizawa, Instability of an Electrooptic Bistable Device with a delayed feedback, IEEE J. Quant. Electr., QE-17(10): :2135 (1981).
3. A. Neyer and E. Voges, Dynamics of Electrooptic Bistable Devices with Delayed Feedback, IEEE J. Quant. Electr., QE-18 (12):2009 (1982).
4. E. Garmire, J. M. Marburger and S. D. Allen, Incoherent mirrorless bistable optical devices, Appl. Phys. Lett., 32:320 (1978).
5. P. W. Smith, Hybrid bistable optical devices, Opt. Eng., 19:456 (1980).

INSTABILITIES AND CHAOS IN TV-

OPTICAL FEEDBACK

G. Häusler and N. Streibl

Physikalisches Institut
der Universität Erlangen
Erwin-Rommel-Straße 1
D-8520 Erlangen

INTRODUCTION

TV-optical feedback systems can simply realize multidimensional (space/time) dynamical systems. Some experiments are shown, demonstrating chaotic and cooperative behavior.

ABOUT TV-OPTICAL FEEDBACK

The simplest TV-optical feedback experiment is performed by a TV-camera looking onto its own monitor (fig. 1). This device can be considered to consist of 250 000 feedback channels in parallel, one for each pixel.

If we introduce some signal processing, such as convolutions, arithmetics (e.g., image subtraction), nonlinearities, etc., those feedback systems become very flexible:

In a first step we used linear devices and we obtained systems for image restoration, iterative image processing and for the solution of partial differential equations in space and time [1].

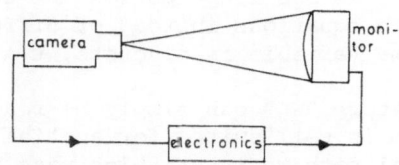

Fig. 1. Principle of a TV-optical feedback system

In a second step the saturation nonlinearity of TV-systems was used to obtain a switching characteristic. This was the basic idea for the construction of flip flop arrays [1], which were reported previously.

NONLINEAR DYNAMICS AND FEEDBACK

In a linear feedback system all the normal modes evolve without mutual interaction: They decrease (or increase) exponentially in space and time [2].

But what happens in more complicated, nonlinear feedback systems? There, generally no independent modes exist.

We observed two kinds of typical behavior: Wild fluctuations in space and time ("chaos") [3] and, the evolution of ordered structures. The latter systems may be called "synergetic", according to Haken [4].

Understanding such phenomena is the goal of "Nonlinear Dynamics". Mathematically, this is the science of coupled nonlinear differential equations, such as the following:

$$(1) \qquad \left. \begin{array}{l} \dfrac{\partial}{\partial t}\, u_1(t) = NL_1(u_1(t), u_2(t), \ldots u_r(t)) \\[2em] \dfrac{\partial}{\partial t}\, u_2(t) = NL_2(u_1, u_2 \ldots u_r) \\[1em] \ldots \end{array} \right\} r \text{ equations}$$

Therein r may vary between 1 and very large numbers exceeding even the Avogadro number. NL_1, NL_2 ... are nonlinear functions of the variables $u_1, u_2 \ldots$

Is it possible, to implement such kind of equations in our TV-optical system?

1. We identify the variables $u_k(t)$ with the spatio-temporal image intensity $I_k(t)$, (of pixel "k"), which is circulating through our TV-optical feedback system. Thus, the number of variables is as large as the number of pixels ($\approx 500^2$ in a TV-picture).

2. The round trip time is one frame period $\tau = 40\text{ms}$. Thus, we modelize difference equations instead of differential equations: our time variable is discrete, $t = \tau, 2\tau, 3\tau \ldots$

3. What kind of operators "NL" can simply be realized? The <u>optical</u> part of the system is well suited for spatial low pass filtering. In the <u>electronical</u> path we insert high pass filters (whose action is equivalent to two-dimensional spatial high pass

filtering). By use of a digital frame-storage, look-up-tables, etc., we realize point-to-point nonlinearity and arithmetics, in TV-real time.

All these operations can be cascaded.

Obviously we are restricted to space invariant systems: The laws of interaction have to be the same overall in space.

On the other hand, the main advantage of such TV-optical systems is that about 250 000 coupled equations can be iterated within steps of $\tau = 40ms$, only.

In our experiments, described below, the following type of equations was implemented:

$$(2) \qquad I_k(t+\tau) = \sum_l a_l NL \left[I_{k-1}(t) \right]$$

$$k = 1,2,\ldots \quad , 250\ 000$$

Here the a_l denote a point spread function (PSF) and NL is a point-to-point nonlinearity.

EXPERIMENTS:

A classical example [3] for chaotic behavior is the iteration of the "logistic" equation

$$(3) \qquad I(t+\tau) = \beta \cdot I(t) \cdot \left[1-I(t) \right] = NL(I)$$

If β exceeds a critical value the output is unpredictable in any practical sense, after a few iterations. In the first experiment, we coupled many of these iterated maps, by optical convolution (simple defocusing).

The temporal power spectra of one pixel (fig. 2) show, with increasing parameter β, typical subharmonics and finally a broad band characteristic. To that extent the temporal behavior of our coupled system is similar to that of the "isolated" system of equ. (3).

The spatial behavior of the system depends on the coupling between adjacent pixels. The results are (pseudo-)random patterns ("texture") whose spatial structure can be influenced by the spatial part of the operator in the feedback branch, in other words, by the "a_l" of equ. (2).

In fig. 3, four kinds of textures have been generated by four different optical PSF's.

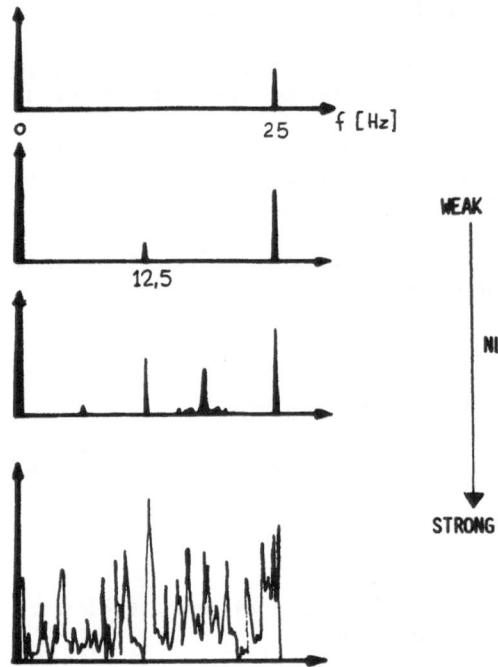

Fig. 2.   Temporal power spectra of one pixel, for different
           strengths of nonlinearity. (Coupled logistic equations.)

Fig. 4 shows an example of "synergetic" or "cooperative"
behavior in the sense, that the range of spatial order is much
larger than the extension of the PSF.

The types of nonlinearity and of PSF are indicated in the
picture. It is remarkable that different strength of the nonlinearity yields different spatial structures.

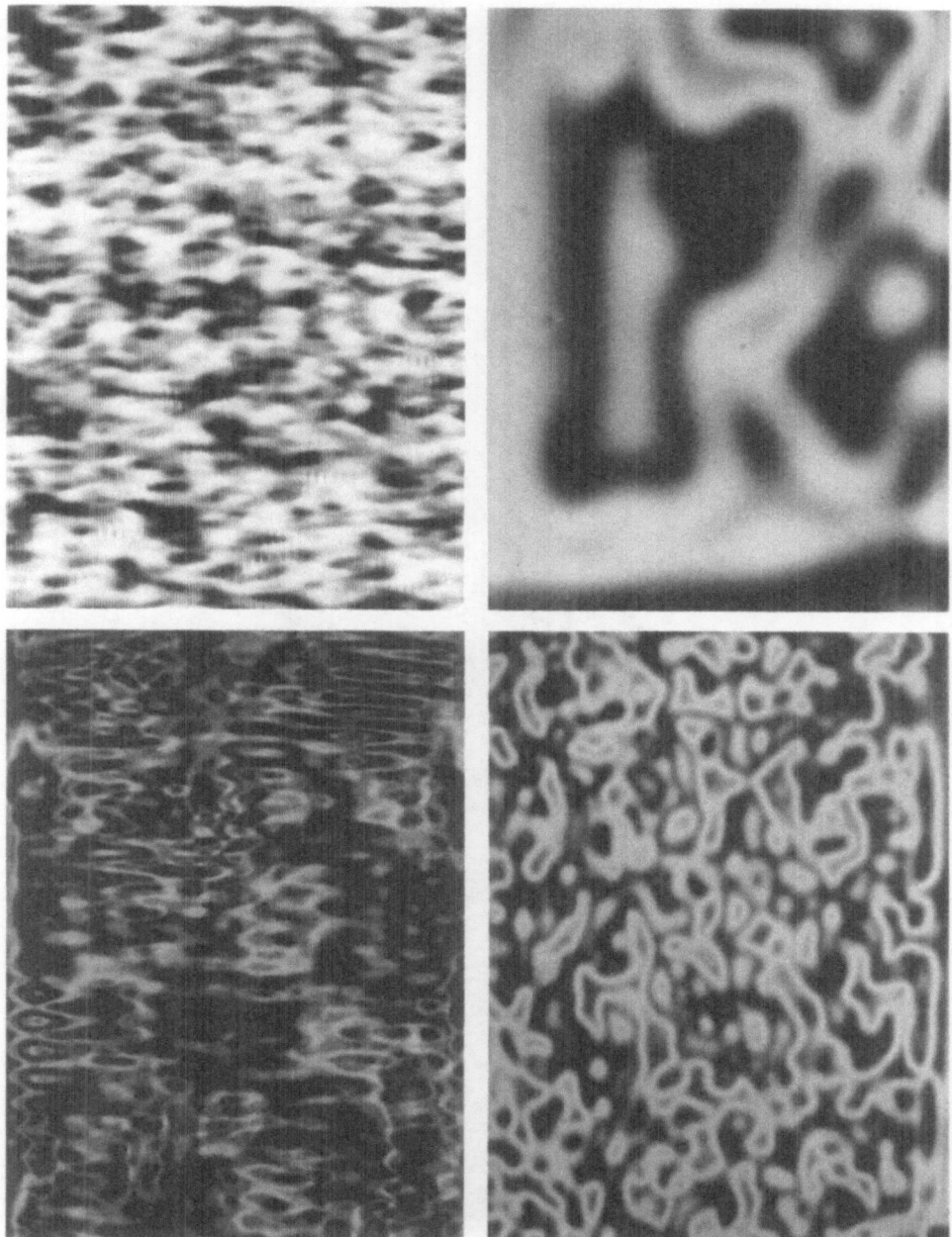

Fig. 3.   Four kinds of (spatial) textures, generated by different
           point spread functions (PSF); in the case of coupled
           logistic equations.

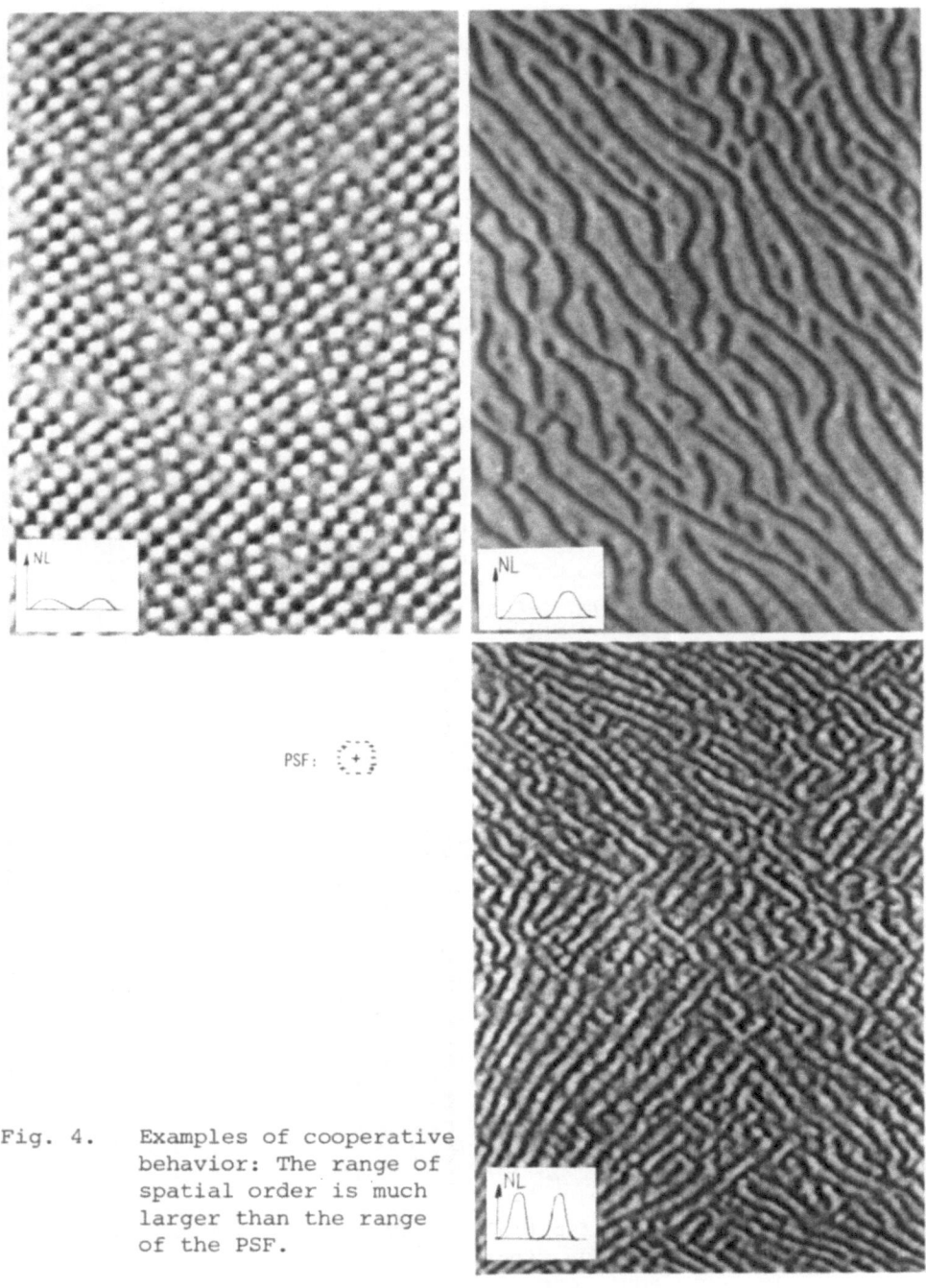

PSF:

Fig. 4.   Examples of cooperative
behavior: The range of
spatial order is much
larger than the range
of the PSF.

CONCLUSIONS

We modelized coupled nonlinear difference equations by TV-optical feedback systems. We could observe chaotic as well as cooperative behavior in the x,y,t domain. It appears that such experiments are quite instructive, because it is possible to observe the evolution of 512^2 pixels in TV-real time. The parameters can be easily changed during the experiments and a large variety of evolving patterns can be "generated". *)

REFERENCES

1. G. Ferrano, G. Häusler, TV-optical feedback systems, Opt. Eng. 19: 442 (1980)
2. G. Häusler, N. Streibl, Stability of spatio-temporal feedback systems, Opt. Acta 30: 171 (1983)
3. R.M. May, Simple mathematical models with very complicated dynamics, Nature 261: 459 (1976)
4. H. Haken, "Synergetics", Springer Verlag, Heidelberg (1978)

---

*) The fascination of these experiments is connected with the dynamical behavior of the "living patterns", evolving from noise, being stable for some time (even in a completely analog system), changing slightly, or switching occasionally to some other structure. Therefore we offer copies of a super 8 film, instead of reprints (for the cost of the copy).

CRITICAL SLOWING DOWN AND MAGNETICALLY-INDUCED SELF-PULSING

IN A SODIUM-FILLED FABRY-PEROT RESONATOR

J.Mlynek, F.Mitschke, and W.Lange

Institut für Quantenoptik, Universität Hannover
Welfengarten 1, D-3000 Hannover 1, FRG

INTRODUCTION

The level degeneracy of the ground state of atoms acting as a nonlinear medium in optical bistability experiments has attracted much attention recently. The well-known process of transverse optical pumping between Zeeman levels has been discussed as a mechanism yielding considerable nonlinearity even at low power densities[1], and the corresponding experiment has been performed on a resonator filled with sodium atoms[2]. The process of longitudinal optical pumping has also been studied[3]. While in Refs. 2 and 3 only a single mode of the radiation field interacts with the atoms, the case of two orthogonally polarized degenerate modes has been dicussed, too. It has been predicted[4] and experimentally established[5] that optical tristability may arise, and it is quite obvious that in the two-mode case the system is capable of displaying a rich variety of behaviour[6-9]. In this paper we report on an expansion of the experiment of Ref. 2 to the study of transients in the one- and two-mode case.

ONE-MODE CASE

The experimental set-up used in the one-mode case is shown in Fig.1. The dye laser, optically isolated from the Fabry-Perot resonator by means of a Faraday rotator (FR), is detuned from the Na D_1 - line by 10 to 20 GHz. The sodium sample ($N \cong 10^{13} cm^{-3}$, buffer gas: argon, p=200 mbar) is placed in a near confocal resonator (beam waist: 120 μm) and can be subjected to a transverse magnetic field B. The intensity of the laser beam can be switched by means of an electro-optic modulator (EOM), thus providing input steps of variable height (power: 5 - 50 mW). The λ/4 - plate serves for establishing a circular pola-

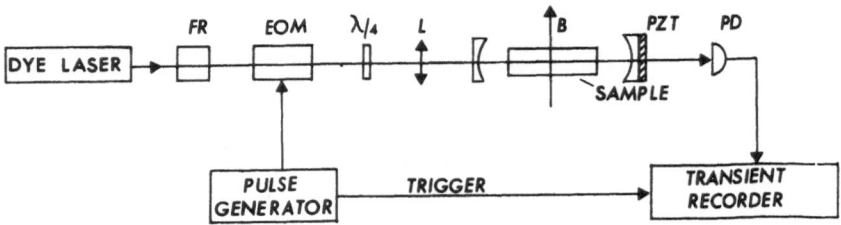

Fig.1. Experimental set-up for the one-mode case.FR: Faraday rotator.
       EOM: Electro-optic modulator.L: mode-matching lens.PZT: piezo
       translator.PD: photo diode.

rization. The Fabry-Perot output intensity is monitored by a photo
diode, and the time dependence is analyzed by means of a transient
digitizer; further details are given in Ref.1o.

    In zero magnetic field, the interaction of the circularly pola-
rized (σ^+) near resonant light beam with the sample would give rise
to a (weak) optical pumping process creating an "orientation" in the
sample. With the direction of propagation of the light beam chosen
as axis of quantization, this process may be regarded as a "longitudinal" optical pumping between Zeeman levels. Evidently this
causes a reduction of absorption and refraction of the sample for a
σ^+ -polarized light beam. In our experiment the the lifetime of the
longitudinal component of the orientation (τ_1) is controlled by
diffusion processes; it is found to be in the µs range. As a consequence, the saturation occurs at very low power densities.

    The transverse magnetic field counteracts the pumping process
and tries to establish the field direction as axis of symmetry. The
corresponding reduction of the strong saturation phenomena observed
in zero magnetic field can be seen in Fig. 2a and 2b, where the
transmission of the sample and its index of refraction are displayed
in dependence on the magnetic field as calculated for the conditions
of our experiment[11].The value of the light power is used as a parameter. In the calculation of Fig.2 the direction of the magnetic
field was used as axis of quantization for convenience. In this frame
of reference the pumping process is called "transverse optical pumping", the orientation is transverse,and the role of differences in
the population density of Zeeman sublevels is taken by the "coherence" between the Zeeman substates. Obviously, the relaxation time
τ_2 of the coherence is controlled by the same diffusion processes
as τ_1 in our experiment, yielding $\tau_2=\tau_1$.

    With respect to optical bistability it can be stated that the
process of optical pumping between Zeeman levels in sodium by circularly polarized light gives rise to strong nonlinearities, which in

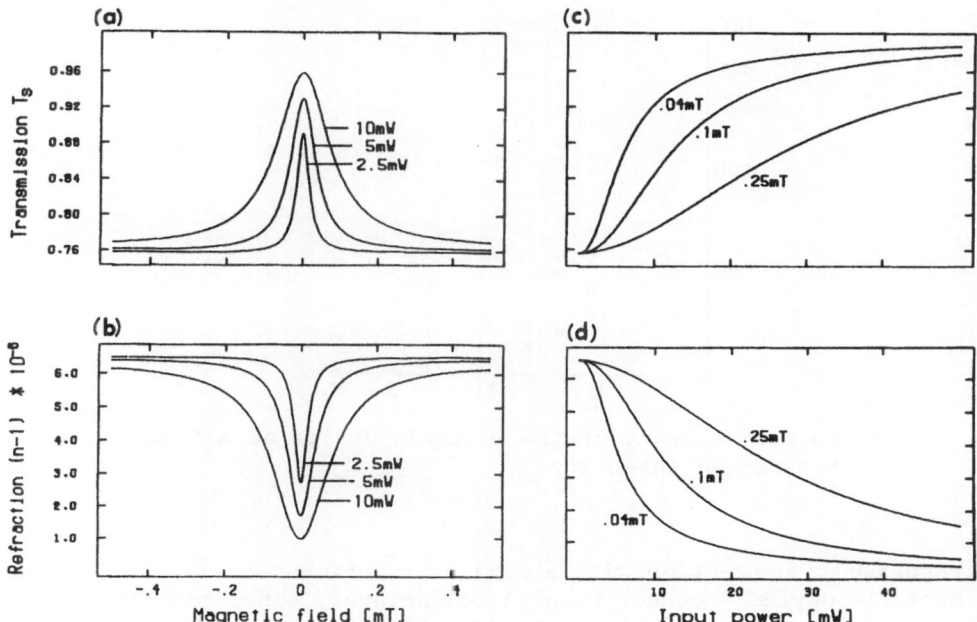

Fig.2. Calculated transmission(a,c) and refraction(b,d) of sodium
       sample in dependence on magnetic field(a,b) and input power
       (c,d).

turn can be controlled by a transverse magnetic field. It may be
helpful to present plots of the transmission and the index of refraction of the sample as a function of light power, with the magnetic field used as a parameter (Fig. 2c and 2d, respectively). It
should be noted, however, that generally the description given here
is sufficient only in stationary or quasistationary experiments.
Under appropriate conditions we could observe coherent Raman-type
oscillatory transients following bistable switching, which can be
attributed to an oscillatory behaviour of the Zeeman coherence.

In our experiment employing the set-up of Fig.1 we observed
the delay τ_D between switching of the input- and of the output-
intensity. τ_D strongly depends on the input intensity I_{in} as soon
as I_{in} is only slightly beyond the critical input intensity I_{cr}
needed for switching. Experimentally determined delays are displayed
in Fig.3 in dependence on I_{in}/I_{cr}. Obviously the system exhibits the
well-known behaviour of "critical slowing down" [10], which up to now
has predominantly been studied in hybrid bistable devices. It should
be noted that all details of the actual behaviour of the system can
be described by a simple computer model. The analysis given very

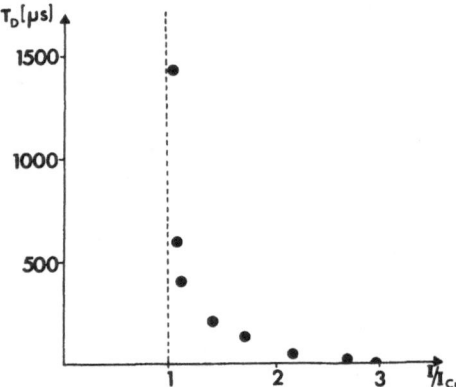

Fig.3. Switching delay of the bistable device as a function of
irradiated intensity.

recently in connection with a similar experiment in an all-optical
bistable device[12] cannot be applied, however, since the diffusion
process governing the time behaviour cannot completely be characterized by a single time constant.

It is clear that the delay time τ_D can also depend very strongly
on the magnetic field B, since the nonlinearity of the system and
thus I_{cr} are controlled by B.

TWO—MODE CASE

The experimental set-up used in the two-mode case is shown in
Fig.4. There are only minor changes with respect to Fig.1 : the
$\lambda/4$ -plate in the input is replaced by a linear polarizer P_1 ,and a
combination of a $\lambda/4$ -plate and a polarizing beam splitter separates
the right- and left-circularly polarized components of the output ;
they are detected by a pair of photodiodes and recorded simultaneously.

In zero or weak magnetic field optical tristability is expected
in this type of set-up.It has been shown in Refs. 4 and 5 that at
low input intensities the σ^+ - and the σ^- -polarized mode of the
cavity are equally excited, giving rise to a linearly polarized
output beam. However, the two modes compete very strongly by optical
pumping, and beyond a critical input intensity either the σ^+ - or the
σ^- -mode will become dominant by spontaneous breaking of symmetry.
Fig.5 shows the time behaviour of the two σ -components following
an input step of linearly polarized light,with I_{in} closely above I_{cr}.
After a certain delay τ_D ,the σ^+ -component jumps to the lower branch

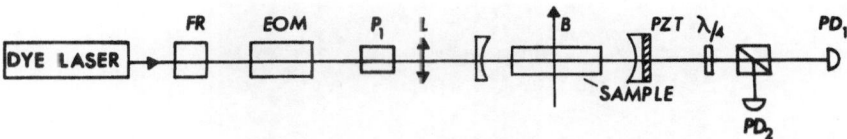

Fig.4. Experimental set-up for the two mode case. P_1:linear polarizer.

of the tristability curve (see Fig.2 of Ref.4), whereas the σ^- -component jumps to the upper branch. Evidently, the two-mode experiment displays a similar type of "critical slowing down" as the one-mode experiment.

When the static transverse magnetic field is increased, the process of mode competition is strongly modified by the Larmor precession affecting the orientation of the atomic sample. Above a critical value B_{cr}, no state is stable, and the system starts to switch between the two high-transmission states (σ^+ -light or σ^- -light transmitted). The intensities of the two polarization components develop into "complementary" regular pulse trains as

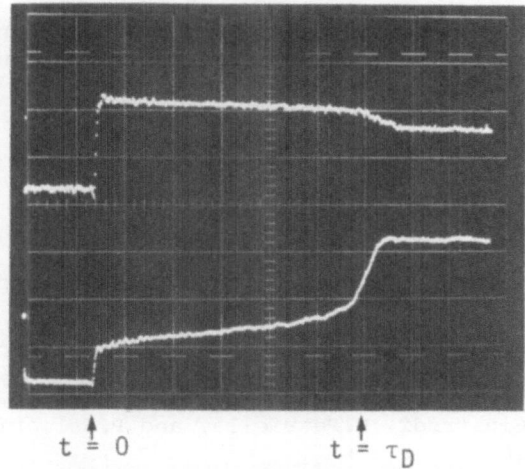

$t = 0 \qquad t = \tau_D$

Fig.5. Transients in optical tristability. Input intensity is switched on at t=o. Upper trace: σ^+ -component. Lower trace: σ^- -component, attenuated xo.5.Time scale 1o µs/div.

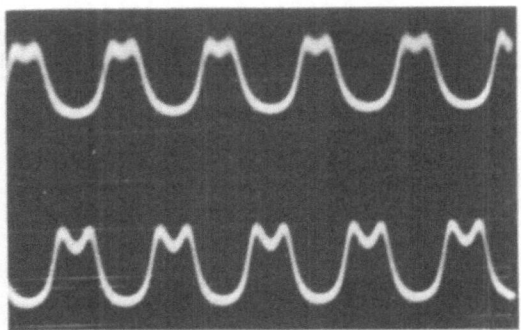

Fig.6. Complementary pulse trains in magnetically-induced self-
       pulsing. Time scale : 2oo ns/div.

shown in Fig.6. These pulse trains last indefinitely in principle;
they were easily observed for minutes. All observations are in excellent agreement with the predictions of Kitano et al.[6] concerning
"magnetically-induced self-pulsing" in optical tristability.Some
details have been given elsewhere[13].

It is quite obvious that within the formal description given
by Kitano et al.[6] there is no place for period-doubling or chaotic
behaviour, since the problem is reduced to the motion of an auto-
nomous system in a two-dimensional phase space. The possible behaviour under modified conditions is presently under study.

The authors gratefully acknowledge the assistance of R.Deserno
and H.J.Schröder.We also thank the Deutsche Forschungsgemeinschaft
for financial support.

REFERENCES

1. D.F.Walls and P.Zoller, Opt.Comm.34 (198o) 26o
   and references therein.
2. J.Mlynek, F.Mitschke, R.Deserno, and W.Lange, Appl.Phys. B28
   (1982) 135
3. F.T.Arecchi, G.Giusfredi, E.Petriella, and P.Salieri, Appl.Phys.
   B29 (1982) 79
4. M.Kitano, T.Yabuzaki, and T.Ogawa, Phys.Rev.Lett. 46 (1981) 924
5. S.Cecchi, G.Giusfredi, E.Petriella, and P.Salieri, Phys.Rev.Lett.
   49 (1982) 1928
6. M.Kitano, T.Yabuzaki, and T.Ogawa,Phys.Rev. A24 (1981) 3156
7. J.A.Hermann and D.F.Walls, Phys.Rev. A26 (1982) 2o85
   and references therein

8. H.J.Carmichael, C.M.Savage, and D.F.Walls, Phys.Rev.Lett.$\underline{5o}$
   (1983) 163.
9. F.T.Arecchi, J.Kurmann, and A.Politi, Opt.Comm. $\underline{44}$ (1983) 421
1o.F.Mitschke, R.Deserno, J.Mlynek, and W.Lange, Opt.Comm. $\underline{46}$
   (1983) 135 and references therein.
11.J.Mlynek, F.Mitschke, R.Deserno, and W.Lange :"Optical bistability
   from three-level atoms using a coherent nonlinear mechanism",
   submitted for publication.
12.G.Grynberg, S.Cribier, and E.Giacobino, in:"Laser Spectroscopy
   VI"(Eds.:H.P.Weber and W.Lüthy), Springer 1983 (in press).
13.F.Mitschke, J.Mlynek, and W.Lange, Phys.Rev.Lett. $\underline{50}$ (1983) 166o.

A CATASTROPHE MODEL OF OPTICAL BISTABILITY

Chun-fei Li and Ai-qun Ma

Department of Physics
Harbin Institute of Technology
Harbin, People's Republic of China

ABSTRACT

A simple catastrophe theory model of optical bistability is proposed, indicating the threshold conditions for attaining bistability and the means of control of the bistable characteristics.

INTRODUCTION

Optical bistability is an example of one kind of catastrophe phenomena, and catastrophe theory has been applied in the past to study optical bistability.[1-3] However, most of these works were related in a complicated fashion to the microscopic atomic processes, making it inconvenient to explain the experimental results. We present here a simple macroscopic catastrophe model for optical bistability in a nonlinear Fabry-Perot etalon. It describes clearly the threshold conditions for attaining bistability and the means of control of the bistable characteristics.

CALCULATIONS AND RESULTS

Most bistable optical devices consist of a nonlinear medium within a Fabry-Perot etalon. Let the input and the output light intensities be indicated by I_i and I_t, respectively. If we neglect the absorption of the medium, the transmittance τ is given by

$$\tau = \frac{I_t}{I_i} = \frac{1}{1 + [(2F/\pi) \sin(\phi/2)]^2} , \qquad (1)$$

167

where F is the etalon finesse and ϕ is the phase shift, which can be expressed in the case of either intrinsic bistability (containing a nonlinear material and optical feedback) or hybrid bistability (containing an electro-optic element and optoelectric feedback) as[*]

$$\phi = \phi_0 + K I_t. \tag{2}$$

Here K is the nonlinear coefficient characteristic of the medium and ϕ_0 is the initial phase shift, which depends on the linear refractive index n of the medium, the etalon length d, and the incident light wavelength λ; namely,

$$\phi_0 = \frac{4\pi}{\lambda} nd. \tag{3}$$

The transmittance τ has a maximum at $\phi = 2m\pi$, as shown in Fig. 1. If the peak is very narrow we can expand $\sin(\phi/2)$ as a Taylor series around $2m\pi$, keeping the first two terms only, so that $\sin(\phi/2) = (-1)^m (1/2) (\phi-2m\pi)$. Let us define the relative phase shift

$$\theta = 2m\pi - \phi_0, \tag{4}$$

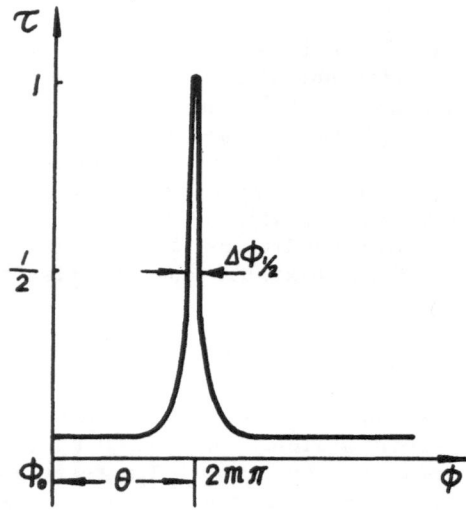

Fig. 1. Etalon transmission as a function of phase shift, where θ is the relative phase shift and $\Delta\phi_{1/2}$ is the full width at half maximum of the transmission peak.

which characterizes the tuning of the resonator relative to the frequency of the incident laser, i.e.

$$\theta = (\nu_m - \nu)\, 4\pi\, \frac{nd}{c}\,, \tag{5}$$

where ν and ν_m are the laser frequency and the resonance frequency of the etalon, respectively. Using (2) we have $\sin(\phi/2) = (-1)^m$ $(1/2)\,|KI_t-\theta)|$; substituting this into (1) we obtain the cubic equation

$$I_t{}^3 + a\, I_t{}^2 + b\, I_t + c = 0, \tag{6}$$

where

$$\begin{aligned}
a &= -2(\theta/K)\\
b &= (\theta/K)^2 + (\pi/KF)^2\\
c &= -(\pi/KF)^2 I_i.
\end{aligned} \tag{7}$$

If we define

$$\begin{aligned}
I_t{}' &= I_t + a/3\\
p &= -(1/3)a^2 + b\\
q &= (2/27)a^3 - (1/3)ab + c.
\end{aligned} \tag{8}$$

then (6) becomes the catastrophe equation, which has one independent variable (I_t') and two parameters (p,q) satisfying the critical condition[5]

$$I_t{}'^3 + p\, I_t' + q = 0. \tag{9}$$

According to catastrophe theory, we can construct a critical surface satisfying (9) in the three-dimensional space (I_t',p,q). But because (9) is equivalent to (6), it is possible to take I_t as the independent variable and I_i and θ as the adjustable parameters, and construct a critical surface in the three-dimensional space (I_t,I_i,θ), as shown in Fig. 2. Note there are two stable regions and one unstable region, with the fold surface corresponding to the bistable region. We can see the bistable loop in the plane parallel to the (I_t,I_i) plane; the negative slope portion is unstable. Projecting the fold surface onto the (θ,I_i) plane, we get a two-branched curve with the bistable region lying in the shaded area between the two branches. The width of the bistable region depends upon θ, i.e. the resonator tuning.

The nature of the roots of equation (9) or (6) depends on the discriminant

$$\Delta = 4p^3 + 27q^2. \tag{10}$$

If $\Delta < 0$, there are three distinct real roots, corresponding to the

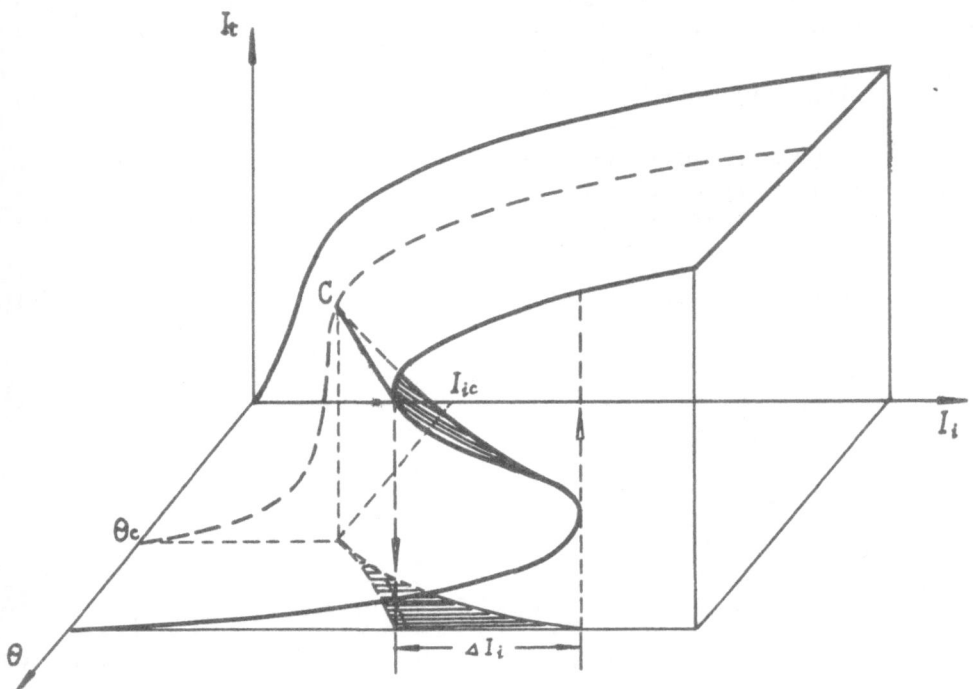

Fig. 2.   Graphical representation of the catastrophe model.   The
fold surface corresponds to the bistable region (shaded area).

presence of bistability.   Using (8) in (10) we get

$$4[-(1/3)a^2 + b]^3 + 27[(2/27)a^3 - (1/3)ab + c]^2 < 0. \tag{11}$$

From (11), it must hold that $-(1/3)a^2 + b < 0$, or using (7),

$$\theta > \sqrt{3} \, \pi/F. \tag{12}$$

Substituting (7) into (11) gives

$$\{[(2/27)(\theta/K)^3 + (2/3)(\theta\pi^2/K^3F^2)] - (2/\sqrt{27})[(1/3)(\theta/K)^2$$
$$- (\pi/kF)^2]^{3/2}\}(kF/\pi)^2 < I_i$$
$$< \{[(2/27)(\theta/K)^3 + (2/3)(\theta\pi^2/K^3F^2)] + (2/\sqrt{27})[(1/3)(\theta/K)^2$$
$$- (\pi/KF)^2]^{3/2}\} \, (KF/\pi)^2. \tag{13}$$

Equations (12) and (13) are the necessary conditions for the occurrence of optical bistability.

When $\Delta = 0$ and $p,q = 0$, two of the real roots are equal for
each branch and we obtain the switching intensities

$$I_{i1,2} = (2/3)[(1/9)(\theta^3 F^2/K\pi^2) + \theta/K]$$
$$\pm (2/\sqrt[3]{27})(KF/\pi)^2[(1/3)(\theta/K)^2 - (\pi/KF)^2]^{3/2}. \tag{14}$$

This represents the two-branched curve in the (θ, I_i) plane in Fig. 2.

The width of the bistable region is

$$\Delta I_i = (4/\sqrt[3]{27})(KF/\pi)^2[(1/3)(\theta/K)^2 - (\pi/KF)^2]^{3/2}. \tag{15}$$

It is obvious that ΔI_i expands with an increase in θ.

When $\Delta = 0$ and $p = q = 0$, all three roots are equal and we obtain the critical point C, at which the bistability begins to disappear ($\Delta I_i = 0$). I_t is a single-valued function of I_i, exhibiting optical triode characteristics as shown in Fig. 2. The coordinate values of the point C in the (θ, I_i) plane are

$$\theta_c = \sqrt{3} \, \pi/F \tag{16}$$
$$I_{ic} = (8/9)(\sqrt{3} \, \pi/KF). \tag{17}$$

When $\Delta > 0$, Eq. (9) or (6) has one real root (and a conjugate pair of complex roots). It is also a single-valued function of I_i without optical bistability.

DISCUSSION

From the above catastrophe model we can draw the following conclusions:

1. The optical bistability depends on the nonlinear coefficient of the medium and the etalon finesse. The threshold conditions for the occurrence of optical bistability are

$$\theta > \sqrt{3} \, \pi/F \approx \Delta\phi_{1/2}$$

$$I_i > (8/9)(\sqrt{3} \, \pi/KF) \approx \Delta\phi_{1/2}/K. \tag{18}$$

To reduce the threshold values of the relative phase shift and the incident light intensity, we must enhance the medium nonlinearity and improve the etalon finesse.

2. The width of the bistable region is dependent on the relative phase shift, i.e. the resonator tuning. Under the condition of sufficient incident light intensity, according to (5), we can adjust

either the laser frequency or adjust the etalon frequency by changing the etalon length or the medium refractive index.   For the
intrinsic device, we can change the position of the lens focus (as a
different intensity within the etalon will result in a different
optical path length) or similarly change the position of the focus
point on the etalon plane; for the hybrid device, we can change the
refractive index via the voltage applied to the crystal, or change
the etalon length by means of the electrostriction effect.

Our results indicate that a good nonlinear medium and etalon
are necessary for attaining optical bistability, and that suitable
adjustment of the incident light intensity and the resonator tuning
are important for controlling optical bistability.

REFERENCES

1.   R. Gilmore and L. M. Narducci, Relation between the equili-
        brium and nonequilibrium critical properties of the Dicke
        model, Phys. Rev. A 17:1747 (1978).

2.   G. P. Agrawal and H. J. Carmichael, Optical bistability
        through nonlinear dispersion and absorption, Phys. Rev. A
        19:2074 (1979).

3.   S. A. Collins and K. C. Wasmundt, Optical feedback and bista-
        bility: a review, Opt. Eng. 19:470 (1980).

4.   Chun-fei Li, The progress of bistable optical devices, Physics
        (China) 11:666 (1982).

5.   Tim Poston and Ian Stewart, "Catastrophe Theory and Its
        Application," Pitman Publishing Limited, London, (1978).

BIFURCATION GEOMETRY OF OPTICAL

BISTABILITY AND SELF-PULSING[*]

D. Armbruster

Institute for Information Sciences
University of Tübingen
Tübingen, F.R. of Germany

In optical bistability, as in most nonlinear systems, the underlying differential equations (here the Maxwell-Bloch-Equations) are
in general not analytically tractable. Usually only partial information about the behaviour of the system in some limiting cases is
available (mean field limit in the pure absorptive or the dispersion-
dominated case). Alternatively, one has to rely on numerical solutions. In either case, one would like to know whether the results so
obtained are in some sense general or exceptional, i.e. whether they
are structurally stable or not. Structural stability is a keyword:
In order that experiments be reproducible, the underlying physics
must be structurally stable. This notion can be made more rigorous
on the level of equations which one uses to model nature: A structurally stable system preserves its quality when its equations are
perturbed: it is insensitive, not only against small changes in the
initial data, but also against small changes in its own specification.

Since in optical bistability we are interested in spatio-
temporal bifurcation phenomena and their interaction, the answer to
the problem of structural stability is given by imperfect bifurcation
theory (developed by Golubitsky & Schaeffer[1]). The aim of imperfect
bifurcation theory is to classify degenerate bifurcation problems.
By bifurcation we mean the change of the topological type of the
solution of a system (p.d.e., o.d.e., algebraic) when a distinguished
physical parameter is varied. This corresponds to the fact that
experimental graphs are usually plotted against a single parameter
(in our case this can be the incident field E_I). Classification is
achieved by means of an equivalence relation, which puts all

---

[*]Work supported by the Stiftung Volkswagenwerk, FRG.

bifurcation problems that behave qualitatively similar (i.e., exhibit the same stability changes) into the same class described by a polynomial normal form. Two classes differ by the number and type of perturbations which change the bifurcation qualitatively. These perturbations are called unfoldings and their number is the codimension of the problem. If partial information on the bifurcation of a system in special cases is available, then the information on the whole system can be completed by determining a singularity (i.e., a degenerate bifurcation point) with finite codimension, which acts as an organizing center for the bifurcation processes. The idea of an organizing center is based on the fact that the apparent global behavior of a system is just a blown up version of its local behavior. If one can find an organizing center, then a full unfolding of it yields all structurally stable bifurcation diagrams which govern the process, thus uncovering new phenomena in addition to those already known. While classical bifurcation techniques tend to avoid degenerate situations (since there the implicit function theorem does not work), we deliberately seek them out.

Using these techniques, Dangelmayr, Güttinger and myself[2] have performed a classification of interacting Hopf and steady-state bifurcations and discussed their imperfection sensitivity. It can be shown that these bifurcation equations have the form

$$
G(x,\lambda,y^2) := \begin{bmatrix} a(x,\lambda,y^2) \\ yb(x,\lambda,y^2) \end{bmatrix} = 0
$$

where x is the amplitude of the steady-state branch (in our problem the amplitude of the cw transmitted field), y is the amplitude of the periodic solution (describing self-pulsing) and λ is a distinguished bifurcation parameter (E_I). G is the result of a Lyapunov-Schmidt reduction or may describe the stationary solutions for the o.d.e. resulting from Haken's slaving or a center manifold reduction. We propose a codimension-4 normal form, into which G can be brought, as an organizing center for the interaction of self-pulsing and optical bistability. This idea sprung up by combining the known experimental (numerical) and analytical facts on optical bistability discovered by Lugiato and others (for a review cf. Bonifacio & Lugiato[3] and Lugiato's article in this volume).

The analytical results are most advanced in the mean field limit of pure absorption where we have the diagram of Fig. 1. It shows for a fixed cavity length,which determines the frequency difference α_1 of the first cavity sidemode and the resonance frequency, those parts of the high transmission branch denoted by X which are stable or unstable respectively. For a point which lies in the region bounded by the full lines, the cw-transmission is unstable to a self-pulsing solution. In the region bounded by the segments AG and AH we find a

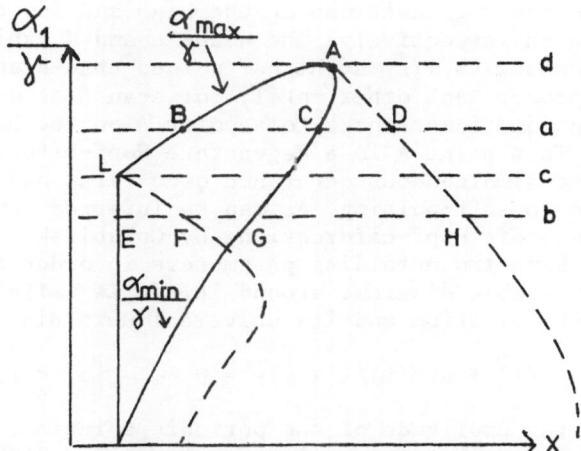

Fig. 1.   Instability regions for self-pulsing in the pure absorptive
          limit. The dashed lines A-H and L-F represent first order
          transitions to self-pulsing and bifurcations to tori respec-
          tively.

coexistence of a stable cw-transmission and stable self-pulsing solution. For details see Lugiato et al.[4] In order to facilitate comparison with the later bifurcation diagrams we show in Figs. 2(a) and
(b) the scans a and b as plots of the transmitted intensity X as a
function of the incident intensity Y. The dashed curve represents the
average intensity of the self-pulsing solution. Note that there are

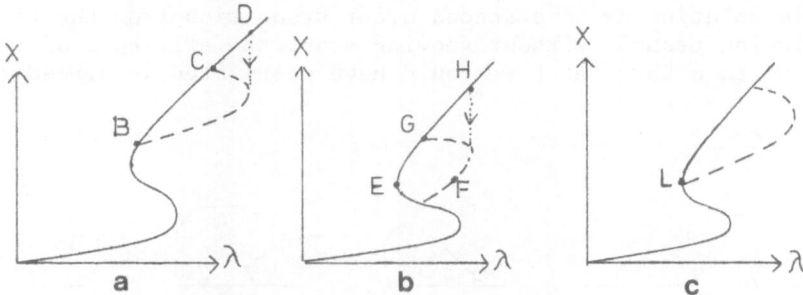

Fig. 2(a-c).   Bifurcation diagrams corresponding to the scans a-c in
               Fig. 1. The dashed lines describe the mean value of the
               self-pulsing intensity while the full line represents
               the cw-transmission.

first and second order transitions at the high and low end of the
self-pulsing branch respectively. The scans c and d exhibit singular
behavior: For scans with α_1/γ at higher values than scan a, the
points B,C,D approach each other until, for scan d at $\alpha_1=\alpha_{min}=\alpha_{max}$,
the self-pulsing solution reduces to a point A on the high trans-
mission branch. This point A is a degenerate Hopf-bifurcation point
and describes the simultaneous occurance of a first and a second
order transition to self-pulsing. As can be inferred from the classification of degenerate Hopf-bifurcations by Golubitsky & Langford[5],
this point must have two unfolding parameters in order to produce
all structurally stable diagrams around it. It is called an H(7)-
degenerate Hopf-bifurcation and its universal unfolding is given by

$$y^5 + 2b\lambda y^3 + (\lambda^2 + \text{sgn}(b)\beta\lambda + \alpha)y = 0 \qquad |b| < 1 \qquad (1)$$

where y denotes the amplitude of the periodic solutions, λ is the
bifurcation parameter, b is a modal parameter and α and β are unfolding parameters. All structurally stable unfolded diagrams for
this degeneracy are shown in Fig. 3. We recognize the two diagrams
for pure absorptive behavior, viz. diagrams 3.1 and 3.3. The other
three diagrams are new and represent two first order or two second
order transitions to self-pulsing, respectively, and an isola solution. Note that for the self-pulsing isola the cw-transmission branch
is not unstable. So the conditions necessary for reaching this stable
pulsing state remain experimentally unclear. We can regard α_1 or the
cavity lenght $\mathcal{L}$ as one of the unfolding parameters. Since there are
no more free parameters in the pure absorptive limit we expect that
cavity detuning Θ and/or atomic mistuning Δ will play the role of the
other unfolding parameter.

Following the scan c in Fig. 2(c) shows the simplest interacting
Hopf and steady-state bifurcation at L. It has codimension 1 and a
slight perturbation of α_1/γ leads either to branching from the
unstable middle sheet of the cw-transmission curve to an unstable
periodic solution or to a second order transition from the high
transmission branch. Without knowing about the existence of the bifurcation to a torus at F we could have postulated it already: Moving

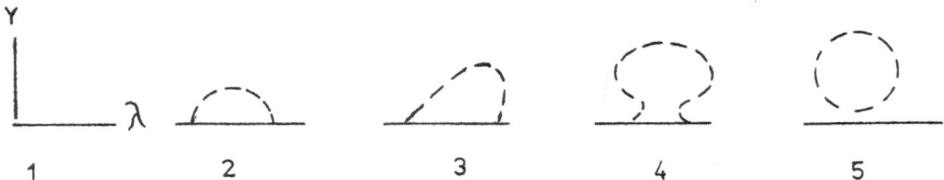

Fig. 3.   Qualitatively different bifurcation diagrams for a full
          unfolding of a H(7)-degenerate Hopf bifurcation.

the secondary bifurcation point B over the limit point L to the point E cannot influence the stable periodic solution around D. Hence to connect E and H there must be a stability change.

In addition to the singularities of the self-pulsing solutions, the cw-transmission curve itself also possesses a singularity, namely a hysteresis point. It has been shown by Agrawal & Carmichael[6] that this point exists for a wide range of C,Δ and Θ. A further partial information about the whole system is the existence of self-pulsing solutions without bistability in the dispersive case (Lugiato[7]). So, although for pure absorption the H(7) point always lies at C > $C_{Hysterese} = 4$, we suggest that the most general interaction of self-pulsing and bistability is governed by an unfolding of an organizing center, which incorporates the existence of an H(7) singularity into a hysteresis point. The corresponding normal form has codimension 4 and its unfolding is

$$G = \begin{bmatrix} -(x^3 + y^2 - \lambda + \beta x + \alpha) \\ -y(x^2 + \lambda^2 + \gamma x + \delta) \end{bmatrix} \tag{2}$$

Figs. 4 and 5 show the most interesting diagrams that are structurally stable and result from G=0. Since stability changes are initiated in the (x,λ)-plane, we only plot the projection of the periodic solution. Stability changes at a limit point of the self-pulsing solution with respect to λ and occurs not necessarily at the maximum pulse amplitude, described by y_{max}. With these pictures it seems possible to fill the gaps between the known diagrams for pure absorption and for the dispersion dominated case giving now an over-all description of the interaction between self-pulsing and optical bistability.

Finally a comment about chaotic behavior in this system is in order. One always looks for Feigenbaum sequences to chaos in optical bistability. We want to point out that there are at least two more routes to chaos which involve bifurcation to tori. The first one is

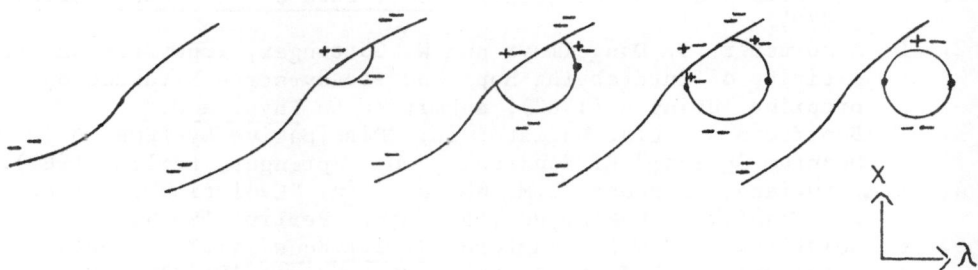

Fig. 4. Bifurcation diagrams without bistability resulting from G=0 of Equ. (2).

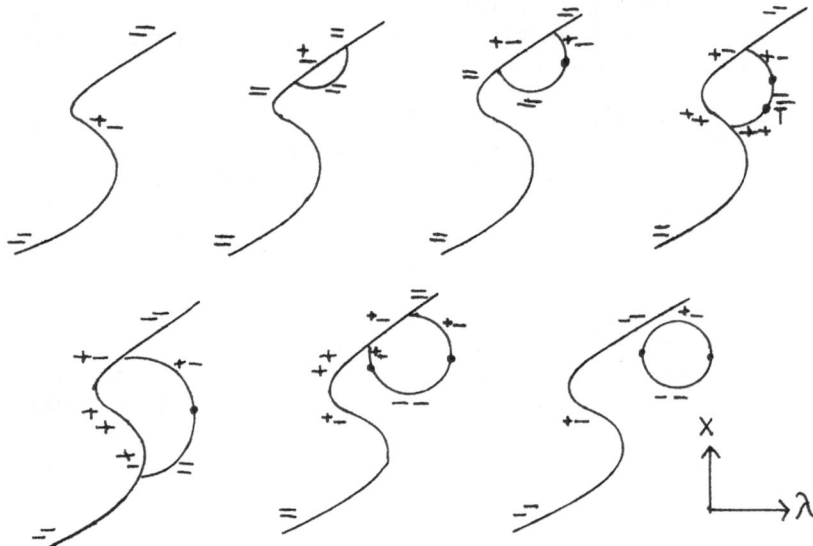

Fig. 5.   Bifurcation diagrams with bistability.

the well-known Ruelle-Takens[8] scenario, which says that the 3-torus
obtained after a further Hopf-bifurcation is structurally unstable
and collapses to a strange attractor. The other one is associated
with Silnikov-bifurcations which occur when the underlying 3-dimen-
sional dynamical system for G is subject to non-symmetric perturbations. Then the 2-torus is structurally unstable and also collapses
to a strange attractor. So it would not be surprising to discover
chaotic behavior in optical bistability, which is not associated with
a Feigenbaum sequence.

REFERENCES

1.   M. Golubitsky and D. Schaeffer, Comm. Pure & Appl. Math. 32:21
       (1979).
2.   D. Armbruster, G. Dangelmayr and W. Güttinger, Imperfection sen-
       sitivity of interacting Hopf and steady-state bifurcations,
       preprint Tübingen (1982), submitted to Physica D.
3.   R. Bonifacio and L.A. Lugiato, in: "Dissipative Systems in
       Quantum Optics," R. Bonifacio, ed., Springer, Berlin (1982).
4.   L.A. Lugiato, V. Benza, L.M. Narducci in: "Evolution of Order
       and Chaos," H. Haken, ed., Springer, Berlin (1982).
5.   M. Golubitsky and W.F. Langford, J.Diff.Equs. 41:375 (1981).
6.   G.P. Agrawal and H.J. Carmichael, Phys.Rev. A 19:2074 (1979).
7.   L.A. Lugiato, Optics Comm. 33:108 (1980).
8.   J.-P. Eckmann, Rev.Mod.Phys. 53:643 (1981).

OPTICAL BISTABILITY WITH FINITE BANDWIDTH NOISE

Axel Schenzle and Thomas Thel*

University of Essen

4300 Essen, West Germany

In this paper we want to demonstrate how the statistical properties of non-linear optical systems change when the fluctuations are generalized from the extreme limit of white noise to fluctuations with a finite bandwidth. While monostable systems seem to be affected only in a quantitative way, multistable processes may undergo qualitative changes when the bandwidth of noise is reduced.

Nonlinear dynamical systems in general may exhibit continuous as well as discontinuous instabilities as a function of an external control parameter, which e.g. may be associated with the energy flux through the system. These models have found widespread interest in the general effort to understand processes far from thermal equilibrium. The field of nonlinear optics and quantum optics provides a large variety of interesting physical examples which can be described by a small number of degrees of freedom. These models are therefore ideally suited for the study of non-equilibrium processes both theoretically and experimentally.

The most prominent example in this context is the continuous bifurcation observed in laser systems when the externally applied energy input exceeds the linear losses in the system (1). Other well known examples can be found in the field of optical parametric processes, where the non-linear response of the medium causes the system to bifurcate as a function of the applied field intensities (2). The phenomenon of optical bistability is an optical example of a discontinuous instability which exhibits a hysteresis type behaviour analogous to the observations in equilibrium first order phase-transitions (3,4,5) Besides the unquestionable practical interest in these non-linear optical devices, they also provide some relatively elementary, but nevertheless realistic physical models for non-equilibrium phenomena.

It is a well known observation that macroscopic systems, when approaching their limits of stability become very sensitive to all sources of external disturbances or noise. It is therefore necessary in order to understand the behaviour in the critical regime or in the regime of so called multistability to include in the description fluctuations even on the macroscopic level. The simplest and most widely used approach to account for noise in a dynamical system is to include a fluctuating term in the equations of motion with zero correlation time and Gaussian statistics. This choice is not only motivated by the simplifications it brings about in the mathematical formulation of the problem, but also by the fact that close to the region of instability the macroscopic dynamics slows down and the time scale of the deterministic evolution may exceed the correlation time of the random fluctuations. In the case of a continuous bifurcation the time scale of the deterministic evolution even grows without limit and the approximation of white noise is always realistic, at least in the immediate neighborhood of the transition point.

The situation is somewhat different in the case of a multistable system where the dynamic evolution, in general, is characterized by more than one time constant. The boundaries of multistable behaviour e.g. are characterized by the divergence of one of these time constants, while the others at at the same time remain finite. It is therefore concievable that situations occur, where the time scales of the noise falls right in between the time scale which govern the deterministic dynamics. It is in this case where we expect qualitative changes in the statistical behaviour when noise with finite bandwidth is considered.

We will now proceed to investigate the role played by finite bandwidth noise, by using the model of absorptive optical bistability (6). As this model is a generalization of the traditional laser model with an additional coherent external field, the results derived here carry over also to the problem of laser action.

The model we want to discuss is characterized as follows:
The field mode in the optical cavity is decribed by a complex amplitude $E(t)$ the time will be measured in units of the cavity damping time and the nonlinear response of the medium is characterized by a material constant $\Gamma^2 = g\,N/V$, g is the dipole matrix element and N/V the atomic density. The strength of the incident laser field is given by E_0 while the transmitted field is proportional to E .

$$\dot{E}(t) = -E - \Gamma^2 E \left(1+|E|^2\right)^{-1} + E_0 + F(t) \qquad (1)$$

$F(t)$ represents all fluctuating influences like pump laser noise or spontaneous emission noise. We assume that $F(t)$ has a characteristic time scale given by Δ^{-1} and is exponentially correlated in time. Therefore $F(t)$ can be represented by an Ornstein- Uhlenbeck process in the form (8)

$$\dot{F}(t) = -\Delta F + \Delta \sqrt{Q}\, \xi(t) \qquad (2)$$

where ξ is the standard stationary Gaussian white noise process

$$\langle \mathcal{F}(t)\, \mathcal{F}(0) \rangle = \delta(t) \qquad (3)$$

The fluctuating force in this model is then described by the following correlation function

$$\langle F(t)\, F(0) \rangle = \frac{1}{2} Q \Delta\, e^{-\Delta t} \qquad (4)$$

The normalization has been chosen in such a way that the time integral over the correlation function is independent of the bandwidth of noise and therefore can be compared with the strength of the associated white noise case, which is contained in the limit $\Delta \to \infty$.

By increasing the two-dimensional phase space of the complex field variable E(t) to four dimensions thereby including also the auxiliary complex variable y which describes the noise, we obtain a formulation of the entire problem in terms of a multivariate continuous Markov-process.   The Fokker-Planck equation for the joint probability distribution P(E,E*,F,F*,t) associated with this stochastic process can be written in the following form:

$$\frac{\partial}{\partial t} P(x,y,t) = -\frac{\partial}{\partial x_i}\left(k_i + \frac{1}{\varepsilon}\, y_i\right) P + \frac{1}{\varepsilon^2}\frac{\partial}{\partial y_i}\, y_i\, P \; + \; \frac{1}{2} Q \varepsilon^{-2}\, \frac{\partial^2}{\partial y_i\, \partial y_i}\, P \qquad (5)$$

where E and F have been decomposed into their real and imaginary parts
$$E = x_1 + i x_2 \qquad \text{and} \qquad F = \varepsilon^{-2}\, (y_1 + i y_2)$$ For simplification of the notation we will use the following abreviations:

$$\Delta = \varepsilon^{-2}, \qquad k_i = -x_i\left(1 + \frac{\Gamma^2}{1 + x_j x_j}\right) + x_0\, \delta_{i1}$$

We assumed without loss of generality that E_0 is real $E_0 = x_0$. The summation convention over repeated indices is implied.

The advantage of this formulation proceeding along the traditional lines of continuous Markov processes is obtained only on the expense that the formulation now contains explicitly the irrelevant noise variable and that the asymmetry in the relative dependence of the variables x and y destroys the property of detailed balance, which was still present in the white noise limit. This formulation also contains more information than there is experimentally accessable i.e. besides containing the properties of the field through e.g. the field auto-correlation function, the joint probability distribution also allows obtaining the cross-correlation function of field x and noise y .   In order to reduce the problem to a lower dimensional and more tractable formulation, which no longer contains the physically irrelevant information about the auxiliary variable y, we attempt to project the dynamics of the entire phase space onto the subspace of the experimentally relevant variable x . Thereby we loose the Markov property of the model which, however, is not of primary importance here as long as we are only interested in single time properties of the process as the stationary correlation function or even the steady state.

A convenient approach in order to arrive at this reduced level of description is , among others, a generalization of the moment expansion method developed by

Wilemsky (9) . The systematic continuation of this expansion in a variety of examples even allows the formulation of the exact dynamics on the reduced phase space in a non-perturbative way (8,10,11) . The steady state distribution on the reduced phase space of the field variable x is obtained by simply integrating over the irrelevant variable y :

$$W_0 = \int P_0(\underline{x},\underline{y}) \, d\underline{y} \tag{6}$$

The conditional probability density of the field variable alone is obtained through the following definition ;

$$W(\underline{x},t/\underline{x}^\circ) = \int d\underline{y}\,d\underline{y}^\circ \, P(\underline{x},\underline{y},t/\underline{x}^\circ,\underline{y}^\circ) \cdot \frac{P_0(\underline{x}^\circ,\underline{y}^\circ)}{\int P_0(\underline{x}^\circ,\underline{y}')\,d\underline{y}'} \tag{7}$$

The average over the initial values $\underline{y}_i^\circ$ is chosen as a conditional average. For the same reason as the joint probability density is conserved in course of time, also the reduced probability $W(x,t)$ is a conserved quantity and consequently has to satisfy the corresponding continuity equation. This equation is easily obtained from the Fokker-Planck equation when inserting the definition eq.(7)

$$\frac{\partial W}{\partial t} = -\frac{\partial}{\partial x_i} \, k_i \, W - \frac{1}{\varepsilon} \frac{\partial}{\partial x_1} \, j_{1,0} - \frac{1}{\varepsilon} \frac{\partial}{\partial x_2} \, j_{0,1} \tag{8}$$

where the matrix-valued probability currents $j_{n,m}(\underline{x},t/\underline{x}^\circ)$ are defined as the conditional time dependent moments of the irrelevant variable y:

$$j_{n,m} = \int d\underline{y}\,d\underline{y}^\circ \, y_1^n \, y_2^m \, P(\underline{x},\underline{y},t/\underline{x}^\circ,\underline{y}^\circ) \, \frac{P_0(\underline{x}^\circ,\underline{y}^\circ)}{\int P_0(\underline{x}^\circ,\underline{y}')\,d\underline{y}'} \tag{9}$$

The evolution equation eq.(8) for the reduced probability has not yet been obtained in closed form i.e. not defined entirely on the variables of the phase space of the field, but it still depends on the dynamics of the irrelevant degrees of freedom through the currents $j_{1,0}$ and $j_{0,1}$. These currents themselves evolve in time according to the Fokker- Planck equation eq.(5) of the joint process. After Laplace transformation their equation of motion reads for arbitrary order n,m:

$$\left[n+m + \varepsilon^2\left(z + \frac{\partial}{\partial x_i} k_i \right) \right] j_{n,m}(z) = j_{n,m}(t=0) - \varepsilon \frac{\partial}{\partial x_1} j_{n+1,m}$$

$$\tag{10}$$

$$- \varepsilon \frac{\partial}{\partial x_2} j_{n,m+1} + \frac{Q}{2} n(n-1) j_{n-2,m} + \frac{Q}{2} m(m-1) j_{n,m-2}$$

This equation actually establishes an entire hierarchy of coupled equations for the currents $j_{n,m}$ which in general can not be solved analytically as the moments $j_{1,0}$ and $j_{0,1}$ in course of time couple to an infinite number of higher order moments.

At this point we have to restrict ourselves to an approximate description in order to truncate this hierarchy at a convenient low order. From hereon we will assume that the bandwidth of the noise $\Delta = \varepsilon^{-2}$ is large compared to the deterministic time scale which in this case is the cavity damping time. This assumption introduces a small parameter ε into the description which then can serve as an expansion parameter in a systematic perturbation series, the lowest order of which is identical to the white noise limit, which is well understood (7,12) .

In lowest non-trivial order in ε^2 the moment hierarchy involves only a finite number of moments according to the relation :

$$ \dot{J}_{n,m} \quad \text{with} \quad n+m \leq 2 \tag{11} $$

This finite number of equations then is easily solved. After inverse Laplace-transformation we insert these moments into the continuity equation eq.(8) which leaves us with the following equation for the conditional probability density on the reduced phase space:

$$ \frac{\partial W}{\partial t} = -\frac{\partial}{\partial x_i} k_i W + \frac{Q}{2} \frac{\partial^2}{\partial x_i \partial x_i} \left(1 - \varepsilon^2 \frac{\partial k_i}{\partial x_i} \right) W $$
$$ + \varepsilon^2 \frac{Q}{2} \frac{\partial^2}{\partial x_1 \partial x_2} \left(\frac{\partial k_1}{\partial x_2} + \frac{\partial k_2}{\partial x_1} \right) W + J(\underline{x}) \cdot \delta(t) \tag{12} $$

It should be mentioned at this point, however, that in spite of the formal analogy of this result to the usual Fokker-Planck equation, eq.(12) does not characterize a Markov process as $W(\underline{x},t/x^\bullet)$ does not satisfy the Chapman-Kolmogorov equation. In view of this fact we will not use the phrase Fokker-Planck equation for eq.(12) . The inhomogeneity in eq.(12) ,abbreviated by $I(x) \cdot \delta(t)$, represents a renormalization of the initial condition which approximates the rapid evolution on the time scale of the noise. As we will not be concerned with this question in the present context, we have not written down this term in all detail here.

Under the assumption of broad but finite bandwidth noise eq.(12) is the desired evolution equation which only makes reference to the probability density and the variables of the relevant phase space of the field. Looking at this result from a more formal point of view it may be interesting to note that, while starting out from a purely additive stochastic process in eq.(1,2) or eq.(5) the reduction of phase space finally introduces multiplicative fluctuations as can be seen from the state dependence of the diffusion matrix. Inclusion of non- white noise is therefore also a mechanism for creating multiplicative fluctuations.

A first physical consequence of the generalization to non-white noise can be derived from the stationary solution of the process. As this model describes the amplitude as well as the phase fluctuations of the bistable device, we are still confronted with a multivariate stochastic process, and already the formulation of the steady state in general is a non-trivial excercise. Only in the special case where the process satisfies the condition of detailed balance (13,14) or the potential condition (15) is the derivation of the steady state solution reduced in a straight forward way to the evaluation of quadratures. In order to find out if the solution can be obtained in analytical form we define the potential

$$W_0(\underline{x}) = exp\ \Phi(\underline{x})\qquad\qquad(13)$$

which guarantees the positive definiteness of the steady state probability $W_0(x)$

When inserting this ansatz into the stationary evolution equation we obtain the following defining equation:

$$\left\{ \frac{\partial\Phi}{\partial x_i} + \frac{\partial}{\partial x_i} \right\} \cdot \left\{ -k_i(\underline{x}) + \frac{\partial k_{ij}}{\partial x_j} + k_{ij}\frac{\partial\Phi}{\partial x_j} \right\}\qquad(14)$$

where $K_{ij}(\underline{x})$ is the diffusion matrix defined in eq.(12). A sufficient but certainly not a necessary condition for integrability is that

$$\Lambda_i(\underline{x}) = K_{ij}^{-1}\left(k_j - \frac{\partial k_{ij}}{\partial x_j} \right)\qquad\qquad(15)$$

derives from a potential, i.e.

$$\frac{\partial\Lambda_i}{\partial x_j} = \frac{\partial\Lambda_j}{\partial x_i}\qquad\qquad(16)$$

Abbreviating $K_{ij}(\underline{x})$ in the form

$$K_{ij}(\underline{x}) = \frac{Q}{2}\left(\delta_{ij} + \varepsilon^2 U_{ij} \right)\qquad\qquad(17)$$

where $\quad k_i = \partial U/\partial x_i \equiv U_i\quad,\quad \partial k_i/\partial x_j = U_{ij}$

For the inverse of the diffusion matrix we obtain

$$K_{ij}^{-1}(\underline{x}) = \frac{2}{Q}(1 - \varepsilon^2 tr\ U_{ij})\begin{pmatrix} 1+\varepsilon^2 U_{22} & -\varepsilon^2 U_{12} \\ -\varepsilon^2 U_{21} & 1+\varepsilon^2 U_{11} \end{pmatrix}\qquad(18)$$

Inserting this expression into eq.(15,16) reveals that in first order in ε^2 the potential conditions are satisfied and the steady state solution is obtained by integrating eq.(15). The steady state density obtained in this way assumes the following form:

$$W_0(\underline{x}) = exp -\frac{2}{Q}\left(U - \frac{\varepsilon^2}{2}\left\{ \frac{\partial U}{\partial x_i}\frac{\partial U}{\partial x_i} + Q\frac{\partial^2 U}{\partial x_i\partial x_i} \right\} \right)\qquad(19)$$

where $U(x)$ is the potential governing the white noise limit (7). When introducing the definition of $U(x)$ into eq.(19) and using polar coordinates $x_1 + ix_2 = r\exp\varphi$ we arrive at the following result

$$\Phi(r,\varphi) = -\frac{1}{Q}\left[r^2 - 2rr_0\cos\varphi + T^2 \ln(1+r^2) \right.$$

$$\left. \varepsilon^2\left(r^2\left(1+\frac{T^2}{1+r^2}\right)^2 - 2rr_0\cos\varphi\left(1+\frac{T^2}{1+r^2}\right) + r_0^2 \right) \right]\qquad(20)$$

where we have assumed for the moment that noise is weak and we can drop the last term in eq.(19) compared to the first one. This expression explicitly shows how the white noise result is modified when the bandwidth of noise is reduced.

The changes which are induced when altering the correlation time of the fluctuations are most easily seen in a plot of the probability density. Unfortunately in this short communication there is not enough space to include also some figures. We therefore briefly describe qualitatively what happens to the probability distribution, when the bandwidth of noise is varied from $\varepsilon = 0.1$ to $\varepsilon = 1.0$. A cut through the distribution along the prefered direction of the phase of the external field shows that the narrow peak shrinks drastically as a function of ε while the wide peak responds barely. In such a picture it becomes obvious that the narrow peak which belongs to the lower or absorptive branch is much more sensitive to the decrease of the bandwidth than the broad peak which represents the transmitting state of the bistable device. This tendency becomes even more pronounced when the distribution is plotted by cutting through the most probable values but in the direction orthogonal to the direction of the phase of the external field. Here an observable amount of shrinking is obtained already for values of ε^2 as small as $\varepsilon^2 = 0.03$ while no modifications are yet seen for the peak corresponding to the upper branch. This means that the field in the cavity is going to be locked more effectively to the phase of the external field when the bandwidth of noise is decreased.

This behaviour, to some extend, can be understood qualitatively when we notice that a bistable device is governed by two different time scales (16) . When we consider the purely deterministic time evolution of the process we can make an estimate on the characteristic time constants by linearizing around the deterministical steady states. Below as well as above the multistable regime there only exists a single stationary solution and consequently only a single time constant γ . Away from the boundaries of multistability these constants approach $\gamma = 1 + T'^2$ for small external field amplitudes and $\gamma = 1$ in the bleached state. As the time scale of the fluctuations is measured in units of the cavity decay time, we will assume in this approximation ε^2 to be small compared to unity. In the realistic experimental situation the parameter T' is large compared to unity and therefore the time scales of relaxation in the limit of weak fields may become comparable or even shorter than the time scale of the fluctuations. This will then result in a narrowing of the already sharp probability distribution of the lower branch. In the limit of strong fields, which satturate the optical transition the dynamical time scale is of order unity and in the approximation used here is always longer than the noise correlation time, it is therefore plausible that only small changes are observed in the transmitting state.

In the bistable regime now both arguments used above apply at the same time for the two coexisting stationary points. As the bandwidth of noise is narrowed, the time scale of the lower branch of the bistable hysteresis cycle is approached first and therefore only the width of the narrow peak is affected, while the peak corresponding to the transmitting state remains roughly unchanged. The diffusion process in the upper branch will finally respond to the increasing time scale of the fluctuations but only when their correlation time is of order one. At the same time the probability distribution of the lower branch has drastically narrowed. However, at this point also the limits of the perturbation expansion are surpassed and the results have to be taken only in a qualitative way. The reduction of the correlation time of noise results in a lowering of the statistical weight of the lower branch, causing the device to be less likely observed in the lower absorbing state. This also affects e.g. the ensemble averages of the field amplitudes. The Maxwell construction, which separates the regions of global stability, however, is not modified.

The knowledge of the stationary state in analytic form opens the possibility of estimating the time scales of global relaxation in the presence of noise. This can be done by using the variational principle for the evaluation of upper and lower bounds to the low lying eigenvalues of the dynamic equation eq.(12) (18) These time constants especially in the tunneling regime are expected to depend sensitively on the bandwidth of noise.

* On leave from the University of Budapest

REFERENCES

1. H.Haken, Rev.Mod.Phys.47,67 (1975)
2. R.Graham, Springer Tracts in Modern Physics Vol.66 (Springer,Berlin ,1972)
3. A.Szoeke, V.Daneu, J.Goldhar and N.A.Kurnit, Appl.Phys.Lett.15,376(1969)
4. S.L.McCall, Phys.Rev.A9,1515(1974)
5. S.L.McCall,H.M.Gibbs and T.N.C.Vencatesan, Phys.Rev.Lett.36,1135 (1976)
6. R.Bonifacio and L.A.Lugiato, Opt.Comm.19,172(1976)
7. A.Schenzle and H.Brand, Opt.Comm.27,485(1978)
8. R.Graham and A.Schenzle, Phys.Rev.A26,1676(1982)
9. G.Wilemsky, J.Stat.Phys.14,153(1976)
10. A.Schenzle and R.Graham, Proceedings of the 5.Rochester Conference on Coherence and Quantum Optics (Plenum Press,New York 1983)
11. A.Schenzle, to be published
12. R.Bonifacio,M.Gronchi and L.A.Lugiato,Phys.Rev.A18,2266(1978)
13. R.Graham and H.Haken, Z.Phys.243,289(1971)
14. R.Graham in Fluctuations, Instabilities and Phase Transitions, ed. by T.Riste (Plenum Press, New York 1975)
15. R.L.Stratonovich, Topics in the Theory of Random Noise (Gordon and Breach, New York 1963)
16. A.Schenzle and H.Brand, Opt.Comm.31,401(1979)
17. A.Schenzle and T.Thel, to be published
18. H.Brand, A.Schenzle and G.Schroeder, Phys.Rev.A25,2324(1982)

STATISTICAL FLUCTUATIONS IN OPTICAL BISTABILITY INDUCED BY SHOT NOISE

S. L. McCall

Bell Laboratories
Murray Hill, NJ 07974

S. Ovadia, H. M. Gibbs, F. A. Hopf and D. L. Kaplan

Optical Sciences Center
University of Arizona
Tucson, AZ 85721

ABSTRACT

Using a hybrid bistable optical devide with a large photon shot noise, we have observed fluctuations and spontaneous transitions between the two states. The experimental results are compared with various discrete and exact models. Although the exact approach to the problem is unsolvable, a good agreement is found with all the approximate models if one allows the counting rates to be adjustable.

INTRODUCTION

The behavior of an optical bistable system in which the fluctuations play the dominant role has been studied on the basis of a Fokker-Planck description by Bonifacio et al.,[1] Schenzle and Brand,[2] Agarawal et al.,[3] and others. However, most of the analyses so far approximate the fluctuations as arising from an intrinsic noise source described by a δ-correlated, Gaussian random process.[4,5] In this paper intensity-dependent shot noise fluctuations of a bistable system are experimentally studied for the first time, using a hybrid bistable optical device, and analyzed using various models.

Optical bistable devices can be extrapolated to devices with a small number of atoms, perhaps 10^3-10^4 atoms.[6] In these devices, statistical fluctuations in the number of photons, or number of excited atoms, will result in spontaneous transitions (glitches) and noisy output. From a practical point of view, the degree of fluctuations in the output is critical for future applications to all-optical signal processing, switching, and parallel computing.

The semiclassical description of absorptive bistability for two-level atoms of peak absorption coefficient α_0 filling a ring cavity of length L and mirror transmissivity T leads to the requirement $\alpha_0 L/T > 8$ for bistability.[7] An optical cavity can have a waist about λ^2 in area. Choosing $T < 0.125$, $\alpha_0 L \simeq 1$ is needed. Now consider a single atom, whose absorption line is only broadened by life-time effects, placed in a cavity of length λ. The absorption cross section is about λ^2 so that α_0, which is the product of cross section and volume density, is about λ^{-1}. Thus $\alpha_0 L \simeq 1$ is achieved. Semiclassically, this single-atom device should be bistable. However, quantum fluctuations would cause the device to switch states spontaneously at a rate of about once per atomic lifetime. Since semiclassical arguments do not rule out devices involving only a few absorbing centers and photons, the study of statistical limitations is especially relevant.

We may estimate the number of atoms required for a suitably small glitching rate. Assume the bit rate is 10^{12}/sec, and that less than one glitch per day, about 10^5 sec, is required. This is an error rate of 10^{-17}. Using Gaussian statistics, this corresponds to 8.48 standard deviations. In the "on" state approximately $\frac{1}{2}N$ atoms are excited, where N is the number of excitable atoms in the device. Setting $(\frac{1}{2}N)^{\frac{1}{2}} = 8.48$, thus N = 144. This is, of course, a gross estimate.

Incidentally, the "limit" set by Keyes[8] for optical switching does not apply to semiconductor devices. He correctly argues that as an absorption line is broadened to make a given device switch faster, that the switching energy behaves as a constant divided by switching time. His evaluation of the constant used highly favorable constants of atomic spectral or0igion. Carriers in semiconductors are not free atoms, and measured results when extrapolated are less than Keyes' limit.

EXPERIMENTAL APPARATUS

In this experiment a hybrid bistable optical device consisting of an electro-optic crystal between crossed polarizers is employed as shown in Fig. 1. The transmission of the modulator as a function of the applied voltage is shown in Fig. 2. The transmitted laser beam was strongly attenuated such that the shot noise in the photon detection process was the dominant noise source; correspondingly, the amplifier gain was increased. This

was achieved by using neutral density and wavelength selective
filters. The quantum efficiency of the photomultiplier-discrimi-
nator system was made very small by using a quite narrow window
in the discriminator, thus effectively eliminating dark noise
problems.

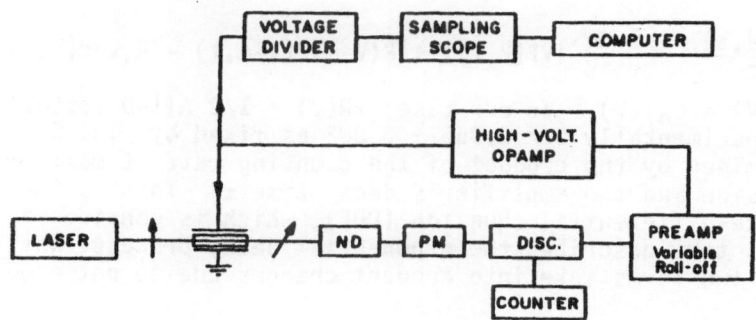

Figure 1. The experimental apparatus.

Data were taken in two modes. In the first, the bistability
operating point is selected within the classical loop limits such
that switching back and forth occurs. By sampling the voltage
repeatedly, steady-state histograms of the amplifier output voltage
times are recorded. In the other mode, the system is forced into
the upper or lower state by an external pulse and the time to
return to the original state is recorded.

"EXACT" MODEL

From Fig. 1, the high-voltage amplifier output can be written
as

$$V(t) = \alpha \sum_i H(t - t_i)\, e^{-(t-t_i)/\tau} \qquad (1)$$

where α is the voltage increment per detected photon, $H(t)$ is the
Heaviside step function, the times t_i denote when a photon was
detected, and τ, the decay time, is set by the preamplifier rolloff frequency. The rate of the detected photons is $C_0 T(V)$ where
C_0 is the counting rate for maximum transmission and $T(V)$ is the
transmission of the modulator at voltage V normalized to maximum
transmission.

From Eq. (1), we have

$$\frac{dV}{dt} = -\frac{V}{\tau} + \alpha \sum_i \delta(t-t_i), \tag{2}$$

a Langevin-type equation for the fluctuating voltage.

Imaging an ensemble of such systems, let $P(V,t)dV$ = probability of V to be in $(V, V + dV)$ at time t. Then, one can show that

$$\frac{\partial P(V,t)}{\partial t} = \frac{1}{\tau}\frac{\partial}{\partial V}[VP(V,t)] + R(V-\alpha)P(V-\alpha,t) - R(V)P(V,t), \tag{3}$$

where $R(V) = C_0 T(V)$. In our case, $\tau R(V) = 1/2\ A[1-D\cos(GV)]$, where experimentally $D = 0.965 \pm 0.002$ as fixed by Fig. 2, and A is determined by the product of the counting rate at maximum transmission and the amplifier's decay time τ. This is the difference-differential equation (DDE), which is considered exact. The first term described the exponential decay process, while the last two terms take into account changes due to detected photons.

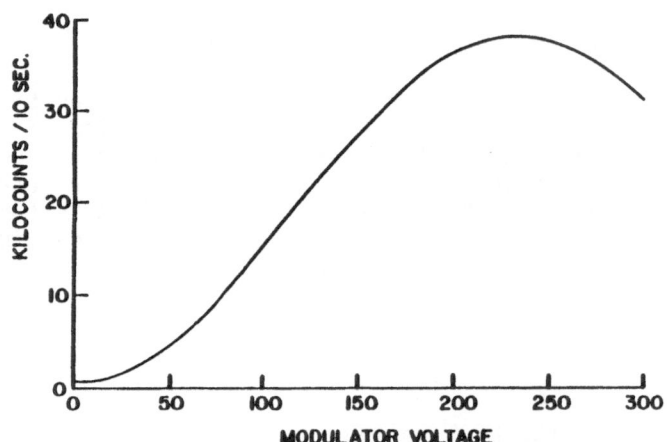

Figure 2.   Count rate vs. modulator voltage.

Expanding the second term to second order in α, one finds

$$\frac{\partial P(V,t)}{\partial t} = \frac{\partial}{\partial V}\left\{\left[\frac{V}{\tau} - \alpha R(V)\right]P(V,t)\right\} + \frac{\alpha^2}{2}\frac{\partial^2}{\partial V^2}[R(V)P(V,t)], \tag{4}$$

the corresponding Fokker-Planck-like equation. The steady-state solution to Eq. (3) is zero for $V < 0$, whereas the solution to Eq. (4) does not have this property.

We will use a class of discrete models, with a limiting case which is identical with Eq. (3). The equation is

$$\frac{dP_N}{dt} = -\frac{N}{\tau} P_N + \frac{N+1}{\tau} P_{N+1} - R_N P_N + R_{N-K} P_{N-K} \tag{5}$$

Here P_N is the probability that exactly N atoms are excited, and the first two terms on the right-hand side represent spontaneous decays. The last two terms represent absorption of a photon. For the third term the number of excited atoms changes from N to N+K; for the last term the change is from N-K to N. Note that there are two sizes of quanta. The decay process involves one step at the time, e.g. N goes to N-1, whereas the absorption process involves K steps, e.g. N goes to N+K. One can convince oneself that as K→∞, Eq. (5) becomes Eq. (3). For any K, a physical model exists. The linewidth of the absorption line must be broad compared to the inverse lifetime, so only on-diagonal elements of the density matrix need be considered. The coefficients R_N are determined then by the mirror-atom arrangement.

We make correspondence with Eq. (3) through $V = \alpha K N$, and $R_N = \frac{1}{2}A(1-D\cos(GV))$.

For K=1, the unnormalized steady-state solution is, with $P_0 = 1$,

$$P_N^{(0)} = \frac{(\tau R_0)(\tau R_1) \ldots (\tau R_{N-1})}{(1)\ (2) \ldots\ (N)} ,$$

so that a bimodal output is obtained if $R_N > N$ in an isolated region. For K > 1, the steady-state solution is not simply expressed.

We may expand Eq. (5) to second order in K. Doing so and substituting so to replace N with V,

$$\frac{\partial P}{\partial t} = \frac{\partial}{\partial V}\left\{\left[\frac{V}{\tau} - \alpha R\right]P\right\} + \frac{\alpha^2}{2}\frac{\partial^2}{\partial V^2}\left\{\left(R + \frac{V}{\alpha\tau K}\right)P\right\} , \tag{6}$$

The same as Eq. (4) with the addition of the last term. As K becomes infinite the last term approaches zero; one should expect that because for finite K, Eq. (5) is a discrete approximation to Eq. (3). For finite K, the last term represents an increase in noise. For finite K, the relaxation process introduces uncertainty, whereas for infinite K and in Eq. (3) the relaxation process is perfectly noise-free. For example, in Eq. (3), if $P(V, t=0)$ is a delta function, and there were no absorption, then $P(V,t)$ is also a delta function. This is not true for finite K. The ratio $V/\alpha\tau K : R$ is about 1/K where it counts, since the peaks of the bimodal distribution occur approximately where $V=\alpha\tau R$, according to the argument of the first derivative term.

Finally, for spontaneous switching measurements, one solves Eq. (5) through an eigenvalue expansion, but only two terms with the smallest two eigenvalues are kept,

$$P_n(t) = P_n^{(0)} + aP_n^{(1)} \exp(-\lambda t/\tau). \tag{7}$$

The smallest eigenvalue is zero, and λ, the smallest nonzero eigenvalue, represents spontaneous switching. If $\tau\!\uparrow$ and $\tau\!\downarrow$ are the mean lifes for the system to stay in the lower and upper states respectively, then it is easily seen that $\lambda = \tau(1/\tau\!\uparrow + 1/\tau\!\downarrow)$ and that $\tau\!\downarrow/\tau\!\uparrow$ is equal to the ratio of steady-state probabilities that the system will be in the upper and lower state.

OBSERVATIONS AND ANALYSIS

In Fig. 3 is presented a histogram of voltages obtained in steady-state. The conditions were $C_0=37$ KHz, $\tau=1.59\times10^{-3}$ sec (100 Hz), D=0.965, and G=.01366. The values of D and G are derived from the data leading to Fig. 2; the value of G is not relevant in our analysis.

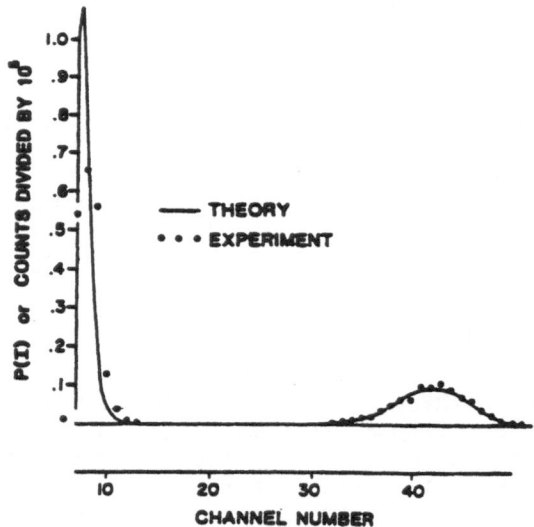

Figure 3. The open circles are the histogram of output voltage occurrences. Zero voltage is about channel 6, the lower state is about channel 8, the upper state about channel 42. The computer-simulated histogram is shown as the solid curve.

The models were fit to the histogram allowing two parameters to be free, A and G in the equation for R

$$R_N = \frac{1}{2}A \; (1-D \cos(GV))$$

where V=N/(AK). The voltage V from here on is thus normalized. The parameter $A=\tau C_O$ experimentally is, from the above, 58.83.

The fitting was done by forcing A & G be such that two experimental parameters were fit, namely the ratio of the peak value of the upper state histogram to the minimum in between, which had an experimental value of 1184, and namely the ratio of times the device spent in the upper state and the lower state, i.e. the ratio of areas of the two peaks. The histogram was fit with K=1,2,3,...10. Values of A thus obtained fit very well the relation A=31.54 + 30.28/K. The value at K=∞ of 31.54 matches poorly the experimental value of 58.83, which means there was twice as much noise in the system as we calculate. Values of G varied from 3.805 at K=1 to 3.756 at K=10. The variation of A with K is precisely what one expects by requiring the coefficient of P in the second derivative term in Eq. (6) to remain constant.

Using models with various K, and also the Fokker-Planck-like equation (3), we may calculate the tunneling times $\tau\uparrow$ and $\tau\downarrow$. In each case, the values of G and A derived by forcing agreement as described above were used. The thusly found tunneling times were essentially model independent. In the discrete model cases, the lowest non-zero eigenvalue thus derived varied from 1.202×10^{-4} for K=1 to 1.186×10^{-4} for K=10. For the Fokker-Planck case, $\lambda=1.156\times10^{-4}$ was found. The point is that if one matches the two parameters by choosing noise and gain levels, then evidently everything else follows correctly, even though different models are used.

In another calculation, A was allowed to vary, but G was determined by the requirement that the ratio of the two mode areas be equal. The model with K=1 was used. Not surprisingly, λ varied exponentially with A, and to well within a factor two, $\lambda=0.1\times10^{-(A/20)}$. For the earlier mentioned rate of 10^{-17}, A∼320 is required.

Figure 4 presents tunneling time data. The theoretical curves used are with the same gain G=3.805 for K=1, and A is adjusted to take into account the change-in counting rate (5.6 instead of 37 KHz) and time constant due to varying bandwidths. The agreement was surprisingly good. Again the point is that once the model is calibrated, everything else follows.

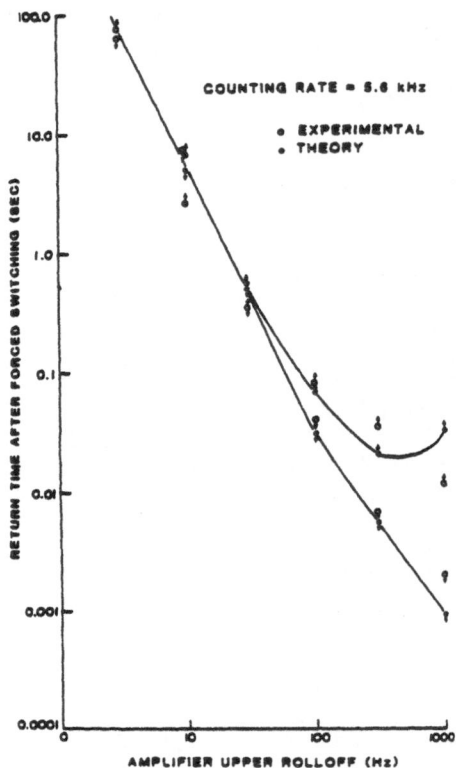

Figure 4. Return time after forced switching to the lower and upper states as a function of the amplifier upper rolloff frequency.

CONCLUSIONS

To fit our data, we used a class of models which were not the correct ones for our system. Fitting the steady-state distribution involves a two-parameter fit, G which alone determined the noise-free operating curve, and A which characterized the amount of noise in the system. A and G were chosen so that the ratio of mode areas in the distribution and the ratio of upper state mode maximum to minimum between the two modes agreed with experiment. G hardly varied from model to model, while A varied in a simply understood way. Resultant eigenvalues were model independent and fit the data. The lack of agreement between thusly derived A and experimental expectation means that the noise was about twice than expected for reasons not presently understood.

As a result of this investigation, we conclude that one can use a Fokker-Planck equation or a simple model such as ours with K=1 to find out what happens, by using an adjusted amount of noise. Of course detail, particularly of the shape of the lower state mode, are incorrect.

ACKNOWLEDGEMENTS

The Arizona portion of this research was supported by the Air Force Office of Scientific Research, The Army Research Office, and the National Science Foundation.

REFERENCES

1. R. Bonifacio, M. Gronchi and L. A. Lugiato, "Photon statistics of a bistable absorber", Phys. Rev. A 18, 2266 (1978).
2. A. Schenzle and H. Brand, "Fluctuations in optical bistable systems", Opt. Comm. 27, 485 (1978).
3. G. S. Agarawal, L. M. Narducci, R. Gilmore and D. H. Feng, "Optical bistability: a self-consistent analysis of fluctuations and the spectrum of scattered light", Phys. Rev. A 18, 620 (1978).
4. R. Graham and A. Schenzle, "Dispersive optical bistability with fluctuations", Phys. Rev. A 23, 1302 (1981).
5. J. Chrostowski, A. Zardecki and C. Delisle, "Time-dependent fluctuation and phase hysteresis in dispersive bistability", Phys. Rev. A 24, 345 (1981).
6. S. L. McCall and H. M. Gibbs, "Conditions and limitations in intrinsic optical bistability", in Optical Bistability, C. M. Bowden, M. Ciftan, and H. R. Robl, eds. (Plenum, New York, 1981).
7. A. Szoke, V. Daneu, J. Goldhar and N. A. Kurmit, "Bistable optical element and its applications", Appl. Phys. Lett. 15, 376 (1969).
8. R. W. Keyes, "Physical limits in digital electronics", Proc. of IEEE 63, 740 (1975).

EFFECT OF DRIVING LASER FLUCTUATIONS IN OPTICAL BISTABILITY

Charles R. Willis

Physics Department

Boston University Boston, MA 02215

The standard model for optical bistability[1],[2] (OB) consists of
a system of N two-level systems in a cavity driven by an external
field of specified amplitude and phase. The only source of fluctuations in the model is the spontaneous emission of the two-level
systems in the cavity. Since the number of photons in the cavity is
large, the spontaneous emission which is roughly inversely proportional to the number of photons in the cavity, is very small. The value
of the dimensionless parameter that measures the ratio of spontaneous
emission to mean field terms in present day OB systems are typically
10^{-6} or less. In a recent paper[3] we showed that driving laser frequency fluctuations cause fluctuations in the phase of the field that
are typically many orders of magnitude larger than the fluctuations
caused by spontaneous emission. In this article, we show the effect
of driving laser fluctuations on the amplitude of the field in the
cavity can be much more important than on the field phase because the
amplitude undergoes critical slowing down and the amplitude fluctuations diverge at the turning points of the OB curve. The increased
fluctuations of the cavity field due to the fluctuations in the
driving laser will cause the system to make a transition from one
branch of the OB curve to the other before the mean field turning
point is reached. In this paper we also include driving laser
amplitude fluctuations. We show that for dispersive OB that the
amplitude of the cavity field is affected by driving laser phase fluctuations. While for absorptive OB the amplitude of the cavity field
is only affected by driving laser amplitude fluctuations. In Ref. 3
we considered only the high Q cavity limit but here we derive the
master equation for the general case where neither the high Q or low
Q cavity limits are valid for both absorptive and dispersive OB. In
this paper we sketch the essential steps in the derivation of the

master equation for OB where the drift terms are the Maxwell-Bloch
equations and the fluctuation terms depend on both driving laser
amplitude and phase fluctuations for both absorptive and dispersive
OB. We obtain the master equation by making the time dependence of
the driving laser field a Gaussian process, and specifying the conditions required on the stochastic time dependence such that the
Stratonovich derivation of the generalized Fokker-Planck equation is
valid.

DERIVATION OF EQUATIONS

The Hamiltonian for N two-level systems interacting with
radiation in a cavity is

$$H = H_m + H_F + H_{int} \tag{2.1}$$

$$H_m \equiv (1/2)\hbar\omega_A \sum_\alpha^N s_\alpha \equiv (1/2)\hbar\omega_A S \tag{2.2a}$$

$$H_F \equiv \hbar\omega_c a^+a \quad , \quad H_{int} \equiv \hbar(\mu P^+a + \mu^* P^- a^+) \tag{2.2b)(2.2c}$$

where $P^\pm = \sum_\alpha e^{\pm i\vec{k}\cdot\vec{X}_\alpha} \sigma_\alpha^\pm$ and the commutation relations for the collective matter operators $[P^+,P^-] = S$ and $[S,P^\pm] = \pm 2P^\pm$. The commuta-
tion relations follow from $s_\alpha = [\sigma_\alpha^+,\sigma_\alpha^-]$ and $[\sigma_\alpha^-,\sigma_\alpha^+]_+ = 1$ and operators from different two level systems commute with each other. For
a complete discussion of the Hamiltonian, see Ref. 5. The frequencies
are ω_A the two-level frequency difference, ω_c the cavity frequency
and $\hbar\mu \equiv \vec{\varepsilon}\cdot\vec{\rho}^*(2\pi\hbar\omega_c/V)^{\frac{1}{2}}$ where $\vec{\varepsilon}$ is the unit vector in the direction
of the cavity field, $\vec{\rho} \equiv <+|e\vec{r}|->$ is the transition dipole moment
between the two states of the two-level system and V is the volume
of the cavity. We can add an external field α to a (where α can be
time dependent) by means of the unitary transformation $D(\alpha)$,

$$D^{-1}(\alpha)aD(\alpha) = a + \alpha \equiv A \quad \text{where } D(\alpha) \equiv \exp(\alpha a^+ - \alpha^* a) \tag{2.3}$$

Since $D(\alpha)$ commutes with matter operators the net effect of $D(\alpha)$ is
to replace a by A in the Hamiltonian Eqs. 2.2b and 2.2c as was first
pointed out in Ref. 6. Using the properties of $D(\alpha)$ we derive by a
straightforward calculation

$$\dot{A} = D^{-1}(\alpha)\dot{a}D(\alpha) + \dot{\alpha} \tag{2.4}$$

where the dot means derivative with respect to time. Consequently,
the net effect of adding a time dependent external field $\alpha(t)$ to the
N two-level systems interacting with radiation in a cavity is to
replace a by A in the Hamiltonian and add the term $\dot{\alpha}(t)$ to the time
derivative of A. The Liouville equation for N two-level systems
interacting with radiation in a cavity is

$$\partial F_N/\partial t + i\hbar^{-1}[F_N,H] = 0 \tag{2.5}$$

In the presence of matter reservoir Λ_A and field reservoir Λ_F we have

$$\partial F_N/\partial t + i\hbar^{-1}[F_N,H] = \Lambda_A F_N + \Lambda_F F_N \tag{2.6}$$

When we transform Eq. 2.5 with the unitary transformation $D(\alpha)$, the Hamiltonian H becomes[6]

$$H = H_o + \Delta H \quad ; \quad H_o \equiv \hbar\Omega(A^+A + (1/2)S) \tag{2.7}$$

and $\quad \Delta H \equiv \hbar\delta_F A^+A + (1/2)\hbar\delta_A S + i\hbar(\mu P^+A + \mu^* PA^+) \tag{2.8}$

The frequencies are: Ω the frequency of the incident laser, $\delta_A \equiv \omega_A - \Omega$ the atomic detuning and $\delta_F \equiv (\omega_C - \Omega)$ the cavity detuning. Going over to the interaction representation using H_o as in Ref. 6 we obtain

$$\partial F_N/\partial t = -i\mathcal{L}F_N + \Lambda_A F_N + \Lambda_F F_N + [\dot\alpha A^+ - \dot\alpha^* A, F_N] \tag{2.9}$$

where

$$\Lambda_F F_N \equiv \kappa[(A-\alpha),F_N(A^+ - \alpha^*)] + h.c. \tag{2.10}$$

κ is the photon inverse lifetime, h.c. is the hermetian conjugate and the last term in Eq. 2.9 which vanishes if α is time independent, yields the $\dot\alpha$ term as required by Eq. 2.4. The radiation reservoir acts only on[2a] the cavity field $a = A - \alpha$. The unitary transformation has no effect on Λ_A. The definition of $\mathcal{L}$ is $[\ \ ,\Delta H]$.

We obtain the equations of motion for $A, A^+, P^\pm$ and S in the self-consistent field approximation by multiplying Eq. 2.9 by each operator in turn, tracing over matter and field variables, and neglecting radiation-matter correlations. The resultant equations are:

$$\langle\dot A\rangle + (\kappa + i\delta_F)\langle A\rangle = \kappa\alpha - i\mu^*\langle P\rangle + \dot\alpha \tag{2.11a}$$

$$\langle\dot P^-\rangle + (\gamma_\perp + i\delta_A)\langle P^-\rangle = i\mu\langle A\rangle\langle S\rangle \tag{2.11b}$$

$$\langle\dot S\rangle + \gamma_\parallel(\langle S\rangle + N) = 2i\{\mu^*\langle A^+\rangle\langle P^-\rangle - \mu\langle A\rangle\langle P^+\rangle\} \tag{2.11c}$$

where $\langle(...)\rangle \equiv Tr(...)F_N$. The equation for $\langle A^+\rangle$ is the complex conjugate of Eq. 2.11a and the equation for $\langle P^+\rangle$ is the complex conjugate of Eq. 2.11b. $\gamma_\perp$ is the inverse polarization relaxation time and $\gamma_\parallel$ is the inverse population inversion relaxation time. With $\dot\alpha$ set equal to zero, Eqs. 2.11 are the Maxwell-Bloch equations which do not contain incoherent spontaneous emission because we neglected radiation matter correlations in Eqs. 2.11. The steady state solution of Eq. 2.11 with $\dot\alpha = 0$ is

$$\langle y\rangle = \langle x\rangle\left\{(1 + \frac{\mathcal{N}}{1+\hat\delta_A^2+|\langle x\rangle|^2}) - i(\frac{\mathcal{N}\hat\delta_A}{1+\hat\delta_A^2+|\langle x\rangle|^2} - \bar\delta_F)\right\} \tag{2.12}$$

where $\hat\delta_A \equiv (\delta_A/\gamma_\perp)$, $\bar\delta_F \equiv (\delta_F/\kappa)$, $\mathcal{N} \equiv (|\mu|^2 N/\gamma_\perp\kappa)$, $\langle x\rangle \equiv 2|\mu|\langle A\rangle/(\gamma_\parallel\gamma_\perp)^{\frac12}$, and $\langle y\rangle \equiv 2|\mu|\alpha/(\gamma_\parallel\gamma_\perp)^{\frac12}$.

In order to obtain a generalized Fokker-Planck equation for our model we first obtain a Langevin description. Since the amplitude and phase noise on the driving laser can arise from different causes, we find it useful to introduce an amplitude and phase variable description for α and the radiation and matter variables. Once we have

Eqs. 2.11 we can drop the average values and take the variables x, x^+, $P^\pm$, and S as classical variables whose statistical properties are determined by the statistical properties of $\alpha(t)$ because we are considering the situations where the driving laser noise is much larger than spontaneous emission. When we substitute the following definitions of the amplitude and phase variables

$$P^\pm \equiv \pm i\wp e^{\pm i\psi} \quad , \quad x \equiv re^{-i\phi} \quad \text{and} \quad y \equiv \mathcal{E}e^{-i\theta}$$

into Eqs. 2.11 we obtain

$$\overset{\bullet}{\xi}_\ell = f_\ell(\xi) + \Sigma g_{\ell m}\eta_m(t) \tag{2.13}$$

where $d\xi_1 = dr$, $d\xi_2 = rd\phi$, $d\xi_3 = d\wp$, $d\xi_4 = \wp d\psi$ and $d\xi_5 = dS$. The definitions of the $f_\ell(\xi)$ are

$$f_1 = -\kappa r + \kappa \mathcal{E}\cos(\theta-\phi) - 2|\mu|^2(\gamma_\parallel \gamma_\perp)^{-\frac{1}{2}} \wp \cos(\psi-\phi) \tag{2.14a}$$

$$f_2 = r\,\delta_F + \kappa\mathcal{E}\sin(\theta-\phi) - 2|\mu|^2(\gamma_\parallel \gamma_\perp)^{-\frac{1}{2}} \wp \sin(\psi-\phi) \tag{2.14b}$$

$$f_3 = -\gamma_\perp \wp - (\gamma_\parallel \gamma_\perp)^{\frac{1}{2}} 2^{-1} Sr\cos(\phi-\psi) \tag{2.14c}$$

$$f_4 = \delta_A \wp - (\gamma_\parallel \gamma_\perp)^{\frac{1}{2}} 2^{-1} Sr\sin(\phi-\psi) \tag{2.14d}$$

$$f_5 = -\gamma_\parallel (S+N) + 2(\gamma_\parallel \gamma_\perp)^{\frac{1}{2}} r\wp\cos(\theta-\psi) \tag{2.14e}$$

The definitions of the $g_{\ell m}$ are

$$g_{11} = \Gamma_u^{\frac{1}{2}}\mathcal{E}\cos(\theta-\phi) \quad , \qquad g_{21} = \Gamma_u^{\frac{1}{2}}\mathcal{E}\sin(\theta-\phi)$$

$$g_{12} = \Gamma_\theta^{\frac{1}{2}}\mathcal{E}\sin(\theta-\phi) \quad , \qquad g_{22} = \Gamma_\theta^{\frac{1}{2}}\mathcal{E}\cos(\theta-\phi) \tag{2.15}$$

where $\mathcal{E} \equiv e^u$, $\eta_1(t) = \Gamma_u^{-\frac{1}{2}}(\dot{\mathcal{E}}/\mathcal{E}) \equiv \Gamma_u^{-\frac{1}{2}}\dot{u}$, and $\eta_2(t) = \Gamma_\theta^{-\frac{1}{2}}\dot{\theta}$. The constants Γ_u and Γ_θ are defined below and are introduced in Eq. 2.15 to make the coefficient of $\delta(t)$ in Eq. 2.17 dimensionless. For Eq. 2.2 to be a Langevin equation we need two sets of conditions on the time dependence of $\dot\theta$ and $(\dot{\mathcal{E}}/\mathcal{E})$. First, the correlation times

$$t_\theta \equiv <\dot\theta^2>^{-1}\int_0^\infty<\dot\theta(\tau)\dot\theta>d\tau \text{ and } t_u \equiv <\dot{u}^2>^{-1}\int_0^\infty<\dot{u}(\tau)\dot{u}>d\tau = t_\mathcal{E} \tag{2.16}$$

must be shorter than the times κ^{-1}, $\gamma_\parallel^{-1}$, $\gamma_\perp^{-1}$, $(\mu\alpha)^{-1}$, and $[|\mu|^2 N(\gamma_\parallel\gamma_\perp)^{-\frac{1}{2}}]^{-1}$. The averages in Eq. 2.16 are over the time dependent Gaussian stochastic processes which determine $\dot\theta$ and $\mathcal{E}$. The second set of conditions[8] require the correlation times t_θ and t_u be small compared with the modulations $\delta_\theta \equiv <\dot\theta^2>^{\frac{1}{2}}$ and $\delta_u \equiv <\dot{u}^2>^{\frac{1}{2}}$, i.e., $\delta_\theta t_\theta \ll 1$ and $\delta_u t_u \ll 1$.

When all the above conditions on the time parameters are met, we can write

$$<\eta_\ell\eta_m(t)> = \delta_{\ell m}\delta(t) \tag{2.17}$$

We can now obtain the generalized Fokker-Planck equation for the probability distribution $\rho(r,\phi,\wp,\psi,S,t)$ because Eq. 2.12 and the condition Eq. 2.17 are the Langevin equations for which the Stratonovich[4] derivation applies. Since the physical noise source is external to the system and is not strictly white noise, then the Stratonovich interpretation is the correct limit[9]. In actual situations, frequently some of the conditions on t_θ and t_u will not be satisfied and a Markoffian description with a Fokker-Planck equation will not be possible. In this article we consider only the cases where all the conditions are satisfied and we have then

$$\frac{\partial \rho}{\partial t} = -\sum_{\alpha=1}^{5} \frac{\partial}{\partial x_\alpha} k_\alpha \rho + \frac{1}{2}\sum_{\alpha,\beta}^{2} \frac{\partial^2}{\partial x_\alpha \partial x_\beta} k_{\alpha\beta}\rho \qquad (2.18)$$

where $k_1 = f_1 + (1/2)\Sigma_m g_{mj}(\partial g_{1j}/\partial x_m) = f_1 + <\mathcal{E}^2>(2r)^{-1}\Big\{\Gamma_u \sin^2(\theta-\phi) +$
$\Gamma_\alpha \cos^2(\theta-\phi)\Big\}$ $\qquad\qquad (2.19a)$

$k_2 = f_2 + (1/2)\Sigma_m g_{mj}(\partial g_{2j}/\partial x_n) = f_2 + <\mathcal{E}^2>(2r)^{-1}\Big\{(\Gamma_\theta - \Gamma_u)\sin(\theta-\phi)$
$\cos(\theta-\phi)\Big\}$ $\qquad\qquad (2.19b)$

$$k_3 = f_3 \quad , \quad k_4 = f_4 \quad \text{and} \quad k_5 = f_5 \qquad (2.19c)$$

and

$$k_{11} = \Sigma_j g_{1j}g_{j1} = <\mathcal{E}^2>\Gamma_u \cos^2(\theta-\phi) + <\mathcal{E}^2>\Gamma_\theta \sin^2(\theta-\phi) \qquad (2.20a)$$

$$k_{22} = \Sigma_j g_{2j}g_{j2} = <\mathcal{E}^2>\Gamma_u \sin^2(\theta-\phi) + <\mathcal{E}^2>\Gamma_\theta \cos^2(\theta-\phi) \qquad (2.20b)$$

$$k_{12} = k_{21} = g_{11}g_{21} + g_{12}g_{22} = <\mathcal{E}^2>(\Gamma_u - \Gamma_\theta)\sin(\theta-\phi)\cos(\theta-\phi) \qquad (2.20c)$$

The inverse relaxation times Γ_θ and Γ_u are

$$\Gamma_\theta = \delta_\theta^2 t_\theta \equiv T_\theta^{-1}, \quad \Gamma_u = \delta_u^2 t_u \equiv T_u^{-1} \quad \text{and} \quad \Gamma_\mathcal{E} = \mathcal{E}^2 \Gamma_u.$$

An alternate form of our Markoff conditions $\delta_\theta t_\theta \ll 1$ and $\delta_u t_u \ll 1$ are $t_\theta T_\theta^{-1} \ll 1$ and $t_u T_u^{-1} \ll 1$.

Eq. 2.18 is too lengthy to write out explicitly and since the drift terms are just the Maxwell-Bloch equations we display only the diffusion terms explicitly

$$\frac{\partial \rho}{\partial t} + \Sigma \frac{\partial}{\partial x_\alpha} k_\alpha \rho = \frac{<\mathcal{E}^2>}{2}\Big\{\frac{\partial^2}{\partial r^2}(\Gamma_u \cos^2(\theta-\phi) + \Gamma_\theta \sin^2(\theta-\phi))$$

$$+ \frac{1}{r^2}\frac{\partial^2}{\partial \phi^2}(\Gamma_u \sin^2(\theta-\phi) + \Gamma_\theta \cos^2(\theta-\phi))$$

$$+ (\frac{1}{r}\frac{\partial}{\partial\phi}\frac{\partial}{\partial r} + \frac{\partial}{\partial r}\frac{1}{r}\frac{\partial}{\partial\phi})[(\Gamma_u - \Gamma_\theta)\sin(\theta-\phi)\cos(\theta-\phi)]\Big\}\rho \qquad (2.21)$$

The normalization of ρ is $\int \rho dr d\phi d\theta d\psi dS = 1$. There are two independent ratios that determine the qualitative and quantitative properties of the diffusion terms in Eq. 2.21, namely, (Γ_u/Γ_θ) and $[\sin(\theta-\phi)/\cos(\theta-\phi)]$. The value of the ratio (Γ_u/Γ_θ) depends on the ratio of driving laser amplitude fluctuations to driving laser phase fluctuations. In absorptive bistability $\cos(\theta-\phi) \gg \sin(\theta-\phi)$ while in dispersive bistability the inequality is reversed.

An important special case of Eq. 2.21 is where $\Gamma_\theta \gg \Gamma_u$, there is some dispersion, and the Γ_θ terms are the dominant diffusion terms on the right had side of Eq. 2.21 which becomes

$$\frac{<\mathcal{E}^2>}{2} \; \Gamma_\theta \left\{ (\frac{\partial}{\partial r} \; \sin(\theta-\phi) \; - \; \frac{1}{r}\frac{\partial}{\partial\phi} \cos(\theta-\phi))^2 \right\} \rho \qquad (2.22)$$

The most important consequence of Eq. 2.22 is that in the practically important case where Γ_θ is large and we have some dispersion, then the cavity amplitude fluctuations caused by the driving laser phase fluctuations Γ_θ are appreciable. In pure absorptive OB the driving laser phase fluctuations have very little effect on the amplitude fluctuations of the cavity field. Thus the combination of driving laser phase fluctuations with some dispersive OB leads to enhanced fluctuation effects which will cause the system to jump from one branch of the OB curve to the other branch before the mean field turning points are reached.

HIGH-Q CAVITY LIMIT

Since the full Eq. 2.21 is so complicated and contains so many parameters we find it necessary to consider various limits in order to get a qualitative appreciation of the effects of driving laser fluctuations. Consequently, we consider the high Q cavity where we can eliminate the matter variables adiabatically if the condition $(\kappa/\gamma_\perp)<<\eta^{-1}$ is satisfied. The adiabatic elimination of the matter variables leads to the following master equation for $R(r,\phi,\tau) \equiv \int \rho(r,\phi,\rho,\psi,S,\tau)d\rho d\psi dS$

$$\frac{\partial R}{\partial\tau} = -\frac{\partial}{\partial r}\left\{ -r + \mathcal{E}\cos(\theta-\phi) - \frac{\eta r}{1+\hat\delta_A^2+r^2} + \frac{<\mathcal{E}^2>}{2r}(\bar\Gamma_u\sin^2(\theta-\phi)+ \right.$$

$$\Gamma_\theta\cos^2(\theta-\phi))\Big\} R - \frac{1}{r}\frac{\partial}{\partial\phi}\left\{ r\delta_F + \mathcal{E}\sin(\theta-\phi) - \frac{\eta\hat\delta_A r}{1+\hat\delta_A^2+r^2} \right\} R$$

$$+ \; \frac{<\mathcal{E}^2>}{2}\left[\frac{\partial^2}{\partial r^2}\left\{ \bar\Gamma_u\cos^2(\theta-\phi) + \bar\Gamma_\theta\sin^2(\theta-\phi)\right\} \right. \qquad (3.1)$$

$$\left. + \frac{1}{r^2}\frac{\partial^2}{\partial\phi^2}\left\{\bar\Gamma_u\sin^2(\theta-\phi) + \bar\Gamma_\theta\cos^2(\theta-\phi)\right\} + \frac{2}{r}\frac{\partial^2}{\partial\phi\partial r} (\bar\Gamma_u-\bar\Gamma_\theta)\sin(\theta-\phi)\cos(\theta-\phi)\right] R$$

where $\tau \equiv \kappa t$ and a frequency with a bar over it has been made dimensionless by dividing by κ. The ϕ drift term of Eq. 3.1 has been cancelled exactly by a corresponding term in the diffusion term. The nonlinear coefficients in Eq. 3.1 make its solution difficult and since Eq. 3.1 does not satisfy detailed balance, we do not know the stationary state. We can get some appreciation of the driving laser fluctuations by solving the linearized form of Eq. 3.1. The details are presented in Ref. 7. The linearized master equation for the phase variable $\chi \equiv \phi-\theta$ obtained from Eq. 3.1 is

$$\frac{\partial R(\chi,\tau)}{\partial\tau} = \frac{\partial}{\partial\chi}\left\{ \lambda_\phi\chi + D_\phi \frac{\partial}{\partial\chi} \right\} R(\chi,\tau) \qquad (3.2)$$

where $\lambda_\phi \equiv [(<\mathcal{E}>/r)\cos\chi]_s = 1 + \eta [1+\hat\delta_A^2 + r_s^2]^{-1}$ and the diffusion coefficient D_ϕ is

$$D_\varphi \equiv (1/2) <\mathcal{E}^2>r_s^{-2}\{\bar\Gamma_u <\sin\chi>_s^2 + \bar\Gamma_\theta <\cos\chi>_s^2\} \tag{3.3}$$

$$= (1/2)\bar\Gamma_u \left[\eta\hat\delta_A[1+\hat\delta_A^2 + r_s^2]^{-1}-\bar\delta_F\right]^2 + (1/2)\bar\Gamma_\theta \left[1+\eta[1+\hat\delta_A^2 + r_s^2]^{-1}\right]^2$$

where r_s is the steady state solution for the amplitude of Eq. 2.12. The steady state solutions for $<\sin\chi>_s$ and $<\cos\chi>_s$ were obtained in Ref. 2d. Since Eq. 3.2 was thoroughly analyzed in Ref. 3 for a special case of D_ϕ, we refer the reader to that reference where all the analysis follows with the more general D_ϕ given by Eq. 3.3. We only remark here that when (D_ϕ/λ_ϕ) becomes of order unity that the cavity field phase fluctuations can cause the system to jump from one branch of the OB curve to the other branch.

The linearized master equation for the amplitude r obtained from Eq. 3.1 is

$$\frac{\partial R(r,\tau)}{\partial\tau} = \frac{\partial}{\partial r}\left\{\lambda_r(r-r_s) + D_r \frac{\partial}{\partial r}\right\} R(r,\tau) \tag{3.4}$$

where

$$\lambda_r \equiv 1 + \frac{\eta(1-r_s^2+\hat\delta_A^2)}{(1+\hat\delta_A^2+r_s^2)^2} + \frac{1}{2}\left\{\bar\Gamma_u \left[\frac{\eta\hat\delta_A}{1+\hat\delta_A^2+r_s^2} - \bar\delta_F\right]^2 + \bar\Gamma_\theta \left[1 + \frac{\eta}{1+\hat\delta_A+r_s^2}\right]^2\right\} \tag{3.5}$$

and

$$D_r \equiv \frac{r_s^2}{2}\left\{\bar\Gamma_\theta \left[\frac{\eta\hat\delta_A}{1+\hat\delta_A^2+r_s^2} - \bar\delta_F\right]^2 + \bar\Gamma_u \left[1 + \frac{\eta}{1+\hat\delta_A+r_s^2}\right]^2\right\} \tag{3.6}$$

The Green's function for Eq. 3.4 is

$$G[(r-r_s),\tau;(r-r_s)_o,0] = Z^{-1}\exp\left\{\frac{-\lambda r}{2D_r}\left(\frac{[(r-r_s)-(r-r_s)_o\exp(-\lambda r\tau)]^2}{[1-\exp(-2\lambda r\tau)]}\right)\right\}$$

where $Z \equiv [2\pi D_r(1-\exp(-2\lambda_r\tau))\lambda_r^{-1}]^{\frac{1}{2}}$ $\tag{3.7}$

and where $(r-r_s)_o$ is the initial value of the deviation of r from the mean value r_s.

The initial value of the amplitude deviation is forgotten on the time scale λ_r^{-1} and the stationary state is reached on the time scale $(2\lambda_r)^{-1}$. If $\lambda_r > D_r$ the steady state is reached before there is an appreciable spread in $r-r_s$ due to fluctuations and the dispersion $\sigma^2 \equiv (D_r/\lambda_r)$ in the steady state is small, thus $r \sim r_s$ with small fluctuations in the steady state. On the other hand if $D_r \geq \lambda_r$ there is an appreciable spread in $r-r_s$ before the steady state is achieved and in the steady state σ^2 is large. We can see the effect of $D_r >> \lambda_r$ most clearly by considering the interval $D_r^{-1}<\tau<\lambda_r^{-1}$ in Eq. 3.7 which becomes

$$G[(r-r_s),\tau;(r-r_s)_o,0] = (4\pi D_r)^{-\frac{1}{2}}\exp[-(r-r_o)^2/4D_r\tau].$$

Thus we see for $D_r \gg \lambda_r$ the amplitude diffuses instead of relaxing to r_s until τ becomes of order λ_r^{-1}. The amplitude variable undergoes critical slowing down and the amplitude fluctuations diverge because $\lambda_r \to 0$ as the turning points of the OB curve are approached. The role of laser jitter is important because D_r due to laser jitter is usually many orders of magnitude larger than spontaneous smission. Consequently $D_r \gg \lambda_r$ is satisfied farther from the mean field turning points and the discontinuous jumps from one branch of the OB curve to the other will occur farther from the mean field turning points. When we calculate the two time amplitude correlation function from Eq. 3.7 we obtain

$$\frac{\langle \Delta r(\tau) \Delta r \rangle}{r_s^2} = \frac{e^{-\lambda_r \tau}}{2\lambda_r} \left\{ \bar{\Gamma}_\theta \left[\frac{n \hat{\delta}_A}{1 + \hat{\delta}_A^2 + r_s^2} - \delta_F \right]^2 + \bar{\Gamma}_u \left[1 + \frac{n}{1 + \hat{\delta}_A^2 + r_s^2} \right]^2 \right\} \quad (3.8)$$

where $\Delta r \equiv r - r_s$. Eq. 3.8 depends on seven dimensionless parameters and will be analyzed in Ref. 7. However, we see the combination of $\bar{\Gamma}_\theta \gg \bar{\Gamma}_u$ and dispersive OB, which are the most frequent occuring experimental situations, combine to lead to large amplitude fluctuations as $\lambda_r \to 0$. We conclude with the observation that we need experimentally determined values or model calculations of the stochastic parameters $\Gamma_\theta, \Gamma_u, t_\theta$ and t_u to determine the magnitude of the effects of driving laser fluctuations induced in OB.

REFERENCES

1. For a general reference see Optical Bistability, edited by C. Bowden, M. Ciftan and H. Robl (Plenum, New York, 1981).
2. (a)R.Bonifacio, M.Gronchi and L.Lugiato, Phys. Rev. A 18, 2266 (1978);(b)P. Drummond and D. Walls, Phys. Rev. A 23, 2563 (1981);(c)G. Agarwal, L. Narducci, R. Gilmore, and D. Feng, Phys. Rev. A 18, 620 (1978);(d)C. Willis and J. Day, Opt. Commun. 28, 137(1979).
3. C. R. Willis, Phys. Rev. A27, 375 (1983).
4. R. L. Stratonovich, Topics in the Theory of Random Noise, (Gordon and Breach, New York, 1963).
5. R. H. Picard and C. R. Willis, Phys. Rev A 8, 1536 (1973).
6. R. Bonifacio and L. A. Lugiato, Lett. al Nuovo Cimento 21, 517 (1978).
7. C. R. Willis to be published.
8. R. Kubo in Fluctuation, Relaxation and Resonance in Magnetic Systems, edited by D. ter Haar (Oliver and Boyd, London, 1962)
9. N. Van Kampen, Stochastic Processes in Physics and Chemistry, (North Holland, Amsterdam, 1981).

SIMULATIONS OF THE STOCHASTIC DYNAMICS OF

SWITCHING FOR OPTICAL BISTABILITY

S.W. Koch, H.E. Schmidt and H. Haug

Institut für Theoretische Physik der
Universität Frankfurt
Robert-Mayer Str.8, D-6ooo Frankfurt-Main
Fed. Rep. Germany

ABSTRACT

Detailed numerical solutions are presented for the dynamics of the fluctuating light-field in a nonlinear resonator showing optical bistability.

The Langevin-equation (1) for the fluctuating light-field E in a semiconductor resonator with a two-photon absorption resonance has been derived in Ref.[1]:

$$\frac{\partial E}{\partial \tau} = \kappa E_0 - E(\kappa + \omega'' |E|^2) + iE(\Delta + |E|^2) + \sqrt{Q/2}\ F(\tau) \ . \qquad (1)$$

All quantities are dimensionless. E_0 is proportional to the amplitude of the incoming field, κ is a damping constant, ω'' is the ratio of imaginary and real part of the dielectric function and Δ is proportional to the detuning. The additive term $F(\tau)$ is the classical noise source representing a Gauss-Markov process. The stationary properties of eq.(1) have been investigated in Ref. [1] and the details of optical bistable operation have been determined. Here, we present numerical solutions of eq.(1) in order to study the dynamic properties of the system, which determine the response to time-dependent incident fields $I_0(\tau) = |E_0(\tau)|^2$. We simulate the noise source $F(\tau)$ using random numbers and obtain the most probable value of e.g. the field intensity in the resonator $<I>$ by averaging over many stochastically independent realizations.[2]

In Figs. (1) and (2) we show as a typical example
the response in intensity and phase to a linearly increasing incident field. Shown are the instantaneous quasistationary solution (dashed line — — in Fig.1) and the
time-dependent solution (dashed lines − − −) of the deterministic part of eq.(1). Furthermore, we plotted a
typical realization of the full stochastic equation (1)
(thin fluctuating curves) and the solution <I> and <φ>
averaged over 1,000 realizations (thick full lines). The
instantaneous solution jumps at τ=32.5 because at this
time the end of the lower branch is reached. The time-
dependent deterministic solution switches even later due
to the inertia of the system. The average intensity <I>
and average phase <φ> show the destabilization of the
off-state due to the fluctuations. Already at τ=25 a
certain percentage of realizations has switched to the
upper branch (one example is shown in Figs. 1 and 2) and
the others follow, after around τ=30 <I> and <φ> reach
their stationary values.

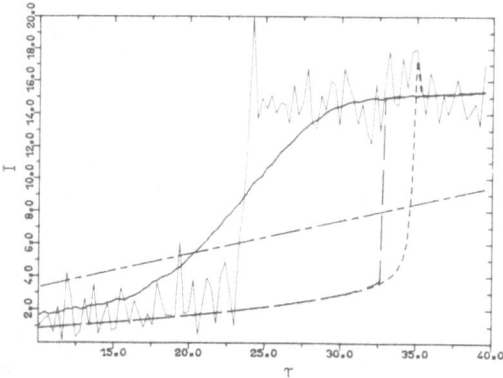

Fig.1: Resulting intensi-
ty for a linearly
increasing inci-
dent intensity I_0
($I_0/100$ is shown
as dashed line
— - —).

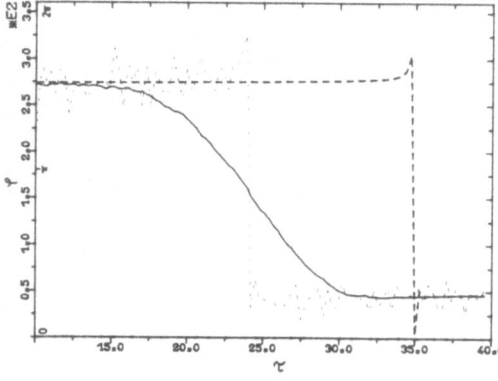

Fig.2: Resulting phase
for a linearly in-
creasing intensity
I_0.

The influence of the noise is also clearly visible in Fig.3.   Here, we plotted the time-development of the field E(τ) as a path in the complex E-plane for an instantaneous switching of I_0 from a value above to a value below the bistable region. The path parameter is the time.   The deterministic solution of eq(1) (dashed line in Fig.3) shows a spiraling approach of the final state, a behaviour which has also been found in Ref.[3].   The inluence of the fluctuations on the spiraling motion is shown by the dotted curve for a single relization.   The full line shows the path averaged over 1000 realizations.   Furthermore, we investigated the stability of the on- and off-state against the noise in the bistable region.   Our results can be fitted by an expression of the form

$$<I(\tau)>I_i e^{-\lambda\tau}+I_f(1-e^{-\lambda\tau}), \qquad (2)$$

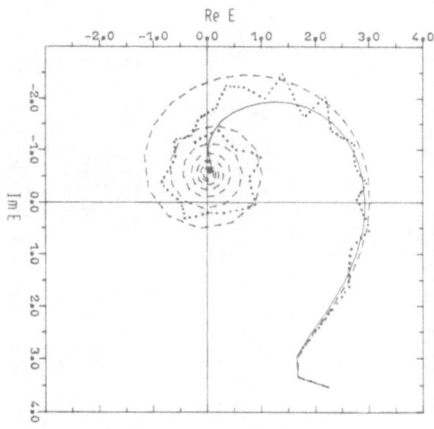

Fig.3: Response to instantaneous down-switching of I_0 (see text)

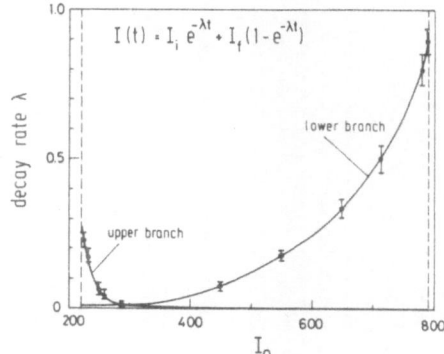

Fig.4: Decay rate λ versus input intensity.

where I_i and I_f are the initial and final values of the intensity in the resonator.   The obtained decay rate of the on- and off-state is plotted in Fig.4 versus the input intensity.   It is seen that under the influence of fluctuations the upper branch is more stable than the lower one due to the fact that the relative strength of the fluctuations in the on-state is much weaker than in the off-state.   For one of these simulations, in which we investigate the decay of a metastable off-state we plot in Fig. 5a,e the time-evolution of the full probability distribution function $f(E,E^*,\tau)$ which has been obtained from 10,000 individual solutions of the stochastic eq.(1). One sees the build-up and decay of the probability current over the potential barrier in the complex E-plane which separates the on- and off-states. Finally, the system has switched to the upper branch and approaches it's stationary solution.

Summarizing our findings, we obtained the detailed time-dependence of the optical bistability described by the stochastic equation (1). The on-state shows a considerably higher stability than the off-state. Thus, the pronounced hysteresis effect predicted by the stationary deterministic solution is not fully destroyed even by relatively large external noise.

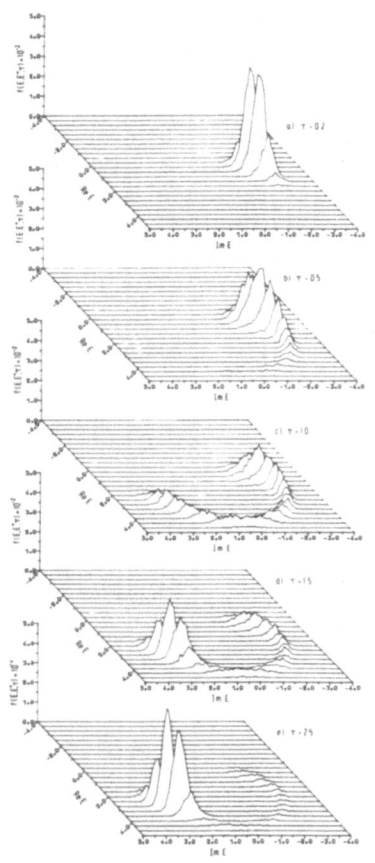

Fig.5: Probability distribution function $f(E, E^*,\tau)$ for different times τ.

References:

1. H. Haug, S.W. Koch, R. Neumann, and H.E. Schmidt, Z. Physik B49, 79 (1982)

2. Fore more details see H.E. Schmidt, S.W. Koch, and H. Haug, Z. Physik B51, 85 (1983)

3. E. Hanamura et al., to be published.

ABSORPTIVE OPTICAL BISTABILITY WITH

LASER AMPLITUDE FLUCTUATIONS

M. Kuś, Institute of Theoretical Physics, Warsaw University, Warsaw 00-681, Poland

K. Wódkiewicz*, Department of Physics and Astronomy
University of Rochester, Rochester, New York 14627, USA

J. A. C. Gallas, Max-Planck Institute of Quantum Optics
D-8046 Garching, Federal Republic of Germany.

It is generally believed that in optical bistability realistic
fluctuations of the injected laser signal can influence the bistable
behavior of the system in an important way. As an illustration of
this general statement, we present in this paper the case of absorptive optical bistability (AOB) with random Gaussian fluctuations of
the driving electric field amplitude.

For such external fluctuations, the microscopic part of the dynamical system can be described by a set of simple macroscopic state
equations which, after a reduction of some of the degrees of freedom, can be converted into a single non-linear state equation which
shows explicitly the bistable behavior of the system.

The theory of AOB in the limit of low-transmission mirrors and
weak enough absorption gives the following well known relation between the dimensionless transmitted x and the incident electric
field y:

$$\frac{dx}{dt} = -(x + \frac{2Cx}{1+x^2}) + y \tag{1}$$

where the time is measured in units of the cavity bandwidth κ and
C is the order parameter of the system.

---

* Permanent address: Institute of Theoretical Physics, Warsaw
  University, Warsaw 00-681, Poland.

A partially coherent laser with a finite bandwidth results in a non-white additive noise in this non-linear macroscopic state equation.

For a laser operating far above threshold the total electric field amplitude $y(t)$ consists of two components: a coherent part A_O and a small fluctuation $\delta y(t)$ which is Gaussian with the following mean value and correlation function:

$$<\delta y(t)> = 0 \qquad <\delta y(t)\,\delta y(t')> = a\ \exp(-\frac{|t-t'|}{b}) \qquad (2)$$

where the dimensionless parameters a and b describe the characteristic properties of the fluctuating laser amplitude. The parameter $b = \kappa \cdot \tau_c$ is the coherence time τ_c of the laser amplitude fluctuations in units of the cavity linewidth and the coefficient a measures the intensity of the noise which in most realistic experiments is $a \leq 0.1$, i.e. a better than 10% stabilization of the amplitude can be obtained.

With such amplitude fluctuations the AOB equation (1) takes the form of the following nonlinear Langevin equation with an additive Orsnstein-Uhlenbeck stochastic process:

$$\frac{dx}{dt} = F(x) + \delta y(t) \qquad (3)$$

where $F(x)$ is the deterministric part of the dynamical evolution 1 given by:

$$F(x) = -\ (x + \frac{2Cx}{1+x^2}) + A_O\ . \qquad (4)$$

In Ref. 1 we have established for the stochastic equation (4) a proper Fokker-Planck equation in the limit of large laser linewidth ($b < 1$) and good stabilization of the amplitude fluctuations ($a < 1$). For times larger than the transients $t \sim b$, this Fokker-Planck equation for the probability distribution of the transmitted field takes the following form:[1]

$$\frac{\partial}{\partial t}\ P = -\ \frac{\partial}{\partial x}\ F \cdot P + D\ \frac{\partial^2}{\partial x^2}\ K \cdot P \qquad (5)$$

where the nonconstant diffusion function $K(x)$ has the following form:

$$K(x) = (1 + b\ F'(x)) = 1 - b - 2bC\ \frac{1-x^2}{(1+x^2)^2} \qquad (5)$$

and where $D = a \cdot b$.

In the white-noise limit, i.e. if $b \to 0$ with $D = ab = $ constant, we have $K \to 1$ and the diffusion term takes the well-known constant form. In this case D plays the role of the diffusion constant. In general, i.e. for $b \neq 0$, the diffusion function 6 depends on the laser linewidth b. This dependence is shown explicitly in Figure 1.

In order to describe a proper Fokker-Planck equation, the diffusion function given by Eq. (6) must be positive. This condition is fulfilled for all values of x only if $b < 1$, i.e. for values of b for which Eq. (5) should hold.

From the Fokker-Planck equation (5) we derive the following form of the stationary solution $(\frac{\partial}{\partial t} P_{st} = 0)$, assuming natural boundary conditions:

$$P_{st} = N \exp \left(- \frac{U(x)}{D}\right) \tag{7}$$

where

$$U(x) = - \int dx \, \frac{F(x)}{1+bF'(x)} + ab \, \ln|1+bF'(x)| \tag{8}$$

and N is a normalization constant.

States of maximal probability are characterized by the absolute minima of the thermodynamical potential $U(x)$. In Figure 7 we have shown the form of this thermodynamical potential for various values of the incident field A_0 and for two values of b. It is clear from these curves that the depth and the width of the bistable minima depend on the laser parameters a and b.

The most probably values of P_{st} given by Eq. (7) lead to the following steady-state relation $(U'(x) = 0)$:

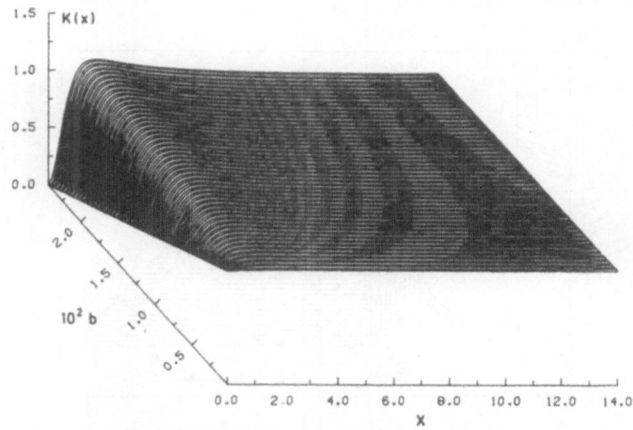

Fig. 1.   Curves of the diffusion function $K(x)$ for various values of x and laser bandwidth b.

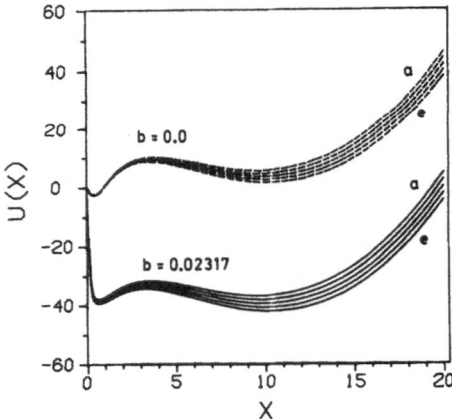

Fig. 2. These two sets of curves show the shape of the thermodynamical potential U(x) for two values of b and for the input
field A_O equal to : 13.7, 13.8, 13,9, 14.0, 14.1 (a-e).
The dotted lines correspond to the white-noise case given
by b=0. All these curves were calculated for C=20 and
a=0.4.

$$A_O = x + \frac{2Cx}{1+x^2} + 4ab^2 C \frac{3-x^2}{(1+x^2)^3} \tag{9}$$

Eq. (9) can be regarded as a generalization of the deterministic bistability condition for the case of laser amplitude fluctuations.

The bistable behavior of the system with laser amplitude fluctuations can be illustrated in Fig. 3 where we have plotted the stationary probability given by Eq. (7) for different values of the coherent laser field A_O.

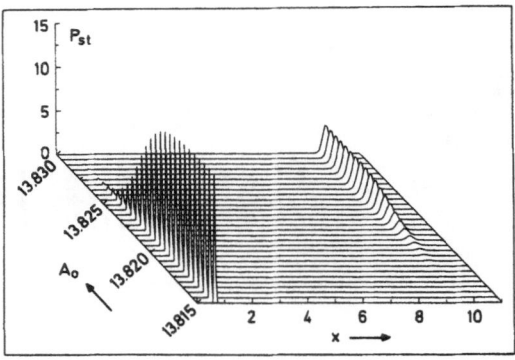

Fig. 3. Development of $P_{st}(x)$ as a function of the injected laser
amplitude A_O for C=20, a=0.4, and b= 0.02317.

As in the case of quantum fluctuations[2] the random amplitude of the laser field leads to a small range of values of A_0 in which the two peaks have a comparable area. Clearly the mean value of the transmitted field will coincide with one of the two deterministic branches except in this narrow transition region where we have large fluctuations.[1]

ACKNOWLEDGMENTS

One of the authors (K.W.) would like to thank Professor H. Walther for his invitation to the Max-Planck Institute of Quantum Optics where a large part of this work has been done. Many discussions with Dr. P. Meystre are also acknowledged.

REFERENCES

1.   M. Kuś, K. Wódkiewicz, and J. A. C. Gallas, Phys. Rev. A 28, 314 (1983).
2.   See P. D. Drummond, this volume and references therein.

ROOM TEMPERATURE BISTABILITY, LOGIC GATE OPERATION, INCOHERENT

SWITCHING AND HIGH SIGNAL GAIN WITH InSb DEVICES

S.D. Smith and F.A.P. Tooley

Physics Department, Heriot-Watt University
Riccarton, Edinburgh EH14 4AS
U.K.

ABSTRACT

We present details of the performance of several devices utilising the giant nonlinear refractive index of InSb (n_2 = 0.1 cm^2/kW at 77 K and 5.5 μm; n_2 = 10^{-4} cm^2/kW at 300 K and 10.6 μm). Optical bistability was observed at 300 K using a CO_2 laser with intensities of order 100 kW/cm^2 and switching times 5-30 ns. At 77 K and 5.5 μm devices could be operated with cw holding beams combined with time-varying switching or signal pulses. The 'all-optical circuit elements' demonstrated are: (i) AND, OR, NAND or NOR gates in which 'switch on' of the function is inferred to be on a picosecond timescale and 'switch off' of the order of carrier recombination time; (ii) the switching of one optical logic gate by another; (iii) switching of a bistable device in both directions with an incoherent beam; (iv) two beam signal amplification in a transphasor of order 1.3 x 10^4.

Recent experiments utilising the giant nonlinear refractive
index of InSb have significantly extended the range of temperature,
wavelength and laser intensity for which operation of devices based
on optically bistable InSb resonators is possible and potentially
useful results for signal processing have been obtained.

Our earliest observations of nonlinear effects[1],[2] in InSb were
at cryogenic temperatures and are shown to be due to band-gap resonant saturation of states immediately above the energy gap[3] where
carrier excitation is presumed to be from band-tail absorption processes. These observations have been extended to 300 K[4] where, with
the band-gap now at 7.3 µm, the CO_2 laser can be used in the 9.6-
10.6 µm region to operate optically bistable devices. Illuminating
a 250 µm thick uncoated parallel-faced crystal with a peak intensity
I ∿ 100 kW/cm^2 from a SLM injection-locked TEA-CO_2 laser produced an
optical thickness change greater than $\lambda/2$ in a dynamic measurement
(time resolution ∿ 1 ns) during a 200 ns pulse, implying that n_2
is of order 10^{-4} cm^2/kW at 100 kW/cm^2. The intensity dependence of
n_2 is consistent with a two-photon carrier excitation process. At
this intensity the absorption coefficient α is of order 10 cm^{-1},
consistent with a carrier density ∿ 1 x 10^{16} cm^{-3}. Optical
hysteresis is shown in Fig. 1, with switch-on times ∿ 5 ns and

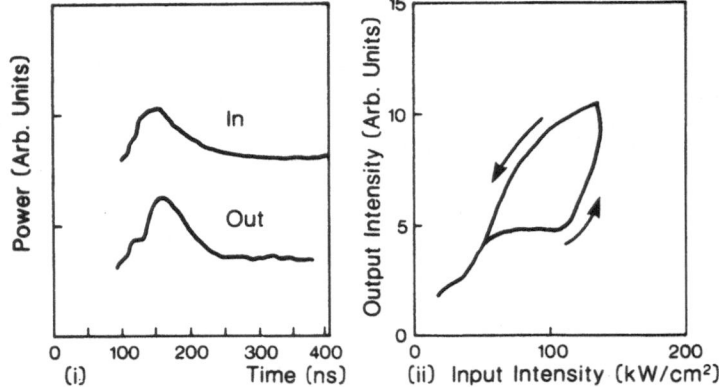

Fig. 1.   (i) Shape of TEA-CO_2 laser pulse with incident intensity
           ≃ 200 kW/cm^2 (upper trace) showing effect of transmission
           (lower trace) through an InSb cavity.
           (ii) Relation between instantaneous incident and transmitted
           intensities showing hysteresis.

switch-off $\sim$ 30 ns, the latter corresponding to carrier lifetime effects. Extension of the experiment to longer pulses ($\sim$ 1 μs) with a pulsed cw CO_2 laser confirms optically bistable switching.

Analysis of the magnitude of n_2 according to equation (3) of reference 4 suggests that, under these conditions, the contribution to n_2 from band-gap resonant blocking of conduction states is greater than that from free carrier plasma effects by a factor $\sim$ 3.

At 77 K with the band-gap at 5.4 μm we have used the CO line at 5.5 μm (1819 cm^{-1}) from an Edinburgh Instruments PL3 laser to demonstrate several 'all-optical circuit elements'. The value of n_2 is $\sim$ 0.1 cm^2/kW and the absorption coefficient $\alpha \sim$ 10 cm^{-1} which facilitates 'cw holding' over a range of input intensities from 10 W/cm^2 to 300 W/cm^2 in a 200 μm thick InSb resonator. Over this range, optically bistable behaviour is observed in three orders, as shown in Fig. 2. This is to be compared with the intensity of order 20 kW/cm^2 previously reported for GaAs or GaAs-GaAlAs device

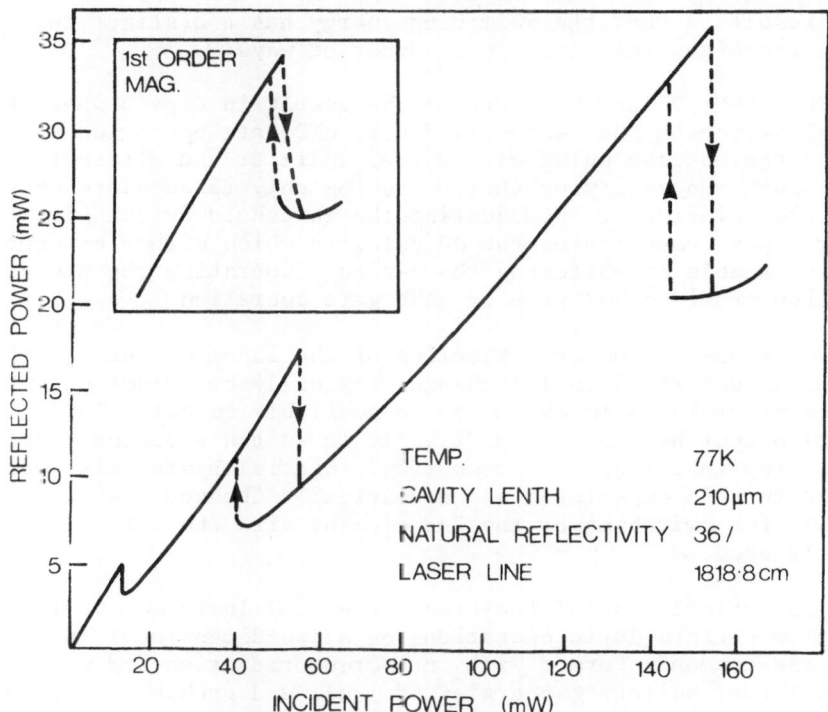

Fig. 2.   Reflected power plotted against incident power from an
          InSb cavity showing three successive interference orders
          displaying bistability.

operation[5] which, combined with larger absorption ($\alpha \sim 10^4$ cm^{-1}), gives consequent difficulty in 'cw holding'.

The InSb/CO laser system seems to be currently unique in that 'cw holding' can be attained indefinitely and the device switched with address or signal pulses down to picosecond time scales.

We have demonstrated logic gate[6] operation in this manner. With a 200 μm thick InSb resonator illuminated over a 200 m ($1/e^2$) diameter the sample was accurately positioned at the input beam waist. Using milliradian parallelism of the faces the resonator could be tuned by translation and thus a convenient initial detuning set to show bistability in 1st (between 11-12 mW input power), 2nd (40-55 mW) and 3rd (145-155 mW) orders. The device could be switched between two stable output states in each order by an external pulse. This was initially provided by Nd:YAG mode-locked laser from which a single switched-out 35 ps pulse was derived. The switching, and single pulse detection, shown in Fig. 3, required an energy of $\sim$ 5 nJ over the 200 μm diameter. When the difference between cw power required to 'switch' and to 'hold' was minimised, the external switching energy was likewise at a minimum. The important experimental result is that the switching energy has a distinct threshold; we have exploited this fact in a number of ways:

(i) AND, NAND, OR and NOR gates - the result in Fig. 3 showing external switching has been extended to AND gate operation by dividing the address pulse with a beam splitter and alternative optical path and verifying that switching only takes place when both pulses are coincident. Readjusting the threshold by changing the holding power demonstrates the OR gate, in which either external pulse is capable of switching the device. Operating the device in reflection provides NOR gate or NAND gate operation.

(ii) The transmission or reflection of the input cw beam in the experiments described in (i) changes typically by 30-40% on switching thus providing $\sim$ 10 mW change in available output. This (transmission) output has been used directly to switch a second logic gate - a first step in all-optical computing. Spatial hysteresis[7] was observed in this experiment and dramatically changed the power available for switching of the 2nd element with the optical system presently used.

(iii) Potentially useful features of optical logic may be its ability to perform logic operations on a two-dimensional image. Since images are commonly formed with incoherent radiation and with the knowledge that switching energies of $\sim$ nJ, and probably much less in future, are practicable, we investigated optically bistable switching using white light incoherent sources. The first source used was a common photographic flash lamp (Sunpak 300) providing a 4 ms long pulse with 80% of the energy radiated during the first 1 ms. With

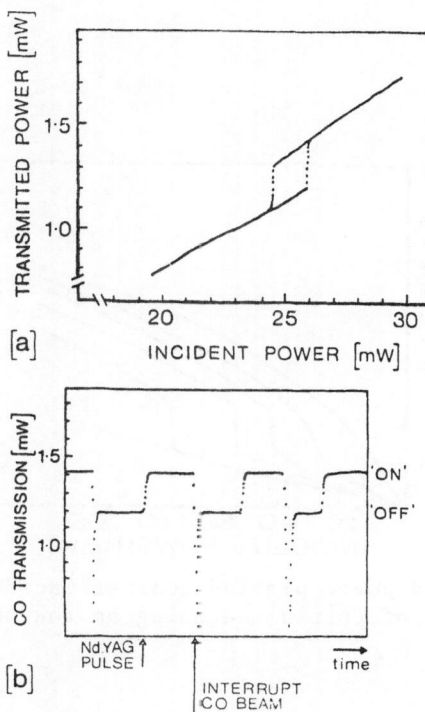

[a]

[b]

Fig. 3.   (a) Transmitted power plotted against incident power for
          a cw CO laser beam (5.5 μm, spot size ∿ 200 μm) passing
          through an InSb cavity at 77 K.
          (b) Transmission of cw CO laser beam showing 'switch-on'
          caused by a single 35 ps pulse. Switch-off is caused
          here by interrupting the CO holding beam.

a colour temperature of 5500 K an effective unfocused portion of the
output amounting to ∿ 1 μJ reached the InSb sample through ZnSe and
perspex windows. With the resonator adjusted by variation of holding
power to be as near as possible to 'switch-up' point, the device could
be reproducibly switched 'on' by one pulse from the flash lamp. On
adjusting the bistable device to be on the upper branch and near
'switch down' point the flash lamp caused the device to switch down.
Since the latter required an increase in refractive index, we can
presume that a heating effect is occurring in contrast to the negative
electronic contribution to n_2. Results were repeated using a Xenon
lamp with the device continually switched up and down using a 30 Hz
chopper and a narrow bistable loop. Thus information impressed in
incoherent light has been transferred purely optically onto a coherent laser beam and incoherent switching of an optically bistable

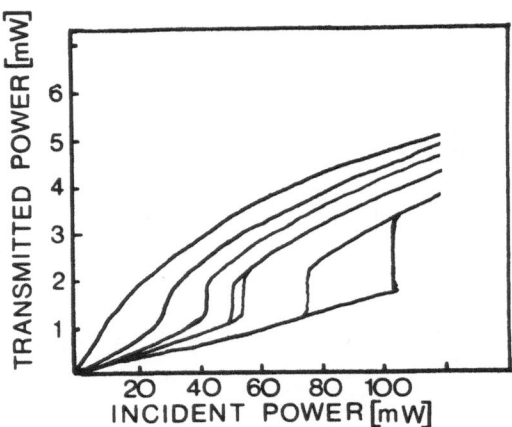

Fig. 4. Transmitted power plotted against incident power showing
         the effect of initial detuning on the observed character-
         istic.

device achieved, to the best of our knowledge, for the first time.

(iv) Transphasor action, in a 4K InSb resonator, was first reported
by workers at this laboratory in 1979. The maximum signal gain
observed was around 10. More recent work, at 77 K, has achieved
spectacular improvements in device operation with the observation
of gain of up to 13×10^4.

An InSb resonator was set with initial detuning adjusted so that
the characteristic contains a differential gain region, where the
output changes by more than the input. Figure 4 shows how the
characteristic changes as the detuning is varied from 0 to π. The
CO laser 'holding' power was set just below the differential gain
region and a second beam introduced, via a beam splitter, along the
same path as the holding beam. This second beam provides a small
modulated signal. When the peak signal power is sufficient, greater
than a few microwatts in practice, the modulation of the signal is
transferred with gain to the reflection and transmission of the

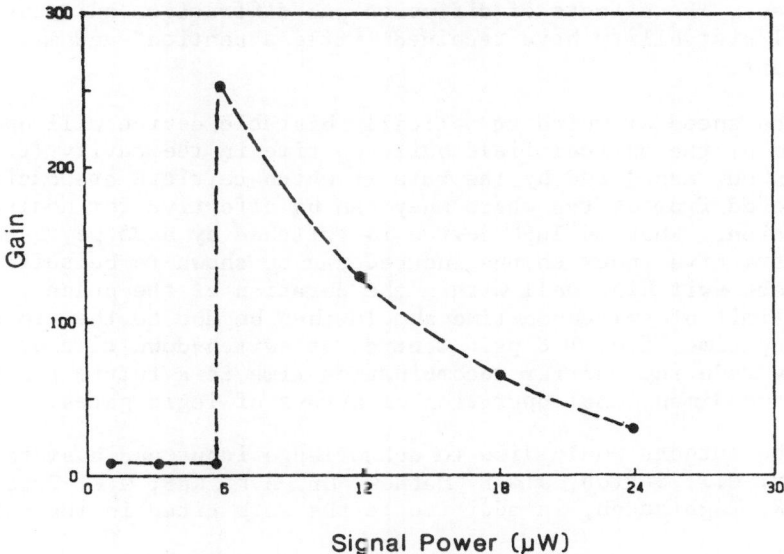

Fig. 5. Gain of an InSb resonator plotted against incident signal
         power.

holding beam from the resonator. Figure 5 shows a plot of this
output as a function of the amplitude of the intensity modulation
of the signal beam. This is a simple method of exploring the shape
of the characteristic, maximum gain corresponding to the steepest
slopes, while limiting action is shown for regions above the 'step'.
By setting the holding power at the midpoint of the differential gain
region and limiting the signal amplitude, we expect this device to
show linear gain and good reproduction of the signal waveform.

     In conclusion, our results show quantitatively how accurately
it is possible to hold a bistable device at any convenient point on
its characteristic. Several novel operations of optical bistable
devices have been demonstrated utilising the ability to hold within
microwatts of the switching points.

     In the above experiments, it has become apparent that the
whole beam is switched or modulated. Since a Gaussian cross-section
holding beam is used, there is a range of intensities across this
beam. In the light of the fact that the observed characteristic
should contain contributions from all parts of the beam, it is
surprising that Gaussian beam optical bistability is so readily
obtainable and that such steep switching and amplifying regions are

obtained. The effects of diffusion and diffraction on Gaussian beam optical bistability have received little attention[8] and may be important.

The speed at which an optically bistable device will operate is limited by the optical field build-up time in the cavity (typically 8 ps in our case) and by the rate at which carriers are excited to or removed from states where they can be effective for nonlinear refraction. When an InSb device is switched by a 35 ps, 5 nJ pulse the refractive index change induced can be shown to be sufficient to effect switching well within the duration of the pulse[6]. The upper limit of switch-on time should then be due to the field build-up time, i.e. $\sim$ 8 ps. Control of switch-down time by artificially reducing carrier recombination time is a future possibility as is two-dimensional operation of arrays of logic gates.

The authors would like to acknowledge input and assistance from Dr. C.T. Seaton, J.G.H. Mathew, Dr. A.K. Kar, M.E. Prise and Dr. M.R. Taghizadeh, in addition to the work cited in the references.

References

1.   D.A.B. Miller, M.H. Mozolowski, A. Miller and S.D. Smith, Optics Commun., 27:133 (1978).
2.   D.A.B. Miller, S.D. Smith and A. Johnston, Appl. Phys. Lett., 35:658 (1979).
3.   D.A.B. Miller, C.T. Seaton, M.E. Prise and S.D. Smith, Phys. Rev. Lett., 47:197 (1981).
4.   A.K. Kar, J.G.H. Mathew, S.D. Smith, B. Davis and W. Prettl, Appl. Phys. Lett., 42:334 (1983).
5.   H.M. Gibbs, S.L. McCall, T.N.C. Venkatesan, A.C. Gossard, A. Passner and W. Wiegmann, Appl. Phys. Lett., 35:451 (1979).
6.   C.T. Seaton, S.D. Smith, F.A.P. Tooley, M.E. Prise and M.R. Taghizadeh, Appl. Phys. Lett., 42:131 (1983).
7.   W.J. Firth, C.T. Seaton, E.M. Wright and S.D. Smith, Appl. Phys. B28:131 (1982).
8.   W.J. Firth and E.M. Wright, Optics Commun., 40:233 (1982); J.V. Moloney and H.M. Gibbs, Phys. Rev. Lett., 48:1607 (1982).

ADVANCES IN GaAs BISTABLE OPTICAL DEVICES

J. L. Jewell, S. S. Tarng, H. M. Gibbs, K. Tai,
D. A. Weinberger, and S. Ovadia

Optical Sciences Center
University of Arizona
Tucson, Arizona 85721
and
A. C. Gossard, S. L. McCall, A. Passner,
T. Venkatesan, and W. Wiegmann
Bell Laboratories, Murray Hill, New Jersey 07974

ABSTRACT

The achievements in GaAs bistable optical devices of room-temperature operation, external switch-on and switch-off, and advances in device fabrication are described.

INTRODUCTION

Bistable optical devices (BOD's) using GaAs as the nonlinear medium are viable candidates for the achievement of small, fast (approximately nanosecond), room temperature,[1] low power (milliwatts), externally controllable[2] optical switches that are easily fabricated[3] and operated. Advances have been made in all of these areas and in our understanding of the physical processes involved.[4] Efforts are in progress to improve performances in ways that are simultaneously compatible.

SWITCHING

When a 1-μs triangular-shaped input pulse is used, both switch-on and switch-off take about 40 ns, whether the sample is bulk GaAs at ≈80K or a GaAs-AlGaAs multiple quantum well (MQW) structure[5] at room temperature. The MQW devices were also investigated with shorter input pulses resulting in faster switching. A 13-ns input yielded subnanosecond detector-limited switch-on, and the switch-off had a fall time of ≈ 2 ns (Fig. 1). This observed fall time is significantly shorter than the probable carrier lifetime τ_c of about 10 to 20 ns.[6] There are at least two explanations. (1) Excited carriers diffuse from the illuminated region in much less than τ_c. This is an interesting possibility because if the diffusion in our MQW's is as large as that measured in bulk GaAs by Olsson,[7] we may be exciting a much larger region than we illuminate. Thus our intensity (optical power divided by focused spot area) requirements could be misleading, appearing too high. Diffusion can also induce crosstalk between switching elements unless inhibited by barriers such as an etched surface or a high density of recombination sites. (2) If the Fabry-Perot device is initially detuned by more than an instrument width, the transmission peak can scan across the laser wavelength in less than τ_c. The observed fall time is merely the time it takes for the peak to scan about one instrument width (high-to-low transmission) and depends heavily on device and operating characteristics.

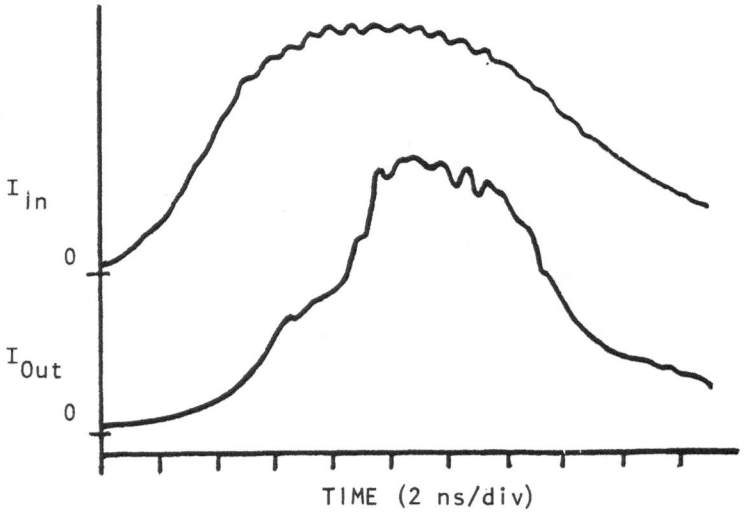

Fig. 1. Subnanosecond switch-on and 2-ns fall time from a 13-ns FWHM E-O modulated pulse. The input and output were synchronized by matching the oscilloscope traces with no GaAs in the beam.

While triangular input pulses are useful for demonstrating bistability and qualitative evaluation of devices, they do not simulate probable working conditions for an optical switch. When the BOD is externally switched, the energy necessary for switch-on can be delivered in 1 ps or less. In the bulk GaAs device at 80K, switch-on occurred faster than the detector limit of 200 ps for a 600-nm, 10-ps, 1-nJ switch pulse. The device was also externally switched off in ≈20 ns by a 7-ns, 600-nm, 300-nJ pulse. This kind of switching off was a brute force technique in that the large pulse energy heated the etalon, shifting the hysteresis loop to higher intensities until the switch-off intensity surpassed the constant input level. Both switch-on and switch-off were performed on a single input pulse[2] as shown in Fig. 2. More recently, only 10 pJ of energy in a ≈10-ps, 800-nm pulse switched on a MQW device at room temperature.

OPTICAL FLIP-FLOP

A triangular input may also yield a somewhat lengthened switch-off since there is still a significant amount of incident light during the transition. To obtain the fastest return to the original "off" state, the input should drop suddenly and stay low until the nonlinear medium has recovered sufficiently for switch-off. A straightforward way to achieve this without affecting the light source is to place another nonlinear Fabry-Perot in front of the BOD (Fig. 3). When initially tuned for high transmission, the negative optical gate (NOG) will transmit the input beam until it is hit by a pulse similar to the BOD's switch-on pulse. In this case, however, the NOG's transmission peak is pushed away from the laser wavelength until the medium recovers. If the "off" time is longer than the recovery time of the BOD, then the BOD will have switched off and can be switched back on any time after the NOG has recovered. Thus the two devices working together form an optical flip-flop. Although this configuration has not yet been tried with GaAs, it was demonstrated with dye etalons[3] and should be realizable for most nonlinear materials.

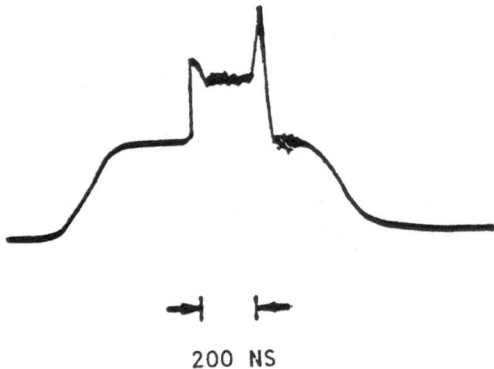

200 NS

Fig. 2. Light-by-light control as a BOD is switched on and off while the input remains constant.

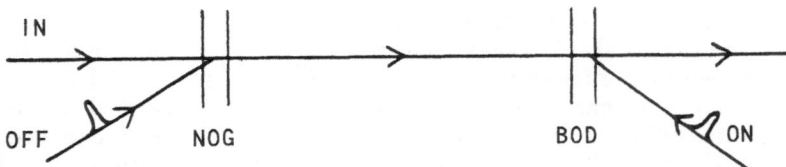

Fig. 3. An optical flip-flop formed from two nonlinear Fabry-Perot etalons in series. The first etalon transmits the input beam acting as a negative optical gate (NOG) in that an incident pulse creates a momentary drop rather than increase in transmission.

ROOM TEMPERATURE

    Room-temperature bistability with modest input power was first
achieved in a MQW structure consisting of 61 periods of GaAs 336 Å
thick and 401 Å AlGaAs layers.[1] Removal of the device from the
liquid nitrogen dewar allowed easier access to the device and
improved focusing with microscope objectives. The contrast and
difference between switch-on (I_{ON}) and switch-off (I_{OFF}) intensities
were then much better than had been observed previously (Fig. 4a).
With this improvement we were immediately able to see room-
temperature bistability in the bulk device although the hysteresis
loop was quite narrow. Wide hysteresis loops from bulk GaAs (Fig.
4b) have been observed recently; however, more power was required to
achieve a response similar to that of the MQW devices. The bulk
device had a 7-μm layer of GaAs compared to only 2-μm in the MQW.
Other factors such as material growth conditions and device
construction help complicate an objective comparison; however the
MQW's have generally looked "better" in terms of performance and
optical spectrum at room temperature.[6]

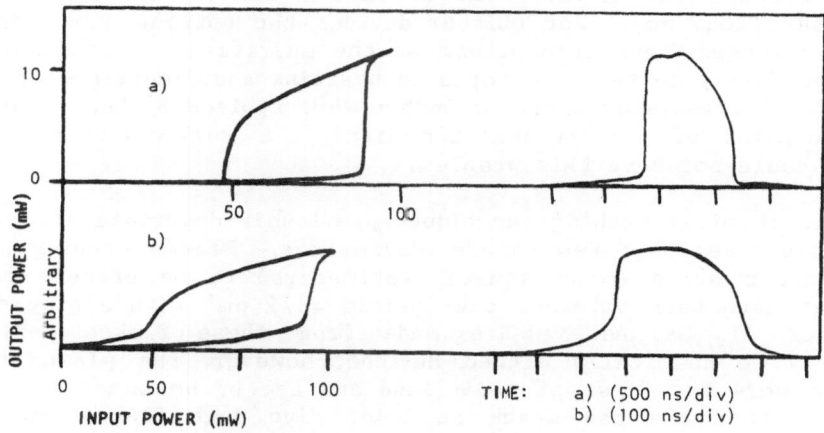

Fig. 4. Room-temperature bistability in a) 336Å MQW and b) bulk GaAs
devices.

FABRICATION

We can now reliably fabricate GaAs BOD's,[3] and there is much room for development in this area. The GaAs-AlGaAs structures are grown on a GaAs substrate by molecular beam epitaxy.[5] This process allows tremendous versatility in material properties as the layer thicknesses and/or doping schemes are varied. As before, a combination of grinding and selective etching removes the substrate leaving the ≈3-μm-thick MQW. Through the use of multiple etch-stop-layers and by etching the entire substrate off each sample, flatnesses of less than λ/4 over a few millimeter square area are obtained. The resulting flake can then be left whole or cleaved into two or three pieces to be studied directly or implemented into a BOD. Many BOD's have been easily built by cementing a flake between dielectric mirrors. These devices, whether the quantum-well thicknesses are 66Å, 152Å, or 336Å, exhibit bistability (room-temperature) with power levels down to around 5 to 10 mW focused to ≈12-μm diameter. They also show regenerative pulsations[9] due to competing excitonic and thermal effects similar to the low-temperature GaAs. Samples thicker than ours (a few micrometers) might be less sensitive thermally to the energy absorbed since it is distributed over a larger volume. A device consisting of three of the 336Å MQW flakes stacked and cemented between 90% mirrors showed forward bistability ($I_{ON} > I_{OFF}$) for on-times up to 300 μs or about 10 times longer than the best such response in single-flake devices. Five flakes stacked between mirrors also showed bistability although with reduced performance. The optimum operating wavelength for a single flake (336Å wells) is at or beyond the current limit of our dye laser (≈885 nm). For thicker devices the optimum wavelength is probably longer (less absorption), so the multiflake devices have not yet been fairly tested. Attempts to heatsink a dielectric-coated MQW device with transistor paste or indium were spoiled by heat generated in absorption of the transmitted light. A well-polished diamond wafer should not have this problem.

The chemical etching techniques previously described are limited to sample sizes of a few square millimeters. Plasma etching allows much larger areas (many square centimeters) to be etched and λ/4 surfaces have been obtained over pieces ≈1/2 cm² with a single stop layer of $Al_{0.3}Ga_{0.7}As$. BOD's made from these flakes performed similarly to chemically etched devices; however, the plasma-etched samples were the first of their kind and are by no means perfected. The fact that one can stack at least five such flakes and still achieve good bistability indicates that the surfaces must make good contact and therefore be quite smooth. With the capabilities of plasma to etch vertical features on a surface,[10,11] one can envision large arrays of circular or rectangular features each being a discrete element with crosstalk virtually eliminated.

There is clearly much room for improvement in GaAs BOD's and when speed, size, power, and temperature are considered, they appear to be promising candidates for practical optical switching.

ACKNOWLEDGMENT

The Arizona part of this work was supported by the Air Force Office of Scientific Research, the Army Research Office, and the National Science Foundation.

REFERENCES

1. H. M. Gibbs, S. S. Tarng, J. L. Jewell, D. A. Weinberger, K. C. Tai, A. C. Gossard, S. L. McCall, A. Passner, and W. Wiegmann, "Room-temperature excitonic optical bistability in a GaAs–GaAlAs superlattice etalon," Appl. Phys. Lett. 41:221 (1982).

2. S. S. Tarng, K. Tai, J. L. Jewell, H. M. Gibbs, A. C. Gossard, S. L. McCall, A. Passner, T. N. C. Venkatesan, and W. Wiegmann, "External off and on switching of a bistable optical device," Appl. Phys. Lett. 40:205 (1982).

3. J. L. Jewell, H. M. Gibbs, A. C. Gossard, A. Passner, and W. Wiegmann, "Fabrication of GaAs bistable optical devices," Matl. Lett. 1:148 (1983).

4. K. C. Tai, J. V. Moloney and H. M. Gibbs, "Optical crosstalk between nearby optical bistable devices on the same etalon," Opt. Lett. 7:429 (1982). See also K. Tai, H. M. Gibbs, J. V. Moloney, S. S. Tarng, J. L. Jewell, and D. A. Weinberger, "Self–defocusing and optical crosstalk in a bistable optical etalon," Topical Meeting on Optical Bistability, Rochester, New York, 1983.

5. A. C. Gossard, "GaAs/AlAs layered films," Thin Solid Films 57:3 (1979).

6. D. A. B. Miller, D. S. Chemla, D. J. Eilenberger, P. W. Smith, A. C. Gossard, and W. T. Tsang, "Large room–temperature optical nonlinearity in $Ga_{1-x}Al_xAs$ multiple quantum well structures," Appl. Phys. Lett. 40:291 (1982).

7. A. Olsson, D. J. Erskine, Z. Y. Xu, A. Schremer, and C. L. Tang, "Nonlinear luminescence and time–resolved diffusion profiles of photoexcited carriers in semiconductors," Appl. Phys. Lett. 41:659 (1982).

8. M. C. Rushford, H. M. Gibbs, J. L. Jewell, N. Peyghambarian, D. A. Weinberger, and C.-F. Li, "Observation of thermal optical

bistability, crosstalk, regenerative pulsations, and external switch-off in a simple dye-filled etalon," Topical Meeting on Optical Bistability, Rochester, New York 1983.

9. J. L. Jewell, H. M. Gibbs, S. S. Tarng, A. C. Gossard, and W. Wiegmann, "Regenerative pulsations from an intrinsic bistable optical device," Appl. Phys. Lett. 40:291 (1982).

10. K. Hikosaka, T. Mimura, and K. Joshin, "Selective dry etching of AlGaAs-GaAs heterojunction," Japan. J. Appl. Phys. 20:L847 (1981).

11. E. L. Hu and R. E. Howard, ""Reactive-ion etching of GaAs and InP using $CCl_2F_2/Ar/O_2$," Appl. Phys. Lett. 37:1022 (1980).

QUANTUM THEORY OF EXCITONIC OPTICAL BISTABILITY

M.L. Steyn-Ross

Physics Department
York University
Toronto,Ontario M3J 1P3

C.W. Gardiner
University of Waikato
Hamilton, New Zealand

1. INTRODUCTION

We present a fully-quantum-mechanical theory of the intracavity interaction of coherent light with semiconductors. It is assumed that the interaction occurs via excitons and discussion is limited to a two-band semiconductor. The interaction between the light field and the semiconductor excites an electron from the filled valence band to the conduction band and thus creates an exciton.

Using a master equation approach we include such effects as exciton-lattice and exciton-exciton interactions. In the two cases of high and low exciton densities, steady-state analysis reveals bistability and hysteresis in the system.

2. MODEL

We consider the intracavity interaction of a coherent driving field with a two-band semiconductor.

In the framework of a second quantised theory we expand the Hamiltonian for the system in terms of field operators,

$$\psi(\underline{x}) = \sum_k a_k \phi_k(\underline{x}) + \sum_k d_{k'} \phi_{k'}(\underline{x}) \qquad 2.1$$

where the $\phi_k(\underline{x})$, $\phi_{k'}(\underline{x})$ are eigenfunctions of the single particle

Hamiltonian; a_k is the fermion destruction operator for a conduction electron with momentum k, and $d_{k'}$ is the fermion destruction operator for a valence hole with momentum k'.

In this manner, we find the Hamiltonian for the light-semiconductor system in momentum space [1],

$$H = \sum_{p_1} E_c(p_1)\, a_{p_1}^+ a_{p_1} - \sum_{p_2} E_v(p_2)\, d_{p_2}^+ d_{p_2}$$

$$+ \tfrac{1}{2}\sum_{\substack{p_1,p_2,q \\ p_1',p_2'}} V(q)\, (a_{p_1+q}^+ \; a_{p_1'-q}^+ \; a_{p_1'} a_{p_1} + d_{p_2}^+ d_{p_2'}^+ \; d_{p_2'-q} \; d_{p_2+q}$$

$$- 2a_{p_1+q}^+ \; d_{p_2'}^+ \; d_{p_2'-q} \; a_{p_1})$$

$$+ \omega_2 \hat{a}^+ \hat{a} + \hbar \Big[\sum_{p_1 p_2} a_{p_1}^+ \; d_{p_2}^+ \; \hat{a}\, g + \text{H.c.}\Big] \qquad\qquad 2.2$$

$$+ i\hbar \Big[E \exp(-i\omega_L t)\, \hat{a}^+ - \text{H.c.}\Big] + \text{excitonic and E.M. damping}$$

terms.

Where $E_c(p_1) = \dfrac{\hbar^2 p_1^2}{2m_e} + Eg$ (m_e is the effective electron mass and Eg the band gap energy).

and $E_v(p_2) = -\dfrac{\hbar^2 p_2^2}{2m_h}$ (m_h is the effective hole mass)

and $V(q) = \dfrac{4\pi e^2}{\varepsilon_o q^2}$ (ε_o is the static dielectric constant).

The first two terms in Eq. (2.2) describe the free Hamiltonian of the two-band semiconductor. The third term describes interparticle interactions within the semiconductor: the first term inside the brackets represents electron-electron interactions, the second term concerns hole-hole interactions and the third term describes electron-hole interactions.

The fourth term of Eq. (2.2) describes the free Hamiltonian for the light field, whose boson operator is given by $\hat{a}$.

The fifth term represents the interaction between light and semiconductor. In deriving this term, we have assumed it is possible to single out a particular mode of the light field; thus g, the coupling constant, does not depend on p_1 and p_2.

The sixth term in the Hamiltonian describes coupling between the external driving field (amplitude E, frequency ω_L) and the field mode.

Finally in the Hamiltonian we refer to damping terms, the form of which will be discussed later.

3. TRANSFORMATION TO BOSON SYSTEM: LOW DENSITY CASE

So far, all discussion has been in terms of electrons and holes. However, as our aim is to formulate a theory of light-exciton interaction we need to express the Hamiltonian in terms of exciton operators. Also, we wish to derive a master equation for the system to systematically describe many-body effects and damping. This requires the Hamiltonian to be written in terms of exciton operators which have a closed set of commutation relations. To this end, we transform fermion pair operators to boson operators, using a modified version of the transformation due to Marumori et al [2],

$$U = |0\rangle \langle 0| \sum_{n=0}^{\infty} \frac{1}{n!} \frac{1}{\sqrt{n!}} \left[\sum_{\alpha\beta} b_{\alpha\beta}^{+} d_{\beta} a_{\alpha} \right]^{n} |0) (0| \qquad 3.1$$

where

$b_{\alpha\beta}^{+}$ = exciton (boson) creation operator, $|0)$ = boson vacuum

a_{α}, d_{β} = electron, hole destruction, $|0\rangle$ = fermion vacuum.
                         operators

The unitary transformation, Eq. (3.1) takes account of antisymmetrisation effects - encountered when mapping from a fermion to a boson subspace. Such effects introduce extra nonlinearities into the Hamiltonian, and ensure the exclusion principle is not violated at high exciton densities. An important case of the transformation is,

$$U \, a_{\alpha}^{+} \, d_{\beta}^{+} \, U^{+} = \left[b_{\alpha\beta}^{+} - \sum_{\gamma\eta} b_{\alpha\gamma}^{+} b_{\beta\eta}^{+} b_{\gamma\eta} \right] \left[1 + \sum_{\alpha\beta} b_{\alpha\beta}^{+} b_{\alpha\beta} \right]^{-\frac{1}{2}} \hat{p} \qquad 3.2$$

where $\hat{p}$ is a projection operator, equal to unity in the boson subspace.

We note that on applying the transformation U to the system Hamiltonian Eq. (2.2), we will generate a term as described by Eq. (3.2) which involves infinite operator expansions. However if we are considering low density systems, higher order terms will become negligible and we may expand Eq. (3.2) to second order only.

Thus, applying Eq. (3.1) to the Hamiltonian Eq. (2.2) and utilizing the transformation,

$$b_{P_1 P_2} = \sum_{\nu k} \frac{1}{\sqrt{\nu}} \, \delta_{k, P_1 - P_2} f_{\nu} \, (\alpha P_1 + \beta P_2) C_{\nu, k} \qquad 3.3$$

[v is the volume of the sytem, ν is the quantum number of the hydrogen-like state described by the wavefunction $f_{\nu}(\alpha p_1 + \beta p_2)$; $\alpha = m_h / (m_e + m_h)$, $\beta = 1 - \alpha$; i.e Eq. (3.3) diagonizes the harmonic part of the excitonic Hamiltonian and $C_{\nu, R}$ is an exciton (boson) destruction operator] the Hamiltonian, Eq. (2.2) becomes,

$$H \rightarrow \tilde{H} \, \hat{p}$$

where

$$\tilde{H} = \sum_{\nu,k} \Omega_{\nu,k} C^+_{\nu,k} C_{\nu,k} + \omega_2 \hat{a}^+ \hat{a} + \sum_{\substack{\mu,k,k'-q \\ \nu,\nu',\mu'}} M_1 C^+_{\mu,k+q} C^+_{\mu',k'-q} C_{\nu,k} C_{\nu',k'}$$

$$- \sum_{\substack{p_2,k,k',\nu,\nu' \\ p_2',\mu,\mu',q}} M_2 C^+_{\mu,k+p_2-p_2'+q} C^+_{\mu',k'+p_2'-p_2-q} C_{\nu,k} C_{\nu',k'}$$

$$+ i\hbar \left[E \exp(-i\omega_L t) \hat{a} - H.c. \right] + \hbar \left[g\hat{a}(r_1 C^+_k - r_2 C^+_k C^+_k C_k) + H.c. \right]$$

$$+ H_{ex-damp} + H_{field-damp} \qquad\qquad 3.4$$

In Eq. (3.4):

$$\Omega_{\nu,k} = E_g - \epsilon_{b,n} + \frac{\hbar^2 k^2}{2M} \quad (\epsilon_{b,n} \text{ is the binding energy of the } n^{th}$$
excitonic state, $M = m_e + m_h$, k is the centre of mass momentum).

r_1 and r_2 describe coupling between the exciton mode and the light field. We have assumed it possible to single out a particular exciton mode k.

The terms M_1 and M_2 describe interactions between all exciton modes [1].

The term $H_{ex-damp}$ describes damping of excitons via coupling to the crystal lattice, i.e. phonons. Assuming these phonons comprise a reservoir in thermal equilibrium weakly coupled to the exciton mode, we may write $H_{ex-damp}$ as,

$$H_{ex-damp} = \sum_i \chi_i C_i \tau_i^+ + H.c. \qquad\qquad 3.5$$

where τ_i represents the reservoir operator.

Finally we assume damping of the field mode occurs through coupling to a reservoir of cavity modes,

$$H_{field\ damp} = \hat{a}^+ \sum_j \chi_4 \tau_j^4; + H.c. \qquad\qquad 3.6$$

where τ_j^4 represents the reservoir mode operators.

4. MASTER EQUATION

We reduce the complexity of the many-body Hamiltonian of the previous section by assuming only one exciton mode is of interest and it couples strongly to the field. All other exciton modes then form a thermal reservoir, weakly coupled to the mode of interest. We may then use the quantum theory of damping [3] to derive a Markovian master equation - an equation of motion for the density operator ρ:

$$\frac{\partial \rho}{\partial t} = -i\delta_1 [c^+ c, \rho] - i\delta_2 [\hat{a}^+ \hat{a}, \rho] + \frac{iM_2}{\hbar} [c^+ c^+ cc, \rho]$$

$$-ig[(r_1 \hat{a}^+ c - r_2 \hat{a}^+ c^+ cc) + (r_1 c^+ \hat{a} - r_2 c^+ c^+ c\hat{a}), \rho]$$

$$+ [E\hat{a}^+ - E^* \hat{a}, \rho] + \frac{\partial \rho}{\partial t}\bigg|_{\text{ex-ex}} + \frac{\partial \rho}{\partial t}\bigg|_{\text{ex-phon}} + \frac{\partial \rho}{\partial t}\bigg|_{\substack{\text{field} \\ \text{damp}}} \qquad 4.1$$

where $\delta_1 = \Omega - \omega_L$, $\delta_2 = \omega_2 - \omega_L$

$$\frac{\partial \rho}{\partial t}\bigg|_{\text{ex-ex}} = K_{1,a}([c\rho, c^+] + [c, \rho c^+]) + K_{1,b}([c^+ \rho, c]$$

$$+ [c^+, \rho c]) + K_{2,a}([cc\rho, c^+ c^+] + [cc, \rho c^+ c^+])$$

$$+ K_{2,b}([c^+ c^+ \rho, cc] + [c^+ c^+, \rho cc])$$

$$+ K_3 ([c^+ c\rho, c^+ c] + [c^+ c, \rho c^+ c])$$

$$\frac{\partial \rho}{\partial t}\bigg|_{\text{ex-phon}} = K_5 \{ (1 + \bar{n}_{ex})[c\rho, c^+] + [c, \rho c^+]) + \bar{n}_{ex}([c^+ \rho, c]$$

$$+ [c^+, \rho c]) \}$$

and

$$\frac{\partial \rho}{\partial t}\bigg|_{\text{field-damp}} = K_4 \{ (1 + \bar{n})([\hat{a}\rho, \hat{a}^+] + [\hat{a}, \rho a^+]) + \bar{n}([\hat{a}^+ \rho, \hat{a}]$$

$$+ [a^+, \rho \hat{a}]) \}$$

where the K's, $\bar{n}$, $\bar{n}_{ex}$ are proportional to coupling strengths between modes and reservoirs and thermal populations of the various reservoirs [1].

From the master equation (4.1) we can use the generalised P-representation [4] to derive a Fokker-Planck equation. In the deterministic limit we neglect noise and find the equations of motion:

$$\dot{\alpha}_1 = -\gamma_1 \alpha_1 - \chi \alpha_1 |\alpha_1|^2 - ig_1 \alpha_2 + ig_2 (\alpha_2^* \alpha_1^2 + 2\alpha_2 |\alpha_1|^2) \qquad 4.2a$$

$$\dot{\alpha}_2 = -\gamma_2 \alpha_2 + E - ig_1 \alpha_1 + ig_2 \alpha_1 |\alpha_1|^2 \qquad 4.2b$$

(and complex conjugate equations).

where $\alpha_1 = \langle c \rangle$, $\alpha_2 = \langle \hat{a} \rangle$

$$\gamma_1 = i\delta_1 - 4K_{2,b} + K_3 + K_{1,a} - K_{1,b} + K_5 = \gamma_a + i\gamma_b$$

$$g_1 = gr_1, \quad g_2 = r_2 g, \quad \chi = 2[K_{2,a} - K_{2,b} - (i/\hbar)M_2] = \chi_a + i\chi_b$$

$$\gamma_2 = K_4 + i\delta_2 = \gamma_c + i\gamma_d.$$

5.   LOW DENSITY OPTICAL BISTABILITY

Equations (4.2) can be solved in the steady state $(\dot\alpha_1 = \dot\alpha_1^* = \dot\alpha_2 = \dot\alpha_2^* = 0)$ to yield the low density state equation:

$$I = n_1 \left\{ \left| \gamma_2^* \left[(\gamma_b + X_b n_1)(g_2 n_1 - g_1) + i(\gamma_a + X_a n_1)(3 g_2 n_1 - g_1) \right] + \right. \right.$$
$$\left. \left. + i(g_2 n_1 - g_1)^2 (3 g_2 n_1 - g_1) \right|^2 \right\} \times \left\{ \left| (3 g_2 n_1 - g_1)(g_2 n_1 - g_1) \right|^2 \right\}^{-1} \quad 5.1$$

$$n_2 = n_1 \left[\frac{(\gamma_b + X_b n_1)^2}{(3 g_2 n_1 - g_1)^2} + \frac{(\gamma_a + X_a n_1)^2}{(g_2 n_1 - g_1)^2} \right] \qquad\qquad 5.2$$

where $I = |E|^2_{ss}$ is the steady-state input field intensity, $n_1 = (|\alpha_1|^2)_{ss}$ is the steady-state exciton intensity and $n_2 = (|\alpha_2|^2)_{ss}$ is the steady-state output field intensity.

Figures 1 and 2 show plots of n_1 vs I and n_2 vs I for small g_2. (The values of n_1 at the points A,B,C,D in Fig.1 correspond to the values of n_1 at the point A,B,C,D in Fig. 2)

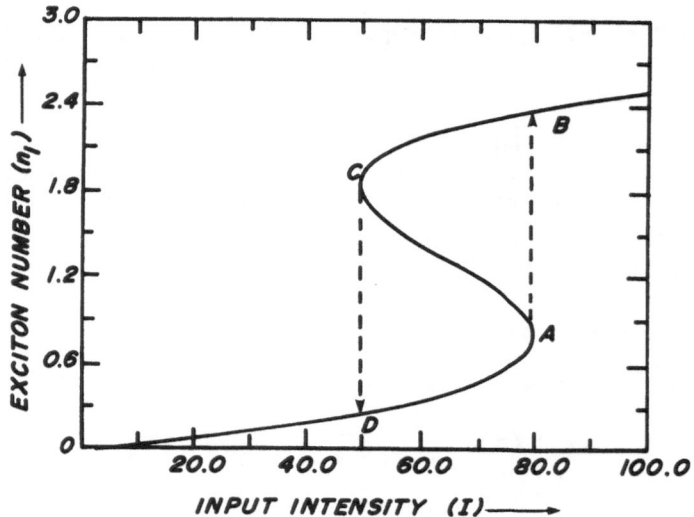

Fig. 1.   Variation of exciton number with input intensity in the
low density limit ($\gamma_a = \gamma_c = \gamma_c = g_1 = 1$, $\gamma_b = -10$ $g_2 = 0.01$,
$X_a = 1$, $X_b = 5$)

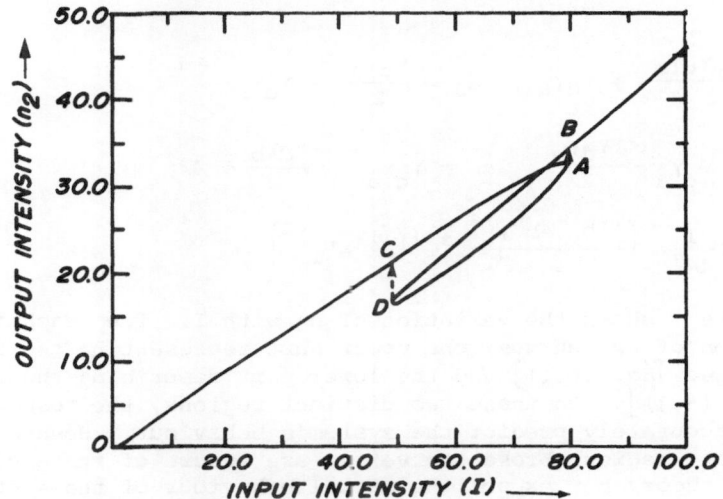

Fig. 2. Variation of output intensity with input intensity in the low density limit (Data as given for Fig. 1)

Dotted arrows indicate expected bistability: a linearised stability analysis shows the middle branch of Fig. 1 to be unstable and the top and bottom branches will be stable [1].

6. HIGH DENSITY OPTICAL BISTABILITY

At high exciton numbers the power series expansion of Eq.(3.2) becomes invalid, and we thus investigate this regime by setting $c = A + \lambda$, $c^+ = A^+ + \lambda*$ (A, A^+ are boson operators, $\lambda = \alpha_1$) and expanding in descending powers of λ, $\lambda*$. The equations of motion become,

$$\dot{\alpha}_1 = -\gamma_1 \alpha_1 - \chi \alpha_1 |\alpha_1|^2 + i\kappa (|\alpha_1|^2)^{-\frac{1}{2}} (3\alpha_2 |\alpha_1|^2 + \alpha_2^* \alpha_1^2) \qquad \text{6.1a}$$

$$\dot{\alpha}_2 = -\gamma_2 \alpha_2 + E + 2i\kappa \alpha_1 (|\alpha_1|^2)^{-\frac{1}{2}} \qquad \text{6.1b}$$

(and complex conjugate equations), where κ is the field-exciton coupling constant.

Equs. (6.1) can be solved in the steady state to yield,

$$I = \frac{(a_1 + a_2 n_1)^2 + (a_3 + a_4 n_1)^2}{4\kappa^2} \qquad\qquad 6.2$$

where

$$a_1 = \frac{\gamma_c \gamma_b}{2} + \gamma_d \gamma_a \ , \quad a_2 = \frac{\gamma_c \chi_b}{2} + \gamma_d \chi_a$$

$$a_3 = \gamma_c \gamma_a - \frac{\gamma_d \gamma_b}{2} \ , \quad a_4 = \gamma_c \chi_a - \frac{\gamma_d \chi_b}{2} + 4\kappa^2$$

and

$$n_2 = \frac{1}{4\kappa^2} \left[\frac{(\gamma_b + \chi_b n_1)^2}{4} + (\gamma_a + \chi_a n_1)^2 \right] \qquad\qquad 6.3$$

Figure 3 shows the variation of n_1 with I. The graph is a combination of two curves; the upper part representing the high density case [Eq. (6.2)] and the lower part describing the low-density case [Eq. (5.1)]. In these two distinct regions, the respective theories accurately predict the system's behaviour. However, in the intermediate region (broken curve) we are unsure of the applicability of either theory but expect the actual behaviour of the system to follow Fig. 3 and that bistability will occur as indicated.

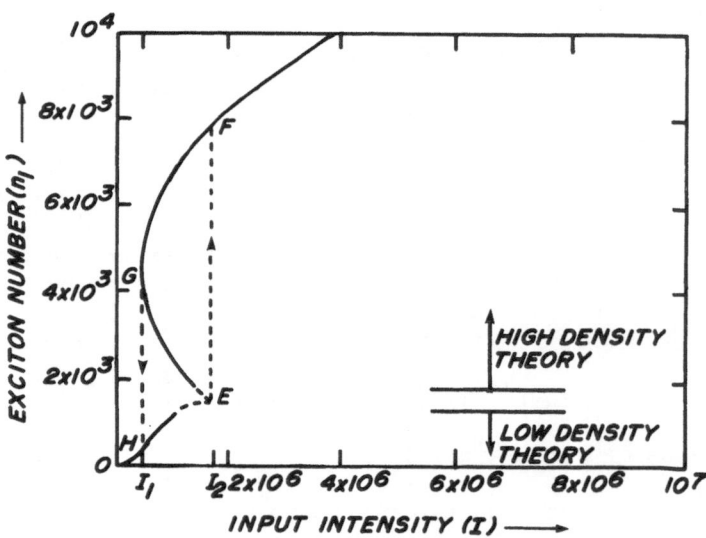

Fig. 3.   Exiton number vs. input intensity, including the effect of
           high density of excitons. ($\gamma_a = 0.5$, $\gamma_b = -5$, $\gamma_c = 1$, $\gamma_d = -5$
           $\kappa = 0.0037$, $g_1 = 34$, $g_2 = 0.0073$, $\chi_a = 10^{-4}$)

Figure 4 shows the variation of n_2 with I. (The values of n_1 at the points E,F,G,H in Fig. 3 correspond to the values of n_1 at the points E,F,G,H in Fig. 4). Upper and lower curves were obtained from the high- and low-density theories respectively, and as before, broken lines indicate regions in which neither theory is valid. Again bistability and hysteresis is displayed.

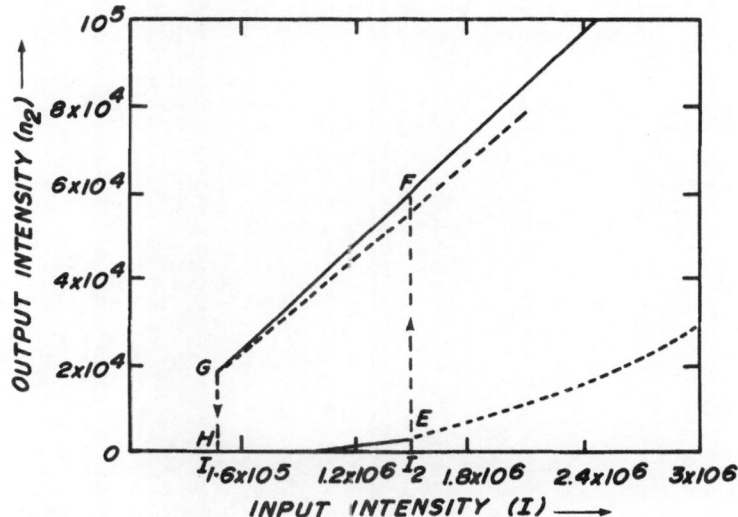

Fig. 4. Output intensity vs input intensity including effects of high exciton density (Data as given for Fig. 3).

REFERENCES

1. M.L. Steyn-Ross and C.W. Gardiner, Phys. Rev. A27 310 (1983)
2. T. Marumori, M. Yamamura and A. Tokunaja, Prog. Theor. Phys. 31 1009 (1964)
3. W.H. Louisell 'Quantum Statistical Properties of Radiation' (Wiley, New York, 1973)
4. P.D. Drummond and C.W. Gardiner, J. Phys. A13 2353 (1980).

OPTICAL BISTABILITY IN SEMICONDUCTORS

C. C. Sung

Department of Physics, University of Alabama in
Huntsville, Huntsville, Alabama 35899, USA

and

C. M. Bowden

Research Directorate, US Army Missile Laboratory, US
Army Missile Command, Redstone Arsenal, Alabama 35898
USA

INTRODUCTION

The first demonstrated optical bistability (OB) in a semiconductor was conducted by Gibbs et al.[1] in GaAs at low temperature in which the OB is primarily dispersive and the nonlinear index arises from light-induced changes in excitonic absorption. Subsequently, and almost coincident with the first GaAs experiment[1] was the work of Miller, Smith and Johnston[2] who observed OB in InSb at 5°K due to a direct bandgap resonance mechanism. Quite recently, Gibbs et al. have observed excitonic OB at room temperature in a GaAs-GaAlAs superlattice etalon[3].

Theoretically, Koch and Haug[4] (KH) have studied OB in the vicinity of the biexciton resonance, and indications are that the intensity-dependent resonance may indeed be the mechanism for the observed OB in GaAs of reference 1. Furthermore, the results of the work of KH relate quantitatively to the two-photon resonance Raman scattering experiments in CuCℓ[5]. Thus, CuCℓ holds particular interest as a candidate for OB.

Recently, it was claimed[6] that incorporation of the complete field exciton-biexciton (FEB) interaction and the local field correction (LFC) in a semiclassical model predicts OB for CuCℓ with

241

the incident field tuned near the two-photon biexciton resonance.
Apparently, the results of reference 6 are intended to be a correction to the nonlinear dielectric function ε in KH. Unfortunately, the results reported in reference 6 for ε cannot be correct
for the main reason that their complex dielectric function ε contains a double pole that does not satisfy the Kramers-Kronig relation[7]. The problem originated from an incorrect treatment of the
FEB as shown later.

The purpose of this paper is to re-examine the model for OB
in excitonic semiconductors incorporating the effects of FEB and
LFC. Since the dielectric function for CuCℓ has been well studied,
we will use a model Hamiltonian for that system to discuss the
problem. In addition, we also incorporate the cavity mirror reflectivity in the calculation to demonstrate its importance combined
with intrinsic nonlinear surface reflectivity and transmitivity
in the interpretation of experimental data.

FIELD DEPENDENCE OF THE DIELECTRIC FUNCTION $\varepsilon(k,\omega)$

The Hamiltonian consists of two parts: the free excitons
and biexcitons, H_O, and the interaction of excitons and biexcitons
with the internal laser field, H'. H_O and H' in the rotating
frame of the external field frequency ω are, respectively,

$$H_O = (\omega_x - \omega)b^+b \;\; + \;\; (\omega_m - 2\omega)B^+B \tag{1}$$

$$H' = ig_1 E^- b^+ \;\; + \;\; ig_2 E^- B^+ b \;\; + \;\; H.c. \tag{2}$$

Here, ω_x and ω_m are the energy levels of excitons and biexcitons and b^+b and B^+B are their collective creation (annihila-
tion) operators and units are chosen such that $\hbar = c = 1$. The
externally-applied field amplitude $E^-(E^+)$ associated with $e^{-i\omega t}$
($e^{i\omega t}$) is treated as a c-number. $g_{1,2}$ is the coupling constant
of the externally-applied field with the induced dipole moments,

$$g_1 = -(N/2V)^{\frac{1}{2}}\vec{e} \cdot \langle \vec{d} \rangle_1 \;\; ; \;\; g_2 = -(N/2V)^{\frac{1}{2}}\vec{e} \cdot \langle \vec{d} \rangle_2 \;\; , \tag{3a,b}$$

where N and V are the number of unit cells and the volume, respectively and $\vec{e}$ is the polarization vector of the external field.
The dipole moment matrix elements $d_{1,2}$ are usually inferred from
the experimental data[4], since it is difficult to obtain the quantitative wave functions of the excitons and biexcitons. Notice also
that the factor $\omega^{\frac{1}{2}}$ in the definition of $g_{1,2}$ is included in $E^{\pm}$.
A short review of the approximations involved in our Hamiltonian
is in order here. First the spatial dependence of the matrix

elements is neglected throughout the paper. As a result of the approximation the ensemble average $\langle B^+B\rangle$ and $\langle b^+b\rangle$ are the number density of biexcitons and excitons, respectively.

We set up the Langevin equations for the b and B by neglecting the fluctuations but introducing relaxation widths γ_x and γ_m

$$-i\,\langle \partial b/\partial t\rangle = -\delta\langle b\rangle - ig_1 E^- + ig_2 E^-\langle B\rangle + i\langle b\rangle\gamma_x \tag{4}$$

$$-i\,\langle \partial B/\partial t\rangle = -\Delta\langle B\rangle - ig_2 E^-\langle b\rangle + i\langle B\rangle\gamma_m \quad, \tag{5}$$

where $\delta = \omega_x - \omega$ and $\Delta = \omega_m - 2\omega$. The dielectric function $\varepsilon^{\pm}$ for $E^{\pm}$ is defined by

$$\varepsilon^{\pm} = 1 + 4\pi \langle P^{\pm}\rangle /E^{\pm} \tag{6}$$

where the polarization $P^{\pm}$ is obtained by writing $H' \equiv -P^+E^- - P^-E^+$ and

$$\langle P^-\rangle = ig_1\langle b\rangle + ig_2\langle b^+B\rangle \quad. \tag{7}$$

The major difference of our work and others in the calculation of $\varepsilon^{\pm}$ is that others have either assumed that[4]

$$\langle b^+B\rangle = 0 \tag{8}$$

or, in Reference 6

$$\langle b^+B\rangle = \langle b^+\rangle\langle B\rangle \quad. \tag{9}$$

Neither is justified as shown here. Instead, we take another Langevin equation with damping constant γ,

$$-i\langle \partial/\partial t(b^+B)\rangle = (\delta-\Delta+i\gamma)\langle b^+B\rangle - ig_1 E^+\langle B\rangle - ig_2 E^-[\langle b^+b\rangle-\langle B^+B\rangle]. \tag{10}$$

The populations $\langle b^+b\rangle$ and $\langle B^+B\rangle$ can be determined by taking another set of Langevin equations as discussed in the next section. In the thermal equilibrium approach, which is used for comparison later, we take

$$\langle b^+b\rangle \approx \langle b^+b\rangle_0 \quad, \qquad \langle B^+B\rangle \approx \langle B^+B\rangle_0 \quad, \tag{11}$$

where $\langle\ \rangle_0$ is the equilibrium value.

The steady-state solution of Eqs. (10), (4), and (5) is used to obtain the dielectric function

$$\varepsilon^- = 1+4\pi\left\{ \frac{g_1^2\Delta'}{\delta\Delta' - |g_2|^2|E|^2} + \frac{g_1^2\,g_2^2\,|E|^2}{(\delta\Delta' - |g_2|^2|E|^2)(\delta-\Delta+i\gamma)} + g_2^2 S_0 \right\}, \tag{12}$$

where $\Delta' = \Delta - i\gamma_m$, $|E|^2 = E^+E^-$, and $\gamma_x = 0$.

$$S_o = (<B^+B>_o - <b^+b>_o)/(\delta - \Delta + i\gamma) \quad , \tag{13}$$

and $<>_o$ is the equilibrium expectation value.

A comparison of Eq. (12) with the earlier works shows that in KH, the last two terms are neglected, which is justified for large width γ or small lifetime of exciton-biexciton pair. In reference 6, S_o is dropped; in addition, the term $\propto g_2^2E^2$ from $<b^+B>$ has a denominator $(\delta\Delta' - g_3^2E^2)^2$ which originated from the unjustified factorization approximation given according to Eq. (9). In principle, one should have another set of equations of motion to determine the expectation values in S_o instead of taking the equilibrium value.[8] This procedure is not taken to avoid more parameters . In practice, $\epsilon^- \to \epsilon_\infty > 1$ as $\omega \to \infty$, which is the contribution from the high frequency resonant states, so that the first term on the rhs in Eq. (12) should be replaced by ϵ_∞ .

LOCAL FIELD CORRECTION (LFC)

The macroscopic field E that appears in Eq. (12) is actually the local field at the exciton or exciton-biexciton position, i.e., it does not include the self-field[9] of each exciton or exciton-biexciton. To discuss the transmitted light as a function of the incident field, the relevant quantity is the total internal macroscopic field E, which satisfies Maxwell's equations.

The standard textbook relations[7] cannot, however, be utilized in this case in a straightforward manner. The reason for this is that the high frequency contribution ϵ_∞ may or may not come from the same local point. Furthermore, the correction is based upon the assumption that the exciton and exciton-biexciton are localized. The justification of this assumption, particularly for the exciton-biexciton, is not obvious.

We summarize the discussion by writing Eq. (12) in terms of the total macroscopic field using the local field correction[9],

$$\epsilon^- = \epsilon_\infty + 4\pi Z_o \left[\frac{g_1^2\Delta'}{\delta\Delta' - g_2^2 E_c^2} + \frac{g_1^2 g_2^2 E_c^2}{(\delta\Delta' - g_2^2E_c^2)(\delta - \Delta + i\gamma)} + g_2^2 S_o \right] \cdot$$

$$= \epsilon_\infty + 4\pi \rho_o \alpha_o \quad , \tag{14}$$

where $Z_o = [1 - \frac{4\pi}{3} \alpha_o \rho_o]^{-1}$, or $Z_o = \frac{\varepsilon_\infty + 2}{3} (1 - \frac{4\pi}{3} \alpha_o \rho_o)^{-1}$, depend-

ing on whether ε_∞ is or is not originated from excitons and excitons and biexcitons. Z_o in Eq. (14) comes from the application
of the LFC to excitons and $E_c = Z_o E$ if the biexciton-exciton pairs
are localized.

We should emphasize that the parameters g_1^2, g_2^2, etc. are inferred from the experimental data on the Raman scattering without
the LFC. A consistent treatment of this effect in all cases is
essential for any quantitatively reliable conclusion. In view of
this remark, the calculation on the LFC presented here shows the
importance of the effect but should not be taken as exhibiting a
quantitatively reliable representation of ε^- unless all material
parameters used are also corrected using the LFC.

EFFECT OF CAVITY MIRROR

In the analysis of the experimental data, it is very important
to include the mirror reflectivity. In this section, we consider
a plane wave incident on a cavity traveling from left to right with
amplitude E_o. By using appropriate boundary conditions at both
ends of the cavity, the transmitted intensity $|E_t|^2$ is determined
as a function of the mirror reflectivity, transmititivity, and the
complex phase shift $-\delta'' + i\delta' = D \sqrt{\varepsilon} \, \omega/c$, where D is the length
of the cavity, and ω/c is the wave number. The result is given by

$$\tau = \frac{|E_t|^2}{|E_o|^2} = \left| 4 \, t_m \, \sqrt{\varepsilon} \left\{ \left[(1 + \sqrt{r_m})\sqrt{\varepsilon} + (1 - \sqrt{r_m}) \right]^2 e^{\delta'' - i\delta'} \right. \right.$$

$$\left. \left. - \left[(1 + \sqrt{r_m})\sqrt{\varepsilon} - (1 - \sqrt{r_m}) \right]^2 e^{-i\delta'' + i\delta'} \right\}^{-1} \right|^2 . \quad (15)$$

DISCUSSION OF RESULTS AND CONCLUSIONS

The contribution of the correction term to the exciton-biexciton contribution, the second term inside the brackets in Eq. (14),
which is neglected in reference 4, is shown in Figure 1. In this
case we have neglected the LFC also, i.e., set $Z_o = 1$. Also,
there are no mirrors or cavity, and reflectivity is due to
reflection at the dielectric boundary which is a nonlinear function
of the internal field. This we call intrinsic reflectivity.

The LFC is applied to the localized excitons, and the results
are shown in Figure 2 for intrinsic reflectivity at the dielectric
boundary.

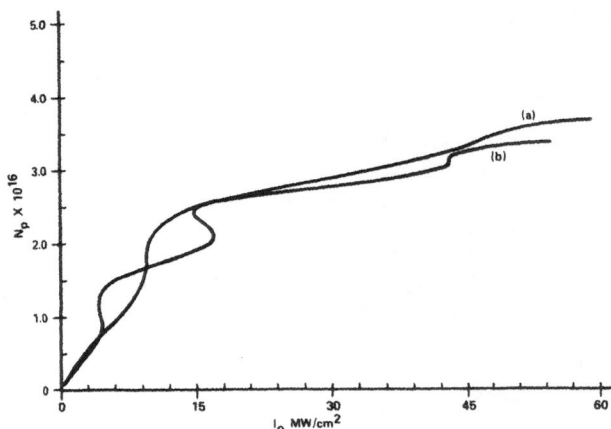

Fig. 1.   Polariton number N_p versus intensity of the incident
          field I_o for the incident field frequency $\omega = 3.18594$ ev,
          and without local field correction. Curve (a): Neglect
          of the FEB correction (see text). Curve (b): Inclusion
          of exciton-biexciton interaction correction with $\gamma_m / \gamma =$
          2.9×10^{-2}. The parameters $g_1^2 = 2.188 \times 10^{-3}$ ev; $g_2^2 =$
          1.572×10^{-22} ev; $\gamma_m = 0.4 \times 10^{-4}$ ev; $\omega_x = 3.2027$ ev; and
          $\omega_m = 6.3725$ ev. In terms of the transmitted intensity I_t,
          $N_p = 2I_t / (\tau \hbar \omega V_g)$ where τ is given by Eq. (15) and $V_g =$
          $c / \text{Re} \left(\sqrt{\epsilon} + \omega \dfrac{d \sqrt{\epsilon}}{d \omega} \right)$ is the polariton group velocity.

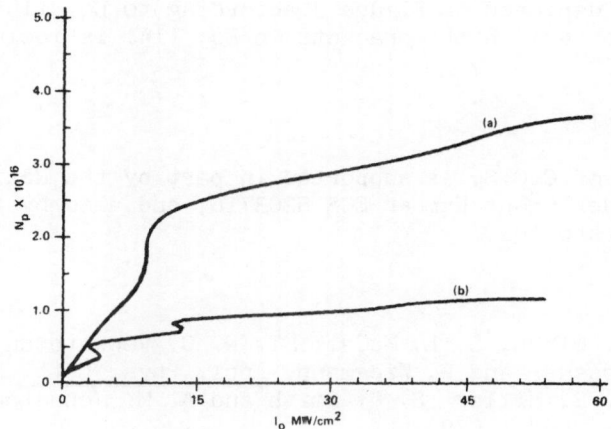

Fig. 2.   Polariton number N_p versus intensity of the incident field
I_O for the incident field frequency ω = 3.18594 ev, and
without the FEB correction. Curve (a): Without the LFC.
Curve (b): LFC applied to the exciton sites. All other
parameters are the same as for Fig. 1.

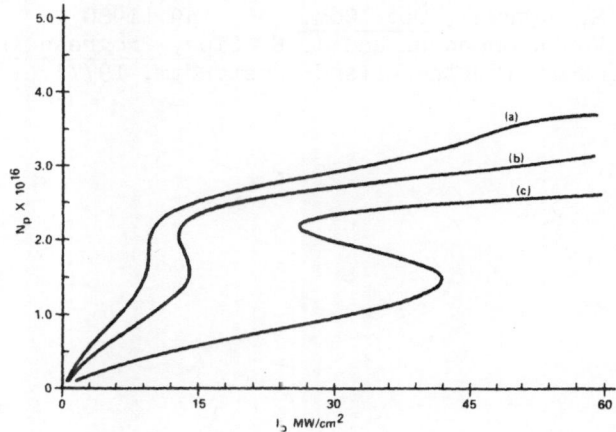

Fig. 3.   Polariton number N_p versus incident field intensity I_O
of frequency ω = 3.18594 ev for a CuCℓ slab 1 μm in length
sandwiched between reflecting films of reflectivity r_m.
Curve (a): r_m = 0. Curve (b): r_m = 0.01. Curve (c):
r_m = 0.10.

The effect of mirror reflectivity, i.e., a cavity, on the OB of CuCℓ is depicted in Figure 3 according to Eq. (15). Here, only the first term in the brackets in Eq. (14) is included and there is no LFC.

ACKNOWLEDGEMENT

The work of C.C.S. is supported in part by the National Science Foundation under Grant Number ECS 8203716, and also by the Battelle Columbus Laboratories.

REFERENCES

1.   H. M. Gibbs, S. L. McCall, T. N. C. Venkatesan, A. C. Gossard, A. Possner and W. Wiegmann, Appl. Phys. Lett. 35, 451 (1979).
2.   D. A. B. Miller, S. D. Smith and A. M. Johnston, Appl. Phys. Lett. 35, 658 (1979).
3.   H. M. Gibbs, S. S. Tarng, J. S. Jewell, D. A. Weinberger and K. Tai, Appl. Phys. Lett. 41, 221 (1982).
4.   S. W. Koch and H. Haug, Phys. Rev. Lett. 46, 450 (1981).
5.   J. B. Grun, B. Honerlage, and R. Levy, "Biexcitons in CuCℓ and Related Systems", in Excitons, edited by E. I. Rashba and M. D. Sturge, North-Holland Publishing Company, (1982).
6.   J. Abram and A. Maruani, Phys. Rev. B26, 4759 (1982).
7.   W. D. Jackson, Classical Electrodynamics, (John Wiley, New York, 1962).
8.   G. S. Agarwal, Opt. Comm. 35, 149 (1980).
9.   J. Van Kronendonk and J. E. Sipe, Progress in Optics XV, edited by E. Wolf (North-Holland, Amsterdam, 1977).

THEORY OF RESONANCE ENHANCED OPTICAL NONLINEARITIES

IN SEMICONDUCTORS

H. Haug

Institut für Theoretische Physik der
Universität Frankfurt
Robert-Mayer-Str.8, D-6ooo Frankfurt-Main
Fed. Rep. Germany

ABSTRACT

   The optical nonlinearities of semiconductors
enhanced by one- and two-photon excitonic resonances
are treated in a unified theory. Illustrations for
the resulting nonlinear absorption and dispersion
spectra are given for InSb, GaAs and CuCl.

   In semiconductors the attractive Coulomb interaction between electrons and holes causes one- and two-
photon resonances, known as excitons and exciton molecules, respectively. Large nonlinearities are observed
in the spectral vicinity of these resonances. These nonlinearities are well-suited for the realization of optical bistability [1,2,3] as has been demonstrated by Gibbs
et al. [1] for GaAs and by Miller et al. [2] for InSb and
very recently by Hönerlage et al. [3] for CuCl. The electron-hole correlations, the screening and the energy renormalization are most conveniently calculated by the
Green's function formalism. The optical dielectric function can be expressed in terms of the retarded photon
self-energy which is also called the polarization function. This connection is not restricted to the linear
response regime.
   In narrow-gap semiconductors the low-intensity optical spectra in the band-gap region are strongly influenced by the electron-hole correlation. For high intensities, the screening of the Coulomb interactions is so
large that these correlations become less important.

In this limit, the optical spectra are those of an electron-hole plasma. In the transition region the optical
spectra change strongly.[4,5] to develop a calculation of
    We have been able to develop a calculation of
the dielectric function which is valid from the low-intensity excitonic limit up to the quasi-metallic high-
intensity limit in which free-carrier screening is the
most important feature. We neglected the contribution
of the excitons to the screening and used a quasi-static
RPA approximation. The integral equation for the complex polarization function has been solved numerically
by matrix inversion. The resulting complex dielectric
function gives immediately both the absorption and the
refraction spectra without using a Kramers-Kronig transformation. Fig.1 shows the resulting spectra for GaAs
for various free-carrier densities. This density can
be related to the light in-
tensity by a rate equation.
For the curve with a free-
carrier concentration of
$n=5\times10^{16}$ cm^{-3} the exciton
energy equals the renorma-
lized band-gap, so that the
exciton is ionized. In this
situation still a considerable excitonic enhancement
exists. At higher densities
a further band-gap shrinkage
and a band-filling is seen,
so that optical gain deve-
lops for $\omega<\mu$ where μ is the
quasi-chemical potential.
Naturally, in a single-beam
experiment μ is always smaller than the laser frequency. The refraction index
is seen to decrease in the
frequency range below the
exciton resonance as the
exciton is ionized.

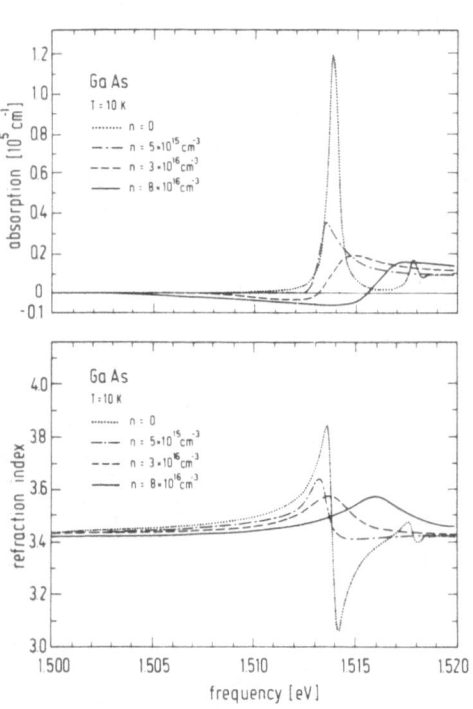

The resulting spectra
for InSb (Fig.2) show no resolved exciton resonance,
because the energy broadening is twice the exciton
binding energy. The resonance enhanced absorption edge
is seen to be bleached by
the increasing light intensity and a corresponding de-

Fig.1: Calculated spectra
of absorption and refrac-
tion for GaAs at various
free-carrier densities
and a temperature of 10 K.

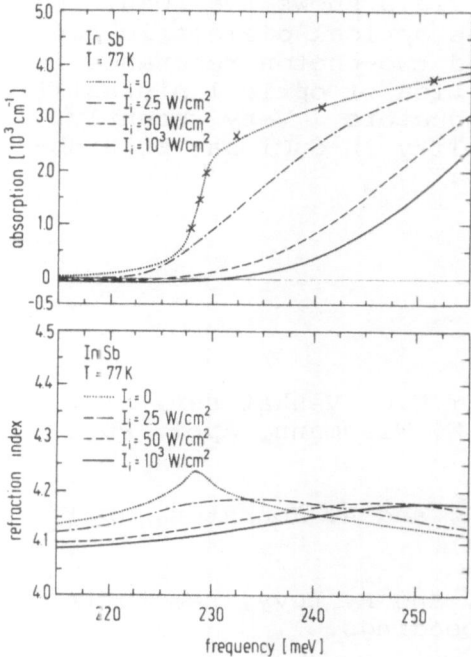

Fig.2: Calculated spectra of absorption and refraction for InSb at various internal excitaiton intensities I at 77 K. The excitation laser frequency was assumed to be ω=225 meV.

crease of the refraction index is obtained. The theoretical results both for GaAs and InSb are at least in good qualitative agreement with all known experimental observations.

In wide-gap semiconductors the static dielectric constant of the unexcited crystal is smaller and therefore the exciton-hole correlations are stronger. In cuprous halides, e.g., one observes not only a one-photon excitonic resonance but also a two-photon resonance due to the exciton molecule. The first photon creates a virtual exciton which fuses with the second photon to yield a real or virtual molecule. The corresponding intensity-dependent photon self-energy can be calculated analytically (Ref. 6 and

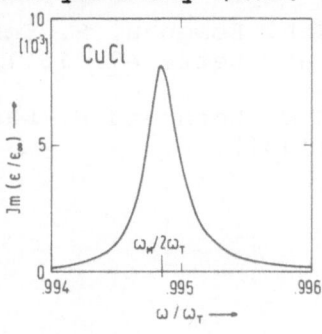

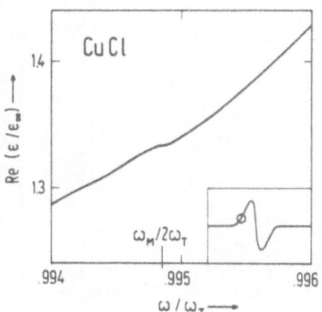

Fig.3: Calculated dielectric function for CuCl at a polarization density of 4×10^{15}cm^{-3}. ω_T and ω_m are the frequencies of the transverse exciton and of the molecule, respectively.

references quoted therein). Fig.3 shows the resulting real and imaginary part of the optical dielectric function for CuCl. The calculated two-photon resonance is strong enough for the realization of optical bistability also with wide-gap semiconductors. Very recently, the predicted optical bistability in CuCl has been observed by Hönerlage et al. [3].

REFERENCES

1. H.M. Gibbs, S.L. McCall, T.N.C. Venkatesan, A.C. Gossard, A. Passner, and W. Wiegmann, Appl. Phys. Lett. 35, 45 (1979).

2. D.A.B. Miller, S.D. Smith, and A. Johnston, Appl. Phys. Lett. 35, 658 (1979).

3. B. Hönerlage, J.Y. Bigot, and R. Levy, see their contribution in this proceedings.

4. S. Schmitt-Rink, J.P. Löwenau, and H. Haug, Z. Physik B 47, 13 (1982).

5. J.P. Löwenau, S. Schmitt-Rink, and H. Haug, Phys. Rev. Lett. 49, 1511 (1982).

6. S.W. Koch and H. Haug, Phys. Rev. Lett. 46, 450 (1981).

OPTICAL BISTABILITY IN CuCl

Bernd Hönerlage, Jean-Yves Bigot, Roland Levy

Laboratoire de Spectroscopie et d'Optique du Corps Solide
(associé au C.N.R.S. n° 232), Université Louis Pasteur
5, rue de l'Université, 67000 Strasbourg (France)

INTRODUCTION

When using a nonlinear medium in a Fabry-Perot interferometer
(F.P.), optical bistability is due to an intensity dependence of
the real or imaginary part of the dielectric function[1,2]. In
solids, its real part defines the well-known polariton dispersion,

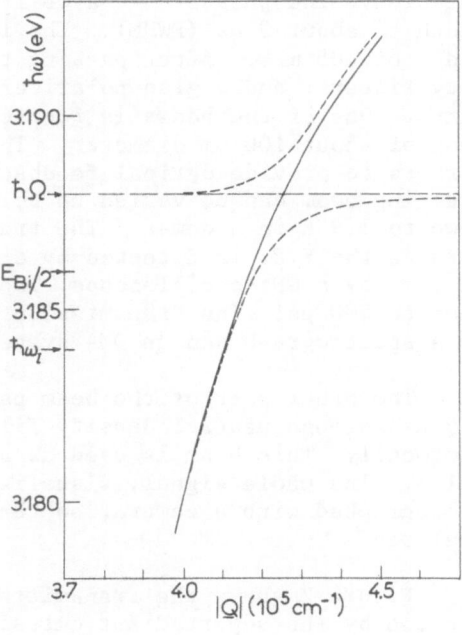

Fig. 1. Lower branch of the
polariton dispersion
with (dotted line) and
without (full line)
renormalization due to
biexcitons.

the lower branch of which is shown in Fig. 1 by the full line[3].
In CuCl, biexcitons with energy E_{Bi} may be created by two-photon
absorption. The absorption coefficient of polaritons with frequency Ω depends therefore on the density of polaritons with frequency ω_ℓ, defined as such

$$\hbar\omega_\ell + \hbar\Omega = E_{Bi} \tag{1}$$

Via Kramers-Kronig relation, the dispersion of polaritons depends
on the absorption process. The dotted lines give the resulting
renormalized dispersion for vanishing damping, using the theoretical model developed by Haug and coworkers[4]. If the laser frequency
ω_ℓ is tuned through the biexciton resonance, the exciting polaritons
are subject to this renormalization effect. Therefore, one might
hope that this process leads to optical bistability if sufficient
feedback is provided. If only virtually created biexcitons are
involved in the process, the switching times of the device will not
depend on the lifetimes of elementary excitations. It may therefore
be very short[5,6].

EXPERIMENTAL RESULTS AND DISCUSSION

In our experimental set-up, given in more detail in Ref.[7], an
excimer laser pumps a dye laser containing a diluted solution of
αNND in ethanol. Great care was taken to keep the intensity of the
superradiant emission of the dye cell small compared to the laser
emission. The pulses have a well-defined shape in time and a half-
width of about 3 ns (FWHM). The laser emission has a spectral
width of .05 meV. After passing through a diaphragm, neutral density filters, and a glan polarizer, the beam is split into two
parts. One of the beams is focused onto the crystal surface to a
spot of about 100 µm diameter. The crystal is placed between two
mirrors to provide optical feedback. The power density of the
exciting beam can be varied up to 60 MW/cm[2] and the F.P. is cooled
down to 1.9 K in a dewar. The transmission of the laser pulses
through the F.P. is detected by a fast photocell and is analyzed
in time by a GHz-oscilloscope. The overall time resolution is inferior to 500 ps. The transmission spectrum may as well be studied
by a spectrograph and an OMA system.

The other part of the beam passes after an optical delay of
5.9 ns through neutral density filters and is detected by the same
photocell. This beam is used as a reference for the shape of the
pulse. The whole signal, visualized on the oscilloscope, is then
photographed with a camera, so that single laser pulses may be
analyzed.

Figure 2 shows the transmitted intensity of the F.P. when
excited by the superradiant emission of the dye cell. The

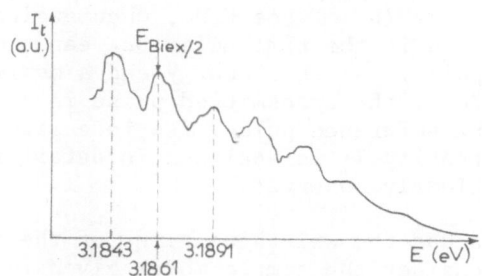

Fig. 2.   Transmission spectrum of the F.P.

thickness of our CuCl sample is about 30 μm, and, as the sample is in direct contact with the mirrors, this is also the length of the F.P.

The F.P. is excited perpendicularly to its surface. The mirrors are thin glass plates, coated with platinium films. Their reflection coefficient is about 90 %. We have chosen platinium, since most other metals react chemically with CuCl. When using natural or directly coated surfaces, no sufficient feedback was obtained. Although the mirrors are highly reflective, only a maximum intensity variation of 40 % is obtained. This is due to the absorption of the crystal near the 1S exciton line and to a lack of parallelism of the crystal surfaces. It is important to notice that the working point of the F.P. close to half the biexciton energy corresponds to a maximum of transmission in the following experiment.

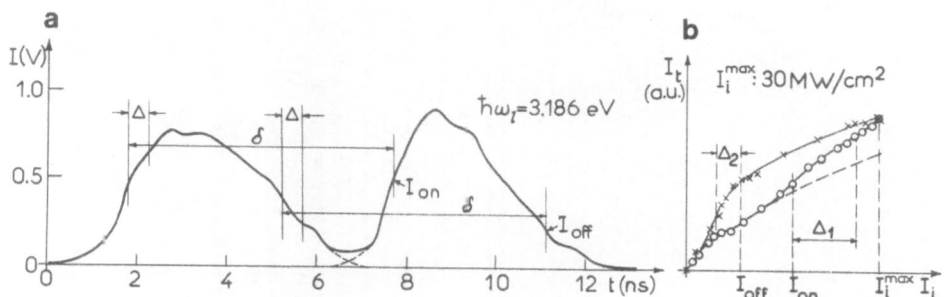

Fig. 3.   Transmitted and reference pulse (a) and transmission
          characteristics (b) of the F.P.

Figure 3 shows the transmitted and the reference laser pulse
as analyzed by the oscilloscope (a) and the corresponding transmission characteristics (b) of the F.P., when excited at a photon
energy of 3.186 eV. δ is the time delay between the transmitted
and the reference pulse. At this energy and a maximum incident
intensity of 30 MW/cm², the transmitted pulse is clearly deformed,
when compared to the reference pulse. Its intensity I_t as function
of the incident intensity I_i is analyzed in detail in Fig. 3(b),
and hysteresis is clearly observed.

Let us now discuss the switching times of the device. For
this purpose, we consider the simple model given in figure 4. We
assume a device, having the transmission characteristics given in
Fig.4(b). Switching is first assumed to be instantaneous (vertical
lines).

If the reference (laser) pulse has a triangular form, the
transmitted pulse will have a different shape, but it will essentially reproduce the laser pulse in three different regions. Our
detection system, however, and the device have a certain time response Δ. Therefore, the transmitted pulse will be deformed in the
way indicated by the dashed dotted line.

If such a pulse is measured, we will determine the transmission characteristics $I_t(I_i)$ given in Fig. 4(b) by the dashed-
dotted lines. Δ_1 and Δ_2 are the regions of deformation due to the
time response Δ. If the time response on the falling part of the
pulse is made longer (e.g. due to a longer switch-off time of the
device), we will find in Fig. 4(b) again a straight line, having
a different slope at the origin than for increasing intensities.

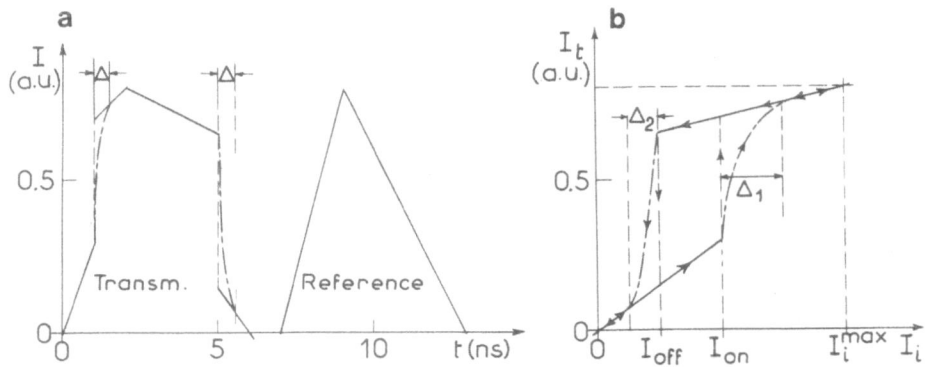

Fig. 4. Model for the transmitted and reference pulse and its
         deformation (a) due to the transmission characteristics
         (b) of the F.P.

In Fig. 3(a), the dashed line gives the (assumed) transmission characteristics without optical feedback. The switching intensities I_{on} and I_{off} are situated at positions where deviations from these characteristics arise. The values Δ_1 and Δ_2 give the regions of deformation, resulting only from the finite time resolution $\Delta = 500$ ps of the detection system. Since no deformation is observed outside this region, we conclude that the switching time has not been resolved. It is therefore inferior to 500 ps for both, the on-to-off and the off-to-on switching.

We have studied the dependence of the hysteresis loop of Fig. 3 on both, the excitation intensity and the excitation frequency. Concerning the excitation intensities, the switching intensities stay at fixed values (I_{on} = 15 MW/cm^2 ; I_{off} = 5 MW/cm^2). If the maximum intensity is below 15 MW/cm^2, no hysteresis is observed. Concerning its dependence on the excitation frequency, we find that it is strongly resonant on half the biexciton energy. Although we did not yet study the switching times in detail, preliminar results show that the switch-off time depends on the energy of the exciting photons. It becomes longer ($\gtrsim$ 4 ns) if the F.P. is excited slightly above half the biexciton energy, and we presume that the finite radiative lifetime of biexcitons then becomes important.

If the working point of the F.P. is changed with respect to the interference fringe pattern of Fig. 2, bistability is still observed[7], but I_{on} and I_{off} are at higher values.

At the end of our experiment, we have allowed the cryostat to heat up to liquid nitrogen temperature ($\gtrsim$ 77 K). As shown in

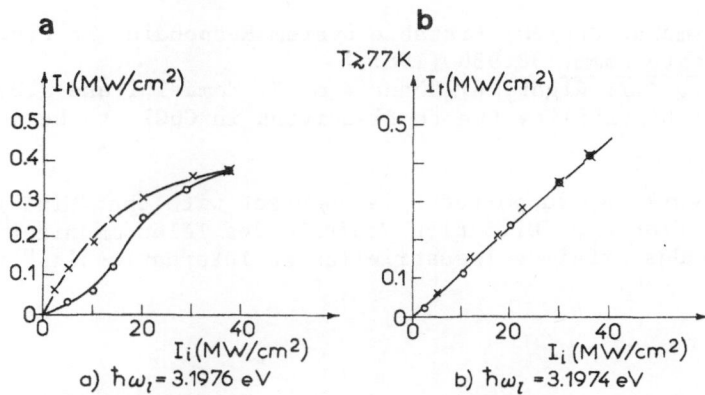

Fig. 5.   Transmission characteristics of CuCl at
          77 K when excited at 3.1976 eV (a) and
          3.1974 eV (b).

Fig. 5, similar optical bistability could then be observed around half the biexciton energy (at 3.1976 eV),

CONCLUSION

We have given experimental evidence of optical bistability due to biexcitons in CuCl at both , 1.9 K and 77 K. We have shown that the phenomenon is strongly resonant at half the biexciton energy, which allows to state that the non-linearity is due to two-photon absorption. Working with virtual transitions, the switching times are shorter than the experimental time resolution which is 500 ps. However, the required input intensity is rather high ($\sim$ 15 MW/cm^2) and it is not clear to us whether this intensity may be diminished considerably.

REFERENCES

1. H.M. Gibbs, S.L. Mc Call and T.N.C. Venkatesan, Differential Gain and Bistability using a Sodium-Filled Fabry-Perot Inter-ferometer, Phys. Rev. Letters, 36:1135 (1976).
2. E. Abraham and S.D. Smith, Optical Bistability and Related Devices, Rep. Prog. Physics, 45:815 (1982).
3. B. Hönerlage, A. Bivas and Vu Duy Phach, Determination of the Excitonic Polariton Dispersion in CuCl by Resonant Two-Photon Raman Scattering, Phys. Rev. Letters, 41:49 (1978).
4. H. Haug, R. März and S. Schmitt-Rink, Dielectric Function for Semiconductors with a High Exciton Concentration, Phys. Letters, 77A:287 (1980).
5. S.W. Koch and H. Haug, Two-Photon Generation of Excitonic Mole-cules and Optical Bistability, Phys. Rev. Letters, 46:450 (1981).
6. E. Hanamura, Optical Bistable System Responding in Pico-Second, Sol. State Comm. 38:939 (1981).
7. R. Levy, J.Y. Bigot, B. Hönerlage, F. Tomasini and J.B. Grun, Optical Bistability Due to Biexcitons in CuCl, to be published.

This work was supported by a contract with the "Ministère des P.T.T." of France - "Direction Générale des Télécommunications" - "Direction des Affaires Industrielles et Internationales".

OPTICAL NONLINEARITY AND BISTABILITY DUE TO THE

BIEXCITON TWO-PHOTON RESONANCE IN CuCl

N. Peyghambarian, H. M. Gibbs, D. A. Weinberger,
M. C. Rushford, and D. Sarid

Optical Sciences Center
University of Arizona
Tucson, Arizona 85721

ABSTRACT

We have investigated the nonlinear dielectric function close to the biexciton two-photon resonance in CuCl, taking into account the background absorption due to the exciton tail in the vicinity of the biexciton as well as the broadening of the biexciton resonance with intensity. This analysis predicts biexcitonic optical bistability with ∿10 MW/cm² intensity for a ∿10-μm-thick CuCl etalon with 90% reflecting mirrors. Our experimental results, obtained with 8- to 12-μm-thick CuCl films sandwiched between 90% reflecting mirrors, verify these predictions. Optical limiting and bistability were observed with ∿15 to 30 MW/cm² intensity. Subnanosecond switching times are limited by the response time of the detection system.

INTRODUCTION

It is well established that biexcitons can be generated directly by two-photon absorption in CuCl. The intensity-dependent nonlinearity arising from this resonance can be used to obtain optical bistability. Recently, Koch and Haug[1] and also Hanamura[2] have predicted the possibility of bistable operation in a CuCl Fabry-Perot etalon. Hanamura used a density matrix formalism and predicted that the switching can respond on a picosecond time scale due to the virtual formation of biexcitons. Koch and Haug obtained the complex dielectric function from a Green's function formalism and suggested bistable behavior for a thin CuCl slab. They used a constant value for the biexciton linewidth, γ_{xx}, and a zero value for the exciton linewidth, γ_x. Their results indicated that a 1-μm-thick CuCl etalon, with natural reflecting surfaces, could give rise to bistability with a switch-on intensity of about 0.1 MW/cm².

The biexciton resonance, however, is located on the tail of the exciton resonance,[3] which acts as a background absorption. Since this background absorption can be fatal to biexcitonic bistability, it should be taken into account by using the proper exciton linewidth. Also, the biexciton absorption line shows a remarkable broadening with input laser intensity,[4] which should also be considered as it can modify the conditions required for achieving bistable operation. The origin of the broadening has been the source of some controversy. Chase et al.[4] and also Itoh et al.[5] have attributed this effect to collisions between excitonic particles. This interpretation is supported by a more recent experiment performed by Peyghambarian et al.[6] in which the two-photon absorption resonance measured with two weak probe beams was broadened by simultaneous injection of biexcitons using an intense pump beam.

CALCULATIONS

We have used the complex dielectric function in the vicinity of the biexciton resonance as given in Ref. 1, together with a model developed for the collision broadening of the biexcitons.[4] This model takes into account liquid-like collisions in which the scattering rate is proportional to the ratio of the mean thermal velocity to the interparticle separation. It yields

$$
\gamma_{xx} = \gamma_{xx,0} + \left\{ \left[(E-E_{xx}/2)^4 + 4(K\sqrt{n_p})^4 \right]^{1/2} - (E-E_{xx}/2)^2 \right\}^{1/2} 2^{-1/2},
$$

where E is the photon energy, K is a constant, $\gamma_{xx,0}$ is the natural biexciton linewidth at low intensities, E_{xx} is the biexciton energy, and n_p is the polariton density.

By choosing $\gamma_x = 0.03$ meV we find that the linear absorption at the tail of the exciton, near the biexciton resonance, is in good agreement with experimental values,[7] as is the calculated value of the group velocity.[8] The light intensity inside the etalon is obtained from

$$
I = n_p E V_E,
$$

where V_E is the energy velocity, which is found to be close to the group velocity away from the biexciton resonance.

The calculated absorption coefficient $\alpha = 2Kn''$ and the change of refractive index $\Delta n' = n'(I) - n'(0)$ are shown in Fig. 1 and are in agreement with recently published experimental results.[5] Figure 2 shows the calculated bistability curves for a 10-μm-thick CuCl slab having a reflectivity $R = 0.9$ and $R = 0.8$ at both surfaces, where $E = 3.177$ eV. One observes that bistability is lost at reflectivities smaller than 0.9 for this case. For a 1-μm-thick sample, bistability is theoretically possible only with extremely high intensities that will destroy the sample, and are therefore impractical to use. When the biexciton broadening is not taken into account, our results indicate that 35 MW/cm² is needed to switch on a 1-μm device with $R = 0.9$ and about 10 MW/cm² is needed for a 10-μm device.

To summarize our calculated results, we show that unlike the previous prediction that a 1-μm CuCl slab can be switched on with 0.1 MW/cm² with uncoated surfaces, bistability can be obtained only with high reflection coatings of the Fabry-Perot etalon for ∿10-μm samples, and with intensities on the order of 10 MW/cm².

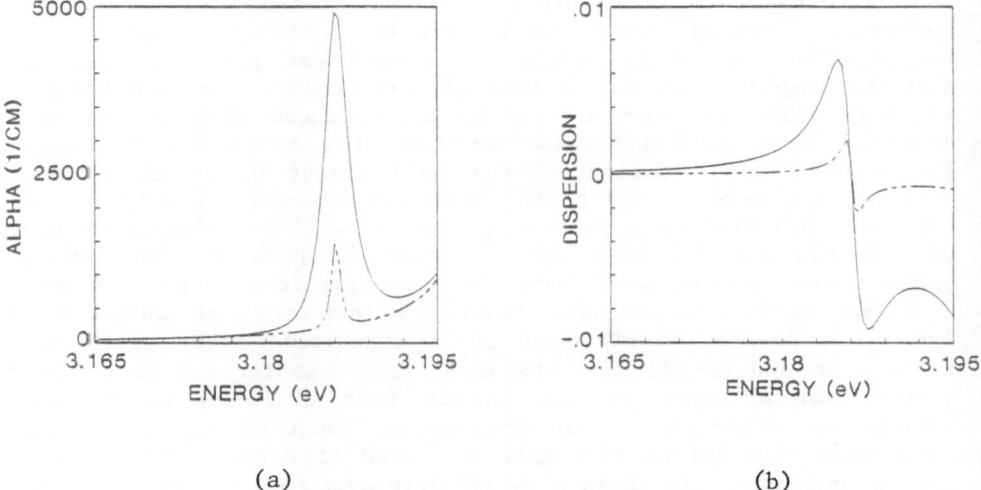

(a)   (b)

Fig. 1.   (a) Calculated α (cm⁻¹) as a function of polariton energy E (eV) for $n_p = 5 \times 10^{16}$ cm⁻³ (solid line) and $n_p = 5 \times 10^{15}$ cm⁻³ (dashed line), which corresponds to ∿10 MW/cm² and ∿1 MW/cm² at the biexciton resonance. (b) Calculated $\Delta n'$ as a function of polariton energy E (eV) for the same parameters.

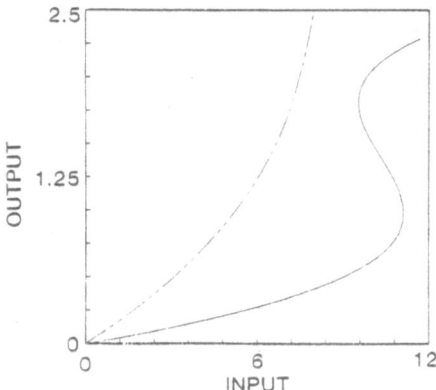

Fig. 2. Calculated output vs input intensities (MW/cm²) for a 10-μm-thick CuCl Fabry-Perot etalon at E = 3.177 eV for R = 0.9 (solid line) and R = 0.8 (dashed line).

EXPERIMENTAL PROCEDURES

High-purity bulk single crystals of CuCl were grown by direct reaction of pure Cu metal and CCl₄ in a sealed and evacuated silica tube.[10] These single crystals were evaporated onto fused silica substrates, which were dielectrically coated yielding 90% reflectivity. A similar dielectric mirror was placed on the CuCl film to complete the fabrication of the etalon. The thickness of the CuCl films (between 8- and 12-μm) produced this way could be measured by a film thickness monitor with a crystal oscillator. The etalons were mounted on the cold finger of an Air Products cryostat and were cooled to liquid-He temperature. The input laser beam was provided by a Hanna type dye laser[11] containing a mixture of 20% BBQ and 80% BPBD dye (to move the peak of the amplified spontaneous emission away from the operating frequencies) pumped by a Molectron UV-400 nitrogen laser. Pulses of 4-ns duration and less than 30-GHz linewidth at 10-Hz repetition rates were easily tuned from 380 to 392 nm. The laser beam was focused to a 20- to 25-μm diameter spot on the sample with a 32-mm focal length microscope objective. An Instrument Technology, Inc. vacuum photodiode with 0.1-ns rise time was used to detect both input and output pulses. The pulses were displayed on a Tektronix 7104 oscilloscope, with an amplifier rise time of 0.6 ns.

Hysteresis loops were obtained with the oscilloscope using the x-y display mode. The input pulse was optically delayed before striking the detector, so that the photodiode signal consisted of a time-resolved output and input pulse pair (time traces as in Figs. 4 through 6). This two-pulse signal was then sent "backwards" through an HP 15104A adder; one output cf the adder was connected directly to the oscilloscope x axis and the other output was cable-delayed and connected to the y axis. The oscilloscope internal time delay was used to synchronize the timing so that the etalon output vs input yielded a straight light in the absence of the sample (see Fig. 6).

EXPERIMENTAL DATA AND DISCUSSION

Figure 3 shows optical limiting action in a 10-μm-thick CuCl etalon at 7K with about 15 MW/cm² intensity. In Fig. 3a the output versus input intensity is plotted, and Fig. 3b shows both output (first pulse) and input (second pulse) versus time. This behavior was observed only in a very small frequency range near the biexciton resonance (√3.186 eV). Changing the frequency by a fraction of a milli-electron volt caused the limiting action to disappear. Optical bistability was not observed in this etalon, presumably because the finesse was too low.

Figures 4 and 5 show optical bistability in a 12-μm etalon at 17K with about 7 to 14 MW/cm² intensity. The operating frequency corresponded to about 3.187 eV (the biexciton resonance, which is 3.1861 eV at 4K, shifts to higher energies at higher temperatures). The laser frequency was changed by a fraction of a milli-electron volt in the sequence of Figs. 4a, 4b, and 4c. Bistability was seen only in a narrow frequency range (Fig. 4b). Detuning the laser by a slight amount to lower frequency (thereby moving it further away from the Fabry-Perot peak) caused the switch-on to occur at higher intensity; detuning it further by a fraction of a milli-electron volt resulted in the disappearance of bistability. Similar bistable

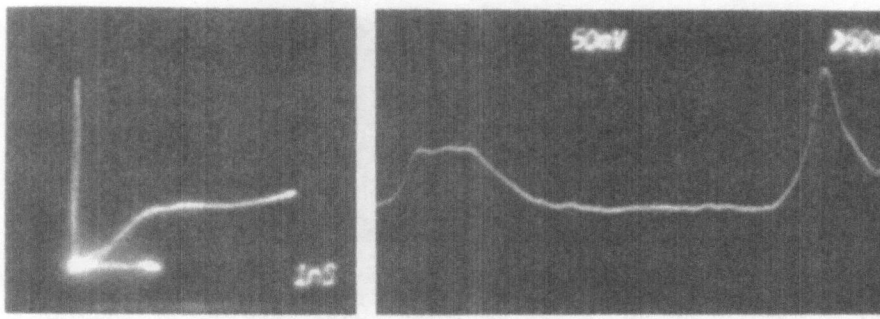

Fig. 3. Limiting action in a 10-μm-thick CuCl etalon. (a) Output vs input intensity. (b) Input pulse (right) and clipped output pulse (left) vs time. Time scale 2 ns/division.

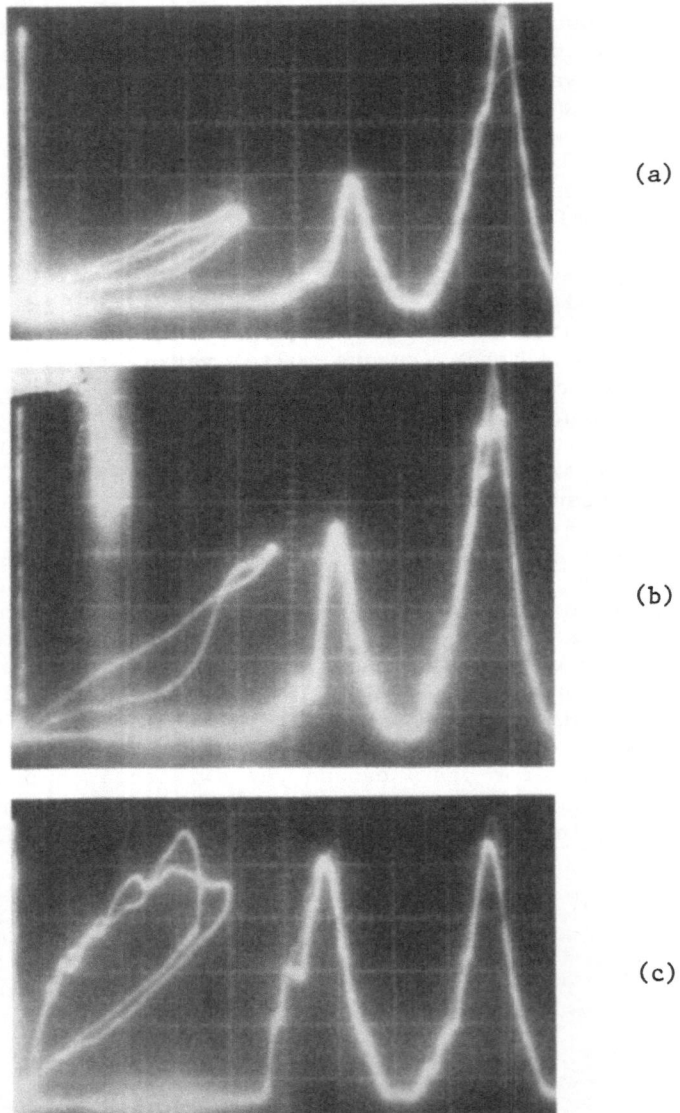

Fig. 4.   Switch-on in a 12-μm-thick CuCl etalon.   The laser
frequency is gradually increased (corresponding to a fraction of a
milli-electron volt change) in the sequence of photos (a) - (c).
Each photo shows hysteresis loop, output pulse (center), and input
pulse (right).   Note only (b) shows switch-on.   Switch-off is not
resolved in (b) as the input dye laser pulse falls too steeply.   The
time scale is 5ns/div.

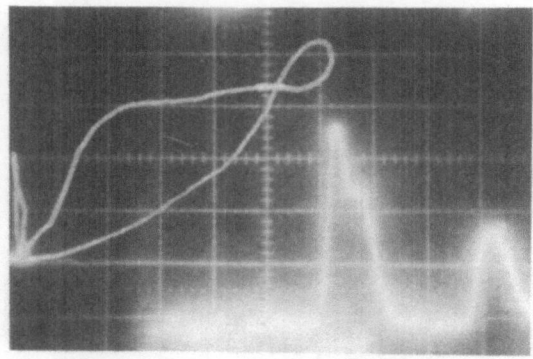

Fig. 5. Switch-off in a 12-μm-thick CuCl etalon. When the input dye laser pulse (right) is adjusted for a slow falloff, switch-off in the output pulse (center) is more convincing. The time scale is 5 ns/div.

behavior was observed when the laser frequency was held fixed and the Fabry-Perot-laser detuning was varied by moving across the sample to regions of differing thicknesses. The Fabry-Perot peak transmission frequency was located on the high-energy side of the laser frequency and both were on the low-energy side of the biexciton resonance. This is consistent for biexcitonic bistability, as the index increases with increasing intensity on the low-energy side of the biexciton resonance (see Fig. 1b). The switch-off is not well resolved in Fig. 4b due to the fast fall time of the input pulse (detection-limited by the oscilloscope rise time). However, when the input pulse was adjusted to fall more slowly, the switch-off was observed as shown in Fig. 5. Figure 6 demonstrates that the output versus input yields a straight line when the sample is removed. Efforts toward measuring the predicted picosecond switching times and the dependence of these times on the detuning from the biexciton frequency are under way. Hopefully, more nearly steady-state hysteresis loops showing sharp switch-on and switch-off will be obtained in the process.

In summary, we present theoretical predictions and corroborating experimental observation of optical bistability using the biexciton two-photon resonance in CuCl.

ACKNOWLEDGMENT

We would like to thank Lawrence Livermore Laboratory for the loan of the nitrogen and dye lasers. We gratefully acknowledge support from the National Science Foundation, the Air Force Office of Scientific Research and the Army Research Office.

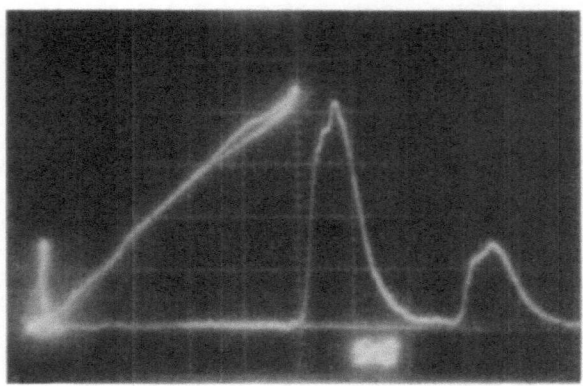

Fig. 6.   Output pulse (center) vs input pulse (right) shows no hysteresis (left) in the absence of a sample.

REFERENCES

1. S. W. Koch and H. Haug, Phys. Rev. Lett. 46:450 (1981).

2. E. Hanamura, Solid State Commun. 38:939 (1981).

3. R. W. Svorec and L. L. Chase, Solid State Commun. 20:353 (1976).

4. L. L. Chase, N. Peyghambarian, G. Grynberg, and A. Mysyrowicz, Opt. Commun. 28:189 (1979).

5. T. Itoh and T. Katohno, J. Phys. Soc. Japan 51:707 (1982).

6. N. Peyghambarian, L. L. Chase, and A. Mysyrowica, Opt. Commun. 42:51 (1982).

7. G. M. Gale and A. Mysyrowicz, Phys. Rev. Lett. 32:727 (1974).

8. N. Peyghambarian, D. Sarid, and H. M. Gibbs, to be published.

9. D. Sarid, N. Peyghambarian, and H. M. Gibbs, to be published in Phys. Rev. B, rapid communication, July 15, 1983.

10. B. Batlog, J. P. Remika, and A. Jayaraman, Bull. Am. Phys. Soc., 25:268 (1980).

11. D. C. Hanna, P. A. Kärkkäinen, and R. Wyatt, Opt. Quant. Elect. 7:115 (1975).

LOW POWER OPTICAL BISTABILITY NEAR BOUND EXCITONS IN CADMIUM SULFIDE

Mario Dagenais and Herbert Winful

GTE Laboratories Incorporated
40 Sylvan Road
Waltham, Massachusetts 02254

INTRODUCTION

Direct gap semiconductors exhibit some of the largest optical nonlinearities measured in condensed media. Using the resonance enhancement of the material nonlinearity near the band gap, very large nonlinear phase shifts can be accumulated over micron size distances and large changes of the absorption coefficient can also be observed at relatively low incident power. For instance, the saturation of free excitons and the associated generation of free carriers following near-band gap excitation in bulk GaAs[1] and InSb[2] or in GaAs/AlGaAs superlattices[1c,3] have been used to demonstrate optical bistability on the 10 mW power scale. The response time of these optical bistable devices is typically tens of nanoseconds and depends directly on the carrier recombination lifetime. Rather than using the extended states (free excitons and free carriers) of a semiconductor, as is usually done, we propose here a completely different approach for demonstrating optical bistability. We would like to use those bound states of a direct gap semiconductor which have a large oscillator strength. In particular, we have studied the saturation behavior of the I_2 bound exciton in cadmium sulfide (CdS), which is an exciton bound to a neutral donor with a binding energy of about 8 meV.[4] Bound excitons have the advantage of having an oscillator strength typically five to ten thousand times larger than the oscillator strength per molecule of the free exciton. This leads to an exceedingly fast radiative lifetime.[5] The I_2 bound exciton in CdS has a radiative lifetime of 500 psec which should permit optical signal processing on a subnanosecond time scale. Furthermore, since the decay is mostly radiative, very little heat is dissipated in the sample. This should allow the demonstration of very high

267

repetitive rates. In addition, the transition linewidth of the I_2
bound exciton was measured to be about 8 GHz at 2K. This is much
narrower than the free exciton linewidth and is an indication that
the power required for seeing large nonlinear effects with the
bound exciton can be substantially less than the power required
with the free exciton. Indeed, in a tight focusing geometry, we
demonstrate here that the I_2 bound exciton in CdS can be saturated
with a cw power of only 3.6μW. This corresponds to a very low
saturation intensity of 58 W/cm^2. We find good agreement with
the predictions of a two-level model, if we make the reasonable
assumption that the line is inhomogeneously broadened. Even for
intensities as large as fifty times the saturation intensity, we
find that thermal effects play a minimal role right at the exciton
resonance. In a more loosely focused geometry, transverse optical
bistability is observed in an uncoated sample with only a milliwatt
of incident power when the laser is tuned near the bandgap. In bistability measurements, the exact role played by the bound exciton has
not yet been fully elucidated. These experiments are done completely cw. Very large hysteresis loops are observed and more
than one hysteresis loop can be seen as the incident intensity
increases. We observe the formation of many rings in the far field.
In our measurements with an uncoated sample, only transverse bistability without power hysteresis is observed. Very regular
self-pulsations are detected on the 10 μsec time scale. Irregular
pulsations are seen at the highest cw intensity used in the
experiment.

SATURATION OF BOUND EXCITONS

     Our saturation measurements on the I_2 bound exciton in cadmium
sulfide were done on a 20 μm thick platelet immersed in liquid
helium. A dye laser operating at about 487 nm was tightly focused
to a 4 μm diameter $(1/e^2)$ spot size. Single mode scans over 30 GHz
range were performed over the bound exciton resonance at different
laser intensities, while monitoring the transmission through the
sample (Figure 1). A remarkably small transition linewidth of
7.9 GHz (FWHM) was observed, which is quite comparable to the
sharp resonances found in atomic physics. We notice that the
width of the resonance is approximately constant up to the largest
intensities used. Furthermore, as the intensity is increased, we
do not observe any shift of the resonance position. Since we can
correlate the position of the peak of the resonance with a certain
sample temperature and since the peak of the resonance does not
appear to shift as the laser intensity is increased, we take this
as evidence that thermal effects play a minor role at the bound
exciton resonance.[6] At low temperature, we expect that the
transition linewidth will be inhomogeneously broadened since
excitons bound at different locations see slightly different
environments. We describe the saturation of bound excitons right
at the peak of the resonance as a saturable, inhomogeneously

broadened two-level system. The following equation describes the
attenuation of a plane wave as it propagates through an inhomogeneously broadened system:

$$\frac{dI}{I} = \frac{\gamma_0}{(1+I/I_s)} \; \tfrac{1}{2} \; dz$$

An excellent fit is obtained if one uses as an adjustable parameter
a saturation power of 3.6 μW, which corresponds to a saturation
intensity of 58 W/cm^2 (Figure 2). The fit describes the experimental results for powers up to 100 times the saturation power. The
linear absorption coefficient at the resonance peak is determined
to be 3.2×10^3 cm^{-1}. A first principles calculation using a two-
level model predicts exactly this power if we assume a dephasing
broadening of 1.8 GHz due to phonon interaction. An independent
experimental study of the dependence of the transition linewidth
on temperature predicts a similar homogeneous linewidth at 2K,
once the inhomogeneous linewidth is subtracted.

OBSERVATION OF OPTICAL BISTABILITY NEAR THE BAND GAP

For studying optical bistability, the laser beam was focused
to a 20 μm diameter spot size (1/e^2) in a high quality plane parallel
cadmium sulfide platelet of 20 μm thickness immersed in superfluid
helium. No reflective coating was deposited on the sample since
the discontinuity of the index of refraction appears to be sufficiently large (n$_{CdS}$ ∼3) to provide the necessary optical feedback.

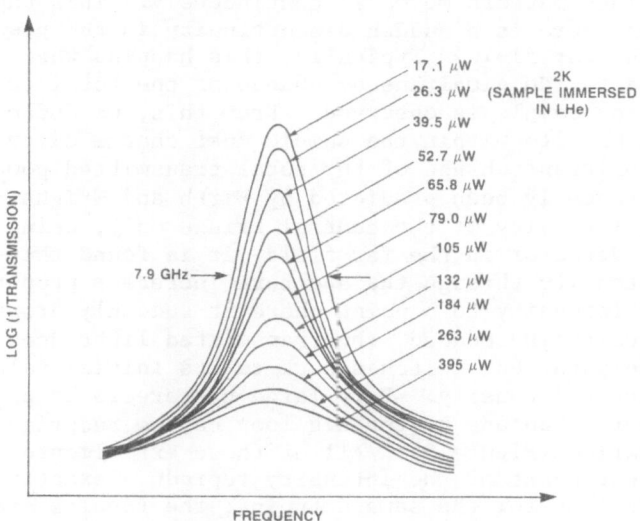

Fig. 1. Saturation of the I$_2$ bound exciton. The incident power
         is given in microwatts.

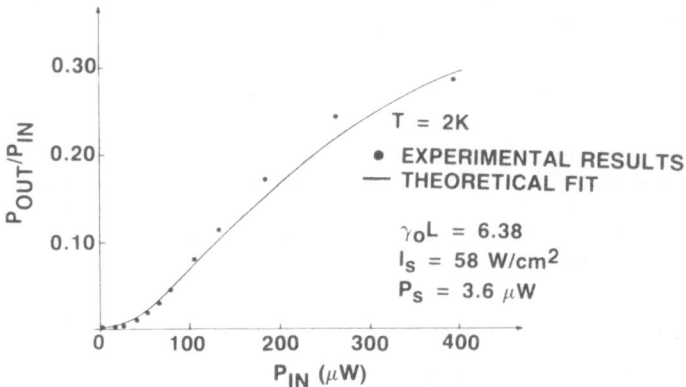

Fig. 2.   Power dependence of the optical transmission right at
the absorption peak of the I$_2$ bound exciton.

At very low intensities, the far field profile of the output beam
from the sample was observed to be Gaussian. As the intensity is
increased, the beam diameter increases and a circular fringe
pattern develops. This effect disappears when the laser is tuned
too far from the free and bound exciton resonances. By increasing
the input field further, the number of rings increases at the same
time as the fringe pattern moves in continuously. This happens up
to a point when there is a sudden discontinuity in the position of
the rings in the far field. Typically, this happens when 3 or 4
rings are present. No simultaneous change of the total transmitted
power through the sample is observed. From this, we deduce that
the laser beam profile within the sample must change discontinuously
without a concomitant change of the total transmitted power. Such
an effect has recently been predicted by Firth and Wright.[7] By
monitoring the intensity of the central fringe only, using an
aperture and a detector in the far field, it is found that the
transmitted intensity through the aperture increases proportionally
with the input intensity to a point where it suddenly drops. As
the input intensity is reduced, the transmitted light does not
follow the same path, but switches back to its initial value at a
much reduced input intensity. Very large hysteresis loops are
observed and more than one hysteresis loop can be recorded as the
intensity is varied (Figure 3). All of these experiments are done
cw and subsequent scans of the intensity reproduce exactly the
same hysteresis loop for the same detuning; the results are reproducible. The formation of rings in the far field is the result
of an induced aberrated lens within the sample. The dependence of
the phase shift on radial position is not simply quadratic as in a
perfect lens, but is proportional to the Gaussian intensity profile

of the incident beam. In our experiment, up to 10 or 12 rings
can be observed at a maximum power less than 10 mW. Using standard
theory,[8] we deduce a change of the index of refraction as large as
$\delta n=0.2$ at the center of the nonlinear lens. As the incident
intensity is increased, the transmitted intensity monitored
through an aperture exhibits highly regular pulsations with a
period on the order of 10 μs (Figure 4). At the highest intensities
used in the experiment, the output appears erratic in time.

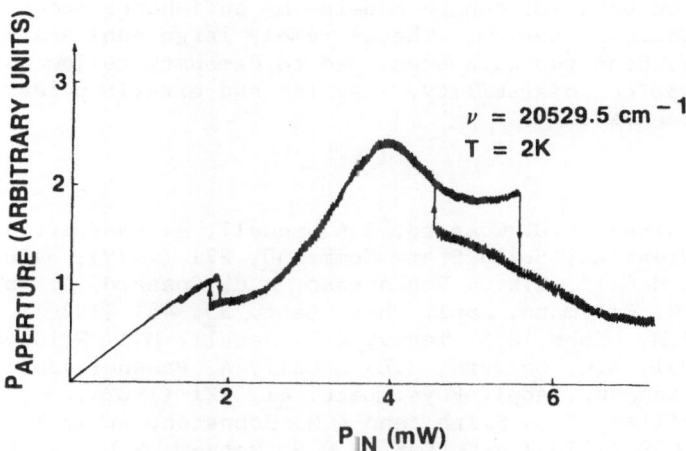

Fig. 3. Observation of transverse optical bistability.

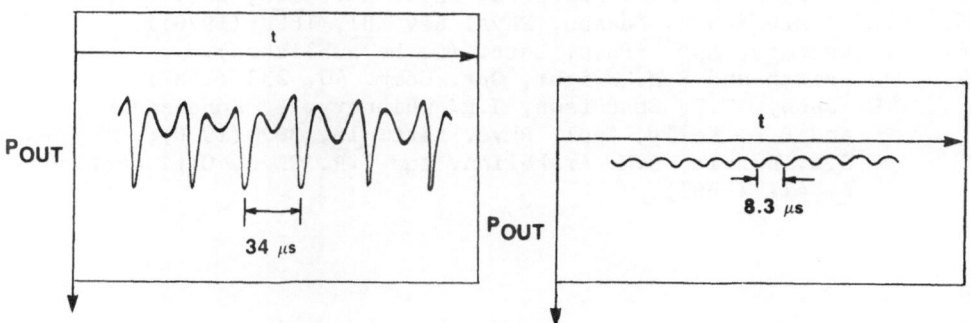

Fig. 4. Observation of self-pulsation.

On the other hand, if the whole transmitted beam is collected with
a large aperture lens, the pulsation tends to be washed out as a
result of averaging across the transverse profile. We do not yet
have a clear picture of the origin of these pulsations or of the

extremely large nonlinearities seen near the band gap responsible for our optical bistability results. Lattice heating does not appear to be significant since there are no substantial shifts in the position of weak absorption lines between the free and bound excitons as the laser intensity is increased.

CONCLUSION

In conclusion, we have been able to saturate the I_2 bound exciton resonance in cadmium sulfide with very low power. The saturation behavior can be modeled by an inhomogeneously broadened two level system. The extremely large nonlinearities seen near the band gap have been used to demonstrate low power transverse optical bistability. Regular and erratic pulsations were also observed.

REFERENCES

1. a) H.M. Gibbs, A.C. Gossard, S.L. McCall, A. Passner, and W. Wiegmann, Solid State Comm. 30, 271 (1979); b) H.M. Gibbs, S.L. McCall, T.N.C. Venkatesan, A.C. Gossard, A. Passner, and W. Wiegmann, Appl. Phys. Lett. 35, 451 (1979); c) H.M. Gibbs, S.S. Tarng, J.L. Jewell, D.A. Weinberger, K. Tai, A.C. Gossard, S.L. McCall, A. Passner, and W. Wiegmann, Appl. Phys. Lett. 41, 221 (1982).
2. D.A.B. Miller, S.D. Smith, and A.M. Johnston, Appl. Phys. Lett. 35, 658 (1979); A.K. Kar, J.G.H. Mathew, S.D. Smith, B. Davis, and W. Prettl, Appl. Phys. Lett. 42, 334 (1983).
3. D.A.B. Miller, D.S. Chemla, D.J. Eilenberger, P.W. Smith, A.C. Gossard, and W.T. Tsang, Appl. Phys. Lett. 41, 679 (1982).
4. D.G. Thomas and J.J. Hopfield, Phys. Rev. 128, 2135 (1962).
5. C.H. Henry and K. Nassau, Phys. Rev. B1, 1628 (1970).
6. M. Dagenais, Appl. Phys. Lett. (to be published).
7. W.J. Firth and E.M. Wright, Opt. Comm. 40, 233 (1982).
8. F.W. Dabby, T.K. Gustafson, J.R. Whinnery, Y. Kohanzadeh, and P.L. Kelly, Appl. Phys. Lett. 16, 362 (1970); S.D. Durbin, S.M. Arakelian, and Y.R. Shen, Opt. Lett. 6, 411 (1981).

ROOM TEMPERATURE OPTICAL NONLINEAR ABSORPTION AND REFRACTION IN GaAs MULTIPLE QUANTUM WELLS

D. A. B. Miller, D. S. Chemla, A. C. Gossard* and
P. W. Smith
Bell Laboratories, Rm 4B-417
Crawfords Corner Road
Holmdel, NJ   07733

ABSTRACT

A brief review is given of the nonlinear optical properties of
room-temperature GaAs multiple quantum wells. These materials show
very large nonlinear absorption and refraction near their exciton
resonances. Furthermore, they are directly compatible with laser
diodes. C.W. nonlinear optics has recently been demonstrated for
the first time with a laser diode as the sole light source.

Progress towards practical all-optical switching and signal
processing devices depends crucially on finding suitable nonlinear
materials. Consequently, various workers have explored so-called
'dynamic' nonlinear optical effects in semiconductors[1] in which
the optical properties change as a result of real excitation of
the material. Semiconductors are particularly attractive for
fabrication as small devices. Unfortunately, while some of these
effects should persist to room temperature, one of the most
attractive, saturation of free exciton resonances[2] is basically
limited to low temperatures in conventional semiconductors because
of thermal broadening of the resonances.

By confining carriers in GaAs quantum wells ∿100Å thick,
spaced apart with GaAlAs barriers, exciton resonances can be
observed at room temperature,[3] primarily because the confinement

*Bell Laboratories, Murray Hill, NJ   07974

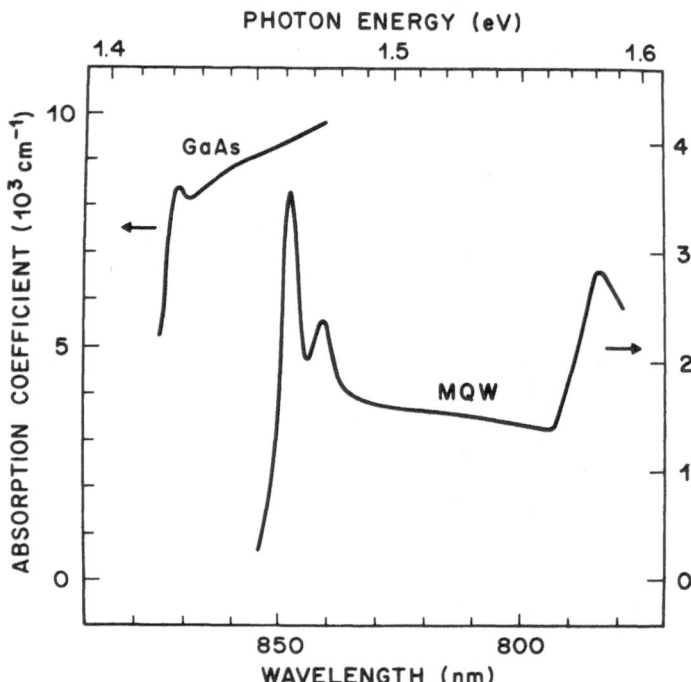

Figure 1. Absorption spectra of GaAs and of Multiple Quantum Well
           (MQW) material.[3] The MQW consists of 102Å GaAs layers,
           each separated by 207Å $Ga_{0.72}Al_{0.28}As$ layers to give a
           total thickness of $\leq$2.4µm. MQW absorption is an under-
           estimate due to thickness loss in etching.

leads to an increase in binding energy, ∿x2 for 100Å layers, whereas
in GaAs the resonance has all but disappeared at room temperature.
This contrast is clearly shown in Figure 1 which shows room-
temperature linear absorption spectra of the two materials. The
GaAs spectrum shows only a very small 'bump' at the band edge (near
870 nm). The quantum well material on the other hand shows two very
clear excitonic peaks near 850 nm. The confinement in thin wells
affects also the interband absorption changing it into a series of
steps, each with an exciton feature. The first step is also
shifted as compared to the band edge in the bulk material; this
shift is primarily responsible for the band edge being at ∿850
rather than ∿870 nm in the quantum well. The peak at ∿780 nm is the
start of the second step. The existence of two peaks at ∿850 nm is
due to the lifting of the degeneracy of the top valence bands by
the confinement. The larger, lower energy peak is the heavy-hole
(hh) exciton, and the smaller, higher energy peak is the light hole
(lh) exciton. In order to observe these effects in absorption, it

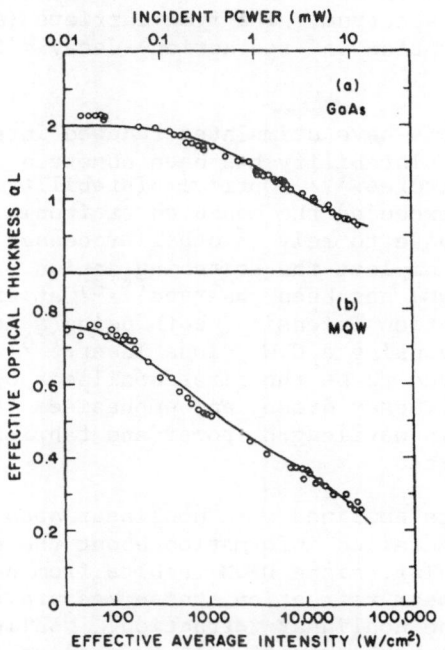

Figure 2.   Absorption saturation of (a) GaAs and (b) MQW at the
            exciton peak[3] (samples as for Fig. 1).

is necessary to grow many layers (e.g., 50-100) to obtain sufficient
absorption length. The separation between the GaAs layers is chosen
sufficiently large that there is little tunnelling between layers,
and the wells therefore behave independently, hence the title
Multiple Quantum Wells (MQW).

     One of the benefits of the clear excitonic resonances in the
MQW is seen in the nonlinear absorption behavior.[3]   Figure 2
shows the comparison of absorption saturation at the exciton peak
(the hh exciton in the MQW) in GaAs and MQW. The nonlinear
absorption starts at ~10 times lower power or intensity in the MQW,
with effective saturation intensities of 200-500 W/cm^2 in the MQW
materials. This absorption saturation appears to recover on a time
scale of ~20 ns in the MQW.

     The mechanism behind the excitonic absorption saturation at
room temperature is thought to be as follows.[3] Optical absorption
near the resonance creates excitons. Within ~0.5 ps these excitons
are ionized by optical phonons into free electrons and holes; this
time is deduced from the measured optical phonon contribution to

the room-temperature line width. This free carrier density inter-
feres with the production of further excitons, thus altering the
excitonic absorption spectrum. The free carriers (and consequently
the changes the absorption and refraction) decay with ∿20 ns
lifetime.

These observations have stimulated renewed interest in MQW
structures. Optical bistability has been observed in room-
temperature MQW structures.[4] Optical bistability observations
have so far greatly exceeded the measured excitonic saturation
intensities and may have to rely on other processes (e.g., interband saturation) to complete the switching action. Degenerate
four-wave mixing (DFWM) has been observed[5,6] at excitation levels
well below the saturation intensity, both using a modelocked
laser[5] and recently using a C.W. diode laser.[6] This latter
observation is believed to be the first nonlinear optics experiment with only a C.W. laser diode, and emphasises the compatibility
of the MQW material in wavelength, power and fabrication technology
with laser diode sources.

DFWM measurements combined with nonlinear absorption results
also give much more detailed information about the nonlinear
behavior. In particular, since DFWM results from both nonlinear
absorption and nonlinear refraction, these measurements when
deconvoluted yield the nonlinear refraction. Nonlinear refraction
at low intensity is otherwise difficult to measure, because the
total nonlinear path length change is small compared to a wavelength; this makes nonlinear interferometry difficult. Figure 3
shows the measured spectra of (a) linear transmission, (b) the
change in transmission of a test beam as the main pump beam is
turned on and off (i.e., nonlinear absorption) and (c) DFWM. The
nonlinear absorption spectrum is complicated, but at 1.455eV for
example where the nonlinear absorption is zero, the entire DFWM
signal arises from nonlinear refraction. The nonlinear refraction
deduced from this can be expressed in three different ways:
$n_2 \sim 2 \times 10^{-4} cm^2/W$; effective $\chi^{(3)} \sim 6 \times 10^{-2}$ e.s.u.; or, arguably the
most meaningful parameter, the change in refractive index for each
excited carrier pair per unit volume, $n_{eh} \sim 2 \times 10^{-19} cm^3$. This is
relatively a very large number and if devices can be designed to
utilize such large nonlinearities, operating powers may be reduced
by orders of magnitude.

Detailed analysis of the nonlinear absorption spectrum[7]
shows that it results entirely from the excitonic parts of the
spectrum at these intensities, with negligible contribution from
interband saturation. This is consistent with the large observed
effects. Furthermore, using the nonlinear absorption data, the
nonlinear refraction spectrum can be calculated using the

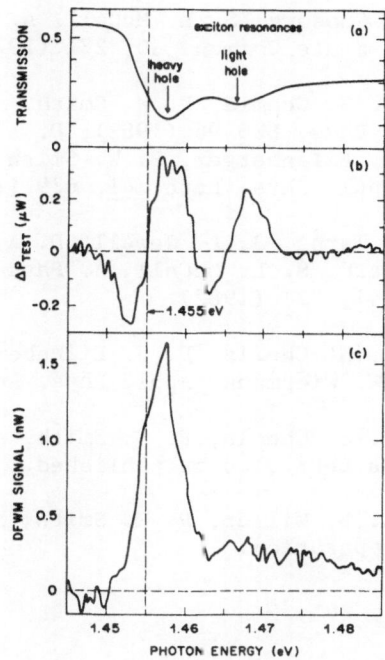

Figure 3. (a) Linear absorption, (b) nonlinear absorption and (c) degenerate four-wave mixing spectra of a sample with 96Å wells.[5] Pump power was ∿300μW, test power ∿60μW in 45 and 30μm diameter spots respectively.

Kramers-Krönig relations. The measured nonlinear absorption and resultant calculated nonlinear refraction can be used to calculate the DFWM spectrum. The result is in extraordinarily good agreement both in form and magnitude with the experimental spectrum,[7] confirming the overall approach to the data analysis.

Work is continuing to characterize these nonlinearities in more detail, and currently these materials offer good prospects for practical room-temperature all-optical switching and signal processing devices compatible with laser diodes.

REFERENCES

1. For recent reviews see A. Miller, D. A. B. Miller and S. D. Smith, Adv. in Phys. 30, 697 (1981); D. A. B. Miller, Laser Focus 19, 61 (1983).

2.   H. M. Gibbs, A. C. Gossard, S. L. McCall, A. Passner and
     W. Wiegmann, Solid State Commun. 30, 271 (1979).

3.   D. A. B. Miller, D. S. Chemla, P. W. Smith, A. C. Gossard and
     W. T. Tsang, Appl. Phys. B28 96 (1982); D. A. B. Miller,
     D. S. Chemla, D. J. Eilenberger, P. W. Smith, A. C. Gossard
     and W. T. Tsang, Appl. Phys. Lett. 41, 679 (1982).

4.   H. M. Gibbs, S. S. Tarng, J. L. Jewell, D. A. Weinberger,
     K. Tai, A. C. Gossard, S. L. McCall, A. Passner and W. Wiegmann,
     Appl. Phys. Lett. 41, 221 (1982).

5.   D. A. B. Miller, D. S. Chemla, D. J. Eilenberger, P. W. Smith,
     A. C. Gossard and W. Wiegmann, Appl. Phys. Lett. 42, 925 (1983).

6.   D. A. B. Miller, D. S. Chemla, P. W. Smith, A. C. Gossard and
     W. Wiegmann, Optics Lett., to be published.

7.   D. S. Chemla, D. A. B. Miller, P. W. Smith, A. C. Gossard and
     W. Wiegmann, in preparation.

NONLINEAR REFRACTION AND NONLINEAR ABSORPTION IN InAs

C.D. Poole and E. Garmire

Center for Laser Studies
University of Southern California
University Park, DRB 17
Los Angeles, CA 90089-1112

Abstract: We report here experimental measurements taken with an HF laser of a large nonlinear index of refraction ($n_2 = 4.5 \pm 2.0 \times 10^{-5}$ cm^2/W) and associated nonlinear absorption near the bandgap in InAs between 6°K and 100°K. These results are shown to be consistent with the bandgap resonant model used to explain the nonlinearities seen in InSb and HgCdTe. Furthermore it is shown that an InAs bistable device operating above liquid nitrogen temperatures should be capable of switching with milliwatt power levels.

Semiconductors operated near their bandgap hold great promise as a nonlinear medium for practical, all optical bistable devices. To date InSb[1] and GaAs[2] are the only semiconductors to exhibit bistability above liquid nitrogen temperatures. Recently a large nonlinear index has been observed in CdHgTe at 175°K but due to rapid saturation of the nonlinearity, bistability may be difficult to realize in this material.[3] In this paper we present the results of measurements of both a nonlinear index and nonlinear absorption in InAs, a III-V semiconductor whose band gap lies between that of InSb and GaAs. Our results give a material figure of merit 4 percent that of InSb. This is consistent with the existing bandgap resonant model used to explain the nonlinearties in InSb and CdHgTe. These results also indicate that optimization of Fabry-Perot parameters should yield bistability in InAs with power levels in the milliwatt range.

The samples consisted of single crystals of n-type InAs ($N_D - N_A = 2.6 - 3.4 \times 10^{16}$cm^{-3}) polished plane and parallel to form Fabry-Perot etalons using the Fresnel reflection from the

faces. The samples were placed in a variable temperature cryostat and the TEM_{00} output of an HF laser was then focused down
onto them with a spot size of 120 μm. The transmitted near field
beam pattern was imaged onto the output detector through a pinhole such that only a 25 μm spot at the center of the near field
beam was detected, allowing plane wave analysis of the transmission. Due to instabilities in the laser discharge, the laser
output was in the form of millisecond pulses providing the variable intensity necessary for the measurements. The input and
output signals were monitored simultaneously to produce signal
out vs. signal in curves which were digitized and sent through a
data acquisition system. Because of the variation of bandgap
with temperature in InAs, we were able to "tune" the bandgap of
the samples to the various HF laser lines, allowing measurements
to be made near the bandgap over a wide temperature range.

Measurements of the nonlinear index in the samples were made
by obtaining nonlinear transfer curves for the Fabry-Perot at
various detunings and comparing the measured curves to theoretical
curves produced using plane wave theory. In generating the theoretical curves the measured values of reflectivity and single pass
loss were used, leaving n_2 as the only fitting parameter. Since
the samples were slightly wedge shaped, detuning of the Fabry-
Perot was achieved by translating the samples perpendicular to
the beam. The absorption remained linear throughout the region
in which the nonlinear index measurements were made ($\alpha \leq 100$ cm^{-1}).
Figure 1a shows a typical set of tranmission curves with their
theoretical counterparts in Fig. 1b. As can be seen, one of the
measured curves came close to demonstrating bistability. Indeed,
the corresponding theoretical curve at a detuning of .63 π radians
demonstrates slight hysteresis.

Figure 2 shows a plot of measured nonlinear index and linear
absorption coefficient versus temperature for two different samples, taken from separate boules. As expected, the absorption
increases with increasing temperature, reflecting the fact that the
bandgap is moving closer to the photon energy. The nonlinear
index also increases with temperature in the same manner as the
linear absorption, corresponding to a roughly constant ratio of
n_2/α. Note that the two samples behave similarly, indicating the
fundamental nature of the effect. The nonlinear index of sample
#1 was also measured at 6°K and was found to be $n_2 \simeq 4 \pm 2 \times 10^{-6}$
cm^2/W with an absorption coefficient of $\alpha \simeq 20$ cm-1. The reason
for the smaller value of n_2 at the lower temperature will be discussed later.

Nonlinear absorption was measured in the samples from 6°K to
100°K. For these measurements, the pinhole in front of the output
detector was removed so that total power out was detected.

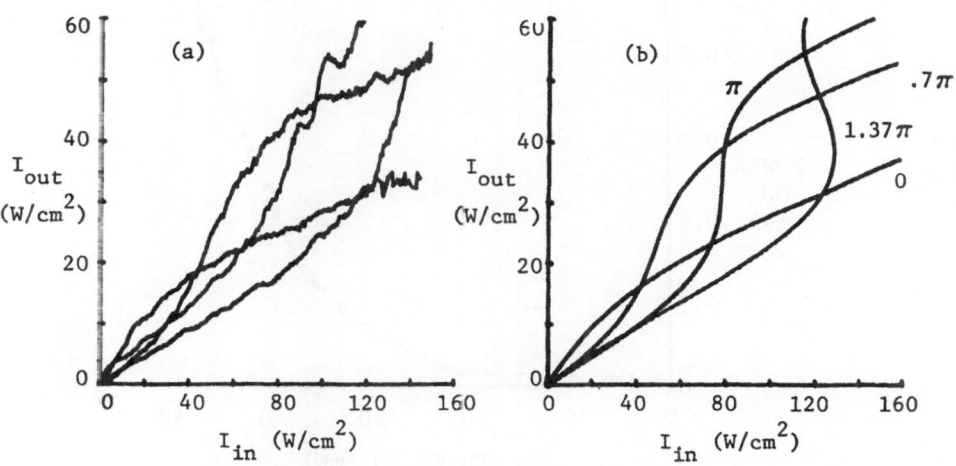

Fig. 1. Intensity out vs. intensity in for NLFP at various detunings. (a) Measured curves for InAs Fabry-Perot etalon at 84°K.
Thickness 170 μm, wavelength 3.096 μm. (b) Theoretical curves
obtained using plane wave theory.
$$R = .3, \ 1 - e^{-\alpha L} = .3, \ n_2 = 4.5 \times 10^{-5} \, cm^2/W.$$

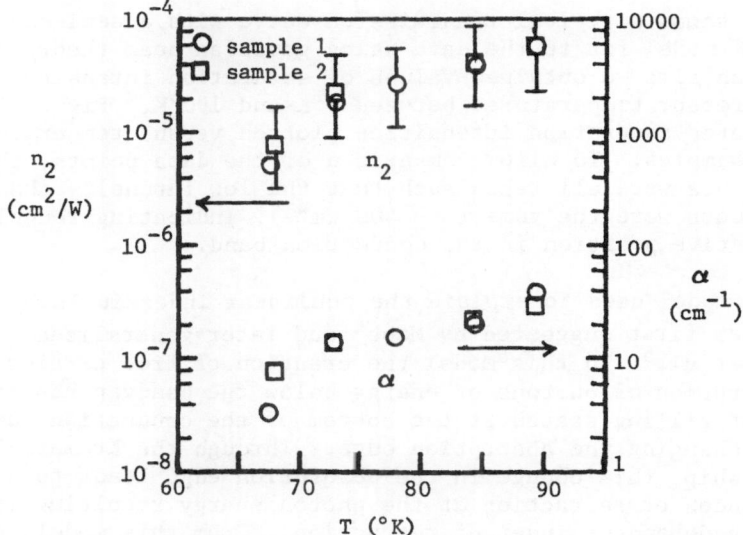

Fig. 2. Measured nonlinear index n_2 and absorption coefficient α
vs. temperature for two different samples of InAs. $\lambda = 3.096 \, \mu m$
Sample 1- thickness 140 μm, $N_D - N_A = 3.4 \times 10^{16} \, cm^{-3}$.
Sample 2- thickness 170 μm, $N_D - N_A = 2.6 \times 10^{16} \, cm^{-3}$.

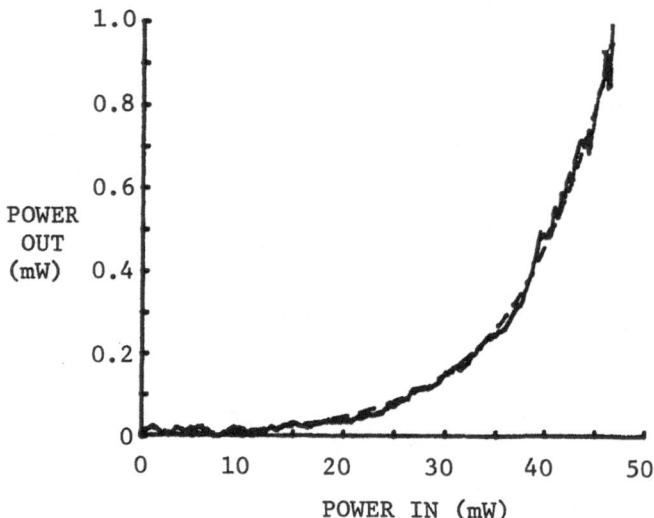

Fig. 3.   Power out vs. Power in for 170 μm thick sample at 18°K.
           Incident spot size is 120 μm, wavelength 2.957 μm.
           Dashed line is a semi-empeirical two-level model fit
           using gaussian beam theory.

Figure 3 shows a typical transmission curve with a semi-empirical
two-level model fit to the data using gaussian beam theory. By
doing such fits we obtained values of saturation intensity for
five different temperatures between 6°K and 100°K. Figure 4 shows
the measured saturation intensities plotted versus temperature for
the two samples. To allow comparison of the data points, the
measurements were all taken such that the low intensity absorption
coefficients were the same ($\alpha \simeq 400$ cm^{-1}), indicating roughly the
same relative position in the conduction band.

The model used to explain the nonlinear index in InSb and
CdHgTe was first suggested by Moss[4] and later generalized by
Miller, et al.[5] In this model the creation of free carriers by
the absorption of photons of energy below the bandgap has the
effect of filling states at the bottom of the conduction band
thereby changing the absorption edge. Through the Kramers-Kronig
relationship, this change in the absorption edge leads to a change
in the index of refraction at the photon energy resulting in an
intensity dependent index of refraction. From this model, one
would thus expect a large nonlinear index in a semiconductor to
be accompanied by low saturation intensities for photons of energy

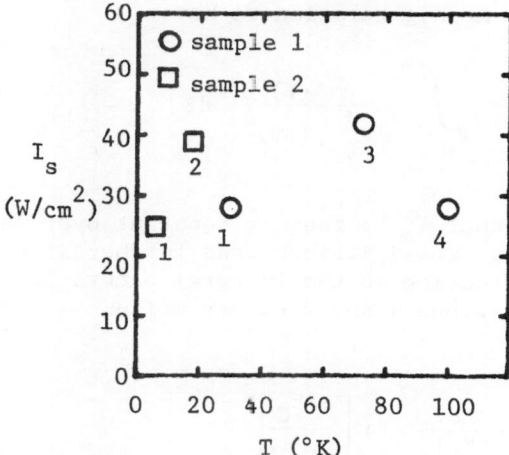

Fig. 4.   Measured saturation intensity vs. temperature in InAs.
Wavelengths and temperatures were selected such that the
low intensity absorption coefficients were the same for
all data pts (α = 400 cm^{-1}). Numbers next to data pts
indicate wavelengths used (1) 2.954 µm (2) 2.957 µm,
(3) 3.046 µm, (4) 3.096 µm.

greater than the bandgap. Indeed low saturation intensities have
been observed in InSb[6,7] and as seen in figure 4, this is also
the case in InAs.

In order to compare our results to the above mentioned model,
we may estimate what the value of n_2/α at photon energies below
the bandgap should be from the measured values of the saturation
intensity at photon energies above the bandgap. If we assume
that every photon absorbed creates a free electron (and hole) and
that a two-level model is adequate to describe the nonlinear
absorption, then it is straight forward to show that the change in
the absorption edge, as expressed by a nonlinear absorption coefficient $\Delta\alpha'(E)$, is related to the linear absorption of photons of
energy less than the band gap by

$$\Delta\alpha'(E) \; = \; \frac{\alpha(h\nu)I}{I_s(E)} \qquad\qquad h\nu < E_G \leq E \qquad\qquad (1)$$

where $\alpha(h\nu)$ is the linear absorption coefficient at the photon
energy and I_s is the saturation intensity for photons of energy
$h\nu' = E$.

From the Kramers- Kronig relation we have

$$\Delta n(h\nu) = \frac{\hbar c}{\pi} \int_{E_G}^{\infty} \frac{\Delta\alpha'(E)}{E^2 - (h\nu)^2} \, dE \quad . \tag{2}$$

Now if we assume that I_s is roughly constant over the range in energy in which $\Delta\alpha'$ is significant and further if we assume that most of the contribution to the integral occurs for energies near E_G then using equations 1 and 2 we may write

$$\frac{n_2}{\alpha} = \frac{\Delta n}{\alpha I} \approx \frac{\hbar c}{2\pi E_G I_s} \, \ell n \left[\frac{2E_G}{E_G - h\nu} \right] \quad . \tag{3}$$

Near 80°K we have $E_G = 0.4$ eV, $I_s \simeq 30$ W/cm^2 and $E_G - h\nu \simeq .005$ eV. The value of $E_G - h\nu$ is obtained from linear absorption measurements at the bandedge and the fact that nonlinear absorption begins at $\alpha \simeq 100$ cm^{-1}. Using these values we then have $n_2/\alpha \simeq 1.3 \times 10^{-6}$ cm3/W. The measured value of n_2/α near 80°K is $1.4 \pm 0.8 \times 10^{-6}$ cm^{-6}/W indicating good agreement considering the crudeness of the model.

The expression derived by Miller et al[5] for the nonlinear index in direct gap semiconductors predicts that the relative value of n_2/α for InAs compared to that for InSb should be

$$\frac{[n_2/\alpha]_{InAs}}{[n_2/\alpha]_{InSb}} \sim \left(\frac{\tau_{InAs}}{\tau_{InSb}} \right) \left(\frac{E_{G\ InSb}}{E_{G\ InAs}} \right)^3 \quad .$$

If we assume the relaxation times are comparable, then the relative figure of merit for InAs goes as the cube of the ratio of the band gaps. Taking values at 80°K of 0.228 eV and 0.400 eV for InSb and InAs respectively we get

$$[n_2/\alpha]_{InAs} \sim 0.2 \, [n_2/\alpha]_{InSb}$$

The measured relative figure of merit is

$$[n_2/\alpha]_{InAs} = .04 \, [n_2/\alpha]_{InSb} \quad .$$

Here we have used $[n_2/\alpha]_{InSb} = 3.5 \times 10^{-5} cm^3/W$ from reference 5.
This represents good agreement considering the uncertainty in the
relaxation times.

The time constant for the nonlinearity in InAs near 80°K can
be estimated to be between 10^{-7} sec. and 10^{-6} sec. based on published free carrier relaxation times.[8,9] It is interesting to note
that near 4°K the carrier lifetimes are considerably shorter. In
reference 8 a lifetime of 5 ns was measured at 5°K, a factor of 20
shorter than at 77°K. This is relevant to our observation of a
slightly lower value of n_2 at 6°K. One might expect the nonlinearity to be larger at the lower temperature since the Fermi distribution would tend to pack the free carriers closer to the bottom
of the conduction band making it easier to fill states. It would
appear, however, that the decreased lifetime at the lower temperature has compensated for this effect resulting in the value of n_2
observed.

As an example of switching intensities that could be achieved
in InAs consider a Fabry-Perot etalon with 90% reflectivity coatings. Using the measured value at 84°K of $n_2 = 4.5 \times 10^{-5} cm^2/W$,
$\alpha = 21 \, cm^{-1}$ and a cavity length of 20 μm, the critical intensity
for the onset of bistability becomes $I_c = 10W/cm^2 = 0.1 \, \mu W/\mu m^2$.

In conclusion we have measured a large nonlinear index and
associated nonlinear absorption in InAs near the bandgap between
6°K and 100°K. A parameter-free calculation shows that the nonlinear index is due to the alteration of the absorption edge by photo-
generated carriers in agreement with the theory of Moss[4] and Miller
et al[5]. A material figure of merit $[n_2/\alpha]_{InAs} \simeq .04[n_2/\alpha]_{InSb}$, was

obtained indicating that bistability should be observable in an
InAs Fabry-Perot with milliwatt powers.

This work was supported by the National Science Foundation
Grant # ECS-8114829.

REFERENCES

1.   D.A.B. Miller, S. D. Smith, and A. Johnston, Appl. Phys.
           Lett. 35, 658 (1979).
2.   H. M. Gibbs, S. L. McCall, T.N.C. Venkatesan, A. C. Gossard,
           A. Passner and W. Wiegmann, Appl. Phys. Lett. 35, 451
           (1979).
3.   J. R. Hill, G. Parry and A. Miller, Optics Comm. 43, 151
           (1982).
4.   T. S. Moss, Phys. Stat. Sol. b 101, 555 (1980).
5.   D.A.B. Miller, C. T. Seaton, M. E. Prise and S. D. Smith,
           Phys. Rev. Lett. 47, 197 (1981).
6.   P. Lavallard, R. Bichard, and C. Benoit a la Guillaume,
           Phys. Rev. B., 16, 2804 (1977).
7.   D.A.B. Miller, M. H. Mozolowski, A. Miller, and S. D. Smith,
           Optics Comm. 37, 133 (1978).
8.   S. S. Li and C. I. Huang, Phys. Rev. B, 4, 4633 (1971).
9.   A.W. Blant-Blachev, L. A. Balagurov, V. V. Karataev, and
           E. M. Omel'yanowskii, Sov. Phys. Semicond. 9, 515
           (1975).
10.   D.A.B. Miller, IEEE J. Quantum Electr., QE-17, 306, (1981).

RESONANT FRUSTRATED TOTAL REFLECTION (FTR) APPROACH TO OPTICAL BISTABILITY IN SEMICONDUCTORS

Bruno Bosacchi

Western Electric
Engineering Research Center
Princeton, NJ 08540

Lorenzo M. Narducci

Drexel University
Department of Physics
Philadelphia, PA 19104

The recent observation of very large nonlinearities and of optical bistability in semiconductor structures at room temperature has heightened the interest in all optical switching and signal processing as a practical prospect in optical information technology. Along with the search for suitable nonlinear materials, some attention is also being paid to the study of various interaction geometries, like the nonlinear interface, several waveguide configurations, and the surface plasmon configuration. Here we consider one further approach of this kind, which makes use of the concept of nonlinear FTR (Frustrated Total Reflection) optical cavity.

The nonlinear FTR optical cavity is a thin film structure into which light is coupled by total internal reflection. At resonance, a very intense electric field builds up in the cavity layer, (enhancement factors up to 2 or 3 orders of magnitude are possible), and a sharp minimum occurs at θ_π in the angular reflectivity spectrum of the structure. θ_m depends on the refractive index n_2 of the cavity layer ($n_2 = n_{20} + \beta \cdot /E/^2$, where n_{20} is the refractive index at low intensity, β is the nonlinear Kerr coefficient, and $/E/^2$ is the optical field inside the cavity layer). Conversely, at a fixed angle of incidence θ_i, appropriately chosen, $/E/^2$ depends on θ_m, being maximum when $\theta_m = \theta_i$. From the simultaneous and reciprocal interdependence of $/E/^2$ and n_2, optical bistability is expected to occur.

We have performed a numerical study of the optical response of a nonlinear optical cavity as a function of the incident power, using the plane wave approximation, and assuming a uniform enhancement of the field in the cavity layer. The results indicate that, with respect to the nonlinear interface configuration, the critical

intensity for switching the system from the high to low reflectivity regime can be lowered of at least one order of magnitude. A
report on this work has just been published,[1] to which the reader
is referred for more information.

REFERENCES

1. B. Bosacchi & L. M. Narducci, Optics Letters <u>8</u>, 324 (1983)

NON-LINEAR FABRY-PEROT TRANSMISSION IN A CdHgTe ETALON

G. Parry, A. Miller and R. Daley

Royal Signal and Radar Establishment
Malvern, Worcestershire, U.K.

ABSTRACT

We extend our study of large nonlinear refractive index changes in CdHgTe to investigate intensity dependent transmission through an etalon. The results provide strong support for a theoretical model which takes into account the dominant Auger recombination process and predicts a refractive index change of $-3.3 \times 10^{-3} \, I^{1/3}$ (W/cm^2).

INTRODUCTION

In a recent paper[1] we reported the results of a theoretical and experimental study of intensity dependent refractive index changes in the alloy semiconductor cadmium mercury telluride at 10.6 μm wavelength and 175K. The changes in refractive index arose through a free carrier or plasma contribution[2] and through a blocking of interband transitions as previously observed in InSb[3,4] and more recently in InAs[5]. By extending the theoretical model due to Miller et al[3] to include Auger recombination we showed that the refractive index of cadmium mercury telluride was strongly intensity dependent and could be modelled in terms of a correction to the linear refractive index proportional to the cube-root of the intensity. This behaviour was confirmed experimentally by studying self-defocussing of a Gaussian laser beam after propagation through a thin crystal. The composition of the alloy, the wavelength of the CO_2 laser and the temperature were such that the absorption in the crystal was sufficiently high to prevent the observation of Fabry-Perot etalon effects, such as those reported in InSb[6] and GaAs[7].

In this paper we report the results of experiments carried out with a different sample of cadmium mercury telluride, at 77K and at

CO_2 laser wavelengths chosen so that sufficiently low absorption occurred in the material to allow Fabry-Perot effects, but at the same time, maintaining sufficient resonance (hence absorption) with the band-edge to retain the large nonlinearity. The combined effect of this nonlinearity with the Fabry-Perot structure should under suitable conditions permit the observation of optical bistability, differential gain, limiting, etc. However, the cube-root dependence of the refractive index on intensity significantly affects the conditions for observation of these nonlinear modes of transmission. We discuss in a separate paper the optimum parameters for bistability under these conditions. In the present paper we report the observation of intensity dependent transmission in the sample and show how the nonlinear Fabry-Perot equations have to be modified to include the cube-root nonlinear dependence.

Preliminary observation of nonlinear Fabry-Perot transmission has been reported by Jain and Steel[8] using an external cavity. They attributed the appearance of random pulsations in the transmitted light to this process. Khan et al[9] have also recently reported nonlinear Fabry-Perot effects using natural reflectivity of a 500 μm thick crystal with a Q-switched CO_2 laser.

EXPERIMENTAL DETAILS

The experiments were carried out using a 2.5 watt line tunable carbon dioxide laser and a sample of cadmium mercury telluride $Cd_{0.23}Hg_{0.77}Te$ at 77K. Radiation from the laser was passed through a variable attenuator and focussed to a beam waist at the sample with a spot diameter ~ 80 μm (1/e points of intensity profile). The total power in the incident laser was monitored using a pyro-electric detector and beamsplitter arrangement. The output power was also monitored using a pyroelectric detector. Both detectors collected all the radiation in the beams allowing direct measurement of the transmission.

The $Cd_{0.23}Hg_{0.77}Te$ sample was approximately 208 μm thick and had been polished plane parallel. The material refractive index was ~ 4.22 so that even without additional coatings it should behave as a Fabry-Perot etalon. Room temperature measurements (when the band-gap energy is much larger than the photon energy) confirmed that the surface quality and degree of parallelism were sufficiently good to observe Fabry-Perot behaviour. Using the value for the band-gap energy E_g calculated by[10] we estimated that the photon energies were ~ 3.6 kT below E_g at 77K.

Particular consideration was given to ensuring that all the transmitted light was collected by the detectors. Previous work[1] had shown that with a Gaussian laser intensity profile significant intensity dependent changes occur to the width of the transmitted beam. Preliminary transmitted far-field beam profile measurements

on the sample using a small area PbSnTe detector showed that intensity dependent beam narrowing occurred. Thus to distinguish nonlinear beam profile changes from nonlinear Fabry-Perot transmission
the total transmitted laser power was measured using a collecting
lens and large area pyroelectric detector.

The transmission was measured over a range of wavelengths and
at a number of points on the crystal. Altering the laser wavelength
changed both the photon energy and the initial tuning of the etalon
structure.

THEORETICAL CONSIDERATIONS

For comparison with experimental results we have taken the
expressions for plane wave Fabry-Perot transmission and included in
these an intensity dependent refractive index of the form $\Delta n \propto I^{1/3}$.
It is important to note that we do not need to include a nonlinear
contribution to the absorption coefficient. No intensity dependent
absorption was observed previously[1]; the nonlinear refractive index
changes for photon energies below the band-gap occur as a result of
generated free carriers through excitation and thermalisation of
electrons via band tail states. Under these conditions the transmission of a Fabry-Perot etalon can be written

$$T = \frac{C}{1+F \sin^2 \delta} \tag{1}$$

where $C = (1-R)^2 e^{-\alpha \ell}/(1-R_\alpha)^2$

and $R_\alpha = R e^{-\alpha \ell}$

Here R is the surface reflectivity, ℓ is the sample thickness and
α is the absorption coefficient. The phase change δ arising from
propagation through a thickness ℓ of the crystal is intensity
dependent and may be written

$$\delta = \delta_0 + \beta I_A^{1/3} \tag{2}$$

where I_A is the averaged etalon intensity in the case of an absorbing cavity[11]. The term δ_0 is the tuning of the cavity at low
intensity and the second term is the cavity intensity dependent
contribution reflecting the cube-root behaviour noted earlier. The
magnitude of the coefficient β varies with temperature, photon
energy and material band-gap and so must be evaluated for the
conditions appropriate to this experiment. In a previous paper[1]
we calculated the refractive index change arising from saturation
of conduction-band states. There is also a significant contribution
due to optically generated free carriers[2] so the total contribution

can be written

$$\Delta n = (\sigma_p + \sigma_s) N_c \tag{3}$$

where σ_p and σ_s are the refractive index changes per electron-hole pair per unit volume arising respectively from free carriers and saturated states and N_c is the number of excited electron-hole pairs. These may be related to material and optical parameters through

$$\sigma_p = - \frac{e^2}{2\epsilon_o m^* n_o \omega^2} \tag{4}$$

and

$$\sigma_s = - \frac{e^2}{2\epsilon_o m^* n_o \omega^2} \left\{ \frac{2}{3\sqrt{\pi}} \frac{m^*}{m} \frac{mP^2}{\hbar^2} \frac{1}{kT} J\left(\frac{\hbar\omega - E_g}{kT}\right) \right\} \tag{5}$$

where e is the electronic charge, ϵ_o the free space dielectric constant, m^* the conduction band effective mass, m is the electron mass, $\hbar\omega$ is the photon energy, P the momentum matrix element[3] and

$$J(a) = \int_0^\infty dx \, x^{\frac{1}{2}} \, e^{-x}/(x-a) \tag{6}$$

For this sample used

$$\sigma_p \sim -1.1 \times 10^{-18} \text{ (cm}^3) \text{ and } \sigma_s \sim -1.2 \times 10^{-18} \text{ (cm}^3).$$

The number of electron-hole pairs in steady state conditions is affected by the Auger nature of the recombination process and for an n-type material with equilibrium electron concentration N_o is governed by the equation

$$N_e(N_o + N_e)^2 = 2N_i^2 \, \tau_i \, \alpha \, I_A/\hbar\omega \tag{7}$$

N_i the intrinsic carrier concentration and τ_i the intrinsic lifetime. From Eq. (7) we deduce that $\beta = -0.41$ rads. $(\text{cm}^2/\text{W})^{1/3}$. Computation of the cavity transmission as a function of incident power involves the rather circuitous procedure of first choosing a density of generated carriers, calculating I_A and hence the transmission of the cavity to finally deduce the input intensity.

RESULTS AND DISCUSSION

Figure 1(a) and (b) show the results of measuring the transmission of the sample at two wavelengths 10.67 μm (P28) and 10.69 μm (P30). They are typical of the results obtained at a number of different wavelengths and positions on the crystal. Different wavelengths give different initial tunings or transmissions of the sample. In (a) the transmission starts at a low value and increases with increasing intensity whereas in (b) the transmission starts at a higher value and decreases with incident intensity. Both experimental curves show clearly a more rapid variation of transmission with intensity at low powers then at higher powers, consistent with the $I^{1/3}$ behaviour expected. The large magnitude of the nonlinearity is also clear from the low input power levels used to observe these effects. In this case a 20:1 chopper was used to minimise thermal effects so the average power levels were only 5% of the levels shown. No time dependent effects were observed on a time scale of 40 μsec associated with the chopped rise time.

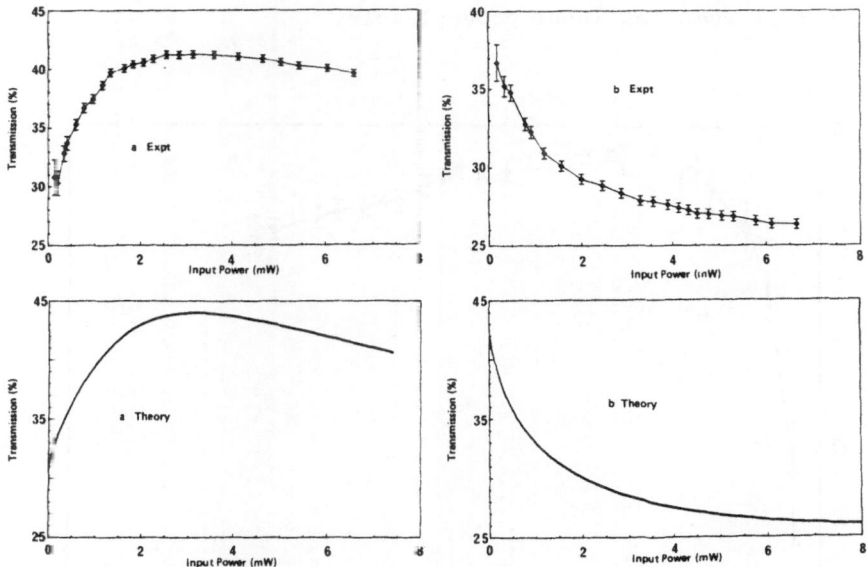

Fig. 1. Experimental and theoretical transmission of the etalon structure using (a) P28 (λ = 10.67 μm) and (b) P30 (λ = 10.69 μm) lines from the CO_2 laser.

Also shown in figure 1 are the calculated values of the transmission as a function of input power based on the plane wave theory and the plasma and band-filling model discussed previously, (Eqn. (1)-(7)). Most of the parameters required for the calculation were known. However, it was necessary to use the reflectivity, absorption and cavity tuning as fitting parameters to obtain approximately the measured peak transmissions. Values of R = 0.18 and 0.22 were used for curves (a) and (b) and α = 30 cm^{-1} for both cases. These reflections are low compared to the calculated Fresnel value of 0.38 probably due to surface quality and degree of parallelism. The optical thickness of the sample varied with wavelength so it was necessary to use the experimentally determined initial cavity tunings δ. The input power scales correspond to spot sizes of 130 μ although the measured spot size was approximately 80 μ. In view of the difference between the Gaussian laser profile used and the plane wave theory this is not unreasonable. The excellent functional agreement between theory and experiment is typical of results obtained over a range of lines from the CO_2 laser and provides strong support for the model for the nonlinearity.

In Figure 2 we show the result of a measurement carried out at a different point on the crystal with an unchopped beam and observed the same highly nonlinear behaviour seen with the chopped beams. We are not aware of any observations of nonlinear Fabry-Perot behaviour at such low input power levels.

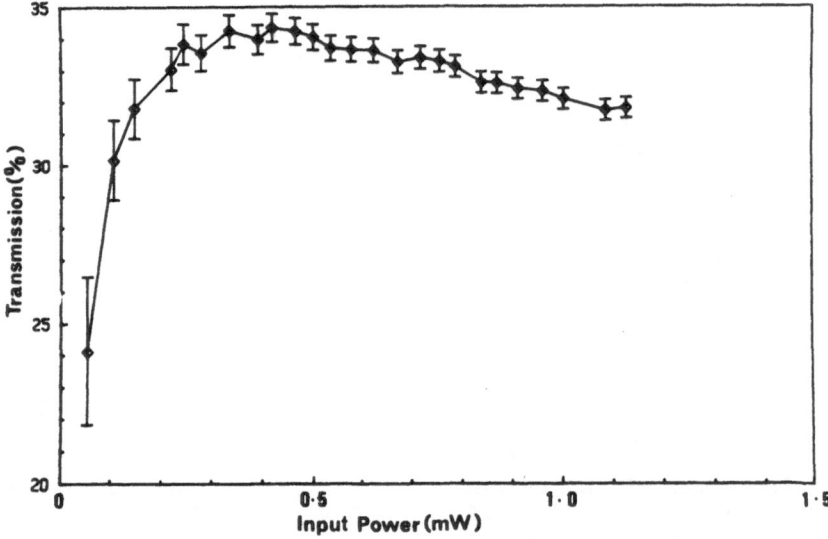

Fig. 2.   Transmission of a low power unchopped beam through the sample.

From transmission versus input power curves it was possible to calculate for each wavelength the variation of induced cavity phase change with output power, which is proportional to the cavity intensity, figure 3. The data measured show the same functional variation with power (for the different wavelength consistent with Auger recombination). The fact the curves appear parallel indicates that the magnitude of the nonlinearity did not vary over the range of wavelengths used (10.51-10.69 µm). This is to be expected when there is a large energy difference between the band-gap energy and photon energy (3.4 kT-3.7 kT), and so provides further support for the theoretical model.

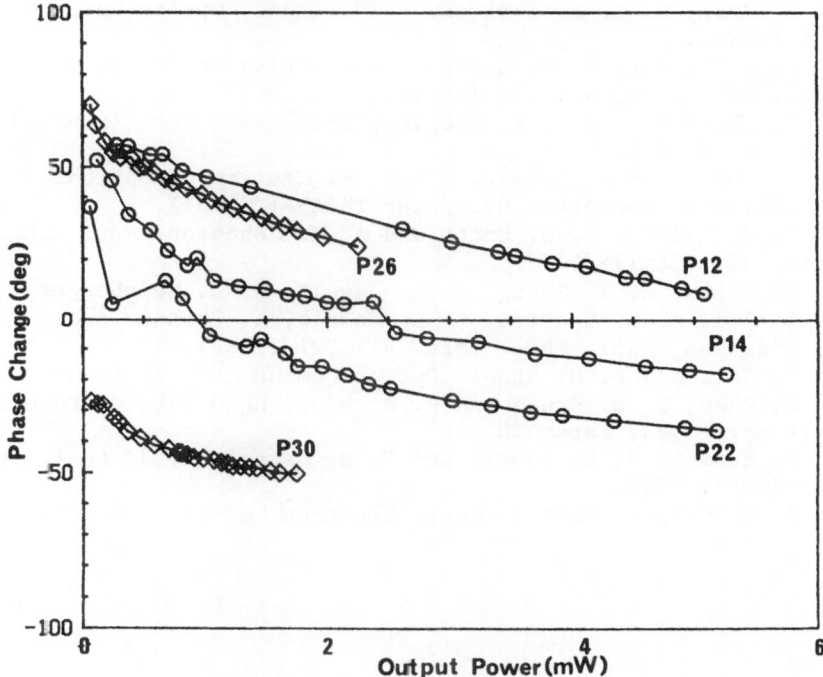

Fig. 3. Plot of intensity dependent phase changes induced in the etalon at five different wavelengths of the CO_2 laser.

CONCLUSIONS

We have demonstrated low power nonlinear Fabry-Perot transmission of a CdHgTe etalon. The results are in good agreement with theory and provide confirmation of the previously reported large nonlinear refractive effect in this material. Transverse effects make a fully theoretical analysis difficult. However, the results reported agree very well functionally with a theory which includes the Auger process as a recombination mechanism. The close agreement between the theoretical and experimental transmission curves also confirms the predicted large magnitude for the nonlinear coefficient.

ACKNOWLEDGEMENT

We would like to thank Mullard Limited for providing the CMT sample and J. R. Hill for stimulating discussions and valuable assistance.

REFERENCES

1.   J. R. Hill, G. Parry and A. Miller, Optics Commun. 43: 151
     (1982).
2.   M. A. Khan, R. L. Bennett and P. W. Kruse, Optics Lett. 6:
     561 (1981).
3.   D. A. B. Miller, C. T. Seaton, M. E. Prise and S. D. Smith,
     Phys. Rev. Lett. 47: 197 (1981).
4.   B. S. Wherrett and N. A. Higgins, Proc. Roy. Soc. (London)
     A379: 67 (1982).
5.   C. D. Poole and E. Garmire, "Topical Meeting on Optical
     Bistability" Rochester NY, paper Th A3-1 (1981).
6.   D. A. B. Miller, S. D. Smith and A. M. Johnston, Appl. Phys.
     Lett. 35: 658 (1979).
7.   H. M. Gibbs, S. S. Tarng, J. L. Jewell, D. A. Weinberger,
     K. C. Tai, A. C. Gossard, S. L. McCall, A. Passner and
     W. Wiegmann, Appl. Phys. Lett. 41: 221 (1982).
8.   R. K. Jain and D. G. Steel, Optics Commun. 43: 72 (1982).
9.   M. A. Khan, P. W. Kruse and R. A. Wood, CLEO 1983 Conference,
     Baltimore 1983, Paper Th H6.
10. G. L. Hansen, J. L. Schmit and T. N. Casselman, J. Appl. Phys.
     53: 7099 (1982).
11. D. A. B. Miller, IEEE J. Quat. Electron. QE-17: 306 (1981).

Footnote: In reference 1 the numerical factor in the expression
            for σ_s (Eqn. (5) here) should have been 1/3 instead of
            the printed 1/2.

BISTABILITY IN EXTERNAL-CAVITY SEMICONDUCTOR LASERS

T.G. Dziura and D.G. Hall

University of Rochester
Institute of Optics
Rochester, N.Y. 14627

INTRODUCTION

Semiconductors have been demonstrated to be promising materials for optical logic devices. Since they possess large absorption co-efficients the optical interaction length can be very small and inter-action times can be made short. They also can have sharp absorption features which provide for large dispersive nonlinearities near a band edge or exciton resonance.

Optical bistability has been observed in intrinsic devices con-structed of ZnS, GaAs[2], InSb[3], and CdS[4]. Picosecond interaction times have been proposed for CuCl[5]. Opto-electronic devices based on satur-able absorption in GaAs (bistable laser diodes) have been built that are capable of high frequency performance[6]. However, the parasitic resistance between absorbing and amplifying sections causes the out-put characteristics to depend on the bias network[7].

In this paper we analyze theoretically the steady-state and dy-namic characteristics of a laser cavity containing two semiconductor diode laser elements. This system provides a unique opportunity to study the effects of the absorber element characteristics on the bi-stable operation of the laser. Our steady-state analysis determines the requirements on the absorber, amplifier, and resonator for bista-bility; our analysis of the transient characteristics predicts switch-on and switch-off with current pulses sent into the absorber or ampli-fier diode, and critical slowing down near the current for switch-on of the output. Bistable operation of such a laser system has been described theoretically[8] and observed experimentally[9].

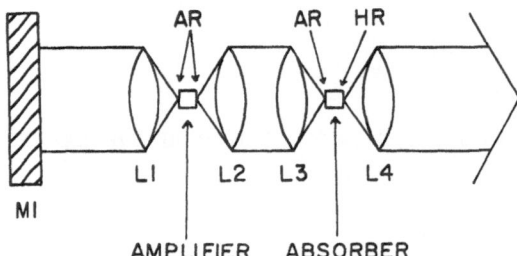

Fig. 1.   Geometry for the bistable semiconductor laser cavity.
L1 – L4 are microscope objectives. M1 is an external
mirror.

STEADY STATE CHARACTERISTICS

The geometry that we are considering is shown in Fig. 1. Two
semiconductor lasers are placed in an external cavity; the first has
both mirror facets anti-reflection (AR) coated, and the second has
one mirror AR-coated and the other mirror high-reflectance (HR)
coated. The external cavity is formed by the external mirror and the
HR-coated facet of the second diode laser.

The rate equations that we will use to model the bistable laser
are

$$\frac{dN_{ph}}{dt} = g_G \left(\frac{L_G}{L}\right) (N_{eG} - N_{oG}) N_{ph} + g_A \left(\frac{L_A}{L}\right)(N_{eA} - N_{oA}) N_{ph} - \frac{N_{ph}}{\tau_{ph}}$$

$$+ \beta \left(\frac{L_G}{L}\right) \frac{N_{eG}}{\tau_{sp}} \tag{1}$$

$$\frac{dN_{eG}}{dt} = \frac{I_G}{eV_G} - g_G(N_{eG} - N_{oG})N_{ph} - \frac{N_{eG}}{\tau_{sp}} \tag{2}$$

$$\frac{dN_{eA}}{dt} = \frac{I_A}{eV_A} - g_A(N_{eA} - N_{oA})N_{ph} - B_r(N_d + N_{eA})N_{eA} - \frac{N_{eA}}{\tau_{nr}} \qquad (3)$$

In the above, N_{ph} is the photon density, N_{eG} is the amplifier carrier density, and N_{eA} is the absorber carrier density. g_G and g_A are gain and absorption coefficients in the assumed linear dependence on carrier density. N_{oG} and N_{oA} are the carrier densities required for transparency in the amplifier and absorber elements respectively. τ_{ph} and τ_{sp} are the photon and carrier lifetimes, and the injection currents in the amplifier and absorber elements are I_G and I_A. L_G, L_A, and L are the lengths of the amplifier, absorber, and resonator respectively, and V_G and V_A are the active volumes of the amplifier and absorber. e is the electronic charge and β is the fraction of spontaneous emission into the lasing mode. B_r is the bimolecular recombination coefficient, N_d is the dopant concentration of the absorber active layer, and τ_{nr} is the nonradiative lifetime for carriers in the absorber element. Bimolecular recombination is chosen to describe the carrier decay in the absorber since the carrier concentration and therefore the carrier lifetime varies over several orders of magnitude.

The steady state solutions to (1) - (3) are shown in Fig. 2. The values of the parameters used in the calculation are $g_G = 2.75 \times 10^{-7}$ cm^3/s, $L_G = L_A = 400$ μm, $L = 15$ cm, $N_{oG} = 2.5 \times 10^{18}$ cm^{-3}, $\tau_{ph} = 120$ ps, $\beta = 5 \times 10^{-5}$, $\tau_{sp} = 2_{sp}$, $V_G = V_A = 2.3 \times 10^{-10}$ cm^3, $B_r = 2 \times 10^{-10}$ cm^3/s, $N_d = 7.5 \times 10^{16}$ cm^3, $\tau_{nr} = 100$ ns, and $I_A = 0$. The amplifier carrier concentration and the amplifier current are normalized to their values at threshold, and the photon density is normalized to the saturation photon density of the absorber, given by (for large N_d)

$$N_{ph,s} = \frac{B_r N_d + 1/\tau_{nr}}{g_A} \qquad (4)$$

It is seen that the area of the hysteresis loop increases with increasing $g_A N_{oA}$, i.e., with increasing small signal absorption coefficient.

Simple expressions for the switching points can be found from (1) - (3) by assuming a constant (i.e., independent of carrier concentration) lifetime for the carriers in the absorber element. The output level N_{ph2}, to which the laser switches from its off state, is given by

$$N_{ph2} = N_{ph,s} [(R/R_{crit}) -1] \qquad (5)$$

where

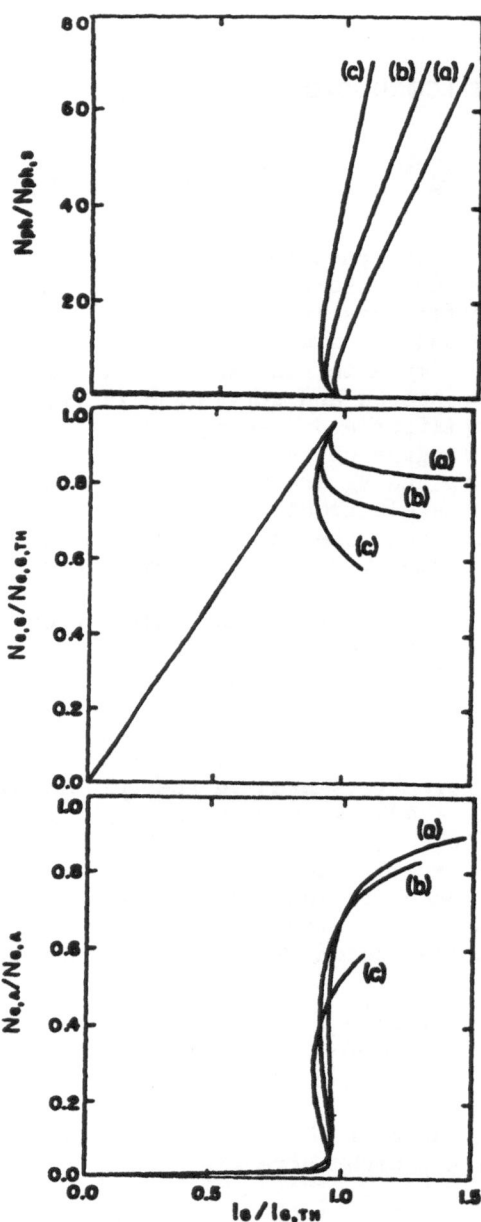

Fig. 2.   Bistable curves for various values of the small-signal absorption at constant saturation photon density. $g_A = 10^{-6}$ cm^3/s and (a) $N_{oA} = 10^{18}$ cm^{-3}; (b) $N_{oA} = 2 \times 10^{18}$ cm^{-3}; (c) $N_{oA} = 10^{19}$ cm^{-3}.

$$R = g_A \, No_A \tag{6}$$

$$R_{crit} = \frac{(L/L_A)(1/\tau_{ph})}{(1/g_G \, \tau_{sp} N_{ph,s}) - 1} \tag{7}$$

We also find for the "on-to-off" switching photon density N_{ph1}

$$N_{ph1} = N_{ph,s} \left[(R/R_{crit})^{\frac{1}{2}} - 1 \right] \tag{8}$$

By comparing (5) with (8) we see that in order to observe bistability the absorption rate R must satisfy

$$R > R_{crit} \tag{9}$$

Equation (7) also places a requirement on the absorber and amplifier elements, for if (1) – (3) are solved for the gain g_G (N_{eG} – N_{oG}) one finds

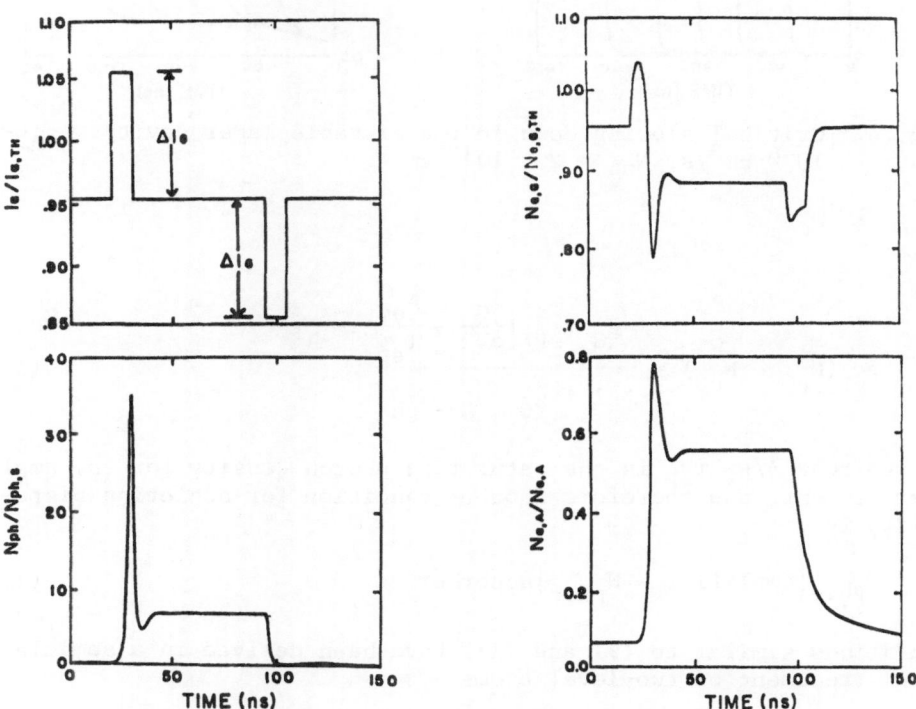

Fig. 3. Turn-on and turn-off of the bistable laser cavity. $g_A = 10^{-6}$ cm^3/s, $N_{oA} = 10^{18}$ cm^{-3}.

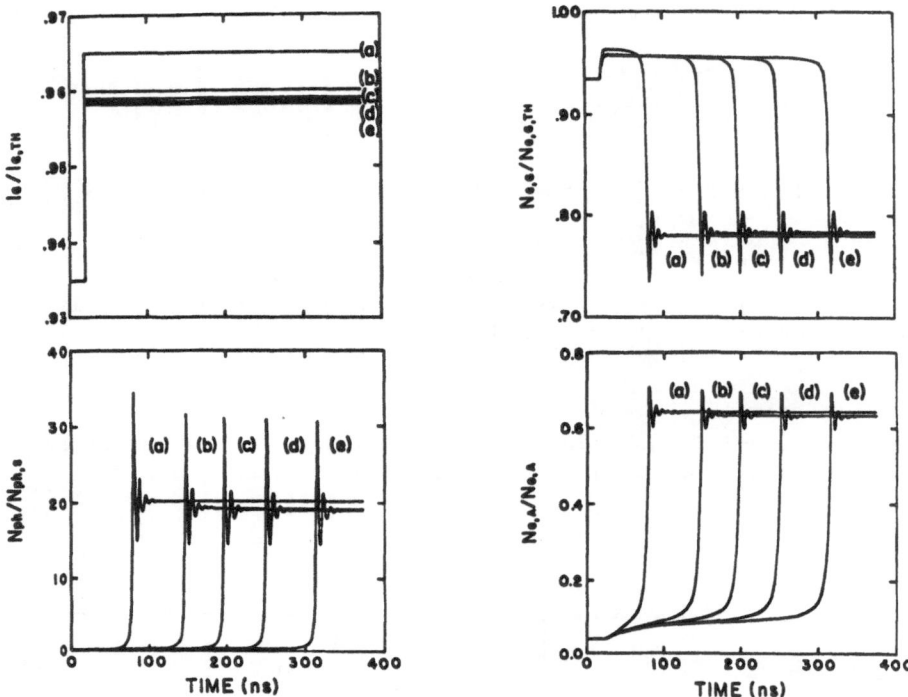

Fig. 4.   Critical slowing down in the bistable laser cavity. g_A = 10^{-6} cm³/s, N_{oA} = 2 x 10^{18} cm⁻³.

$$g_G (N_{eG} - N_{oG}) = \frac{g_G \tau_{sP} (\frac{I_G}{eV_G} - \frac{N_{oG}}{\tau_{sp}})}{1 + g_G \tau_{sp} N_{ph}} \qquad (10)$$

We see that $1/g_G \tau_{sp}$ is the saturation photon density for the amplifier element, and therefore another condition for achieving bistability is

$$N_{ph,s}|\text{amplifier} > N_{ph,s}|\text{absorber} \qquad (11)$$

Conditions similar to (9) and (11) have been derived in a semiclassical treatment of two-level atoms[10].

DYNAMIC CHARACTERISTICS

The transient behavior of the bistable laser cavity was investigated by solving (1) - (3) numerically using a GEAR routine[11]. In Fig. 3 we show the turn-on and turn-off of the output by a current pulse sent into the amplifier. Overshoot and ringing, which occurs in the turn-on of the laser, has been observed experimentally in dispersive devices[12]. The delay and risetime in turn-on depends on the amplitude of the current pulse and the initial bias current. As the switching current plus bias current amplitude reach the current for turning on the laser, the delay time becomes much longer than any time constant of the system, while the risetime remains fairly constant. This is illustrated in Fig. 4, where step function current pulses sent into the amplifier are shown switching on the cavity. As the final pulse amplitude nears the switch-on current the delay time increases to several hundred nanoseconds, while the final risetime remains the same. This phenomenon has been observed experimentally in bistable laser diodes[13] and is an important consideration in the design of practical, high speed logic devices.

We also predict similar behavior when the absorber element is used to turn on the laser. However, this method of switching is fundamentally different from switching via the amplifier; the absorption is reduced directly by the injection of carriers, whereas in amplifier switching the absorption saturates due to the emission of photons by the amplifier element.

CONCLUSION

We have used standard, single-longitudinal mode laser rate equations to analyze the steady-state and dynamic behavior of a bistable semiconductor laser resonator. Our steady-state analysis reveals that a minimum absorption rate, which depends on the cavity lifetime and the relative saturability of the absorber and amplifier elements, is necessary for bistability. The absorption must also be easier to saturate than the gain. Our analysis of the dynamic behavior predicts critical slowing down near the switch-on current, and overshoot and ringing with fast risetime pulses.

ACKNOWLEDGEMENTS

This work was supported by the sponsors and participants of the Laser Fusion Feasibility Project of the Laboratory for Laser Energetics.

REFERENCES

1. F.V. Karpushko and G.V. Sinitsyn, J. Appl. Spect. USSR 29,
    1323 (1978).
2. H.M. Gibbs, S.S. Tarng, J.L. Jewell, D.A. Weinberger, K. Tai,
    A.C. Gossard, S.L. McCall, A. Passner, and W. Wiegmann, Appl.
    Phys. Lett. 41, 221 (1982).
3. D.A.B. Miller, S.D. Smith, and A. Johnston, Appl. Phys. Lett.
    35, 658 (1979).
4. M. Dagenais, J. Opt. Soc. Am. 72, 1835A (1982).
5. E. Hanamura, Solid State Commun. 38, 939 (1982).
6. Ch. Harder, J.S. Smith, K.Y. Lau, and A. Yariv, Appl. Phys.
    Lett. 42, 772 (1983).
7. Ch. Harder, K.Y. Lau, and A. Yariv, IEEE J. Quant. Elect.
    QE-18, 1351 (1982).
8. T.G. Dziura and D.G. Hall, IEEE J. Quant. Elect. QE-19, 441
    (1983).
9. W.A. Stallard and D. J. Bradley, Appl. Phys. Lett. 42, 858
    (1983).
10. L.A. Lugiato, P. Mandel, S.T. Dembinski, and A. Kossakowski,
    Phys. Rev. A18, 238 (1978).
11. A.C. Hindmarsh, Lawrence Livermore Laboratory, Rep. UCRL-51186,
    (1972).
12. T. Bischofberger and Y.R. Shen, Appl. Phys. Lett. 32, 156
    (1978).
13. K.Y. Lau, Ch. Harder, and A. Yariv, Appl. Phys. Lett. 40, 198
    (1982).

INTRINSIC INSTABILITIES IN HOMOGENEOUSLY BROADENED LASERS

Lloyd W. Hillman, Robert W. Boyd, Jerzy Krasinski, and
C.R. Stroud, Jr.

The Institute of Optics
University of Rochester
Rochester, New York 14627

We have studied the behavior of a cw homogeneously broadened
ring dye laser operating far above threshold. We have found that
although the laser operates in a single mode when it is near threshold, it is intrinsically unstable when pumped far above threshold.
The onset of the instability is characterized by an abrupt increase
in the power circulating in the cavity. In addition, the spectral
output of the laser changes. Below the onset of the instability
lasing occurs at gain center, while above this threshold, lasing
occurs simultaneously at two frequencies symmetrically displaced from
gain center. The splitting of the two lasing frequencies follows a
parabolic power dependence. At still greater pump powers, higher
order splittings and instabilities were observed. These instabilities exhibit hysteresis about the critical pump powers.

If one models a continuously pumped, homogeneously broadened
ring laser with conventional rate equations for the inversion and
photon density, one is led to the conclusion that the laser will operate in only a single longitudinal mode at a time. Only if there is
inhomogeneous broadening or spatial hole burning does such a theory
predict multimode operation. However, Risken and Nummedal[1] showed
that if one treated the problem with what are now called the Maxwell-
Bloch equations, the predictions are different. These equations show
that single-mode operation is intrinsically unstable when the laser
is pumped far above threshold. When the initial single mode becomes
intense, it is strongly coupled to two modes whose frequencies are
symmetrically displaced from the initial mode by approximately the
Rabi frequency. There has been recent theoretical progress on this
problem.[2,3] This work shows that these side modes are not independent
but satisfy an eigenvalue equation which relates their amplitudes and
phases. When the laser is operated in a single mode well above

305

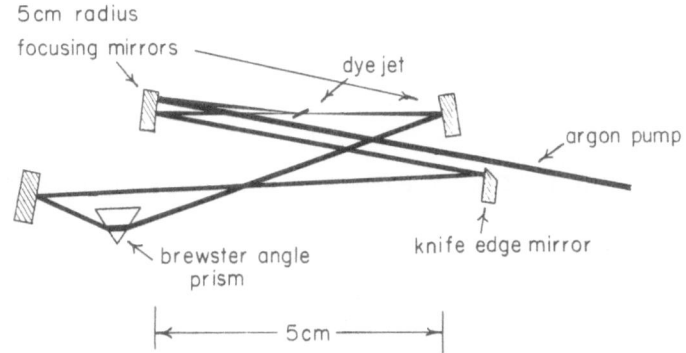

5 cm radius
focusing mirrors

dye jet

argon pump

brewster angle
prism

knife edge mirror

|← 5cm →|

Fig. 1.   Schematic drawing of the high-Q ring dye laser cavity

threshold, the induced gain of the side modes can exceed threshold. Thus, the single mode laser becomes unstable and switches to multimode operation with a mode separation dependent on the intracavity power.

The experimental observations were made on a rhodamine 6G ring dye laser pumped by an argon ion laser. A high-Q ring cavity of length 25 cm was constructed using broadband high reflectors for all mirrors and a weakly dispersive prism as shown in Fig. 1. Cavity losses were so low that we observed more than 80 watts of circulating dye laser power corresponding to an intensity of greater than 50 MW/cm^2 at the focus in the dye jet using only 2 watts of pump power. The circulating power was determined by measuring the intensity of Rayleigh scattering from the intracavity laser beam by nitrogen.[4] The dye laser was adjusted so that it operated at the center of the dye gain curve in the lowest order (Gaussian) mode. The spectrum of the laser was measured using a one-meter spectrometer with a vidicon camera located at the back focal plane. The video signal was either recorded with a video cassette recorder for storage, or digitized and analyzed directly by computer.

Figure 2 shows the observed dependence of the dye laser power on the argon pump power. As the pump power is increased from zero, the dye laser first reaches threhsold (marked A in Fig. 2), and operates at gain center with its power increasing linearly with pump power. A further increase in the pump power results in a discontinuous increase in the power of the dye laser. At this point, the dye laser switches from operation at a single frequency at gain center (B) to a laser operating simultaneously at two frequencies symmetrically dis-

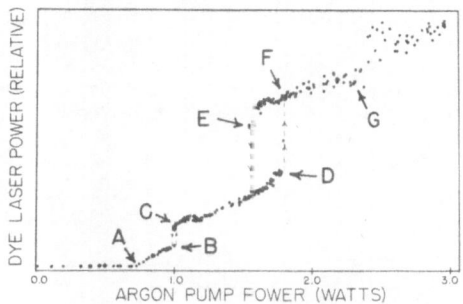

Fig. 2.   Dye laser power versus argon laser pump power. The circulating power in the dye laser cavity exhibits discontinuous jumps as the pump power of the argon laser is varied.
The lasing threshold is marked A. The single frequency
operation of the dye laser becomes unstable at B. Points
D and G indicate the locations of higher-order instabilities. Hysteresis loops exist around points B and C, and
D, F and E.

placed from gain center (C). The transition from a one-frequency
laser to a two-frequency laser is analogous to a first-order phase
transition. Figure 2 also shows that this instability is bistable,
exhibiting a small hysteresis loop about the critical power. The
numerical work on spontaneous mode locking of a homogeneously broadened laser by Risken and Nummedal[5] predicted a similar hysteresis.

As the pump power is further increased, we again find a nearly
linear dependence of the dye laser power on the pump power (from C to
D in Fig. 2). However, the spectral output shows a strong power dependence as illustrated in the left column of Fig. 3. The two lasing
wavelengths split further apart as the cavity power is increased. We
have observed splittings as large as 70Å. At the top of the left column of Fig. 3, we have plotted the wavelengths of the two lasing frequencies as a function of the relative power in the dye laser. We
find that the dependence of the wavelength splitting on the cavity
power is fit closely by a parabola.

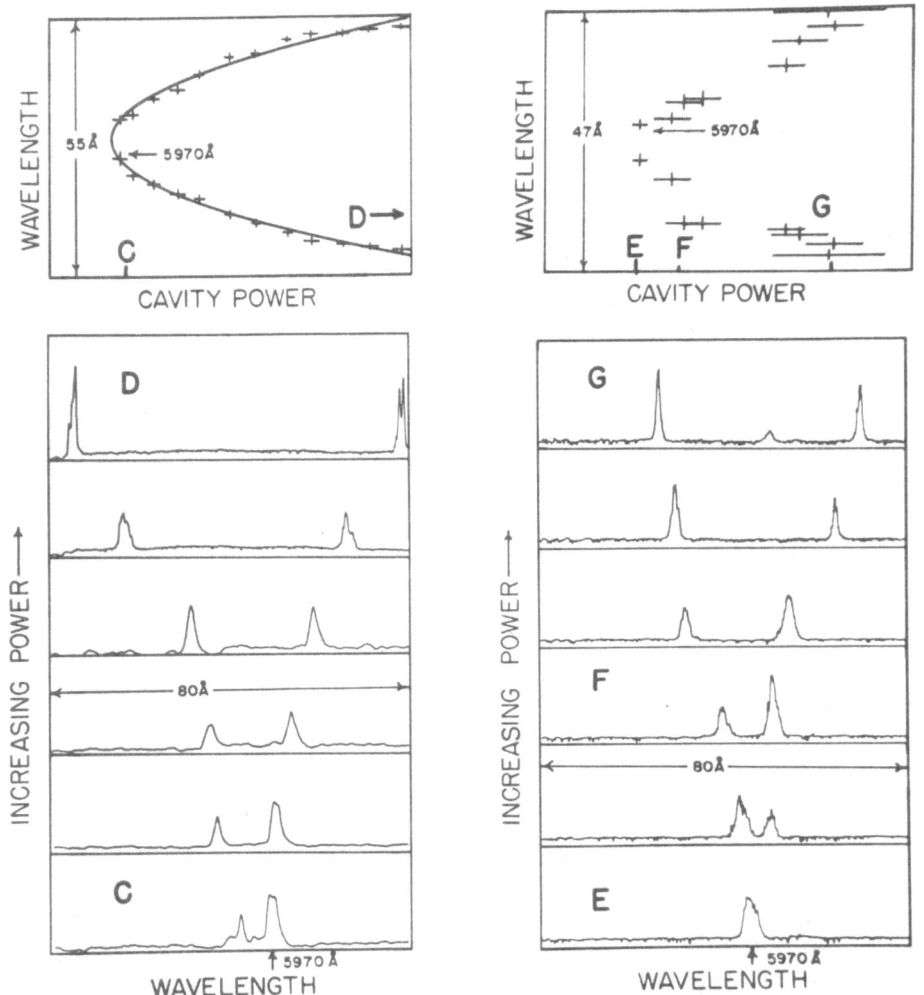

Fig. 3.   Power dependence of the dye laser spectrum. The letters
           mark positions corresponding to those similarly marked on
           Fig. 2.   The two curves in the top row show the splitting
           between the two laser frequencies as a function of power in
           the cavity.   The figures in the left column refer to the
           region of the first splitting (C to D), and those on the
           right to the second splitting (E,F,G).

        At still higher pump powers, another discontinuous jump in the
dye laser power was observed. This was preceded by the growth of a
mode at the center of the gain curve. This mode rapidly grew and

split causing the outer two modes to be extinguished. This higher
order stability exhibited a large hysteresis loop (D,E,F in Fig. 2).
The spectral output of the dye laser along the F,E,G branch is illustrated in the right column in Fig. 3. Beyond G further power jumps
(including negative jumps), hysteresis, and spectral splitting were
observed, which led rapidly to a seemingly chaotic state.

This paper reports the first observation, to our knowledge, of
the laser instability predicted by Risken and Nummedal. A splitting
of the lasing wavelengths which was parabolic with laser power was
observed. This is consistent with the dependence produced by Rabi
oscillations which strongly couple the two fields. The single frequency operation of the laser became unstable because there is strong
parametrically coupled gain for two side modes which are displaced by
the Rabi oscillation frequency on either side of the lasing mode.[2]
The observation of power dependent wavelength splittings in the higher
order instabilities suggests that there are sub-Rabi-frequency resonances in the gain profile which couples the two strong modes. Such
subharmonic parametric resonances have been predicted by Tsukada and
Nakagama[6] for two saturating lasers interacting with a two-level
atomic resonance.

The homogeneously broadened ring dye laser is an interesting
system to study since it displays a series of instabilities, only
the lowest order of which was previously predicted. The appearance
of new subharmonic frequencies at each higher order instability is
suggestive of a new route to chaos.

This work was supported in part by the U.S. Army Research Office,
and by the Office of Naval Research under contract number N00014-76-
0001.

REFERENCES

1.   H. Risken and K. Nummedal, "Instability of off resonance
      modes in laser," Phys. Lett. 26A, 275-276 (1968).
2.   L.W. Hillman, R.W. Boyd, and C.R. Stroud, Jr., "Natural
      modes for the analysis of optical bistability and laser
      instability," Opt. Lett. 7, 426-428 (1982).
3.   S.T. Hendow and M. Sargent III, "The role of population
      pulsations in single-mode laser instabilities," Opt. Comm.
      40, 385-390 (1982), and "Effects of detuning on single-mode
      laser instabilities," Opt. Comm. 43, 59-63 (1982).
4.   L.W. Hillman, J. Krasinski, J.A. Yeazell, and C.R. Stroud,
      Jr., "Intracavity power measurement by Rayleigh scattering,"
      submitted to Applied Optics (1983).
5.   H. Risken and K. Nummedal, "Self-Pulsing in Lasers," J.
      Appl. Phys. 39, 4662-4672 (1968).

6.   N. Tsukada and T. Nakagama, "Modulation of optical bista-
      bility by an additional laser beam," Phys. Rev. A 25,
      964-977 (1982).

BISTABLE INJECTION LASERS

Ch. Harder, K. Y. Lau and A. Yariv

Applied Physics 128-95
California Institute of Technology
Pasadena, CA 91125

Semiconductor lasers with inhomogeneous current injection have been proposed nearly twenty years ago [1] as highly compact and efficient bistable devices. Recently, we demonstrated that a semiconductor laser with a segmented contact, as shown in Fig. 1, displays bistability without pulsations. The key to the proper design of such a bistable laser is the electrical isolation between the two segments which requires that the parasitic resistance (R_p in Fig. 2) between the two contacts be as large as possible. This can be achieved by doping the top cladding layer only slightly p-type.

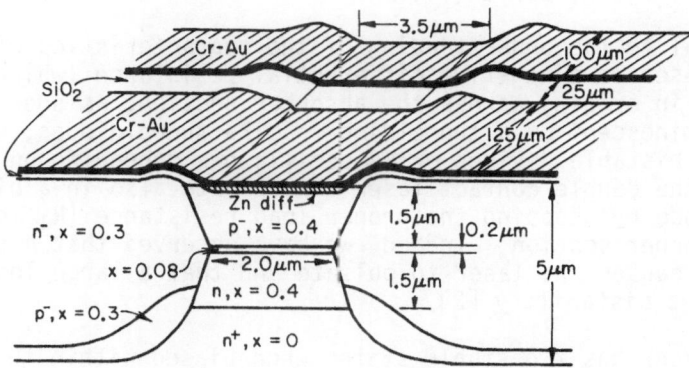

Fig. 1. $Ga_{1-x}Al_xAs$ buried heterostructure laser with a segmented contact.

311

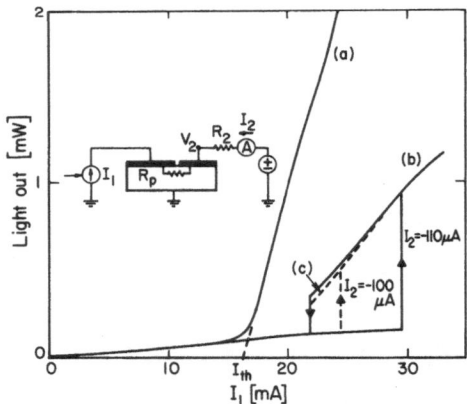

Fig. 2.   Measured static current light characteristic of the two
          segment contact laser. For homogeneous pumping: curve (a)
          and for inhomogeneous pumping: curve (b) and (c). Curve (b)
          corresponds to the case of a large saturation intensity of
          the saturable absorber and curve (c) to the case of a small
          saturation intensity.

     When both sections of the laser are pumped equally, the light
current characteristic is linear as shown in Fig. 2, curve (a). For
bistable operation, the absorber section is pumped with a very small
negative constant current I_2 (to introduce the saturable absorption)
and the light output as function of the current through the gain
section displays a giant hysteresis (Fig. 2, curve (b)). The amount
of saturable absorption can be easily adjusted by changing the current
I_2 through the absorber section resulting in a different size of the
hysteresis (Fig. 2, curve (c)).

     Crucial to the understanding of the characteristics of this
bistable laser is a negative differential resistance (which is opto-
electronic in origin) across the absorber section, as shown in
Fig. 3, reminescent of a tunneldiode. Like tunneldiodes, which can
operate as bistable devices or as oscillators depending on the bias
resistor, the double contact laser can operate also in a bistable or
unstable mode by choosing the proper load resistance (R_2 in Fig. 2)
at the absorber section. Indeed, we have observed that a small load
resistance causes the laser to pulsate and that a large load resistance causes bistability [2].

     The laser has two stable states when biased within the hysteresis
(i.e. I_1=25mA; I_2=-110μA in Fig. 2), an off state (the laser emits
only spontaneous radiation) and an on state (the laser radiates
stimulated emission).   A small positive current pulse superimposed
on I_1 increases the gain momentarily, thus bleaching the saturable

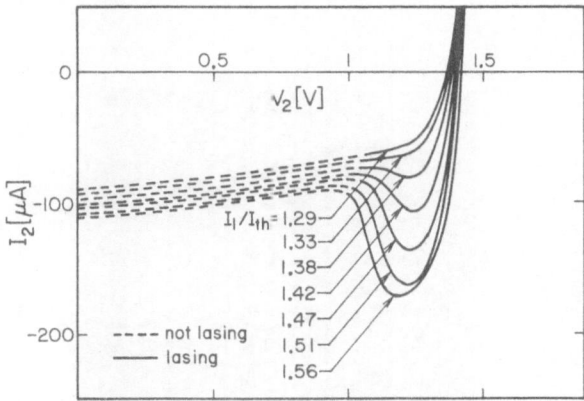

Fig. 3. Current voltage characteristic of the absorber section for
         different pump currents in the gain section I_1/I_{th}.
         I_2, V_2 and I_{th} are defined in Fig. 2.

absorption and the laser switches on. Switching off is achieved by
a subsequent negative current pulse as shown in Fig. 4. While the
switching off is usually fast, the switching on dynamics depends
critically on the trigger pulse amplitude. Measurements of the
technologically important delay between trigger pulse and switching
are shown in Fig. 5 as a function of the trigger pulse amplitude.
For small trigger pulse amplitudes these switching delays are very
long but they can be reduced by increasing the trigger pulse to
typically 5 ns at power delay products of 100 picojoules.

A simple model explains the main features of an injection laser
with inhomogeneous current injection. A set of three rate equations
is used, one for the density of the carriers in the gain section, one
for those in the absorbing section and one for the density of photons
in the lasing mode. The calculated light current characteristic is
shown in Fig. 6. It is a unique feature of semiconductor lasers that
the quasi-Fermi levels of the inverted population can be acessed
directly since the externally applied voltage is equal to the difference of the quasi-Fermi levels. Many interesting effects in injection lasers are a direct consequence of this fact [2]. Therefore,
for a complete description, two equations relating the voltage across
the absorber section and the gain section to the respective carrier
densities have to be added to the three rate equations. The
calculated current voltage characteristics of the absorber section
displays a negative differential resistance as shown in Fig. 7.

It has been shown that a bistable laser with a saturable absorber
is an example of a system with a first order phase transition. As
far as device applications are concerned one of the most important

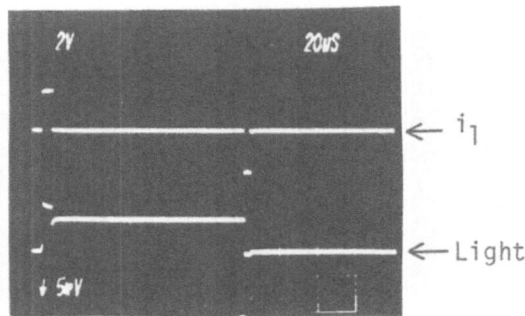

Fig. 4. Gain-switching of the bistable laser. Top trace:
Current I_1 through the gain section. Lower trace:
Light output, Hor.: 20µs/div.

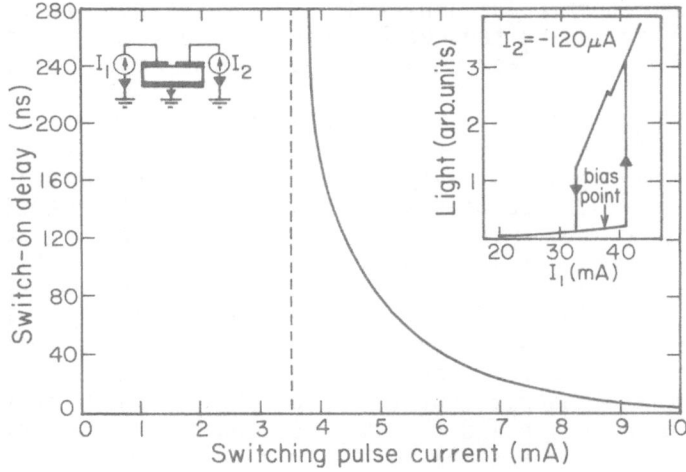

Fig. 5. Switch-on delay time as function of the trigger pulse
amplitude. The critical trigger pulse amplitude is
3.5mA.

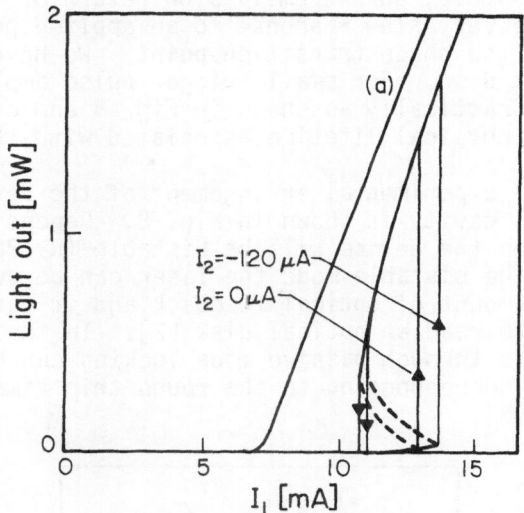

Fig. 6.   Calculated static light current characteristic.   For
          homogeneous pumping: curve (a) and for inhomogeneous
          pumping: curve (b) and (c).

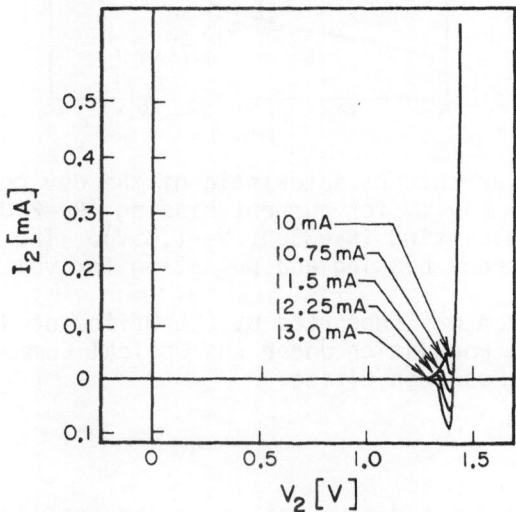

Fig. 7.   Calculated current voltage characteristic of the absorber
          section for different pump currents in the gain section I_1.

predictions gained from this point of view is that of the critical slowing down, that is, an extremely slow return of the system to the equilibrium or a very slow response to an applied perturbation in the vicinity of the phase transition point. We have observed this critical slowing down. For small trigger pulse amplitudes the delay time increases drastically as shown in Fig. 5 and can be very long compared to any physical lifetime associated with the system.

The typical experimental arrangement of the device in an external optical cavity is shown in Fig. 8. Depending on the biasing condition the device will be bistable ($R_2=200k\Omega$) or pulsating ($R_2=330\Omega$). In the bistable mode the laser can be switched on and off by varying the amount of optical feedback and it can be used as an optical stylus to read an optical disk [2]. In the pulsating mode ultrashort pulses through passive mode locking can be generated at a repetition rate corresponding to the round trip time in the external optical cavity [3].

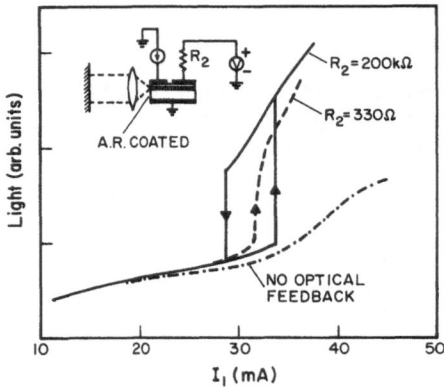

Fig. 8.   Light current characteristic of the device in an external optical cavity for current biasing ($R_2=200k\Omega$, V=-20V) and voltage biasing ($R_2=330\Omega$, V=-1.05V). The device is bistable for current biasing and pulsating for voltage biasing.

This research was supported by the Office of Naval Research, the National Science Foundation under the Optical Communication Program and by the Army Research Office.

REFERENCES

[1] G. J. Lasher, Solid State Electron., 7:707 (1964).
[2] Ch. Harder, Kam Y. Lau, and Amnon Yariv, IEEE J. Quantum Electronics, 18:1351 (1982).
[3] Ch. Harder, John S. Smith, K.Y. Lau and Amnon Yariv, Appl. Phys. Lett., 42:772 (1983).

PRECISE MEASUREMENTS ON OPTICAL BISTABILITY AND PASSIVE Q- SWITCH

IN A CO_2 LASER WITH SATURABLE ABSORBER

E. Arimondo[+], B.M. Dinelli[++] and E. Menchi[+++]

(+) Istituto di Fisica Sperimentale, Università di Napoli, Italy and Gruppo Nazionale di Struttura della Materia del C.N.R., Pisa, Italy (++) Istituto Ricerca Onde Elettromagnetiche (C.N.R.), Firenze, Italy   (+++) Istituto di Fisica dell'Università di Pisa, Italy

INTRODUCTION

The laser with intracavity saturable absorber (LSA) is an interesting system in quantum optics because by the choice of the gain and the absorption of the intracavity media, several different non linear phenomena can be realised. Optical bistability has been investigated quantitatively by Ruschin and Bauer in a CO_2 laser with SF_6^7. Recently a large interest arose owing to experimental observations and the development of precise theoretical models. Passive Q-switch (PQS) consists of an operation with undamped spikes in the output power, a self pulsing behaviour produced by the instability of the continuous wave state and is a phenomenon of growing interest from an experimental as well as theoretical point of view[2].

In general the existing papers are concerned either with PQS or with bistability and not with both together. In recent works[3,4] the simultaneous occurrence of PQS and bistability has been observed and reported and the possible combinations that can be found in LSA experiments have been classified. At present a large effort is made in the precise experimental determination of the LSA operation parameters in regions of bistability and instabilities, in the comparison with various theoretical models and the development of more suitable ones. In this work we report the measurements on bistability and PQS in a CO_2 laser containing a SF_6 absorber. We make a quantitative comparison between a theoretical description of optical bistability and the experimental results. In the course of the analysis we concluded that the available models of bistability were not satisfactory and a more precise description was developped. The experiment concerns the infrared CO_2 laser and molecular absorption transitions where the single mode laser operation is easily realised and molecular parameters are well determined. As the theoretical models do not include the frequency pulling effects produced by the mismatch of amplifier and absorber frequency, we have set up an experiment to determine the influence of laser frequency on the optical bistability. As far as PQS is concerned, recent theoretical analysis have derived the essential features of the phenomenon but a complete model to compare with experimental data does not exist. We report a set of measurements on the pulse period and the pulse intensity as a function of the parameters specifying the LSA, including frequency.

THEORETICAL MODELS

The evolution equations of a laser with or without a saturable absorber have been treated by many authors and various models have been proposed to describe the behaviour of the amplifying and absorbing media and to determine the conditions of stability and instability. The following fundamental steps have been introduced in the theory of the LSA and will concern the analysis of the present work: i) PQS and bistability in a two level model with the assumption of small losses in the cavity[5,6] ii) LSA in a four level model with small losses[3] iii) the laser Rigrod theory without absorber but with the assumption of non negligible losses[7] iiii) an extension of the Rigrod model to a laser with a non saturable absorber[8]. The four level model has been introduced to include the effects of other energy levels on the resonant ones, both in the active and in the passive medium, and the multiplicity of vibrational and rotational levels present in the complex molecules interacting with the field. Before the outline of our model we present here a brief discussion of the theoretical results of i) and ii) mentioned above, in relation to both bistability and PQS.

Two level model

The state equation of the stationary solution in a ring laser is:

$$I\left(1 - \frac{A}{1+I} + \frac{\bar{A}}{1+aI}\right) = 0 \qquad [1]$$

where I = intensity normalized to the saturation intensity I_a of the absorber; A = normalized gain; $\bar{A}$ = normalized absorption; a = ratio of the saturation intensities of the amplifier and absorber. The stability analysis of the $I=0$ and $I=I_+$ solution of equation [1] shows that for $a>1$ (saturable absorber) and for $X_+ < aA < a(\bar{A}+4)$ where $X_+ = a + \bar{A} - 1 + 2\sqrt{\bar{A}(a-1)}$ bistability occurs. A further discussion of the instability leads to the following relation for the passive Q- switch $A\gamma^2 < \bar{A}a\bar{\gamma}^2$ where $\gamma, \bar{\gamma}$ are the relaxation rates in the amplifier and absorber. However this condition is not satisfied by the experimental observations on weak absorbers. This suggests that a two level model is not a suitable approximation for PQS.

Four level model

This model leads to a state equation similar to [1] where the parameters $A, \bar{A}, a$ are defined in a slightly different way. By means of a rather laborious calculation and through the Routh - Hurwitz theorem, the stability conditions were investigated. For I_+ negligible, the instability condition reduces to:

$$A < \zeta_a \bar{A}$$ [2]

where ζ depends on the relaxation rates of the four level model. This relation allows an analysis of the PQS region in the $A, \bar{A}$ plane and a comparison with the domain of existence of the trivial solution $I = 0$ and the solution $I = I_+$. On the basis of this model, for instance, the possible outcomes of experiments on the output intensity as a function of the pump parameter A can be investigated, keeping the other parameters fixed. Typical results, relevant to the experimental observations here presented, are shown in fig. 1.

Theory with large gain and losses

The bistability observations of this paper were quantitatively analysed by extending the Rigrod model to describe the LSA[9]. Thus we have properly taken into account the mirror losses and the exponential growth and attenuation in the amplifier and absorber respectively. On the contrary a mean field approximation has been introduced to describe the standing wave operation: the gain and absorption coefficient have been calculated through an average over the wavelenght of the laser radiation. The steady state laser output power has been determined graphically from the crossing points of two curves describing the laser operation. Altough an homogeneous broadening was introduced to describe the low pressure inhomogeneously broadened absorber, the agreement between theoretical fit and the experimental results is very good.

The intensities per unit area, normalized to the saturation intensity I_a of the amplifier, of the forward and backward propagating waves are denoted by I_f and I_b respectively. The gain coefficient of the amplifier is expressed by:

$$g(z) = g_0 / 1 + I_f + I_b$$ [3]

where g_0 is the unsaturated gain coefficient.

As the gain coefficient is istropic:

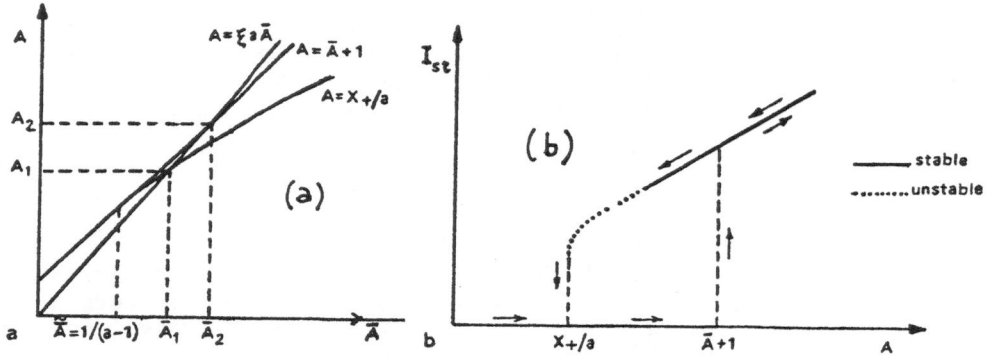

Fig. I.   (a) Phase plane $(A,\bar{A})$ of the system. (b) Plot of emitted
           light I vs. pump parameter A.

$$g(z) = \frac{1}{I_f}\frac{dI_f}{dz} = -\frac{1}{I_b}\frac{dI_b}{dz} \qquad [4]$$

and it results that $I_f \cdot I_b = const = C$ [5]
We can introduce similar equations for the absorber, i.e.

$$\alpha(z) = -\alpha_0/1 + \partial I_f + \partial I_b < 0$$

where ∂ is the ratio of the saturation intensity of the amplifier
and absorber. If we introduce the effective reflectances r_1, r_2 at the
laser boundaries, by integration of equation [4] over the length ℓ
of the amplifier and of a similar equation for the absorber over its
lenght d and making use of equation [5] we derive the state equation for the LSA:

$$\beta I \left[exp(\eta I) + \gamma \right] = I \, exp(-\eta I) + \delta/\eta \qquad [6]$$

where $\eta, \beta, \gamma, \delta$ are defined as a function of $A, \bar{A}, \partial, r_1, r_2, \ell, d$ -
Furthermore the following definitions have been introduced:

$$A = -2g_0 \ell / \ell n(r_1 r_2) \quad ; \quad \bar{A} = -2\alpha_0 d / \ell n(r_1 r_2) \qquad [7]$$

that in the limit of $1-r_1, 1-r_2 \to 0$ reduce to the amplifier and
absorber parameters introduced in the treatment of Lugiato et al.[6]
With proper approximation, a state equation similar to [1] but valid
for a cavity with r_1 and r_2 losses, is obtained:

$$1 - 2g_0\ell/(2-r_1-r_2)(1+I) + 2\alpha_0 d/(2-r_1-r_2)(1+\partial I) = 0 \qquad [8]$$

     The state equation [6] allows to determine the intensity I
and the output power P_{out} from the end mirror with transmittance T
is $P_{out} + TII_a$ S where S is the laser beam across area. If no
absorber is present in the cavity, i.e. $d \to o$ the state equation of
Rigrod is obtained:

$$I = \sqrt{r_2} \left[(A-1) / (\sqrt{r_1} + \sqrt{r_2})(1 - \sqrt{r_1 r_2}) \right] \cdot \ell n \left(1/\sqrt{r_1 r_2} \right) \qquad [9]$$

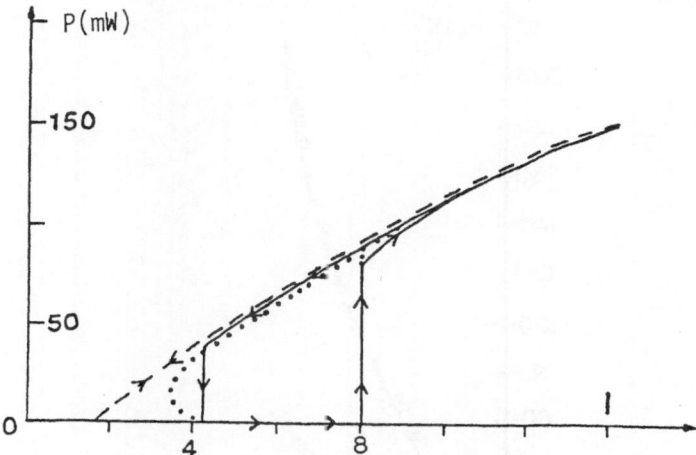

Fig. 2. Experimental and theoretical curves for the optical bistability. The dashed line represents the observed behaviour
for the laser without absorber, the continuous line reports
the behaviour recorded with the intracavity absorber, and
the dotted line is a theoretical fit.

In the case of an unsaturable absorber, i.e. a → o another relation
similar to that used by Kaufman and Oppenheim, and useful for ana-
lysing experimental data can be derived:

$$I = \sqrt{r_2}\, \exp\left(-\varkappa_o d\right) \frac{(A - \bar{A} - 1)\, \ln\left(1/\sqrt{r_1 r_2}\right)}{\left[\sqrt{r_1} + \exp\left(-\varkappa_o d\right)\sqrt{r_2}\right]\left[1 - \exp\left(-\varkappa_o d\right)\sqrt{r_1 r_2}\right]} \qquad [10]$$

EXPERIMENT

The apparatus, typical of an intracavity experiment, was composed by a single mode CO_2 laser operating on the IOP(I6) line for
bistability measurements and on the IOP(22) line for PQS. The amplifier length was ℓ=76 cm., absorber length d = 36 cm., output mirror
and grating with 3 m. radius of curvature. A special effort was made
to have a very stable operation on a long time scale, to avoid the
effects of thermal drifts and hysteresis. To observe the frequency
dependence of bistability and PQS we used a CO_2 waveguide laser and
monitored the beating signal of the LSA output against the reference
laser.

Bistability

The dashed line of Fig.2. reports the measured output power
(in mW) of the CO_2 laser versus the amplifier current(mA) in absence

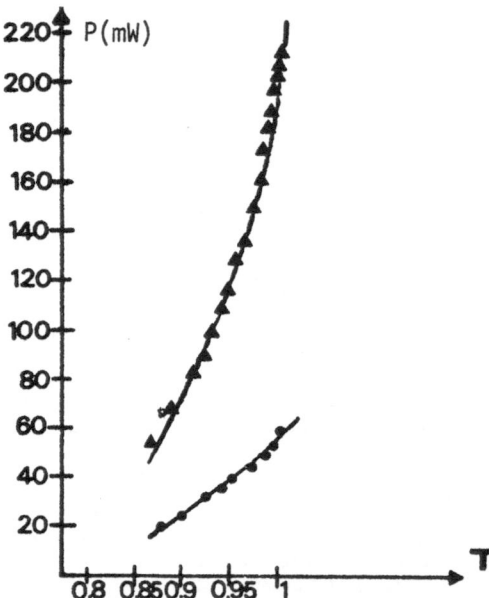

Fig.3. Output laser power (in mW) versus the transmission of the
NH$_3$-He mixture (in the 1:5) ratio) introduced in the absorber
cell. Upper and lower curves have been obtained at different
apertures of the intracavity iris, 7. and 5.9 mm respectively.
The continuous lines are theoretical fits obtained for the
reflectances reported in the text and amplifier saturation
parameters I$_a$=50(5)^W/cm^2 and I$_a$=20(5)^W/cm^2 respectively.

of the intracavity absorber. The continuous line represents the
output power when 25 mTorr pressure SF$_6$ gas was introduced in the
absorver cell, with a bistable behaviour extending over a large range
of the monitored region. The laser parameters were determined looking
at the output power with the absorber cell filled with ammonia diluted in helium, that behaves as an unsaturable absorber in the range
of intracavity power. Typical results are shown in Fig. 3. Making
use of the equation [10] to fit this laser operation, we have derived
the true reflectances of the laser end mirrors r$_1$=0.95(3);r$_2$=0.82(3)
and the amplifier saturation parameters. For the fit of bistability
it may be noticed that also in our model the upper point of the Bistability cycle is given by the relation A=Ā+1. From the experimental
data the absorption coefficient for SF$_6$ resulta 0.15 cm^{-1}. Torr^{-1}
in agreement with the 0.55 cm^{-1}. Torr^{-1} measured for the laser peak
small signal absorption in previous work. The theoretical curve (dotted
line) drawn in Fig2. corresponds to the Ā value derived above and

the saturability parameter a = 60 that provides the best fit of the experimental results and in good agreement with the value measured previously. As far as the frequency influence is concerned we have observed that in absence of the absorber a laser frequency shift smaller than 5 MHz occurs near threshold in the 2-3 mA range, but no shift occurs at larger currents. When the absorber is introduced no laser frequency shift occurs over the bistability region.

Passive Q- switch

In a LSA the PQS appears when the laser operates at a low pumping parameter near the threshold value, i.e. at low current or pressure in the amplifier cell. At larger values of the pumping parameter, as reached increasing the amplifier current or pressure, the Q-switching regime terminates and the c.w. operation is recovered. Because in recent theoretical analysis the Q-switching pulse has been specifically investigated[2] we measured width, intensity and period of the pulses for the IOP(22) line, in presence of SF_6 intracavity absorption with pressures in the 20 - I20 mTorr range. The pulse width did not change with the pumping inversion in the amplifier and with the laser frequency, in agreement with the prediction that the pulse width is equal to the inverse of the photon decay rate in the cavity.For the pulse height it resulted that laser frequency and pumping inversion produced changes smaller than I0% and 5% respectively, while the iris diameter affected the pulse height up to 25%. For the pulse period it resulted that a dramatic rise occurred whenever a transition from the passive O- switching to laser off operation is approached, i.e. a typical slowing down phenomenon at a bifurcation with an unstable solution.Such a behaviour is illustrated in Figs. 4. 5. for the dependence of the period pulse or the amplifier current, i.e. the pumping inversion, and the laser frequency detuning respectively. In Fig. 4.the critical slowing down at low currents is observed for different SF_6 pressures, while the extension for the Q- switching region changes with the pressure. In Fig. 5. an increase in the pulse period is observed on both sides of the frequency tuning, with a longer period for laser tuning approaching the center of the SF_6 absorption at $\Delta f = + 50$ MHz. The pulse period is of the order of the decay time of the amplifier excited state but experimental observations show that when the decay time decreases increasing the amplifier pressure, the pulse period does not decrease because the amplifier inversion is simultaneously increased.

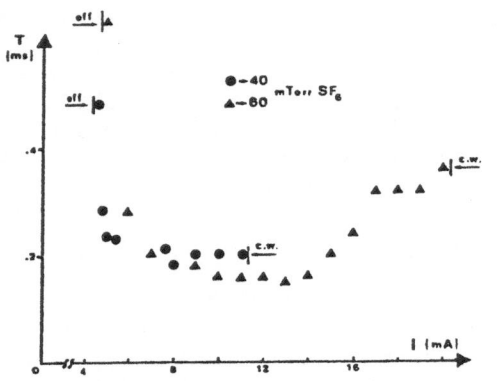
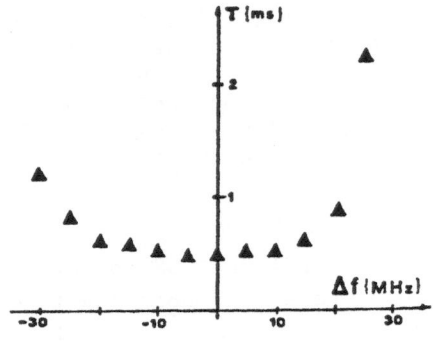

Fig. 4. Dependence of the PQS period on the amplifier current for the CO_2 10P(22) line with SF_6 absorber.

Fig. 5. Dependence of the PQS period on the laser detuning for the CO_2 10P(22) line with 60 mTorr SF_6 pressure.

CONCLUSION

We have shown that the bistable behaviour observed in a CO_2 laser containing a saturable absorber can be described properly in our theoretical treatment were large gain and losses have been introduced. Any remaining discrepancy concerning the value of the absorption coefficient included in the fit could be solved by including the transverse spatial gaussian distrubution of the laser beam and the inhomogeneous broadening of the absorber in the theoretical mode. Preliminary measurements show that the influence of the frequency detuning on the optical bistability is negligible. However observations on different laser lines and different absorbers are required to make a complete test of the frequency dependent phenomena. The experiments on the PQS provide additional information on the behaviour of the LSA. A wider range of observations has to be conducted with a special effort to analyse the behaviour below threshold and in region near the PQS occurrence.

REFERENCES

1. S.Ruschin,S.H.Bauer, Chem.Phys.Lett. 66(1979)100; Appl.Phys. 24(1981)45.
2. J.C.Antoranz,J.Gea,M.G.Velarde, Phys.Rev.Lett. 47(1981) 1895. J.C. Antoranz,L.L.Bonilla,J.Gea,M.G.Velarde,Phys.Rev.Lett.49(1982)35.
3. E.Arimondo,F.Casagrande,L.Lugiato,P.Glorieus,Appl.Phys.B30(1983)57.
4. A.Jacques,P.Glorieux, Opt.Comm. 40(1982)455.
5. H.T.Powell, G.J.Wolga, IEEE.J. QE7 (1971) 213.
6. L.Lugiato,P.Mandel,ST.Dembinski,A.Kossakowski,Phys.Rev.A18(1978).
7. W.W.Rigrod, J.Appl.Phys. 36(1965)2487.   1145.
8. Y.J.Kaufman, U.P.Oppenheim, Appl.Opt. 13(1974)374.
9. E. Arimondo, B.M.Dinelli, Opt.Comm. 44(1983)277.

OPTICAL BISTABILITY IN CHOLESTERIC LIQUID CRYSTALS

Herbert G. Winful

GTE Laboratories Incorporated
40 Sylvan Road
Waltham, Massachusetts 02254

INTRODUCTION

Cholesteric liquid crystals possess an intrinsic helical structure which leads to some rather striking linear optical properties.[1] These properties include a very strong rotatory power and the Bragg reflection of circularly polarized light of the appropriate helicity and wavelength. In this paper we consider nonlinear optical propagation in cholesterics. For wavelengths in the Bragg regime, we find that the intense light can change the pitch of the helix and thus yield a bistable reflection characteristic.[2] In the regime of wavelengths much shorter than the helix pitch (the so-called Mauguin limit) where there are no Bragg reflections, we show that the polarization state of the transmitted light exhibits bistability as the incident intensity is varied. In this limit one may place the cholesteric liquid crystal between crossed polarizers and observe an intensity bistability which does not rely on either Bragg reflections or mirror feedback.

EQUATIONS FOR THE DIRECTOR AND LIGHT FIELD

To sum up the analysis, we begin with the free-energy of the liquid crystal in the presence of the light field. Minimization of the free-energy yields an Euler-Lagrange equation for the director which describes the orientation of the elongated liquid crystal molecules. The light wave couples to the local dielectric anisotropy to exert orienting torques on the molecules whose orientation in turn affects the propagation of the wave. A simultaneous solution of the coupled Maxwell and Euler-Lagrange equations yields the self-consistent field and director distributions. An analytic solution is obtained by assuming the presence of only two counterpropagating (Bragg regime) or co-propagating (Mauguin limit) modes and by employing the slowly varying envelope approximation.

   Consider a cholesteric slab of length L whose helix axis is oriented along $\hat{z}$ (Fig. 1). The average orientation of the aniso-tropic liquid crystal molecules is described by the director $\hat{n}(z)$ whose components are:

$$n_x = \cos\theta(z), \quad n_y = \sin\theta(z), \quad n_z = 0. \tag{1}$$

The director rotates about the z axis, and in the absence of ex-ternal perturbations the angle θ is given by $\theta = q_o z$, where q_o is the unperturbed wave number of the helix whose pitch is $p = 2\pi/q_o$. In the presence of an intense light wave propagating along the z-axis the cholesteric director assumes a new configuration which may be found by minimizing the total free energy [1]

$$F = \int d^3r \left\{ \frac{1}{2} K_{22} \left(\frac{\partial\theta}{\partial z} - q_o \right)^2 - \frac{\langle \bar{E} \cdot \bar{D} \rangle}{4\pi} \right\}. \tag{2}$$

Here K_{22} is a Frank elastic constant for twist deformations. The electric displacement is related to the electric field E through

$$\bar{D} = \epsilon_\perp \bar{E} + \epsilon_a \hat{n}(\hat{n} \cdot \bar{E}), \tag{3}$$

where $\epsilon_a = \epsilon_{\|} - \epsilon_\perp$, $\epsilon_{\|}$ and $\epsilon_\perp$ being the dielectric constants parallel and perpendicular to the local director. The angular brackets in Eq. (2) signify a time average over a few optical cycles.

   Minimization of the free-energy and use of the Maxwell equations leads to the following set of equations for the director and the electric field:[1-3]

$$\frac{d^2\theta}{dz^2} = \frac{\epsilon_a}{4\pi K_{22}} \left[\mathrm{Re}(E_+ E_-^*)\sin 2\theta - \mathrm{Im}(E_+ E_-^*)\cos 2\theta \right], \tag{4}$$

$$\frac{-d^2E_+}{dz^2} = k_o^2 (E_+ + \delta E_- e^{i2\theta}), \tag{5}$$

$$\frac{-d^2E_-}{dz^2} = k_o^2 (E_- + \delta E_+ e^{-i2\theta}), \tag{6}$$

where $k_o^2 = (2\pi/\lambda)^2 \epsilon$, $\delta = \epsilon_a/2\epsilon$, and $\epsilon = (\epsilon_{\|} + \epsilon_\perp)/2$. The electric field vector is taken as

$$\bar{E} = \mathrm{Re}\left\{ [E_-(z)(\hat{x}+i\hat{y})/\sqrt{2} + E_+(z)(\hat{x}-i\hat{y})/\sqrt{2}]e^{-i\omega t} \right\},$$

where $E_{\pm} = (E_x \pm iE_y)/\sqrt{2}$.

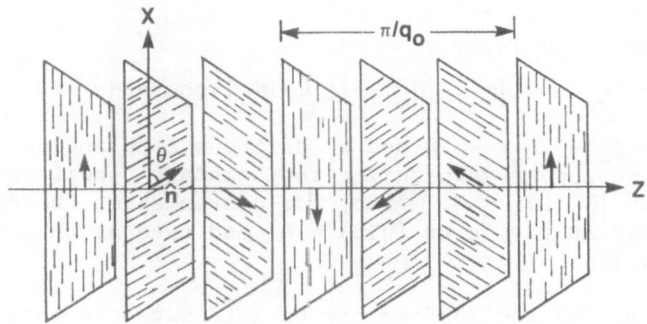

Fig. 1. Schematic of a cholesteric liquid crystal.

Equations (4), (5), and (6) completely describe the steady state nonlinear propagation of light in a cholesteric liquid crystal. They contain the rapid variations of the electric field which occur on a wavelength scale, as well as the detailed evolution of the director θ. To make analytical progress, we now use the slowly varying envelope approximation to remove the rapid electric field variations. It turns out that in real liquid crystals, the dielectric anisotropy δ is a small quantity of order 0.01. The error made in using the slowly varying envelope approximation can be shown to be of order δ^2.

OPTICAL BISTABILITY IN THE BRAGG REGIME

It is convenient to define a reduced wavelength given by $\lambda' = \lambda/p\sqrt{\varepsilon}$ where λ is the vacuum wavelength of the incident light and p is the pitch of the helix. Bragg reflection then occurs in a band of width 2δ centered about $\lambda'=1$ for circularly polarized waves whose helicity matches that of the cholesteric. Let us assume that a right circularly polarized wave with $\lambda' \sim 1$ is incident on a right-handed cholesteric. The field in the medium may be taken as a sum of counterpropagating right circularly polarized waves of the form

$$E_\pm = |\mathcal{E}_\pm(z)|\exp[i\phi_\pm(z) \pm iq_o z], \tag{7}$$

where the plus and minus signs refer to forward and backward waves, respectively. Then in the slowly varying envelope approximation Eqs. (5) and (6) become

$$d|\mathcal{E}_+|/dz = \kappa|\mathcal{E}_-|\sin\Psi \ , \tag{8a}$$

$$d|\mathcal{E}_-|/dz = \kappa|\mathcal{E}_+|\sin\Psi \ , \tag{8b}$$

$$d\Psi/dz = 2q_o + \kappa(|\mathcal{E}_-|/|\mathcal{E}_+| + |\mathcal{E}_+|/|\mathcal{E}_-|)\cos\Psi - 2d\theta/dz \ . \tag{8c}$$

Here $\Psi = \phi_+ - \phi_- + 2q_o z - 2\theta$ and we define a coupling constant $\kappa = \delta k_o/2$. We assume zero-reflection boundary conditions on the field so that $|\mathcal{E}_-(L)| = 0$. Also, for convenience, the director is constrained at the input $[\theta(0) = 0]$ and free at the exit $(d\theta/dz|_L = q_o)$.

Equations (4) and (8) can now be integrated to obtain explicit expressions for the field and director distributions.[2] In particular, the relation between incident (I) transmitted (J) intensities is

$$I = u_3 + \frac{u_2 - u_3}{1 - (u_1-u_2)(u_1-u_3)^{-1} sn^2[2\kappa L/g,k]} \ , \tag{9}$$

where $u_1 > u_2 > u_3 > u_4$ are the roots of the quartic

$$Q(u) = (u-J)[u-(u-J)^3] \ , \tag{10}$$

and sn is a Jacobi elliptic function with $g = 2/[(u_1-u_3)(u_2-u_4)]^{\frac{1}{2}}$ and $k = [(u_1-u_2)(u_3-u_4)]^{\frac{1}{2}} g/2$. The intensitities are normalized by $|\mathcal{E}_c|^2 = \pi K_{22}(\omega/c)^2(\varepsilon_a/\varepsilon)$. As shown in Fig. 2 , the relation between transmitted and incident intensities is multivalued. The transmitted intensity will exhibit hysteresis and bistability as the incident intensity is varied. Physically, the effect of the intense field is to distort the cholesteric helix and increase its pitch. A light wave which at low intensity satisfies the Bragg condition for the structure may not suffer Bragg reflections at higher intensities. Thus a transition from high reflection to high transmission occurs as the incident intensity is increased beyond a critical value. The physics of this bistability is similar to that predicted for distributed feedback structures.[4]

OPTICAL BISTABILITY IN THE MAUGUIN LIMIT

The Mauguin limit is determined by the condition $4(\lambda'/\delta)^2 \ll 1$. In the limit where the light wavelength is much shorter than the helix pitch ($\lambda' \ll 1$) Bragg reflection may be safely neglected and, except for normal dielectric reflections, an incident wave will enjoy total transmission. The field may thus be resolved into two co-propagating circularly polarized components of the form

$$E_+ = |\mathcal{E}_+(z)|\exp[i\phi_+(z) + ik_o z] \ , \tag{11}$$

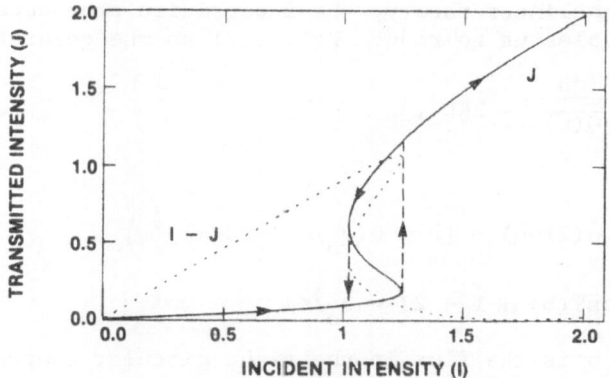

Fig. 2. Transmitted (solid line) and reflected (dashed line)
         intensities vs incident intensity for $\kappa L = 2$.

where the plus and minus signs refer to right and left circular
components, respectively. Using this field in Eqs. (5) and (6)
and making the slowly varying envelope approximation we obtain

$$d|\mathcal{E}_+|/dz = \kappa |\mathcal{E}_-| \sin\Psi , \tag{12a}$$

$$d|\mathcal{E}_-|/dz = -\kappa |\mathcal{E}_+| \sin\Psi , \tag{12b}$$

$$d\Psi/dz = \kappa (|\mathcal{E}_-|/|\mathcal{E}_+| - |\mathcal{E}_+|/|\mathcal{E}_-|)\cos\Psi - 2d\theta/dz , \tag{12c}$$

where $\Psi = \phi_+ - \phi_- - 2\theta$ and $\kappa = \delta k_o/2$. Physically, Eqs. (12a) and
(12b) describe the exchange of energy between the right and left
circular modes which are coupled through the dielectric anisotropy
δ. They can be integrated to yield $|\mathcal{E}_+(z)|^2 + |\mathcal{E}_-(z)|^2 =$ constant,
which expresses the conservation of power. An arbitrary polariza-
tion state of the field may be completely determined by specifying
one of the amplitudes (say $|\mathcal{E}_+|$) and the relative phase $(\phi_+ - \phi_-)$
which is twice the angle between the major axis of the polarization
ellipse and the x-axis. Equation (12c) describes the evolution of
the orientation of the electric field relative to the director.
The director in turn is driven by the electric field as shown in
Eq. (14).

     In the linear theory $(d\theta/dz = q_o)$ the solution of Eqs. (12)
shows that the transmitted mode amplitudes $|\mathcal{E}_+(L)|$ and $|\mathcal{E}_-(L)|$ are
sinusoidal functions of sample length. The polarization state
(orientation and ellipticity) of the transmitted light thus varies
periodically with L.

In the nonlinear theory, the integration procedure outlined in Ref. 4 enables us to reduce Eqs. (12) to the quadrature

$$\int_{I}^{J} \frac{du}{\sqrt{Q(u)}} = 2\kappa L , \qquad (13)$$

where

$$Q(u) = u(2I-u) - [\Gamma + u(q_o/\kappa + 2J-u)]^2 , \qquad (14)$$

and

$$\Gamma = I(\cos\Psi(\theta) + I - 2J - q_o/\kappa) .$$

The variable u is the flux in the right circular component normalized by $|\mathcal{E}_c|^2$; I and J are its values at the input and output, respectively.

Since Q(u) is a quartic polynomial, Eq. (13) is an elliptic integral which can be inverted to yield explicit expressions for the the field and director distributions. The analytic expressions, however, are cumbersome[5] and only graphical results will be shown here. Figure 3 shows the transmitted intensity in the right circular component as a function of sample length for an incident beam of intensity 0.8 (I = 0.4) linearly polarized at 45° to the x-axis. The parameters used are $\lambda' = 0.006$, $\delta = 0.03$, $q_o/\kappa = 0.4$. It can be seen that $|\mathcal{E}_+(L)|^2$ (and hence the polarization state) is a multiple valued function of the sample length. This is a result of the distortion of the cholesteric helix by the light wave.

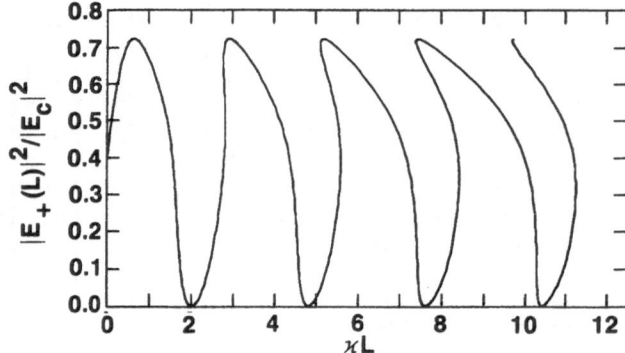

Fig. 3.   Transmitted intensity in right circular component as a function of κL for I = 0.4.

For example, at $\kappa L = 8$ we find three values of $|\mathcal{E}_+(L)^2|$ each of which corresponds to a different spatial distribution of the director and the polarization. The total transmitted intensity $|\mathcal{E}_+(L)|^2 + |\mathcal{E}_-(L)|^2$ is, of course, constant.

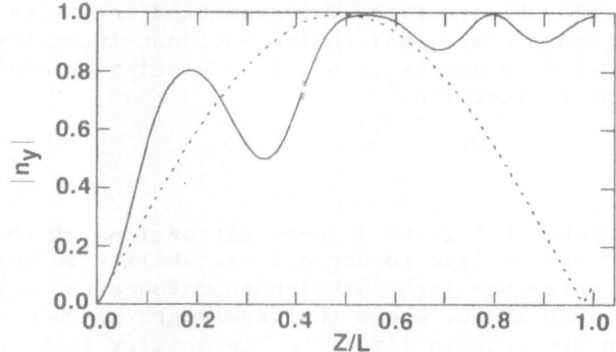

Fig. 4a.   The y component of the director for zero field (dashed line) and for I = 0.4 (solid line). $\kappa L = 8$, and J is on the lower branch of transmission curve in Fig. 3.

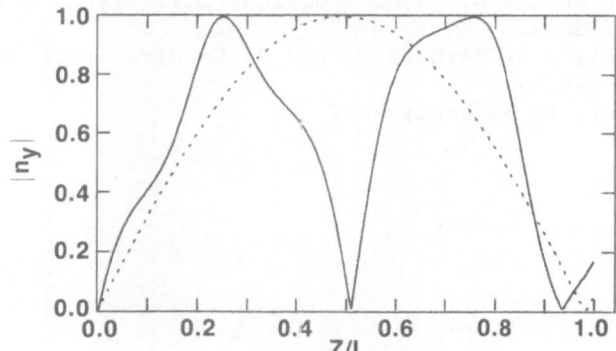

Fig. 4b.   The y component of the director for zero field (dashed line) and for I = 0.4 (solid line). $\kappa L = 8$, and J is on the upper branch of transmission curve in Fig. 3.

In Figs. 4a and 4b the dotted curve shows the y-component of the director in the absence of an applied field. The solid curves show the director distributions for the lowest value of J (Fig. 4a) and the highest value of J (Fig. 4b) at $\kappa L = 8$. On the lower branch, the intense field tends to unwind the helix while on the upper branch it tends to coil up the helix. For a fixed value of κL, switching between the different distributions may be obtained by varying the input intensity. Assuming that these steady states are stable, the structural bistability induced by the applied field manifests itself through a polarization bistability of the transmitted field while the total transmitted intensity remains single valued and proportional to the incident intensity. An intensity bistability can be observed by placing an analyzer at the exit of the cholesteric.

CONCLUSIONS

We have shown that light induced distortions of the helical cholesteric structure lead to optical bistability in both the Bragg regime, where the light wavelength matches the helix pitch, and in the Mauguin limit, where the wavelength is much shorter than the pitch. In the Mauguin limit the bistability is in the polarization state of the transmitted light. No external reflectors are needed for these effects.

REFERENCES

1. P.G. deGennes, "The Physics of Liquid Crystals," Clarendon, Oxford, (1974).
2. H.G. Winful, Phys. Rev. Lett. 49:1179 (1982).
3. N.V. Tabiryan and B. Ya. Zel'dovich, Mol. Cryst. Liq. Cryst. 69:19 (1981).
4. H.G. Winful, J.H. Marburger, and E. Garmire, Appl. Phys. Lett. 35:379 (1979).
5. H.G. Winful, to be published.

OPTICALLY-INDUCED BISTABILITY IN A NEMATIC LIQUID CRYSTAL

Hiap Liew Ong and Robert B. Meyer

Department of Physics
Brandeis University
Waltham, MA   02254

ABSTRACT

The structural bistability in an optical field of intensity ~ 200 Watts/cm^2 in a nematic liquid crystal is discussed.

Since 1979, considerable effort has been expended by many researchers in analyzing the purely optical-field-induced molecular reorientation in a homeotropically oriented nematic liquid crystal (NLC) by a normally incident light wave.[1-4] Recently, the exact solution which is consistent with Maxwell's equations for describing the orientation of the molecule was obtained by Ong.[5] In this paper, we shall discuss the exact solution describing the orientation of the director. In particular, by examining the deformation angle near the threshold, the criterion for the existence of a first-order Freedericksz transition is obtained.

We consider a homeotropically oriented NLC cell of thickness d confined between the planes z=0 and z=d of a Cartesian coordinate system. The NLC director always lies in the xz-plane and in the absence of a light beam, the directors are parallel to the z-axis everywhere. $\theta(z)$ denotes the angle between the director and the z-axis. Then the director can be described by $\hat{n}(\vec{r})$ = (sin θ, 0, cos θ). A harmonic time-dependent light beam is normally incident on the NLC medium with the polarization parallel to the plane of incidence, which is the xz-plane.

By taking the time average of the energy flow $\partial F_{opt}/\partial t$ + div $\vec{S}$ = 0, we find that div $\langle \vec{S} \rangle$ = 0, where F_{opt} is the electro-

333

magnetic energy density and $\vec{S}$ is the Poynting vector of the optical field. Consequently the time-average of the z-component of the Poynting vector is a constant throughout the medium. This conclusion agrees with the geometrical optics approximation[3] but disagrees with the infinite-plane-wave approximation[4] in which the magnitude of the Poynting vector is a constant. If the scattering loss in traversing the medium can be neglected, then $\langle S_z \rangle$ is equal to the intensity of the incident beam, I.

It can be shown that for a normally incident wave, $F_{opt} = S_z n_p/c$ with $n_p = n_o n_e / (n_o^2 \sin^2 \theta + n_e^2 \cos^2 \theta)^{1/2}$, where n_o and n_e are the ordinary and extraordinary refractive indices respectively at the optical wavelength λ.[5] The free energy density of the NLC can then be written as

$$F = \frac{1}{2} k_{11} (\text{div } \hat{n})^2 + \frac{1}{2} k_{22} (\hat{n} \cdot \text{curl } \hat{n})^2$$
$$+ \frac{1}{2} k_{33} (\hat{n} \times \text{curl } \hat{n})^2 - \frac{I}{c} n_p(\theta) \qquad (1)$$

where k_{11}, k_{22}, and k_{33} are the splay, twist and bend elastic constants. By the symmetry of the problem, we look for solutions which are symmetrical w.r.t. the $z = d/2$ plane, i.e. $\theta(z) = \theta(d-z)$. Then minimization of the total free energy $\int F \, d^3\vec{r}$ leads to the Euler equation which has a solution of the form

$$z = (\frac{ck_{33}}{2I})^{1/2} \int_0^{\theta} [\frac{1 - k \sin^2 \theta}{n_p(\theta_m) - n_p(\theta)}]^{1/2} d\theta \qquad (2)$$

for $d\theta/dz \neq 0$ and $0 < z < d/2$, where $k = (k_{33} - k_{11}) / k_{33}$ and $\theta_m = \theta(z=d/2)$. In obtaining Eq. (2), we have made use of the rigid boundary conditions at the two interfaces, i.e. $\theta(z=0) = \theta(z=d) = 0$.

We compute the solution for the tilt angle up to and including terms $\sim \theta^2$ and obtain the following equation for the tilt angle:

$$\theta \sim \theta_m \sin (\pi z/d) \qquad (3)$$

where $\theta_m^2 = (\sqrt{I/I_{th}} - 1) / B$, $I_{th} = c k_{33} (\varepsilon_\parallel/n_o \varepsilon_a) (\pi/d)^2$, where I_{th} is the threshold intensity, $B = (1-k-9u/4)/4$, $u = \varepsilon_a/\varepsilon_\parallel$, $\varepsilon_\perp = n_o^2$, $\varepsilon_\parallel = n_e^2$, and $\varepsilon_a = \varepsilon_\parallel - \varepsilon_\perp$. As a comparison, infinite-plane-wave approximation predicts that $I_{th} = ck_{33} (\varepsilon_\parallel^2/n_o \varepsilon_\perp \varepsilon_a) (\pi/d)^2$ and $B = (1-k-9w/4-3u/4)/4$ where $w = 1-(\varepsilon_\perp/\varepsilon_\parallel)^2$.[4,5]

From Eq. (3), we see that as $B > 0$, the transition is second

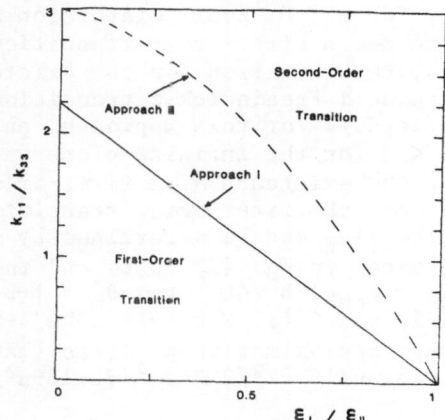

Figure 1. The criterion for the existence of a first-order
optically-induced Freedercksz transition, where this approach is
referred to as approach I and the infinite-plane-wave approximation is referred to as approach II.

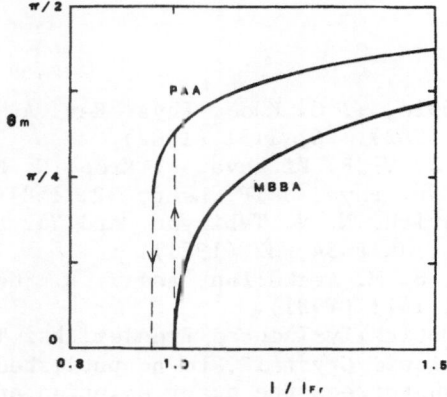

Figure 2. The maximum deformation angle as a function of the
pump beam intensity for MBBA and PAA. For MBBA, we put λ = 6328
Å, n_0 = 1.544, n_e = 1.758, k_{11} and k_{33} = 6.95, 8.99 $*10^{-7}$ dyn
respectively. The Freedericksz transition is second-order with
B=0.06, G=0.06, and I_{Fr}=120.6 W/cm^2 for a cell of 250 μm thick.
For PAA, we put λ=4800 Å, n_0=1.595, n_e=1.995, k_{11} and k_{33} = 9.26,
18.10 $* 10^{-7}$ dyn respectively. The Freedericksz transition is
first-order accompanied by hysteresis with B = -0.08, G = 0.07,
I_{Fr} and I'_{Fr} = 149.0 and 142.8 W/cm^2 respectively for a cell of
250 μm thick.

-order. However, for $B < 0$, small distortion are not stable and the transition becomes a first-order transition accompanied by hysteresis. Thus, the criterion for the existence of the first-order optically-induced Freedericksz transition is given by $B<0$, i.e. $k_{11}/k_{33}+9\varepsilon_\perp/4\varepsilon_\parallel<9/4$ for this approach, and $k_{11}/k_{33}+3\varepsilon_\perp/4\varepsilon_\parallel + (3\varepsilon_\perp/2\varepsilon_\parallel)^2 < 3$ for the infinite-plane-wave approximation. The criterion for the existence of a first-order transition is shown in Fig. 1. For the first-order transition, the lower threshold intensity I'_{th} can be determined by $dI/d(\theta_m^2) = 0$. By computing the integral in Eq. (2) up to and including terms $\sim\theta^4$, we find that $I'_{th} = I_{th}(1-B^2/4G)^2$ and $\theta_m^2 = [-B+(B+4GF)^{1/2}]/2G$, where $F=(\sqrt{I/I_{th}} -1)$ and $G=(11/2-k+9u/4+63ku/4-9k^2/2-261u^2/32)/96$. Infinite-plane-wave approximation predicts that $G = (11/2-k+9w/4 +3u/4+63kw/4+3ku+153wu/16-9k^2/2-261w^2/32-189u^2/32)/96$.[5]

By examining the material parameters from known NLC's, we notice that the bistability could be observed experimentally for PAA (p-azoxyanisole) in a temperature range of 110 to 130°C and a pump beam of intensity $\sim$150 Watts/cm^2 for a cell of 250 μm thick. Figure 2 shows the predicted maximum deformation angle as a function of the intensity for PAA and MBBA [N-(p-methoxybenzylidene-p-butylaniline)]. For MBBA, the transition is second-order but for PAA the transition is first-order accompanied by hysteresis. This hysteresis, if it is confirmed experimentally, may be exploited in bistable display systems.

REFERENCES

1. See, for example, I. C. Khoo, Phys. Rev. A. 25, 1040 (1982);
        25 1637 (1982); 26, 1131 (1982).
2. A. S. Zolot'ko, V. F. Kitaeva, N. Kroo, N. N. Sobolev, and L.
        Chilag, Sov. Phys. JETP. Lett. 32, 158 (1980).
3. B. Ya. Zel'dovich, N. V. Tabiryan, and Yu. S. Chilingaryan,
        Sov. Phys. JETP 54, 32 (1981).
4. S. D. Durbin, S. M. Arakelian, and Y. R. Shen, Phys. Rev.
        Lett. 47, 1411 (1981).
5. H. L. Ong, "Optically-Induced Freedericksz Transition in a
        Nematic Liquid Crystal", to be published. A detailed
        comparison between the exact solution and the solutions
        obtained from the geometrical optics approximation[3] and
        the infinite-plane-wave approximation[4] is made.

THEORETICAL AND EXPERIMENTAL STUDY OF OPTICAL BISTABILITY
IN FOUR-LEVEL NONRADIATIVE DYES

Zhen Fu Zhu* and E. Garmire

Center for Laser Studies, DRB 17
University of Southern California, MC 1112
Los Angeles, CA 90089-1112

INTRODUCTION

We consider bistability which occurs in materials which may be
described by a multilevel energy spectrum for which bistability is
caused by an intensity-dependent nonlinear refraction which is
derived from the saturation of absorption. Such a model for a nonlinear refractive index was first introduced by Liao and Bloom to
explain degenerate four-wave mixing in ruby.[1] Since, in this system, optical bistability occurs at intensities near or higher than
the saturation intensity, the typical power series expansion of the
susceptibility is not valid. Correct analysis requires use of the
functional form of the susceptibility. We provide here such an
analysis for a four-level nonradiative dye with a relatively large
unsaturable absorption loss. In this case optical bistability has
been observed at intensities near or higher than the saturation
intensity.

FIELD EQUATIONS FOR NONLINEAR FABRY-PEROT CAVITY FILLED WITH A SATURABLE ABSORBER

At intensities near or higher than the saturation intensity
the power series expansion of the susceptibility is not valid, and
one has to use the functional form of the overall susceptibility
in the wave equation, $\chi_{NL}(I)$. We may generalize the analysis of
McCall and Gibbs[2] to include a nonlinear refractive index and a
linear loss. Following their analysis, we obtain an expression for
the nonlinear round-trip phase change as a function of position
inside the cavity. In this case, because of the linear loss, their

*Visiting Scholar from Beijing Institute of Environmental Features,
 The People's Republic of China.

equation 5.14 becomes an integral, given by

$$\phi = 2\pi k \int_0^Z \left[2\chi(0) + |\chi(2k)| \left(\left|\frac{E_F}{E_B}\right| + \left|\frac{E_B}{E_F}\right| \right) \right] dz' \tag{1}$$

where E_F and E_B are the forward and backward fields, respectively. We have

$$\chi(0) = \frac{1}{2\pi} \int_0^{2\pi} \chi_{NL}(I)\, d\phi \quad ; \quad \chi(\pm 2k) = \frac{1}{2\pi} \int_0^{2\pi} \chi_{NL}(I)\, e^{\mp i\phi} d\phi \tag{2}$$

where we write the intensity

$$I = \left|E_F\right|^2 + \left|E_B\right|^2 + E_F E_B^* \, e^{-i\phi} + E_F^* E_B \, e^{i\phi} \tag{3}$$

with $\phi = 2kz$.

We may write

$$|\chi(2k)| = \frac{1}{2\pi} \int_0^{2\pi} \chi_{NL}\left(|E_F|^2 + |E_B|^2 + 2|E_F|\,|E_B|\cos\phi \right)\cos\phi \, d\phi \tag{4}$$

as shown by McCall and Gibbs,[3] eq. 7.

This phase change is used in the usual expression for the transmission of a lossy Fabry-Perot interferometer[4]:

$$T = \frac{I_t}{I_i} = \frac{(1-R_1)(1-R_2)e^{-\alpha L}}{(1-R)^2} \quad \frac{1}{1+F\sin^2\left(\frac{\beta+\phi}{2}\right)} \tag{5}$$

where $R = \sqrt{R_1 R_2}\, e^{-\alpha L}$, $F = 4R/(1-R)^2$, $\phi = \phi_B(L) - \phi_F(L)$

is given by Eq. (1) and $\beta = \dfrac{4\pi n_0 L}{\lambda} - 2m\pi$ (m = 0, 1, 2 ...), is detuning parameter of the Fabry-Perot cavity.

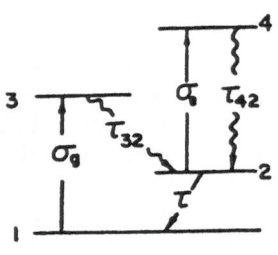

Fig. 1 Four energy-level model for saturable dye with an unsaturable absorption loss.

POLARIZATION CHARACTERISTICS IN FOUR-LEVEL NONRADIATIVE DYES

It is well known that a nonradiative dye is a multilevel saturable absorber. Research on saturable absorption of organic dyes shows that in order to explain the unsaturable absorption, a four-level model with two transitions absorbing the laser radiation is required[5]. Consider a four-level model such as that shown in Fig.1 in which the absorption occurs for both the $1\rightarrow3$ and $2\rightarrow4$ transitions with cross sections σ_g and σ_e, respectively. In nonradiative dyes, such as those used for Q-switching and mode locking, the lack of fluorescence implies that the $3\rightarrow2$ and the $4\rightarrow2$ transitions are so fast ($\tau_{32}<\tau_{42}<<\tau$) that the steady-state populations N_3 and N_4 in levels 3 and 4 respectively, are near zero. For this system the steady-state populations in levels 1 and 2 may be given by rate equations

$$N_1 = N(1 + p)^{-1} \qquad\qquad N_2 = N-N_1 = Np(1 + p)^{-1} \qquad (6)$$

where N is the total density of absorbers and $p = I/I_s$. $I_s = h\nu/\sigma_g\tau$ is the saturable absorption intensity at photon energy $h\nu$ for a dye with relaxation time τ and I is the intensity inside the cavity.

The total steady-state susceptibility can be given by

$$\chi = \chi_{\circ} + N_1\chi_1 + N_2\chi_2 = \chi_{\circ} + N\chi_1(1+p)^{-1} + N\chi_2 P(1+P)^{-1} \qquad (7)$$

where $\chi_{\circ}$ is the linear susceptibility of solvent, χ_1 and χ_2 are the molecular susceptibilities in levels 1 and 2. Clearly, the linear susceptibility (Let p = 0) is $\chi_L = \chi_{\circ} + N\chi_1$. Thus, the nonlinear susceptibility is

$$\chi_{NL} = \chi - \chi_L = N(\chi_2 - \chi_1)p(1 + p)^{-1} \qquad (8)$$

Thus, for nonradiative dyes there is a nonresonant nonlinear susceptibility proportional to the metastable state population. Therefore, there is a large nonlinear phase change due to the population redistribution in nonresonant levels.

From the form of the nonlinear susceptibility, eq. 8 , the nonlinear refractive index, n_2, can be determined. Writing $\chi_{NL} = \chi_3 |E|^2$ and using $I = (n_{\circ}c/8\pi)|E|^2$ we obtain

$$\chi_3(esu) = \frac{N(\chi_2-\chi_1)n_{\circ}c}{8\pi I_s} (1 + I/I_s)^{-1} \qquad (9)$$

and[6]

$$n_2 \ (cm^2/w) = \frac{\chi_3(esu)}{76n_{\circ}^2} = \frac{N(\chi_2-\chi_1)c}{608 \ n_{\circ}\pi I_s} (1 + I/I_s)^{-1} \qquad (10)$$

Thus, for nonradiative dyes the nonlinear refractive index is proportional to the dye concentration. For example, for BDN dye with $N = 5.1 \times 10^{17} \text{cm}^{-3}$, we will show that $\chi_2 - \chi_1 = 3.8 \times 10^{-22} \text{esu}$, $I_s = 9 \times 10^{12}$ erg/cm^{2}sec so that we obtain $n_2^1 = 2.3 \times 10^{-10}$ cm^{2}w^{-1}.

Once the functional form of nonlinear susceptibility is known, the roundtrip phase change can be easily estimated by using Eqs.(1) and (8). Substituting Eq. (8) into Eq. (4) and integrating gives

$$\chi(0) = N(\chi_2 - \chi_1) [1 - A] \tag{11}$$

$$|\chi(2k)| = -N(\chi_2 - \chi_1) \ b^{-1} [1 - (1 + p_F + p_B)A] \tag{12}$$

where

$$A = [1 + 2(p_F + p_B) + (p_F - p_B)^2]^{-1/2}$$

$$b = 2 \ \sqrt{p_F p_B}$$

p_F, p_B are forward and backward intensities, p_t is the transmitted intensity of Fabry-Perot interferometer, all in units of the saturation intensity.

Substituting Eqs. (11) and (12) into Eq.(1) and integrating gives

$$\Delta\phi = r\alpha_o L \ \sigma_I \tag{13}$$

where $r = \dfrac{8\pi^2 (\chi_2 - \chi_1)}{n_o \lambda \sigma_g}$ is a constant, independent of the dye concentration, and

$$\sigma_I = L^{-1} \int_0^L \left\{ 1 - A - \frac{p_F + p_B}{4 p_F p_B} [1 - (1 + p_F + p_B) A] \right\} dz' \tag{14}$$

Here $\alpha_o = N\sigma_g$ is the low-intensity absorption coefficient of the dye, and σ_g is the cross-section for linear absorption.

THEORETICAL RESULTS AND COMPARISON TO EXPERIMENT

In Fig. 2 we plot the nonlinear refractive index of a nonradiactive dye (BDN) as a function of the intensity, using experimentally measured parameters for BDN dye dissolved in dichloroethane. These parameters were determined by measurements of the absorption as a function of intensity for several dye concentrations. From these measurements, we determined that the saturable absorption intensity $I_s = 9 \times 10^5$ W/cm^2, the ground and excited state absorption cross sections are $\sigma_g = 1.94 \times 10^{-16}$ cm^2 and $\sigma_e = 2.3 \times 10^{-17}$ cm^2;

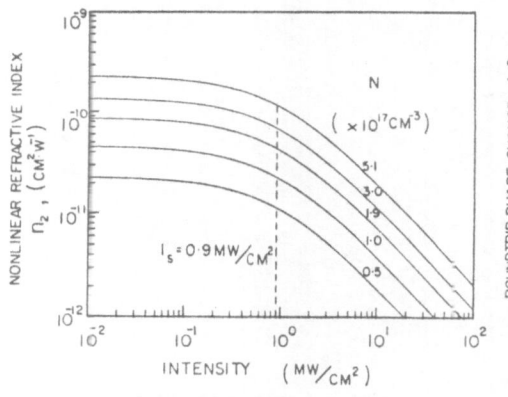

Fig. 2 Change of nonlinear refraction index with the intensity for
BDN dye. Numbers indicate the dye
concentrations. σ_g=1.94x10^{-16}cm^2;r=1.

Fig. 3 Saturation of roundtrip phase change with increasing intensity for BDN
dye.

the unsaturable absorption loss is 12%.

Fig. 2 shows that the nonlinear refractive index n_2 of BDN dye
is proportional to the dye concentration; i.e., the higher the concentration, the larger the nonlinear refractive index. For low intensity, the nonlinear refractive index is a constant, while for
I > I_s, the nonlinear refractive index is decreased. This is a
characteristic of the nonlinear refractive index derived from the
saturation behavior.

Fig. 3 shows the characteristics of roundtrip phase change of
BDN dye as a function of the intensity inside the cavity. Numbers
in Fig. 3 are the initial optical densities used in our experiment.
From Fig. 3 it can be seen that the roundtrip phase change is also
proportional to the dye concentration; the higher the concentration,
the larger the roundtrip phase change. For low intensity ($I/I_s \ll 1$),
the roundtrip phase change is proportional to the intensity. For
the intensities near the saturation intensity the roundtrip phase
change dramatically increases as the intensity is increased. For
high intensity ($I/I_s \gg 1$), the roundtrip phase change tends to
saturation and remains a constant for a given dye concentration.
Usually, for bistable operation due to nonlinear refraction, the
laser intensity must be large enough to change the roundtrip phase
change more than π. For this reason, in Fig. 3 the optimum working
region to observe the bistability should be in such a region that
working intensities are near the saturation intensity and the dye
concentrations are higher than a certain value. For six dye concentrations and working intensity (~15MW/cm^2) used in our experiment, Fig. 3 shows that there is a bistability only for three lar-

Fig. 4 Calculated input-output
   characteristics curves for BDN
   dye in a Fabry-Perot interferometer at various detuning parameters. Numbers indicate the
   cavity deturning (modulo 2π)
   in radians. R=0.925, $\alpha_o L$=5,
   η=0.12 and r=1. $p_i = I_i / I_s$
   $p_s = I_t / I_s$. Both nonlinear refraction and absorption are
   taken into account, as well as
   the effects of standing waves
   inside the cavity.

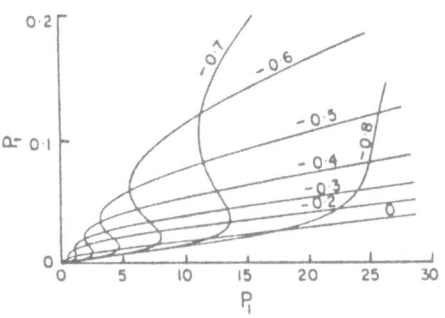

ger dye concentrations, in agreement with our experimental observation to be described later.

     In Fig. 4 we plot numerical evaluations of Eq. (5), using
experimentally measured parameters for BDN in dichloroethane. From
Fig. 4 it can be seen that there is only a range of detunings for
which hysteresis curve is a sizeable fraction of the incident intensity. For the case given here, the optimum detuning is about -0.6.
This nonzero value of optimum detuning indicates that nonlinear
refaction is the dominant mechanism for bistability in this case.
And note that bistability disappears for detuning either too large
or too small.

     Fig. 5 shows how the switch-on and switch-off intensities which
limit the bistable region depend on the initial optical density of

Fig. 5 Normalized inten-
sities for switch-on and
switch-off as a function
of the initial optical
density of BDN dye in a
Fabry-Perot with detuning
β(modulo 2π in radians)
as the parameter. Solid
curves are for switch-on
intensities and dashed
curves are for switch-
off intensities. Circles
represent points at which
bistability was observed
experimentally and crosses
represent points at which
bistability was not observed.

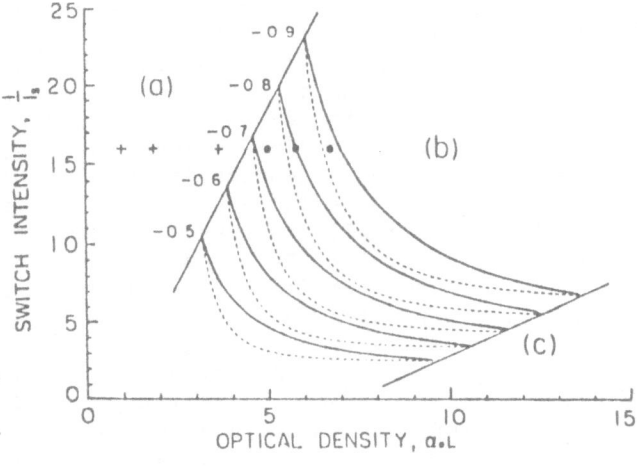

BDN dye and the detuning parameter of the interferometer. It is
interesting to note that the whole region is divided into three
parts. We can expect a bistable behavior only when the parameters
chosen are in region (b). Region (c) represents the well known fact
that for a given optical density there is a threshold in intensity
below which bistability can not be observed, for any detuning. Region (a) occurs because for large detunings the width of the hysteresis curve becomes an increasingly small fraction of the switching
intensity. So, for a given initial optical density, both the detuning and switch intensity must be chosen for operation within
region (b).

    We made measurements of optical bistability in BDN dye dis-
solved in dichloroethane. Using a Fabry-Perot with mirrors of reflectivity of 0.9 and 0.95, and separated by 1mm. The laser was a
single-mode Q-switched Nd:YAG with a pulse width of 20ns and a peak
power of 5 MW. Bistability was observed for certain dye concentrations and incident intensities by measuring a change in the time-
dependent behavior of the output, as determined from the oscilloscope traces. Transfer curves, determined from point-by-point
analysis, demonstrated switch-on behavior and hysteresis. Six dye
concentrations were studied, three of which demonstrated bistability.
In fig. 5 we have marked with o's the dye concentrations and peak
intensities for which bistability was observed and with +'s those
values at which bistability was not observed. It can be seen that
there is excellent agreement between the results of the theoretical
prediction and the experimental results.

CONCLUSION

    Using Maxwell equations, we have discussed the optical bistabi-
lity behavior of the saturable absorbers with linear absorption and
nonlinear refraction due to the saturation behavior. Because in this
system, the nonlinear effects occur at the intensities near or higher
than the saturation intensity, the typical power series expansion of
the susceptibility is not valid, we used the functional form of the
susceptibility. As an example, we have shown experimentally that our
results obtained with nonlinear Fabry-Perot filled with a nonradia-
tion dye (BDN) are in excellent agreement with the theoretical prediction.

    We have also derived the nonlinear characteristics of Fabry-
Perot interferometer from a Four-level model of a saturable absorber,
and discussed in detail these nonlinear behaviors in the condition of
different intensities, especially the saturation behavior in intensities near or higher than the saturation intensity. The result of
our experiment can be described by the theory.

ACKNOWLEDGMENTS

    The authors gratefully acknowledge the assistance of Craig Poole,
as well as the loan of equipment from Professor M. Bass.

This work was sponsored in part by NSF.

REFERENCES

1.   P.F. Liao and D.M. Bloom, "Continuous-wave backward-wave gen-
       eration by degenerate four-wave mixing in ruby," Opt.
       Lett., 3, 4 (1978)

2.   S.L. McCall and H.M. Gibbs, "Optical Bistability" Dissipative
       Systems in Quantum Optics, R. Bonifacio, ed., Springer
       Verlag, Berlin, 1982

3.   S.L. McCall and H.M. Gibbs, "Standing wave effects in optical
       bistability," Opt. Commun., 33, 335 (1980)

4.   D.A.B. Miller, "Refractive Fabry-Perot bistability with linear
       absorption" IEEE J. Quantum Electr. QE-17,306 (1981).

5.   L. Huff and L.G. DeShazer, "Saturation of optical transitions
       in organic compounds by laser flux," J. Opt. Soc. Amer.,
       60, 157 (1970)

6.   R.K. Jain, "Degenerate four-wave mixing in semiconductors:
       application to phase conjugation and to picosecond-
       resolved studies of transient carrier dynamics," Opt.
       Eng., 21, 199 (1982)

OBSERVATION OF THERMAL OPTICAL BISTABILITY, CROSSTALK, REGENERATIVE PULSATIONS, AND EXTERNAL SWITCH-OFF IN A SIMPLE DYE-FILLED ETALON

M. C. Rushford, H. M. Gibbs, J. L. Jewell,
N. Peyghambarian, D. A. Weinberger, and C. F. Li*

Optical Sciences Center
University of Arizona
Tucson, Arizona 85721

ABSTRACT

Continuous-wave optical bistability was easily achieved with the single-wavelength outputs from a cw visible Ar laser, rhodamine B or 6G dye lasers, and a He-Ne 632.8-nm laser. Input powers ranged from 5 mW to 2 W. A switch-on time as short as 20 µs was achieved with 100 mW of cw dye laser input. Using two adjacent or overlapping bistable transmission regions on the same etalon, we observed light by light control, with possible applications in optical signal processing. With two dye etalons in series we produced an optical flip-flop, permitting external switch-off as well as switch-on of an intrinsic bistable device. The simple etalon construction and flexibility in choice of lasers should encourage widespread use in testing various bistable operations, as well as in demonstrations and student labs.

INTRODUCTION

Thermal optical bistability can be defined as dispersive optical bistability[1-3] in which the intensity dependence of the optical path length arises from a temperature change of the intracavity medium. For example, off-resonance absorption of 10% in an etalon made of a Corning glass filter resulted in thermal changes in both the physical length and the refractive index of the

---

*Visiting Scholar from Harbin Institute of Technology, Harbin, People's Republic of China

etalon.[4]   In GaAs[5] and Si,[6] band-to-band absorption provides the heating mechanism.   Bistability has been seen in interference filters,[7-9] and, at least some of the time, the mechanism is thermal.[10]

Perhaps the simplest device exhibiting thermal optical bistability is that described herein, namely dye solution between two mirrors.   The bistable etalon construction is shown in Fig. 1. The dye solvent is held by capillary attraction in the 10- to 100-μm space between the Fabry-Perot plates.   The plates are 90%-reflective dielectric-coated 1 mm-thick glass, flat to λ/2 over 0.5 mm.

The simplicity of the dye-filled etalon is attractive since a large-area etalon can be obtained using standard flat Fabry Perot plates by placing the dye and solvent, via capillary attraction, in the thin plate spacing.   This is encouraging for the development of techniques for two-dimensional image processing, and studies of crosstalk.

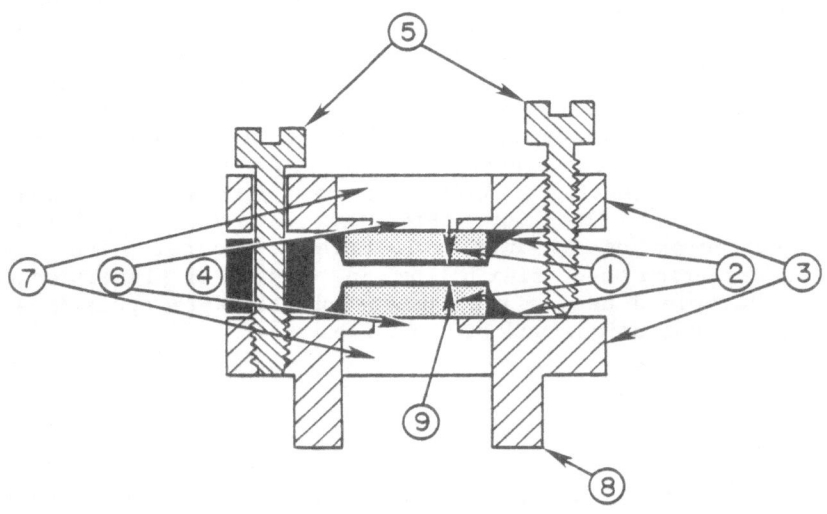

Fig. 1.   General dye etalon construction.   (1) Fabry-Perot plates, 90%-reflective dielectric, (2) glue or wax, (3) Al metal holder, (4) rubber compressed 50% to 80% of its uncompressed thickness, (5) set of three each, push and pull screws, spaced 120° apart, (6) hole size 80% to 90% of mirror size, (7) hole size larger than microscope objective, (8) holding surface for NRC mount, (9) 10- to 100-μm space for dye solvent to be held by capillary attraction.

OBSERVATION OF THERMAL OPTICAL BISTABILITY
IN A DYE-FILLED ETALON

The addition of rhodamine B dye at a concentration of 10^{-3} molar to ethylene glycol or other solvents lowers the bistability turn-on intensity compared with that for solvent alone. The dye aids by increasing the absorption, heating the solvent, and thereby lowering its index of refraction.[11] Ethylene glycol (EG) has a change of index of about $-0.00025/°C$.[12] The index change going from "off" to "on" in bistability for RB dye in EG was estimated to be typically -0.015 or less from the Fabry-Perot peak formula $2nd \cos\theta = m\lambda$ and the observed fringe angular shift of <0.1 radians. The typical temperature increase was about 60°C, achieved with an input power of 70 mW at 600 nm. Switch-on times were from 200 to 20 μs, strongly dependent upon the focus parameters of the input light. The most effective focusing was with a 10X microscope objective, giving a spot size of 5 to 10 μm in diameter.

Using these experimental values and an estimated dye absorption α of 2% per pass, and the thermal constants of EG, we calculate the turn-on time (heating time) to be

$$t_{on} = \frac{\rho \, C_p \, \Delta T \, V}{\alpha \, P} = 60 \text{ μs,}$$

where $\rho = 1.1134 \text{ gm/cm}^3$, $C_p = 2.34 \text{ J/gm °C}$, $\Delta T = 60°C$, $V = 5 \times 10^{-10} \text{ cm}^3$, and $P = 70 \times 10^{-3}$ W.

A typical transmission waveform and loop are shown in Fig. 2, where the entire transmitted beam was detected to avoid complications from radial switching[13] and spatial rings. The high-state far-field intensity profiles were similar to those of Weaire et al.[14]

REGENERATIVE PULSATIONS

In general, if the laser input lies within a strong dye absorption band, then the dye-filled etalon regeneratively pulsates.[15] For LDS 820 dye in EG and 30 mW of 632.8-nm laser input, or similarly for R6G in EG and 50-mW single Ar laser line input, the oscillation period is from a few milliseconds to seconds, where no cw bistability was generally possible. However, when LDS 820 is replaced with cresyl violet perchlorate, cw bistability is possible with 5 mW of 632.8-nm laser input. The mechanism for these pulsations has not been identified, but it may be due to absorption by triplet state dye molecules. Identification and control of the mechanism of regenerative pulsations could lead to controlled switching in a single etalon.

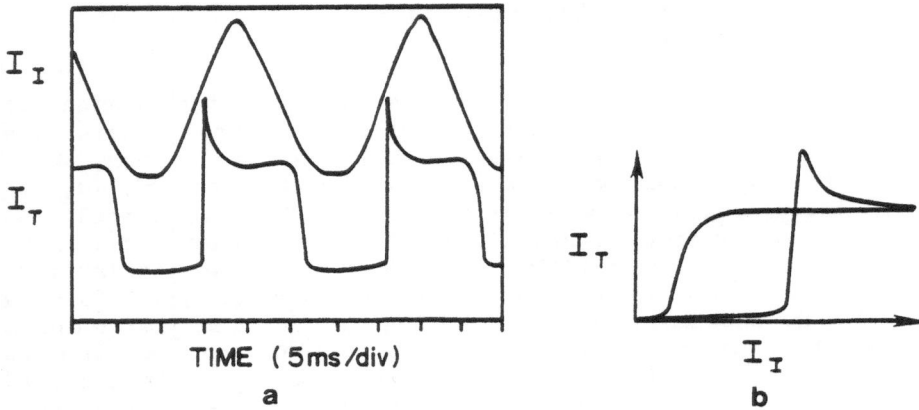

Fig. 2. Thermal optical bistability in the dye etalon of Fig. 1. (a) Time dependence of input intensity I_I and output intensity I_T. (b) Bistability loop showing overshoot and limiting.

Pulsations also occurred when the input wavelength corresponded to the lasing range of RB dye in EG. The focused laser spot was placed a few focal spot diameters from the edge of an air bubble in the etalon containing RB in EG. It may be that an asymmetric heat diffusion made one side of the focal spot region hotter, forming an effective wedge which detuned the etalon and caused switch off.

CROSSTALK

When two beams at holding intensity were focused within two spot diameters on the same etalon, switching one beam from a low to a high transmission state caused the other beam to switch on also. It was possible to see independent bistability with the two focused beams as close as three to four spot diameters apart; this is consistent with a diffractive coupling theory,[16] but the crosstalk here is via thermal diffusion. Thus, arrays of independent devices on the same etalon can be contemplated for two-dimensional information processing.

On the other hand, deliberate crosstalk between beams of different wavelengths results in control of one beam by another. In one experiment, the focused spots of red and yellow laser beams were made to overlap. If the red beam's transmission was in its "on" state, then when the yellow, more-powerful beam's transmission was turned on, the red beam's transmission was forced off. This occurred because the narrowband etalon could not transmit simultaneously at the red and yellow wavelengths. When the yellow beam was turned off, the red beam's transmission turned back on.

CONTROLLED SWITCHING

Two dye etalons were used to form a light-by-light-controlled flip-flop. The idea (by Jewell) was to put two etalons in series, using the transmission of one to control the transmission of the second. Figure 3 shows the setup as used to demonstrate the flip-flop. The first etalon (NOG) was as described in Fig. 2; the second (BD) was piezoelectrically tuned in thickness for convenience. A yellow beam was tuned in wavelength to the transmission peak of the first etalon. The transmitted light fed the second etalon and its thickness was adjusted for high-state bistability. A heating pulse from a 514.5-nm wavelength Ar laser was used to shift the transmission peak of the first etalon, causing a decrease in transmission. The decreased intensity on the second etalon allowed it to cool and thus switch off. After the first etalon cooled and its transmission rose again, a heating pulse from the same Ar laser was used to switch the second etalon to its high state again. The first etalon, a negative optical gate (NOG), could easily produce a 70% pulsed reduction of its transmission, thus switching off a large variety of bistable loops in the second etalon. The system waveform is shown in Fig. 4.

This technique differs from that of Szöke[17] in that dispersive rather than saturable absorptive bistability is used, and the first etalon operation does not depend on the second etalon. Thus, in the Jewell flip-flop, the spacing between the two etalons is not critical. Note that this flip-flop arrangement enables complete external control of an intrinsic bistable etalon.

In conclusion, most of the phenomena associated with dispersive optical bistability have been demonstrated using a simple dye etalon. This device should find applications in demonstrations, student labs, and bistability studies including parallel optical processing, crosstalk, and new device configurations.

We gratefully acknowledge useful discussions with K. Tai and S. S. Tarng and for support from the Air Force Office of Scientific Research, the Army Research Office, and the National Science Foundation, and for equipment loaned by Lawrence Livermore National Laboratory, and Professors William Wing and Stephen Jacobs.

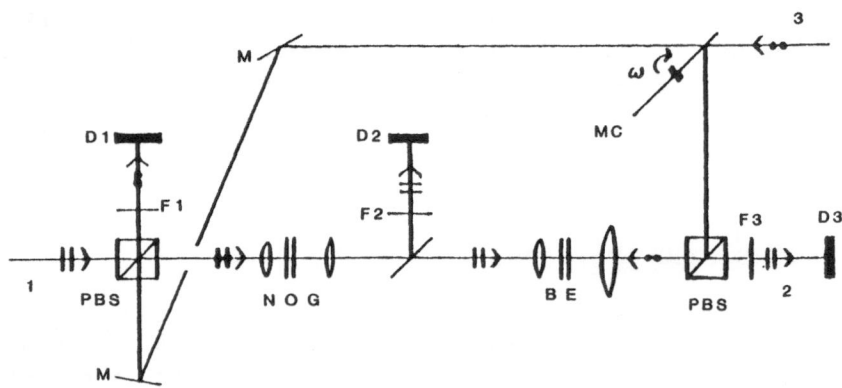

Fig. 3.   Optical flip-flop.   (1) 100-mW cw horizontally polarized
yellow input beam from dye laser.   (2) 4-mW transmission of yellow
beam through bistable etalon (BE) switched on or off by (3) 1 W of
vertically polarized cw Ar laser at 514.5 nm, directed by rotating
mirror (90% reflective)/chopper MC.   Filter F1 allows 514.5-nm beam
to hit detector D1 for scope triggering.   Filter F2 allows yellow
beam to hit detector D2 for negative optical gate (NOG) monitoring.
Filter F3 allows yellow beam output to hit detector D3 for on or
off state monitoring.   Yellow and green beams are combined by
polarization beamsplitters (PBS).

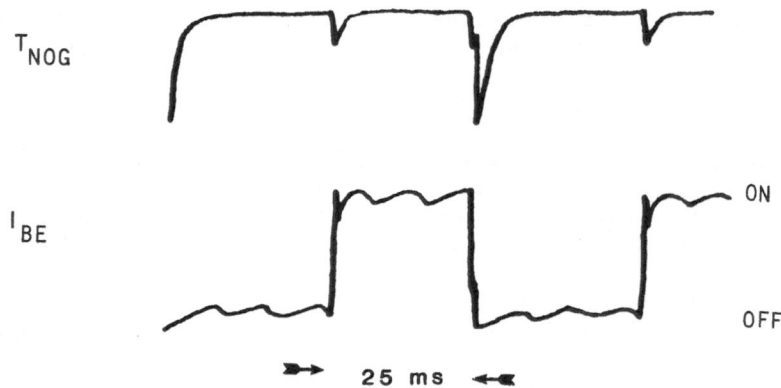

Fig. 4.   Response of optical flip-flop.   Top trace is transmission of
NOG monitored by D2.   The large pulse switches BE off; the small
pulse is timing reference from mirror/chopper transmission and BE
transmission affecting the NOG.   Bottom trace is on or off intensity
monitored by D3.   The on-to-off intensity ratio is $\leq 10^2$.   The ac
ripple is due to room lights.

REFERENCES

1. H. M. Gibbs, S. L. McCall, and T. N. C. Venkatesan, " Optical bistability," Opt. News 5:6 (1979).

2. H. M. Gibbs, S. L. McCall, and T. N. C. Venkatesan, "Optical bistability devices: the basic components of all-optical systems," Opt. Eng. 19:A63 (1980).

3. E. Abraham and S. D. Smith, "Optical bistability and related devices," Rep. Prog. Phys. 45:815 (1982).

4. S. L. McCall and H. M. Gibbs, "Optical bistability via thermal effects in a glass filter," J. Opt. Soc. Am. 68:1378 (1978).

5. H. M. Gibbs, S. L. McCall, T. N. C. Venkatesan, A. C. Gossard, A. Passner, and W. Wiegmann, "Optical bistability in semiconductors," CLEA 1979 and Appl. Phys. Lett. 35:451 (1979). These papers emphasize excitonic bistability, but thermal bistability is mentioned.

6. H. J. Eichler, F. Massmann, and C. Zaki, "Modulation and compression of Nd:YAG laser pulses by self-tuning of a silicon cavity," Opt. Commun. 40:302 (1982).

7. F. V. Karpushko and G. V. Sinitsyn, "Spectral characteristics of nonlinear interferometers in a strong field," J. Appl. Spectrosc. USSR 26:204 (1977).

8. F. V. Karpushko and G. V. Sinitsyn, "An optical logic element for integrated optics in a nonlinear semiconductor interferometer," 29:1323 (1978).

9. F. V. Karpushko and G. V. Sinitsyn, "The anomalous nonlinearity and optical bistability in thin-film interference structures," Appl. Phys. B 28:137 (1982)

10. D. A. Weinberger, H. M. Gibbs, C. F. Li, and M. C. Rushford, "Room-temperature optical bistability in thin-film interference filters," Annual Meeting of the Optical Society of America, Tucson, Arizona, October, 1982.

11. F. P. Schafer, ed., 1973, "Dye Lasers," Springer-Verlag, New York, pp. 3-4.

12. S. Leutwyler, E. Schumacher, and L. Woste, "Extending the solvent palette for cw jet stream dye lasers," Opt. Commun. 19:197, (1976).

13. H. M. Gibbs, J. L. Jewell, J. V. Moloney, M. C. Rushford, K. Tai, S. S. Tarng, D. A. Weinberger, A. C. Gossard, S. L. McCall, A. Passner, and W. Wiegmann, "Room temperature optical bistability and self-defocusing in semiconductor etalons," IQEC '82 Postdeadline.

14. D. Weaire, B. S. Wherrett, D. A. B. Miller, and S. D. Smith, "The effect of low power nonlinear refraction on laser beam propagation in InSb," Opt. Lett. 4:331 (1979).

15. J. L. Jewell, H. M. Gibbs, S. S. Tarng, A. C. Gossard, and W. Wiegmann, "Regenerative pulsations from an intrinsic bistable optical device," Appl. Phys. Lett. 40:291 (1982).

16. K. Tai, J. V. Moloney, and H. M. Gibbs, "Optical crosstalk between nearby optical bistable devices on the same etalon," Opt. Lett. 7:429 (1982).

17. A. Szöke, "Bistable optical device," United States Patent 3,813,605, May 28, 1974.

RADIATION PRESSURE INDUCED OPTICAL BISTABILITY AND MIRROR CONFINEMENT

A. Dorsel[1], J.D. McCullen[2,3], P. Meystre[2], E. Vignes[2], and H. Walther[1,2]

[1] Sektion Physik, Universität München
Am Coulombwall 1, D-8046 Garching, West-Germany

[2] Max-Planck-Institut für Quantenoptik
D-8046 Garching, West-Germany

[3] On leave of absence from the Department of Physics
University of Arizona, Tucson, Az 85721, U S A

1. INTRODUCTION

Over the last few years, considerable progress has been made in the development and understanding of optical bistable devices. A number of nonlinear systems have been shown to exhibit bistability, ranging from atomic vapors to semi-conductors. Switching times down to fractions of nanoseconds have been reached. The contributions presented at this meeting show that optical bistability is on the way to becoming applicable in data processing. Although the major thrust of research is now aimed towards this goal, optical bi-stability still remains a subject of fundamental interest. Of particular importance are the search for new nonlinearities and a better understanding of the dynamics of nonlinear interferometers.

In this paper, we discuss what may well be the simplest all-optical system leading to optical bistability, and show that its dynamical behaviour leads to a novel effect, that we call "mirror confinement": Under appropriate conditions, the length of an opti-cal cavity can be dynamically stabilized by the combined effects of gravitation (weight) and radiation pressure.

In Section 2, we discuss the principle of radiation pressure induced optical bistability, and show that it can be reached for very reasonable values of the incident laser intensity. In Section 3, we present our experimental set-up, and discuss two types of experiments performed on this system. Finally, Section 4 is a summary and conclusion.

2:   THEORY

In dispersive optical bistability, the optical length of an interferometer is changed nonlinearly by the intensity-dependent index of refraction of a dispersive medium. In the case of a Kerr medium, the nonlinear index is proportional to the intensity inside the resonator, and so is the change in optical length. Combined with the feedback provided by the resonator mirrors, this leeds to bi- and multistability[1].

Since the only role of the medium is to change the cavity length, one can imagine eliminating it by introducing nonlinearities in the <u>physical</u> cavity length. Radiation pressure is capable of doing just this.

Consider a plane Fabry-Perot resonator in which one of the mirrors is suspended to swing as a pendulum, the other one being fixed. The equation of motion of the moving mirror is

$$\ddot{x} + \gamma\dot{x} + \Omega^2 x = 2\kappa RTI_i / [1 + R^2 - 2R\cos(\phi_0 - \phi)] \quad , \tag{1}$$

where x measures the mirror displacement. Here, R and T are the mirror reflection and transmission coefficients, and I_i is the incident intensity in Watts/m^2. The pendulum motion is characterized by its damping rate γ and eigenfrequency Ω, and

$$\kappa = \frac{S}{mc} \, ,$$

where S is the moving mirror irradiated surface and m its mass.

The pendulum damping γ is on the order of sec^{-1} and much smaller than the resonator linewidth. Thus, in Eq. (1) we have adiabatically eliminated the field and used the well-known relation

$$I = TI_i / [1 + R^2 - 2R\cos(\phi_0 - \phi)] \tag{2}$$

between incident intensity and intensity inside the resonator. The phase $(\phi_0 - \phi)$ is given as usual by

$$\phi_0 - \phi = 4\pi L(1/\lambda - 1/\lambda_c) \tag{3}$$

where λ is the laser wavelength, λ_c a cavity wavelength ($\lambda_c = 2L/n$, n integer), and L the length of the resonator. Writing $L \cong L_0 + x$, where L_0 is the cavity length in absence of light, yields readily

$$\phi_0 = 4\pi L_0 (1/\lambda - 1/\lambda_c) \tag{4}$$

and

$$\phi = 4\pi x/\lambda \quad , \tag{5}$$

where $\lambda_c^\circ = 2L_0/n$. Noting that at steady-state, $x = 2\kappa RI/\Omega^2$, we find a one-to-one correspondance between this situation and usual dispersive bistability in the presence of a Kerr medium. The only difference is that the nonlinear phase ϕ is now due to a physical change of length of the resonator, instead of an effective change. Thus, the usual argument leading to dispersive bistability can readily be applied to this case. In particular, for small phases, the steady-state solution is, with $I_t = TI$,

$$I_i = I_t [1 + R^2 (\phi_0 - \frac{8\pi\kappa R}{S\Omega^2 \lambda T} I_t)^2 / T^2] \tag{6}$$

The threshold condition for bistability yields readily

$$I_i = 4I_t/3 = \lambda\Omega^2 T^2/\pi\kappa\lambda 3R \sqrt{3R} \quad . \tag{7}$$

Taking for instance $\lambda = 5000$ Å, $\Omega \cong 1$ sec^{-1}, $T = 1-R = 2 \times 10^{-2}$, and $S = 1$ cm^2 yields I_i(Watt) $\cong 0.4m$, where m is the moving mirror mass in kilograms.

Because of a number of problems to be discussed, this value should not be taken too seriously, and may be off by two to three orders of magnitude. But it indicates that optical bistability induced by radiation pressure can be achieved for very low input intensities and should readily be observable experimentally.

3.   EXPERIMENTS

The interferometer constructed for the experimental studies consists of a plane massive mirror and a small quartz plate, mass 60 mg. Both mirrors are dielectrically coated, with a reflectance of 99 % for the fixed mirror and 92 % for the moving one. The plate is suspended by 2 tungsten wires to hang about 0.8 mm from the fixed mirror. The resulting interferometer has a finesse of about 15. The incident laser is an Ar$^+$ system, which can operate single or multimode, and provides powers between a few mW and 5W.

The experiments were done both in air and in reduced pressure. The interferometer was acoustically isolated in an evacuable chamber which is mechanically tied to bedrock. Filling with air leaves

open the question of radiometric effects on the mirror; the observed response times, on the order of .1 to .3 second, indicate however that radiometer forces are not a major contributor to the
effects seen. Mechanical motion and drift of the system somewhat
limit the ability to observe the bistability. Observations of this
effect could only be made at night when civilization-caused ground
noise was minimal. A separate seismometer monitored the ground
noise in order to help identify peaks in the spectral response not
due to radiation-driven mirror motion.

This lead us to consider two types of experiments: in the
first, the incident light intensity was held fixed, and the audiofrequency spectrum of the transmitted intensity was measured using
a Fast-Fourier-Transform spectrum analyzer.

In presence of mechanical noise, Eq. (1) becomes

$$\ddot{x}+\gamma\dot{x}+\Omega^2 x = 2\kappa RTI_i/[1+R^2-2R\cos(\phi_0-\phi)] + F(t)/m \quad . \quad (8)$$

In the numerical results discussed here, the noise $F(t)$ is
simulated by a sequence of short impulses of random strength regu-
larly distributed in time. A full scale noise analysis is underway
and will be presented elsewhere.

Depending on the value of I_i, three regimes of operation were
identified, characterized by mirror motions small, comparable, and
large compared with the width of the Fabry-Perot resonance.

In the small displacement limit, the audiospectrum consists of a
single peak about the effective "resonant" frequency Ω_{eff} of the
driven pendulum. For larger mirror displacements, we observe the
appearance of a sequence of harmonics of Ω_{eff}. This is because the
interferometer length can now change sufficiently to produce
significant variations in transmission. Theoretical and corresponding experimental spectra in this regime are shown in Fig. 1a. (Note
that depending on I_i and ϕ_0, the mirror displacement $x(t)$ may
become quite complex. This point, which will be discussed elsewhere,
has however little influence on the audio-spectrum.)

For appropriate choices of I_i, ϕ_0, and $F(t)$, the spectrum changes
from a single pattern of harmonics to one with the major frequency
split. It is easy to convince oneselve that this occurs in the
bistable region. The radiation field, together with the gravitational restoring force, can be thought of as forming a potential
presenting minima at the stable operating points. Under the action
of noise, the system can hop between the first two wells, with two
different "resonant" frequencies Ω_{eff}, leading to a splitting of
the audio-spectrum (Fig. 1b).

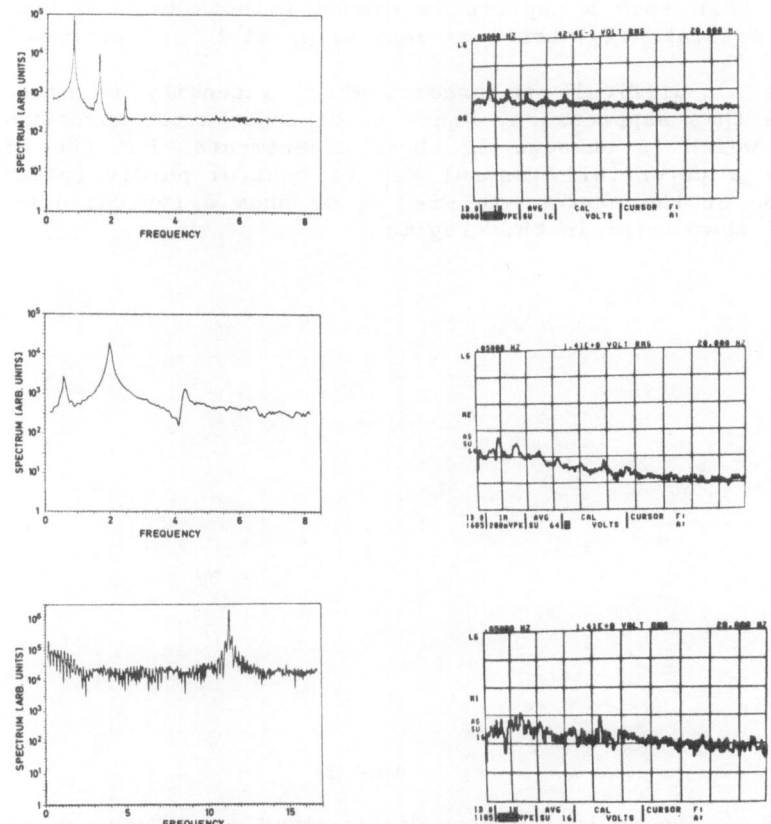

Fig. 1   Theoretical (left) and experimental (right) audiospectra
showing the appearance of harmonics of the pendulum frequency Ω_{eff} (up), frequency splitting (middle), and
mirror confinement (down). The frequencies in the theoretical curves are in arbitrary units, and 2 Hz/division in
the experimental ones.

As the intensity I_i is further increased, still another characteristic solution type appears. Here, the morror is no longer
particularly sensitive to the driving force, but damps into a
"steady-state" mode in which it is weakly driven about one of the
higher order equilibrium points indicated by the static soution
(Fig.1c). That such a solution is possible is not surprising; the
potential wells corresponding to higher order resonances are increasingly narrower, and therefore have resonant frequencies Ω_{eff}
quite different from Ω.   The mirror, once captured in

the well, no longer responds to a driving force of frequency about
Ω. But that such a capture is common is not obvious. However, it
occurs persistently when a minimum value of I_i is attained.

It is difficult to predict which intensity is necessary to
achieve this suppression, since it depends on the magnitude of the
noise, which is unknown in these experiments. But the effect is
clearly a potentially useful way to control purely optically the
position of the mirror. In Fig. 2, we show a typical displacement
x(t) of the mirror in this regime.

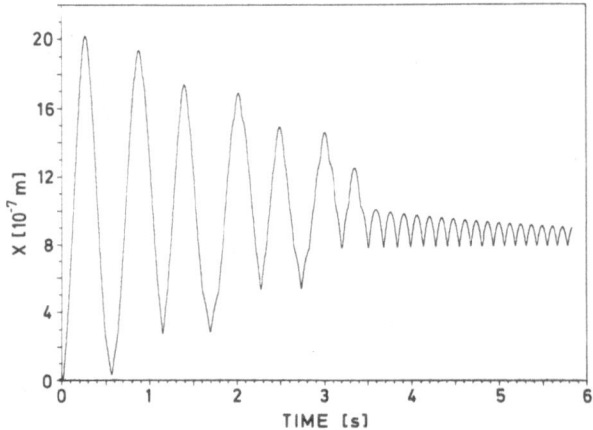

Fig. 2   Theoretical mirror displacement as a function of time,
            exhibiting a transition to the confined regime.

This first series of experiments having convinced us that we
had the system rather well under control, we improved further the
mechanical isolation of the interferometer and managed to observe
static optical bistability. Here, the laser intensity was slowly
scanned across the bistable region, from feeble intensity to high
intensity, and back. Scanning cycle times ranged from two to five
minutes, times long compared to the damping time of the mirror. The
result of one such scan is shown in Fig. 3. The hysteresis cycle
typical of bistable behaviour is clearly evident.

It would be tempting to associate Fig. 3 with the regime of
Fig. 1b, where the audiofrequency spectrum is split. This, however,

cannot be done unambiguously: the possibility of noise "washing-out" the structure of the first two, rather shallow, potential minima, cannot be excluded, and it is quite possible that the observed hysteresis involves a higher order Fabry-Perot resonance. This might also explain the discrepancy between the input powers levels of about 200 mW required to observe the splitting and the higher powers used in Fig. 3.

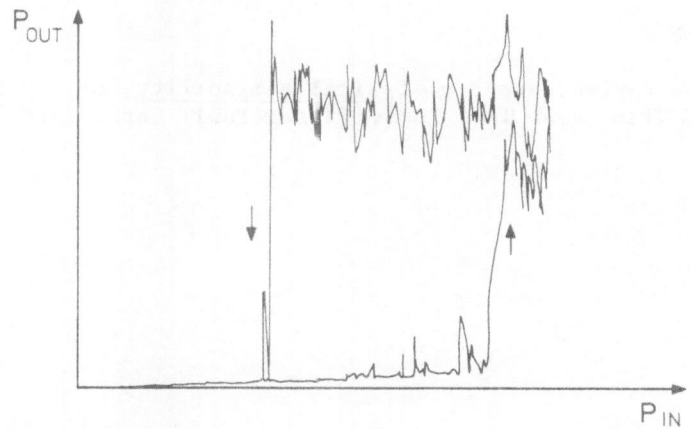

Fig. 3   Experimental hysteresis cycle. In this example, the
          switching-up power is about 2.2 Watt and the switching-
          down power about 1.1 Watt. The transmitted powers are
          on the order of 5 mW in the upper branch and 0.2 mW on
          the lower branch.

4.   SUMMARY AND CONCLUSION

These preliminary results demonstrate with certainty that bistability may be achieved without the use of non-linear media. The general agreement of calculations using radiation pressure strongly indicate that this is the source of the non-linear behavior. Perhaps the most interesting feature of these experiments, however, is the discovery of the mechanical stabilizing effect at higher laser power levels. An understanding of the nature and limits of this effect is now being gained both experimentally and theoretically, and will be reported in detail in the near future.

Radiation pressure bistability and optical mirror confinement open the way to a number of novel applications where the measurement of very small displacements is desired. Our system can operate as an extremely sensitive transducer between mechanical and optical signals, and could be used in laser stabilization, optoacoustic spectroscopy, e.g. of surfaces, and photoelectric effect studies: the momentum of the photoelectrons ejected by a magnetic surface is about three orders of magnitude larger than that of the incident photons. Further possible applications include laser stabilization, an all-optical microphone, and seismometers, to list a few.

REFERENCES

1.   For a review, see e.g. Optical Bistability, Ed. by C.M. Bowden, M. Ciftan, and H.R. Robl, Plenum Publ. Corp. (New York) 1981.

MIRRORLESS OPTICAL BISTABILITY USING THE LOCAL FIELD CORRECTION

Fred A. Hopf

Optical Sciences Center, University of Arizona, Tucson,
Arizona 85721, USA

and

Charles M. Bowden and William H. Louisell (deceased)

Research Directorate, US Army Missile Laboratory, US Army
Missile Command, Redstone Arsenal, Alabama 35898, USA

INTRODUCTION

Recently, there has been considerable interest in the problem of optical bistability[1]. Most experimental and theoretical work has concentrated on Fabry-Perot devices[2]. In addition, there has been an interest in devices using self-focusing concepts[3]. In this paper we consider the question of whether one can dispense with optical paraphenalia altogether, and make an optically-bistable device out of the material system alone.

This idea of making "mirorless" bistable devices[4], including using the local field correction (LFC)[5] to make the bistability is not new. Earlier calculations have fallen into three categories. Some treat the medium as a single particle[5]. Aside from the inconsistency of using a many-body effect like the LFC in a single-particle calculation, this procedure leaves open serious questions as to whether the bistability is possible unless proper boundary conditions are imposed. Some of these calculations have yielded quite unrealistic results, predicting large nonlinearities at very weak powers[5]. Other calculations use the Dicke model[4], which neglects the propagation of the electromagnetic field. Since the propagation of optical fields can be important, this limits the results to qualitative interpretation. Finally, there are calculations in which non-unique molecular states are invoked to make mirrorless bistability, such as anharmonic oscillators (e.g., HF)[6]. In such cases, bistability is exhibited by each molecule independently. Since there are no strong intermolecular forces, each molecule fluctuates between its bistable states independently of all others. Under such circumstances, the bistable condition decays (called a "glitch") rapidly in time due to fluctuations.

The first-principles aspects of the LFC are considered here, although not in detail. The point of a development from first principles is to indicate where the fundamental assumptions of the treatment lie. Our primary purpose in this development is not to demonstrate in any absolute sense that optical bistability is possible using the LFC, but rather to develop the problem in a systematic fashion so that the basic issues are at least clear. We thus consider this paper to be a preliminary exercise. We hope it will result in more attention to a problem that we believe is of both fundamental and practical interest.

We formulate the problem in terms of a self-consistent solution of the macroscopic, semiclassical density-matrix and Maxwell equations of two-level atoms interacting with a radiation field. We show that bistable operation is restricted to situations in which slowly-varying solutions are invalid. Hence we have developed techniques for numerically evaluating the exact Maxwell equations. The semiclassical solution does not predict bistability as such. In the regime in which there are non-unique steady states, the number of steady-state solutions are semi-infinite. The question of the stability of all these solutions is beyond the scope of this work, and may require quantum electrodynamics for proper evaluation.

FORMULATION OF THE PROBLEM

In this development, the slowly-varying approximation is made only with respect to time, and not in the usual fashion with respect to space. This is exact in the Maxwell's equations, and is exact in Schrodinger's equation for circular polarization (for linear polarization we take the rotating-wave approximation to be valid). The convention for defining amplitudes of macroscopic fields (all other fields have the same convention) reads

$$E = \frac{1}{2} (E \exp(-i\nu t) + cc.) \quad , \tag{1}$$

$$P = \frac{1}{2} (P \exp(-i\nu t) + cc.) \quad , \tag{2}$$

and the slowly-varying amplitude of the off-diagonal density-matrix element is

$$\rho_{ab} = \frac{1}{2} (iR_{ab} \exp(-i\nu t) + cc.) \quad . \tag{3}$$

In each case, ν is the frequency, ρ denotes the 2X2 density matrix, t denotes time, $i = \sqrt{-1}$, and cc. denotes the complex conjugate. Assuming homogeneity of the medium, the macroscopic polarization is,

$$P = i\mu N R_{ab} \quad , \tag{4}$$

where μ is the dipole matrix element, and N is the density of atoms. The local field amplitude, E_{loc}, which drives each atom is given by[7],[8]

$$E_{loc} = E + i\zeta\mu N R_{ab} \quad , \tag{5}$$

where $\zeta = 4\pi/3 + s$, and s is a constant which is due to correlations with respect to atomic positions that occur in crystals[7],[8]. The density-matrix equations of the two-level atoms whose upper (lower) states are labeled a(b) read

$$\frac{dR_{ab}}{dt} = -(i\delta + \gamma_{ab})R_{ab} + \frac{\mu}{2h} E_{loc}n \quad , \tag{6}$$

$$\frac{dn}{dt} = -\gamma(1+n) + \frac{\mu}{h} (E_{loc}^* R_{ab} + cc.) \quad , \tag{7}$$

where the inversion n is

$$n = \rho_{aa} - \rho_{bb} \quad , \tag{8}$$

and ρ_{aa} and ρ_{bb} are the diagonal elements of the density matrix. In the above formulation, it is assumed that the lower state decay is zero, so that $\gamma = \gamma_a$. The decay of the off-diagonal element is γ_{ab}, δ is the detuning of the atoms from the field, and h is Planck's constant divided by 2π.

Upon substitution of Eq. (5) into Eq. (7) one obtains

$$\frac{dn}{dt} = -\gamma(1+n) + \frac{\mu}{h} (E^* R_{ab} + cc.) \quad . \tag{9}$$

Thus the local field drops out of the equation for the inversion. The equation for the off-diagonal element can then be rearranged using Eq. (5), to read

$$\frac{dR_{ab}}{dt} = -(i (\delta - B\gamma_{ab}n) + \gamma_{ab})R_{ab} + \frac{\mu}{2h} En \quad , \tag{10}$$

where

$$B = \frac{\zeta N\mu^2}{2h\gamma_{ab}} \tag{11}$$

One therefore sees that the local field correction causes an inversion-dependent frequency shift. It can be shown that when taken off resonance, this shift leads to the Clausius-Mosatti relationship. To obtain a magnitude for B, it is useful to recall the formula for the on-resonance absorption coefficient α which is

$$\alpha = \frac{4\pi\nu N\mu^2}{ch\gamma_{ab}} \quad . \tag{12}$$

If we take s = 0, then α can be expressed in terms of B as

$$\alpha = \frac{8B\pi^2}{\lambda\zeta} \quad . \tag{13}$$

Here we have used $\nu/c = 2\pi/\lambda$, where λ is the wavelength of light in vacuum. The numerical factors that relate B to α are appreciable. Since B is the shift in the central frequency relative to linewidth, it follows that for any appreciable shift, the absorption coefficient is of the order of or much greater than $1/\lambda$. Since bistability occurs near resonance, the slowly-varying approximation in space is not usable.

MAXWELL'S EQUATIONS

As noted previously, the large values of the absorption coefficient needed for bistable operation preclude the use of the slowly-varying approximation in the spatial coordinate to compute answers. On the other hand, since the media are opaque, one is not interested in large propagation distances, and hence it is feasible to numerically integrate the exact equations over distances short compared to an optical wavelength. There has been no work, to our knowledge, on how to solve an exact, nonlinear propagation problem in this limit, so we outline our technique in some detail.

In the steady-state, plane-wave case, Maxwell's equations reduce to

$$\frac{d^2E(z)}{dz^2} - \frac{\nu^2E(z)}{c^2} = \frac{4\pi\nu^2P(z)}{c^2} \quad , \tag{14}$$

where the amplitudes are independent of time, and hence E is described by an ordinary differential equation. To obtain the solution to Eq. (14), it is necessary to apply boundary conditions on the amplitudes. We denote by z the position in a medium extending from z = 0 to z = L. The total field to the left of the medium reads

$$E = \frac{1}{2} (E_i \exp\ i(kz - \nu t) + E_R \exp\ i(-kz - \nu t) + cc.), \quad z \leq 0, \tag{15}$$

from which one obtains the amplitudes and derivatives as

$$E = E_i \exp (ikz) + \Gamma_R \exp (-ikz), \quad z \leq 0 , \tag{16}$$

$$\frac{dE}{dz} = ik [E_i \exp (ikz) - E_R \exp (-ikz)] \quad z \leq 0 . \tag{17}$$

For fields to the right of the medium we have

$$E = \frac{1}{2} (E_T \exp (i(kz-\nu t) + cc.) , \quad z \geq L , \tag{18}$$

from which one obtains the amplitudes and derivatives as

$$E = E_T \exp (ikz) , \quad z \geq L , \tag{19}$$

$$\frac{dE}{dz} = ikE_T \exp (ikz). \quad z \geq L . \tag{20}$$

Here the subscripts T and R denote the transmitted and reflected field. We assume normal incidence. In this case the important boundary conditions are the continuity of E and H (the magnetic field) at each boundary. At z = 0, Eqs. (16) and (17) provide the boundary conditions.

$$E = E_R + E_i , \quad z = 0 \tag{21}$$

$$\frac{dE}{dz} = ik(E_i - E_R) , \quad z = 0 \tag{22}$$

to integrate Eqs. (14), (9) and (10) across the medium, provided the reflected field is given. Eqs. (19) and (20) reduce to

$$\frac{dE}{dz} = ikE , \quad z = L . \tag{23}$$

Eq. (23) is a criterion that must be met in order that the solution be valid.

Before discussing how the solution is obtained it is useful to define the following standard notation

$$r = \frac{E_R}{E_i} , \quad R = |r|^2 , \quad T = \frac{|E_T|^2}{|E_i|^2} , \tag{24}$$

$$A = 1 - T - R . \tag{25}$$

Here R and T are the usual reflection and transmission coefficients, A is the absorptance, i.e., the fraction of optical energy that is dissipated in the medium, and r is the amplitude reflection coefficient.

The difficulty with the solution is that E_R, i.e., r is un-known. The solution method involves guessing r until one finds a result that meets the criterion of Eq. (23). When that is found, then one has a valid solution of the problem insofar as all the boundary conditions are met. The iterative procedure is discuss-ed elsewhere[11].

RESULTS

In the calculation we assume that the atoms are always locally in the same superposition of atomic states. To assure this, we specify, in advance of the calculation, the state that the atom will be in (the "upper" or the "lower") if a non-unique atomic state is encountered. The "upper" state is the one with the largest value of ρ_{aa}, and the "lower" state has the smallest value of ρ_{aa}. The former defines the "upper" bistable branch, and the latter defines the "lower" branch. These branches would be reached, in practice, by the usual methods, the lower branch by adiabatically increasing a weak field, and the upper by adi-abatically decreasing a strong field. The upper and lower branches have the highest and lowest rates of dissipation of optical ener-gy (i.e., as measured by the absorbance). By this measure, the terms upper and lower have the same significance as in a cavity. However, the transmission characteristics of the two states are reversed from the bistability in a cavity, in that the lower branch has the higher transmission.

In Figure 1 the transmission function is shown as a function of detuning δ for three values of B, each with two or three values of E_i. The case B = 4 and E_i = 3 was not finished due to extreme difficulties in numerical convergence, probably because it is approaching the condition of an optical transistor. For B < 6, bistability is not expected, and is not found. Note the shift in the resonance frequency as the atoms become saturated. Hopefully this shift could be measured experimentally, which would be an additional verification that the ideas presented here are valid. For B = 6, $\delta \cong -2$, and $E_i \cong 2$, a non-unique absorption is seen. Note that the curves for the upper and lower branches join con-tinuously, so that the states taken by the atoms as the tuning is varied, are indeterminate in a steady-state analysis.

In Figure 2, a bistable loop is shown for B = 6, δ = -1.5, and E_i variable over a range from 1 to 3. Both the absorptance A and transmission T are shown. The absorptance is the same for this and for bistability in a cavity, insofar as in both cases a larger amount of optical energy is dissipated if the atoms are in the upper branch. The condition on transmission then reverses the normal one. The lower branch has a higher transmission than the upper branch.

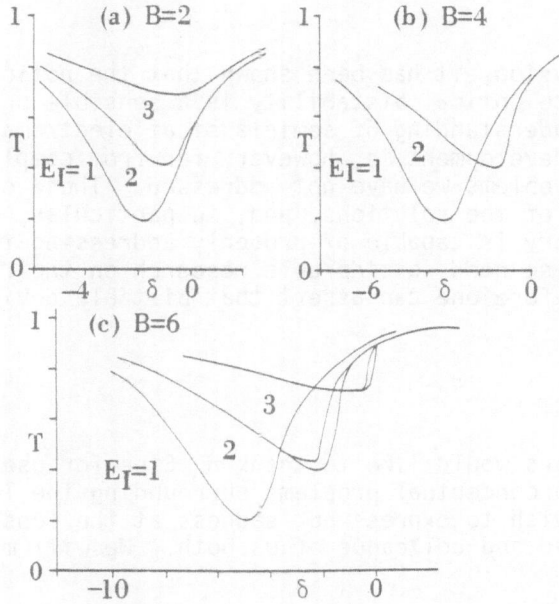

Fig. 1.   Transmission T vs. detuning δ for: (a) B = 2 and E_i = 1,2,3.   (b) B = 4 and E_i = 1,2.   (c) B = 6 and E_i = 1,2,3.

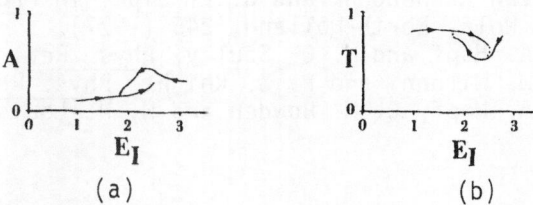

Fig. 2.   (a) Absorptance A.   (b) Transmission T vs. input field E_i for δ = -1.5 and B = 6.

CONCLUSION

        In conclusion, it has been shown that the notion of using the
LFC to generate optical bistability is a sensible one, insofar as
the present understanding of semiclassical electromagnetism is concerned. The development is, however, far from complete. There are
a number of problems we have not addressed. These chiefly involve
the stability of the solutions, and, in particular, whether semi-
classical theory is capable of properly addressing the stability
question. These need considerable research on their own, and must
be answered before one can assert that bistable devices of this kind
are possible.

ACKNOWLEDGEMENT

        The authors would like to thank J. Sipe for useful conversations
in clearing up conceptual problems surrounding the local field correction. We wish to express our sadness at the loss of Bill Louisell.
He was a friend and colleague of us both. We will miss him greatly.

REFERENCES

        1. C. M. Bowden, M. Ciftan and H. R. Robl, eds., Optical Bi-
stability, Plenum, New York, 1981.
        2. E. Abraham and S. D. Smith, Rep. Prog. Phys. 45, 815 (1982).
        3. A. E. Kaplan, Sov. Phys. JETP. 45, 896 (1977); Opt. Lett.
6, 360 (1980); IEEE J. Quant. Electron. QE-17, 336 (1981).
        4. C. M. Bowden and C. C. Sung, Phys. Rev. A19, 2392 (1979);
C. M. Bowden, XI IQEC, IEEE Cat. 80CH1561-0,589 (1980); C. M. Bowden,
in Optical Bistability, edited by C. M. Bowden, M. Ciftan and H. R.
Robl, Plenum, NY, 1981, p. 405.
        5. I. Abram and A. Maruani, Phys. Rev. B26, 4759 (1982).
        6. C. Flytzanis, private communication.
        7. J. D. Jackson, Classical Electrodynamics, second edition,
John Wiley, NY, 1962, Chapt. 4.
        8. J. Van Kranendonk and J. E. Sipe. in Progress in Optics XV,
edited by E. Wolf, North-Holland, 245 (1977).
        9. F. A. Hopf and M. O. Scully, Phys. Rev. 179, 399 (1969).
        10. P. W. Milonni and P. S. Knight, Phys. Rev. A10, 1096 (1974).
        11. F. A. Hopf, C. M. Bowden and W. H. Louisell, to be published.

NONLINEAR OPTICAL INTERFACES

P. W. Smith and W. J. Tomlinson

Bell Laboratories
Holmdel, NJ, USA   07733

In this paper we review the results of previous theoretical
and experimental studies of nonlinear interfaces, and report new
experimental results obtained using a liquid suspension of dielectric particles[1] as the nonlinear medium. This medium has a
sufficiently large nonlinearity to permit nonlinear interface
experiments to be performed with low-power CW laser beams, and
thus simple, direct experimental verification of theoretical
predictions is now possible for the first time.

An example of a nonlinear interface is shown in Fig. 1.
Light is incident from a linear medium (refractive index n_0)   on
an interface with a nonlinear medium of index $n = n_0 - \Delta + n_2 I$.
Note that the index depends on I, the intensity of the light in
the medium, and the nonlinear coefficient n_2 is a measure of the
strength of this dependence.

Early theoretical work on such nonlinear interfaces predicted
bistable behavior for incident plane waves[2], as illustrated in
Fig. 2. If a plane wave is incident so that ψ is less then the
critical angle $\psi_c \left(= \sqrt{2\Delta/n_0} \right)$, it will be totally reflected if I
is low. As I is increased, however, a switch to transmission will

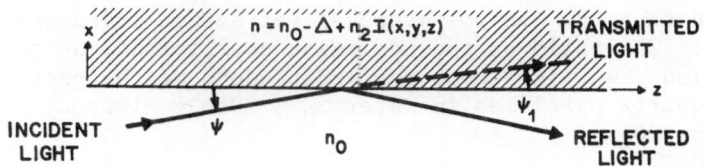

Figure 1. A nonlinear interface

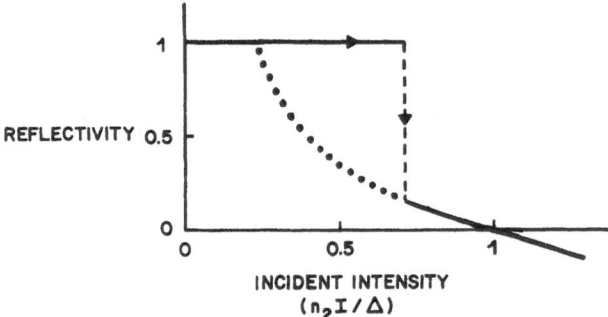

Figure 2. Reflectivity vs. intensity for a nonlinear interface:
plane-wave theory.

occur at a certain critical I. The plane wave theory further
predicts that, upon lowering the incident intensity, the reflectivity will follow the dotted path in Fig. 2 and optical
hysteresis will occur. The first experimental results appeared
to exhibit optical bistability although the detailed behavior did
not conform to the plane-wave predictions.[3]

We recently developed a numerical technique to study the
behavior of a (two-dimensional) Gaussian light beam at a nonlinear
interface.[4] The results of this analysis show several new
features. The most dramatic deviations from the plane-wave
results, however, are the prediction of a second reflectivity
'step' as the input power is increased, and the absence of
hysteresis upon subsequent reduction of the input power. These
results[5] are shown in Fig. 3. The parameters for this calculation were chosen to correspond to the experiments described
below. The critical angle for this calculation was $\psi = 9.4°$.

Our experiments were performed using an aqueous suspension of
700A° quartz particles as the nonlinear medium. As shown in
Fig. 4, a particle with refractive index n_a, surrounded by a
liquid with refractive index n_b, behaves as a dipole with an
effective polarizability, p, which depends on the radius, r, of
the particle and the refractive indicies. The force on this
dipole depends on the gradient of the (light) intensity ∇E_0^2, and
can be written as the gradient of a potential ϕ. The particle
density, and thus the (volume average) refractive index depends on
exp $[-\phi/kT]$. The ratio of the nonlinear coefficient, n_2 to the
scattering loss, α_0, is a constant independent of the particle
size. For quartz particles in water $\alpha_0 \sim 20cm^{-1}$ gives $n_2 \sim 10^5\times$
the value for CS_2.

We performed experiments using a CW argon ion laser beam with

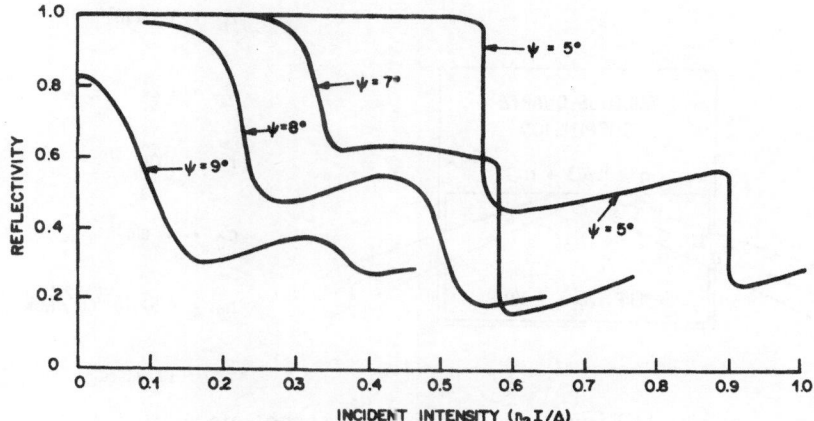

Figure 3. Reflectivity ms. intensity for a nonlinear interface:
two-dimensional Gaussian beam theory.

a LiF crystal (n=1.39) as the linear medium and an aqueous suspension of 700 Å quartz particles (n=1.37) as the nonlinear medium.
The experimental set-up is shown in Fig. 5 which also lists the
pertinent parameters. The experimental conditions were chosen to
correspond to the parameters used for the numerical analysis. An
experimental curve of reflectivity vs (normalized) incident

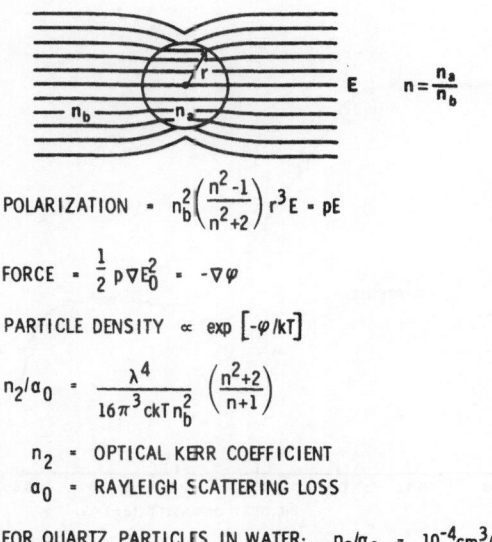

Figure 4. Nonlinear optical properties of a liquid suspension of
dielectric particles

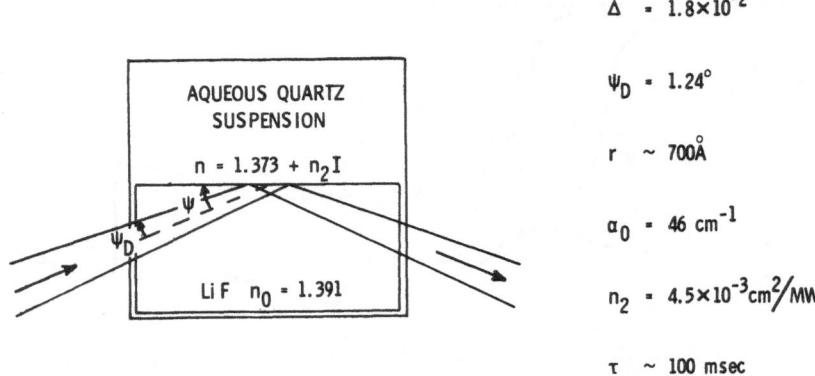

$\Delta = 1.8 \times 10^{-2}$

$\Psi_D = 1.24°$

$r \sim 700 \overset{\circ}{A}$

$\alpha_0 = 46 \text{ cm}^{-1}$

$n_2 = 4.5 \times 10^{-3} \text{cm}^2/\text{MW}$

$\tau \sim 100 \text{ msec}$

Figure 5. Experimental set-up

intensity is shown in Fig. 6. It can be seen that there is
qualitative agreement with the predictions of the Gaussian beam
analysis (Fig. 3) and that its behavior does not conform to the
plane-wave predictions (Fig. 2). Note, however, that the reflectivity 'steps' are smaller than the (two-dimensional) Gaussian
beam predictions. Figure 7 shows that the critical intensity for
the first and second reflectivity 'steps' is in good quantitive
agreement with the results of our analysis. Note that the agreement is absolute: <u>there are no adjustable parameters</u>.

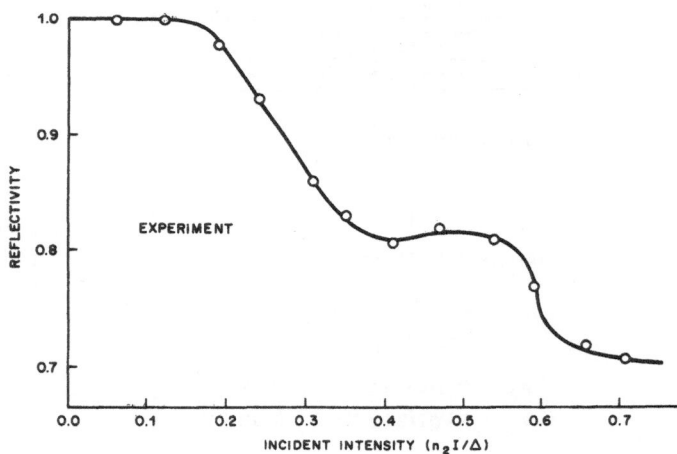

Figure 6. Experimental plot of reflectivity vs. incident intensity.

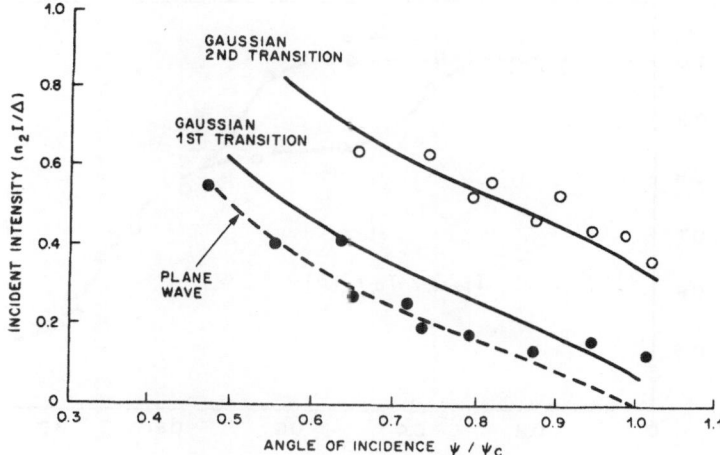

Figure 7. Critical intensity for reflectivity 'jumps' as a
function of angle of incidence

For some experimental conditions, hysteresis was observed.
Only the upper branch of the reflection hysteresis characteristic
was stable, however. The lower branch would decay to the upper
branch with a time constant of ~10^3 × the response time of the
nonlinear medium.

It is interesting to take a new look at the experimental data
for the glass – CS_2 interface[3] and to reinterpret it in terms of
our present understanding of nonlinear behavior. The original
experimental data has been replotted in Fig. 8 to show that at
intensities substantially above the first reflectivity 'step'
there is a second 'step'. These results appear to be in qualitative agreement with our numerical analysis. The hysteresis
observed in these early experiments is also not inconsistent with
our prediction of an unstable lower state, for the measurement
time was of the order of 300× the response time of the nonlinearity (2ps) and our current experiments show that this is of the
order of the stability time of the lower state.

What are the implications of our results for device applications? Nonlinear interface devices are attractive because of
their simplicity, and their potential for sub-picosecond response
(with suitably fast nonlinear materials). The quasi-stability of
the lower branch could cause problems. Gaussian beam devices do
not have a high contrast between "on" and "off" states. Because a
high contrast is predicted from the two-dimensional Gaussian beam
analysis, we would expect that the contrast could be much improved
by using an eliptical input beam. Further work will be necessary

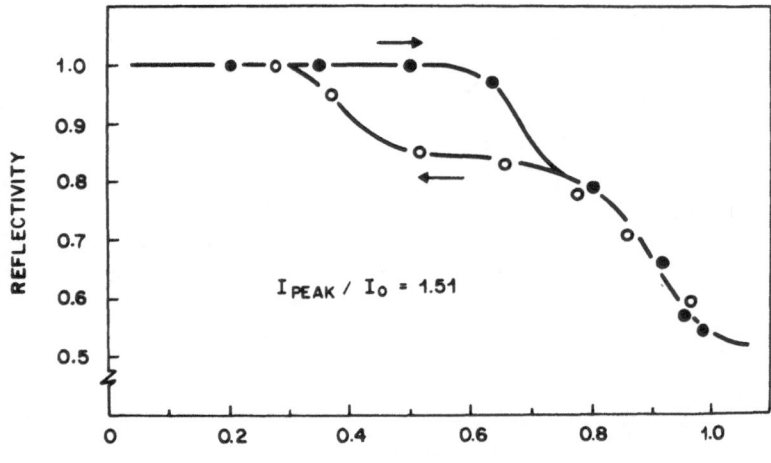

Figure 8. Reflectivity vs. intensity found from an analysis of the
glass/CS_2 data reported in Ref. 3.

to fully determine the practical applications of nonlinear
interface devices.

REFERENCES

1. See, for example, P. W. Smith, P. J. Maloney, and A. Ashkin,
    "Use of a Liquid Suspension of Dielectric Spheres as an
    Artificial Kerr Medium" Opt. Lett. 7, 347 (1982).

2. See, for example, A. E. Kaplan, "Theory of Hysteresis
    Reflection and Refraction of Light by a Boundary of a
    Nonlinear Medium" Sov. Phys., JETP 45, 896 (1977).

3. P. W. Smith, W. J. Tomlinson, P. J. Maloney, and J. P. Hermann,
    "Experimental Studies of a Nonlinear Interface" IEEE
    J. Quantum Electron QE-17, 340 (1981).

4. W. J. Tomlinson, J. P. Gordon, P. W. Smith, and A. E. Kaplan,
    "Reflection of a Gaussian Beam at a Nonlinear Interface",
    Appl. Opt. 21, 2041 (1982).

5. P. W. Smith and W. J. Tomlinson, "Nonlinear Optical
    Interfaces: Switching Behavior", submitted to IEEE
    J. Quantum Electron.

OBSERVATION OF BISTABILITY IN A JOSEPHSON DEVICE

Barry Muhlfelder and Warren W. Johnson

Department of Physics and Astronomy
University of Rochester
Rochester, N.Y., 14627

INTRODUCTION

We have been testing a certain type of <u>double</u> Josephson junction Superconducting Quantum Interference Device (SQUID) for use as a very low noise amplifier.[1] We have found that this device is capable of displaying quasi-bistable behavior that appears to be similar to certain types of optical bistability. We have good evidence that the cause of the observed bistability is the interaction of an electromagnetic resonator (a length of transmission line) with nonlinear elements (the Josephson junctions).

The discovery of this effect prompted Drummond to suggest to us that a <u>single</u> junction device would also display similiar effects and would be easier to analyze. This work is reported in the following paper.[2]

DESCRIPTION OF THE DEVICE

The physical arrangement of this SQUID is shown in Figure 1. It was fabricated at NBS-Boulder, using a thin film Pb alloy technology.[3] A square loop or ring of a thin film superconductor (labeled TL) is broken and the ends reconnected by two Josephson junctions. The loop TL and the ground plane underneath are separated by a thin layer of dielectric, and so form the upper and lower conductors of a transmission line (of the stripline type). The device has two input parameters: the bias current I_b, and the magnetic flux Φ_{ex} that threads the loop TL. The output parameter is $<V(t)>$, the time average voltage across the device.

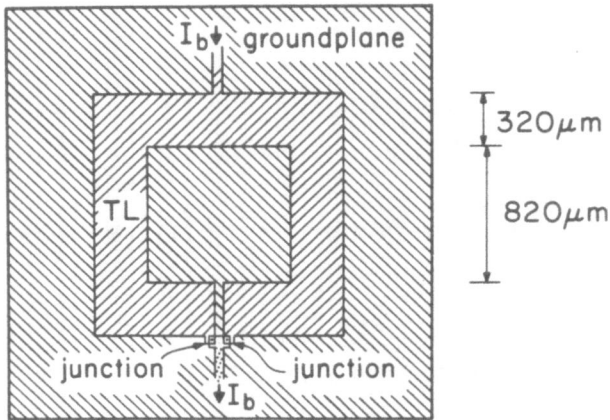

Figure 1.   The arrangement of the device, as seen from above.

QUASI-BISTABILITY

The major observation is that the output voltage will spontaneously switch back and forth between two distinct values, even though the input parameters are fixed.   An oscilloscope photograph is shown in Figure 2.   It looks very similiar to the quasi-bistability observed in ring dye lasers.[4]

The two distinct values of the output voltage must correspond to two different frequencies of oscillation, as required by one of the Josephson equations.   This naturally suggests that the junctions are interacting with either of two electrical resonances.   Our hypothesis is that the junctions are phase-locking to either the first or the third half-wave resonance of the loop TL.

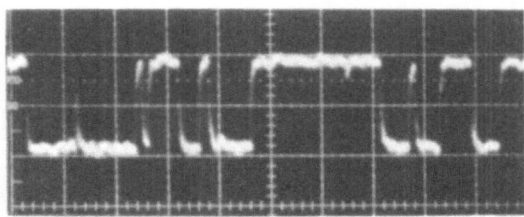

Figure 2.   The output voltage of the SQUID as a function of time
           at some point in the quasi-bistable region. The input
           parameters are fixed. The vertical scale is 50 µV/div.
           The zero of voltage is near the horizontal axis. The
           horizontal axis is 0.5 ms/div.

The first evidence for this hypothesis comes from estimating the voltage values that would correspond to transmission line resonances. If we suppose that the effective length ℓ of the stripline is the mean distance around the loop ($\ell=5.0$mm), if we use the propagation velocity $c/\sqrt{\varepsilon}$ for a transmission line, if we use $\varepsilon = 10.6$ from the estimated thicknesses of the two dielectrics which make up the insulating layer, and if we use a Josephson equation to relate frequency and voltage, then we expect half-wave resonances to occur at;

$$V = \frac{h}{2e}\left(\frac{n}{2}\right)\frac{c}{\ell\sqrt{\varepsilon}} \approx 38.\left(\frac{n}{2}\right)\mu V. \tag{1}$$

The values of V for n=1,2, and 3 are indicated in Figure 3. The one-half and three-half resonances roughly correspond to the two observed voltage states.

The absence of a two-half resonance is explained by two constraints on this transmission line. The first is that the upper and lower currents must be equal in magnitude and opposite in direction; this follows from the perfect diamagnetism of the groundplane. The second is that the time varying current joining

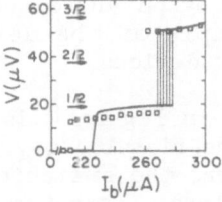

Figure 3.   The output voltage of the SQUID as a function of the bias current for a typical device. The data are shown as solid lines. Where the device is quasi-bistable, the two possible output values are connected by vertical lines. The arrows show the output voltages that are estimated to correspond to half-wave resonances of the transmission line. The squares show solutions found by computer simulation; there is a small region where two solutions are found for the same bias current.

the two junctions must be equal; this follows from the first
constraint and because the lower conductor for this geometry is
joined together at the two ends. These two contraints then require that the current distribution in the transmission line be
an _even_ function of distance from its center. All even numbered
half-wave resonances have currents that are _odd_ functions of
distance; therefore they cannot be present in this geometry.

Further evidence for this hypothesis comes from simulating
the system on a computer, incorporating the constraints. By
modeling the transmission line as a set of discrete inductors and
capacitors, it is possible to integrate forward in time the
complete system's equations of motion. We assume that I_b can be
taken to be time-independent. There are some values of input
parameters for which the simulation will converge to either of
two different stable solutions, the choice dependent on initial
conditions. These results are shown as squares in Figure 3. The
model parameters were adjusted, within their estimated errors, to
give the best qualitative fit to the data. The two solution
region in the simulation is displaced and smaller than the data,
but is close enough to convince us that the hypothesis is
correct. These simulations do not include noise sources, so we
do not expect them to show spontaneous switching.

SWITCHING

While we have identified the main factors that cause bistability, we do not yet have any clear understanding of the switching process. At this time we must content ourselves with simply
describing various experimental observations.

Figure 4 shows the region where the output is double-valued
for one particular example of the device. We have observed
similiar behavior on four devices.

Although the region in Figure 4 is labeled bistable, in no
case have we observed true bistablility, i.e. spontaneous switching never ceases. However, the switching rate is a strong function of the input parameters. Near the center of the region the
switching is less than one Hz. Near the edges of the region it
excedes 10^4 Hz. Switching rates much greater than 10^4 Hz would
not be detectable with our present apparatus, and the output
would appear to be single valued. The observed switching rates
are always very much less than the Josephson frequency, which is
greater than 10^{10} Hz.

There is no conclusive evidence, but we think it unlikely
that the switching is caused by external noise sources. Judging
by its appearance, the switching was a random precess. We took

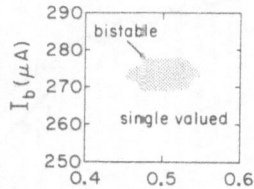

Figure 4.   The region in the input parameter space for which two
            output values are observed.

precautions to exclude external noise, including electrical
sheilding, and we used very quiet and stable current supplies.
When other values of the input parameters were chosen, these
precautions were sufficient to allow observation of the (very
low) noise levels that were expected from the effect of Johnson
noise in resistors that shunt the junctions. We therefore think
that the external noise power available to induce switching
cannot be much greater than the internal noise, and is perhaps
much less. It may be that the thermal noise power available from
these resistors is sufficient to cause the switching.

We have also measured the power spectral density of the
output voltage when the device is spontaneously switching. The
result, for the conditions of Figure 2, are shown in Figure 5.
These measurements were stimulated by a recent paper on hopping
phenomena in driven non-linear systems[5], which predicts a low
frequency power spectral density proportional to $1/f$. We see no
sign of such behavior but instead observe that the spectral
density falls as $1/f^2$ for f larger than the mean switching rate.
We are not yet able to say if this system satisfies the assumptions made in Ref. 5.

CONCLUSION

We have discovered a type of quasi-bistability that differs
qualitatively from the type which currently is the basis for a

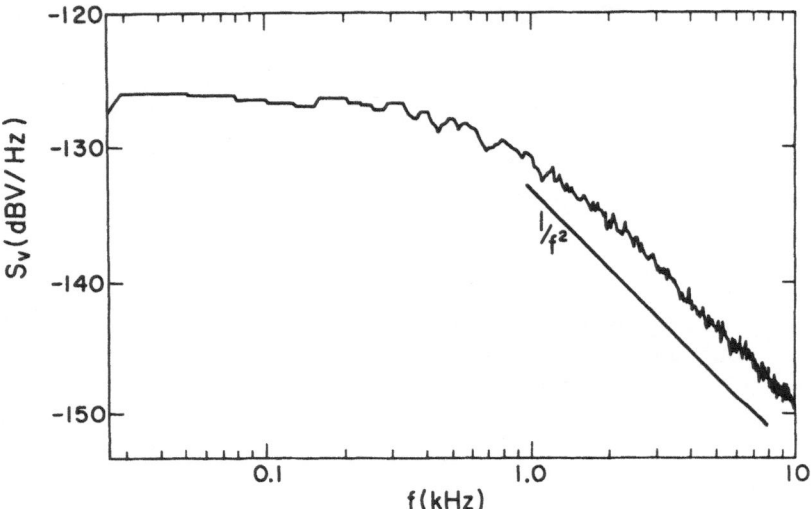

Figure 5. The spectral density of the output voltage when the
           device is spontaneously switching between the two
           output values. The data were taken under the same
           conditions as in figure 2. We see no evidence for a
           1/f region in this spectrum.

digital computer technology.[5] It has a strong analogy to optical
bistability because it also involves a multi-mode electromagnetic
resonator. The spontaneous switching observed is not yet
understood.

ACKNOWLEDGEMENTS

     This work would not have been possible without the collabor-
ation and good counsel of M.W.Cromar.

REFERENCES

1.   B. Muhlfelder, W. W. Johnson, and M. W. Cromar, IEEE Trans.
     Mag. 19, 303, (1983).
2.   P. D. Drummond, B. Muhlfelder, and W. W. Johnson, following
     paper, this conference.
3.   R. H. Havemann, C. A. Hamilton, and R. E. Harris, J. Vac.
     Sci. Technol. 15, 392 (1978).
4.   L. Mandel, Rajarshi Roy, and Surendra Singh, "Optical
     Bistability", Plenum Press, N.Y. (1981) (previous conf.)
5.   F. T. Arrecchi and F. Lisi, Phys. Rev. Lett., 49, 94 (1982)

MULTISTABILITY IN JOSEPHSON JUNCTION DEVICES

Peter D. Drummond, Barry Muhlfelder,
and Warren W. Johnson

Department of Physics and Astronomy
University of Rochester
Rochester, N.Y., 14627

INTRODUCTION

Some of the major features of the bistability reported in the preceeding paper[1] were tentatively explained by the interaction of the non-linear elements (the Josephson junctions) with a "cavity" (a length of transmission line). That type of interaction is superficially similiar to one that causes optical bistability, so that it seems interesting to pursue its analysis with similiar theoretical techniques. However, that system has so many variables that a detailed analysis would be rather laborious.

We therefore consider a simplified version of that type of system. We calculate the input-output relation for this system to lowest order in a singular perturbation technique. We predict that this system can have a whole "ladder" of multistable states. Computer simulations agree well with this analytic technique, except for cases in a certain limit. Experiments intended to test this prediction were performed shortly before this conference and the preliminary results are in good qualitative agreement.

A schematic of the model system is shown in Figure 1. A single idealized Josephson junction, with critical current I_c, is driven by a constant current I_b (the control parameter) and is shunted by: 1) its parasitic capacitance C, 2) a damping resistor R, and 3) a length of transmission line. (A transmission line is a waveguide in which we need consider only the TEM transverse mode.) This transmission line will support a set of modes distinguished only by their longitudinal wavelength.

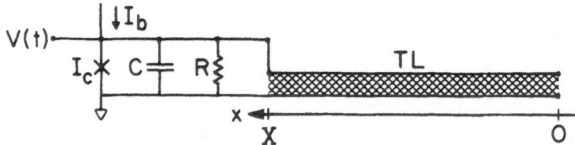

Fig. 1. The electrical schematic of the model system: a single
         junction shunted by a resistor R, capacitor C, and
         transmission line, TL.

There is a significant body of literature discussing a
related system. That system is a long narrow Josephson junction;
It can be viewed as a transmission line shunted by junctions
along its entire length, rather than at just one end. Under some
conditions it has steps in its input-output relation (I_b vs $\langle V \rangle$)
which are called Fiske steps. Each step is associated with exciting a low order mode of the transmission line. (Further
discussion and references to the literature can be found in a
recent text.[2])

A most important distinction between this work and the
earlier work on Fiske modes lies in which parameter (the "bias
parameter") is to be considered to be externally controlled and
time independent. In the earlier work the system was considered to be (nearly) voltage biased; here we wish to consider the
current biased case. This causes Fiske-like steps to become
hysteresis loops in the input-output relation.

RENORMALIZED PERTURBATION TECHNIQUE

The nonlinearity in the model system comes from the single
Josephson junction. The basic equations for the idealized junction are taken to be

$$V(t) = \frac{\phi}{2\pi} \frac{d}{dt}\theta(t), \quad I(t) = I_c \sin\theta(t) \qquad (1,2)$$

where V is the instantaneous voltage difference across the junction, $\phi = h/2e$ is the flux quantum, θ is the order parameter
phase difference across the junction, I is the instantaneous
super-current through the junction, and I_c is the critical current of the junction. Eq.(2) has often been found to give a good
account of experimental data even though it is only an approximation to the complete Josephson theory.

The unperturbed case is taken to be a current biased idealized junction shunted only by a resistor R. Then the scaled (i.e. dimensionless) equation of motion is:

$$\frac{d}{d\tau} \theta + \sin\theta = J \tag{3}$$

where $\bar{J} \equiv I_b/I_c$, $\tau \equiv t/t_s$, $t_s \equiv \phi/(2\pi I_c R)$. We remark that this equation occurs in many areas, including laser gyro theory and super fluorescence.

Its solution is well known. When $J<1$, the steady state solution is the static one (i.e. θ = constant, or "the D.C. Josephson effect"). When $J>1$, the solution for θ is a linear ramp plus a nonsinusoidal periodic function. Using some identities recently found for the solution,[3] it is possible to find an explicit solution of the form:

$$\theta(\tau) = \nu\tau - \phi_0 + \sum_{n\neq 0} c^{(n)} \exp(in[\nu\tau - \phi_0]) \tag{4}$$

whose Fourier coefficients are:

$$c^{(n)} = \frac{1}{in} \left[\frac{1 + \frac{n}{|n|} i\nu}{J(J+\nu)} \right]^{|n|} \tag{5}$$

The dimensionless fundamental frequency ν and the dimensioned period t_p are found to be:

$$\nu = \sqrt{J^2 - 1}, \quad t_p = 2\pi t_s/\nu \tag{6,7}$$

The measurable output variable is the time averaged voltage

$$\langle V(t) \rangle = \phi/t_p. \tag{8}$$

In the perturbed case we consider the addition of an arbitrary linear network shunting the junction. We want to determine how the network's presence changes the input-output relation calculated above, and to determine if the system has more than one solution for the same bias parameter. Although chaotic solutions may exist, we will restrict ourselves in this paper to finding periodic solutions. The advantage of using singular perturbation theory from the known resistively shunted behavior is that all the structure of the exact resistive solution is already included.

The new equation of motion can be written in the form

$$\frac{d}{d\tau} \theta + \sin\theta = J + \sum_{n\neq 0} c^{(n)} \exp(in\nu\tau) L(n\nu) \tag{9}$$

where the summation is the current from the network. We have

made use of Eqs.(1&4). Each Fourier coefficient $L(n\nu)$ is the
dimensionless network admittance at the indicated frequency multiplied by a frequency factor. We restrict ourselves here to
networks that do not allow a D.C. current, so $L(0)=0$.

The singular perturbation technique is as follows. We wish
to find a solution to Eq.(9) of the form

$$\theta(\tau) = \varepsilon(\tau) + \theta_0(\tau) \tag{10}$$

such that

$$\langle \varepsilon(\tau) \rangle = 0 \tag{11}$$

Here θ_0 indicates the exact solution to Eq.(3), for some renormalized current J_o and phase θ_0. We note that allowing the parameter J_o to be renormalized to remove secular terms so that
$\langle \varepsilon(\tau) \rangle = 0$ is entirely similar to the classical technique of
Linstedt, who renormalized the frequency of an harmonic oscil-
lator that was perturbed. Here, the parameter J_o is renormalized, which defines the frequency implicitly. We also choose
to expand in terms of a perturbation amplitude $\varepsilon(\tau)$, rather than
in terms of a parameter like $L(n\nu)$, as is done in some perturbation techniques. The result of this choice is that the first
iteration of this technique will contain implicitly terms up to
infinite order in $L(n\nu)$. This allows us to obtain multistable
results and to handle resonant behavior.

Substituting Eq.(10) into Eq.(9), and expanding to first
order in $\varepsilon(\tau)$, gives:

$$\frac{d}{d\tau}\theta_0 + \sin\theta_0 = J_o \tag{12}$$

$$\frac{d}{d\tau}\varepsilon + \varepsilon\cos\theta_0 = J - J_o + \sum_{n\neq 0}[C_o^{(n)} + \varepsilon^{(n)}]L(n\nu)\exp(in\nu\tau) \tag{13}$$

where $\varepsilon^{(n)}$ are the Fourier coefficients of $\varepsilon(\tau)$, and $C_o^{(n)}$ is
Eq.(5) with $J=J_o$ and $\nu = \sqrt{J_o^2 - 1}$. Eq.(11) requires that $\varepsilon^{(0)}=0$,
and we consider here only networks that have $L(0) = 0$. Now it
can be shown that the n-th Fourier coefficient of $\cos\theta_0$ is
$|n|\nu C_o^{(n)}$. Hence

$$in\nu\varepsilon^{(n)} + \sum_{n\neq m}\varepsilon^{(n-m)}C_o^{(m)}|m|\nu$$

$$= (J-J_o)\delta(n) + L(n\nu)[C_o^{(n)} + \varepsilon^{(n)}] \tag{14}$$

In general, this requires the solution of an infinite number of simultaneous equations for $\varepsilon^{(n)}$, since each coefficient is coupled to all the others. Luckily, it is easy to see from Eq.(5) that the terms become small for large n, so that a reasonable approximation to truncate the series with only three terms $(n=0,\pm1)$.

We can now solve for the three coefficients by a straightforward iterative procedure. Performing only the first iteration we find:

$$\varepsilon^{(\pm 1)} = \frac{L(\pm\nu)c_0^{(\pm 1)}}{\pm i\nu - L(\pm\nu)} \tag{15}$$

$$J - J_0 = 2\nu|c_0^{(1)}|^2 \; Re(\frac{L(\nu)}{i\nu - L(\nu)}) \tag{16}$$

The second equation amounts to a condition that fixes J_0. Combining these gives the lowest order result for the input-output relation:

$$J = \sqrt{1+\nu^2} + \frac{2\nu}{(\nu + \sqrt{1+\nu^2})^2} \; Re(\frac{L(\nu)}{i\nu - L(\nu)}) \tag{17}$$

MULTISTABILITY IN THE MODEL SYSTEM

Next, we wish to analyze the case of a current-biased junction shunted by a resistor, a capacitor, and a length of transmission line. The equations of the coupled system are:

$$\beta_c \; \frac{d^2}{d\tau^2}\theta + \frac{d}{d\tau}\theta + \sin\theta = J + \frac{1}{I_c}I(t,X) \tag{18}$$

$$\frac{\partial}{\partial x} I(t,x) = -c \; \frac{\partial}{\partial t} V(t,x) \tag{19}$$

$$\frac{\partial}{\partial x} V(t,x) = -\ell \; \frac{\partial}{\partial t} I(t,x) \tag{20}$$

where: the dimensionless capacitance is $\beta_c = 2eI_cR^2C/\hbar$; $V(t,x)$ and $I(t,x)$ are the A.C. voltage and current on the transmission line at time t and position x, X is its length, and where ℓ and c are its inductance and capacitance per unit length. Expanding V and I in a Fourier series, using the open termination boundary condition at x = 0, and using Eq.(1) for a boundary condition at x = X, we find that

$$\frac{1}{I_c} I(t,X) = r \sum c^{(n)}n \; \tan(n\nu\pi/\nu_0) \; \exp(in\nu\tau) \tag{21}$$

where

$$r \equiv R/\sqrt{\ell/c} \, , \qquad \nu_o \equiv \pi t_s/(X\sqrt{\ell c}) \qquad\qquad (22,23)$$

This implies that the network response function is now:

$$L(n\nu) = \beta_c(n\nu)^2 + n\nu r \, \tan(n\nu\pi/\nu_o) \qquad\qquad (24)$$

Substituting this into Eq.(17) is our final result.

A typical result of this calculation is plotted as the thick solid line in figure 2. For some values of input current J there are three possible values for the output voltage ; if similiar situations are a guide, then (at least) the negative slope solution should be unstable and (at most) two different voltages will be observable for the same input current. There are several ranges in J where this can happen, so that there should be a set of overlapping steps in the output as a function of the input.

By inspecting equation (17), we can interpret this calculation as predicting that the shunting network tends to reduce J, when viewed as a function of ν, except where the response function has a zero. There results one peak in J for each zero of $L(n\nu)$. When we turn the graph sideways to find ν as a function of J, we will find that most of the possible values of ν will be close to these peak values, and if the sides of the succesive peaks overlap, we should expect multistability. For the open end transmission line the peaks come at integer multiples of ν_o, which is equivalent to Eq.(1) of Ref.(1). Thus with this geometry we should get all the half-wave resonances, not just the odd ones, and the steps in the output should be evenly spaced.

We have also confirmed the main features of this calculation by computer simulation of the system; those results are shown as points in Fig. 2. In the simulation we replaced Eqs. (19) and (20) by a discrete set of inductors and capacitors. A second order Runge-Kutta algorithm was used to integrate the equations of motion; the output parameter was calculated when a stable limit cycle had been found. By incrementing the input parameter both up and down, and by using the preceding solution for the initial conditions, it was possible to find several bistable regions. The lowest order perturbation calculation is in good agreement with the simulation, except for ν less than 0.5, which is evidently where we need to keep higher order terms.

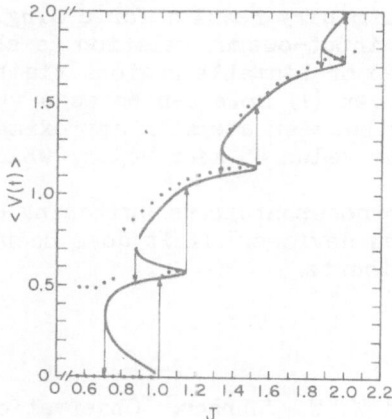

Fig. 2. A comparision of the lowest order perturbation calcula-
         tion (thick solid line) and the computer simulations
         (points). The horizontal axis is the dimensionless
         input current J, and the vertical axis is the dimension
         -less output voltage ν (which is incorrectly labeled
         as $\langle V(t)\rangle$). The thin solid lines show where we suspect
         that switching should occur.

EXPERIMENT

     Stimulated by the preceding calculations, we were able to
fabricate at NBS-Boulder an approximation to the model system.
The parameters were estimated to be I_c = 155 μA, R = 0.40 Ω, C =
0.5 pF, X = 2.75 mm, c = 10.9 nF/m, and ℓ = 6.1 nH/m. This
implies that the dimensionless parameter β_c = .04, and that the
steps should be spaced 46 microvolts apart.

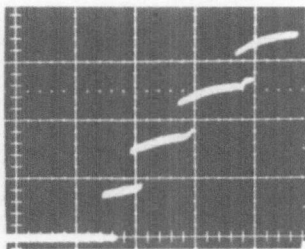

Fig. 3. The measured input-output relation for the first device
         tested. Input current is the horizontal axis (100 micro-
         amps per division) and output voltage is the vertical
         axis (50 microvolts per division).

We have two preliminary results for a single device:

1) The measured input-output relation is shown in Fig. 3. There are four overlap or bistable regions visible in the photograph, and at least seven (!) more can be seen when the range is expanded. The spacing between steps is approximately equal, as predicted, and has the value 45 microvolts, which we regard as very good agreement.

2) We have seen no spontaneous switching of the type seen in the double junction devices.[1] If it does occur its mean rate is less than 50 millihertz.

REFERENCES

1. B. Muhlfelder, and W. W. Johnson, "Observation of Bistability in a Josephson Device", preceding paper, this conference.
2. A. Barone and G. Paterno, "Physics and Applications of the Josephson Effect", Wiley, New York (1982).
3. J. D. Cresser, W. H. Louisell, P. Meystre, W. Schleich, M.O. Scully, Phys. Rev. A25, 2214 (1982).

DISTRIBUTED FEEDBACK BISTABILITY IN CHANNEL WAVEGUIDES

G. I. Stegeman and C. Liao
Optical Sciences Center and Arizona Research Labs
University of Arizona
Tucson, AZ 85721

H. G. Winful
G.T.E. Laboratories Inc.
40 Sylvan Road
Waltham MA 02254

INTRODUCTION

The phenomenon of optical bistability holds considerable promise for application to all-optical logic and signal processing. For applications that will ultimately require high packing densities, i.e. multiple operations in series or parallel, it is desirable to minimize both the material volume and required power per element. Such requirements will probably necessitate using material nonlinearities versus a hybrid approach. The cross-sectional area and hence both the volume and switching power can be minimized by using integrated optics waveguides. The feedback required for bistability can be achieved with distributed feedback gratings. In this paper we analyze the power, waveguide, and grating parameters needed for optical bistability in a two-dimensional waveguide.

CHANNEL WAVEGUIDES

Channel waveguides consist of a rectangular region of a high refractive index material (n) surrounded by media of lower refractive index. Waves are guided by the high refractive index region by total internal reflection at each channel boundary with the result that the fields decay in an evanescent fashion into all of the bounding media. For the typical fields shown in Fig. 1, we write

$$(\mathbf{E,H}) = \frac{1}{2} (\mathbf{E_0, H_0})\ f(x,y)\ \exp[i(\omega t - \beta_0 z)]\ a_+(z) + cc$$

for a wave of amplitude $a_+(z)$ traveling along the z axis. Here $f(x,y)$ is normalized so that $a_+(z)^2$ is the guided wave power in watts. Marcatili[1] has obtained approximate forms for $f(x,y)$ that are valid for waves strongly confined to the channel. There are two types of modes, E^x (with dominant fields E_x and H_y) and E^y (dominant fields E_y, H_x). For example, $f(x,y)$ for the E^y modal field in the channel is[2]

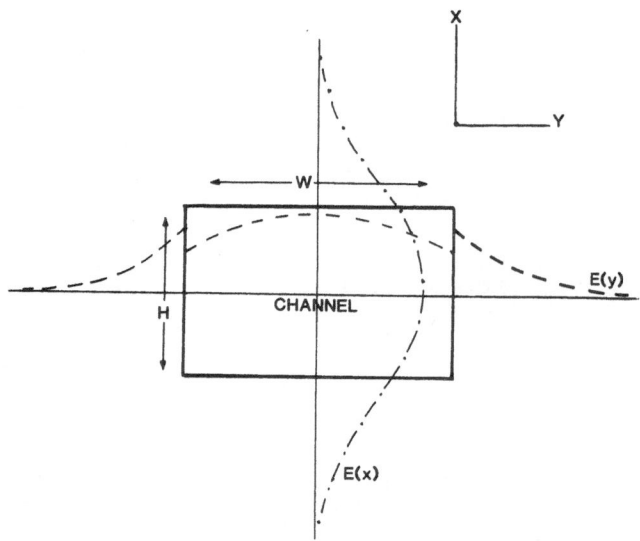

Fig. 1. Cross section of a rectangular waveguide of width W and height H. Typical E^y field distribution variation with both x and y coordinates is shown.

$$E_y = iB\ \frac{\beta_0{}^2 + \kappa_x{}^2}{\beta_0}\ \cos[\kappa_x(x+\zeta)]\ \sin[\kappa_y(y+\eta)]$$

$$H_x = -iB\varepsilon_0 c\ n_c{}^2 k\ \cos[\kappa_x(x+\zeta)]\ \sin[\kappa_y(y+\eta)]$$

$$E_z = B\kappa_y\ \cos[\kappa_x(x+\zeta)]\ \cos[\kappa_y(y+\eta)].$$

Similar expressions[2] are obtained for the fields in the bounding media. The dispersion relations are derived from the continuity of the fields at each channel boundary and can be found in reference 2. The constant B is evaluated from the normalization condition

$$1 = \frac{1}{2} \int E_y H_x\ dx\ dy,$$

where the integration includes the fields in the regions surrounding the channel. The dispersion relations are solved on a computer to yield allowed values of β_0 for a given channel of width W and height H. Minimum channel dimensions are typically about one wavelength across the narrowest dimension of the channel. Hence cross-sectional areas of one wavelength squared are possible.

GRATING REFLECTION

A guided wave can be reflected back along the -z axis by a grating fabricated along one of the channel interfaces. We consider here a buried channel (Fig. 2a) and a ridge waveguide (Fig. 2b). From coupled mode theory,[3] for a sinusoidal grating the amplitude of the reflected wave, $a_-(z)$, is given by

$$\frac{d}{dz} a_-(z) = -i\Gamma \, e^{-i\Delta\beta z} \, a_+(z)$$

where[4]

$$\Gamma = \frac{\omega\varepsilon_0}{8} \, u_0 \int [n^2(x = x_+') - n^2(x=x_-')] \, [\mathbf{E}_i(x=x_+')\cdot\mathbf{E}_r(x=x_-')] \, dy$$

$$\Delta\beta = 2\beta_0 - \kappa.$$

The grating has a wave vector κ that lies in the plane x = x' and has a peak-to-peak displacement $2u_0$. The subscripts "i" and "r" refer to the incident and reflected waves.

The backward wave can also be reflected into the forward wave leading to

$$\frac{d}{dz} a_+(z) = i\Gamma \, e^{i\Delta\beta z} \, a_-(z).$$

(a)

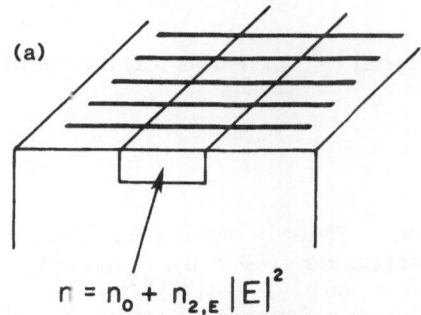

$$n = n_o + n_{2,\varepsilon} |E|^2$$

(b)

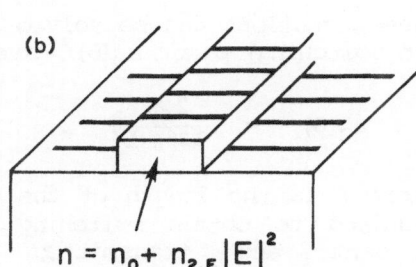

$$n = n_o + n_{2,\varepsilon} |E|^2$$

Fig. 2. Two distributed feedback channel waveguides investigated numerically.

INTENSITY–DEPENDENT PROPAGATION WAVE VECTOR

If the channel medium is characterized by an intensity-dependent refractive index, the propagation wave vector β depends on the guided wave power.[5] Writing the intensity dependent refractive index as

$$n_c(I_0) = n_c + n_2 I_0,$$

where I_0 is the intensity of a plane wave,

$$\beta = \beta_0 + \Delta\beta_0 \, a_+(z)^2$$

for a single guided wave. Using the plane wave approximation for the waveguide fields (and neglecting temporarily the effects of the grating), coupled mode theory gives[5]

$$\frac{d}{dz} a_+(z) = -i\Delta\beta_0 \, a_+(z)^2$$

with

$$\Delta\beta_0 = \frac{\omega\varepsilon_0^2 c}{4} \iint n_c^2 \, n_2 \, E_i^4 \, dx \, dy.$$

DISTRIBUTED FEEDBACK BISTABILITY

Inside the grating, there are both forward and backward traveling waves. It can be shown that the combination of both the grating and the intensity–dependent effects leads to[6]

$$i \frac{d}{dz} a_+(z) = \Gamma e^{-i\Delta\beta z} a_-(z) + \Delta\beta_0[a_+(z)^2 + 2a_-(z)^2]a_+(z)$$

$$-i \frac{d}{dz} a_-(z) = \Gamma e^{i\Delta\beta z} a_+(z) + \Delta\beta_0[2a_+(z)^2 + a_-(z)^2]a_-(z).$$

These equations can be solved analytically[6] in terms of an incident (or switching) power $a_c(0)^2$ given by

$$a_c(0)^2 = \frac{2\beta_0}{3\Delta\beta_0 kL},$$

where L is the length of the grating. This is typically the power required to obtain switching on resonance ($\Delta\beta = 0$). Writing the incident and transmitted power as $I = a_+(0)^2/a_c(0)^2$ and $J = a_+(L)^2/a_c(0)^2$, then for $\Delta\beta = 0$ the grating transmission is given by

$$T = J/I = 2(1 + nd\{2\sqrt{[(\Gamma L)^2 + J^2]};[1+(J/\Gamma L)^2]^{-1}\})^{-1}.$$

Here $nd(u;m)$ is a tabulated Jacobian elliptic function. As shown in Fig. 3, switching occurs for $a_+(0)^2 \sim a_c(0)^2$. In order to obtain bistability, the grating feedback term ΓL must be of the order of unity, or larger.

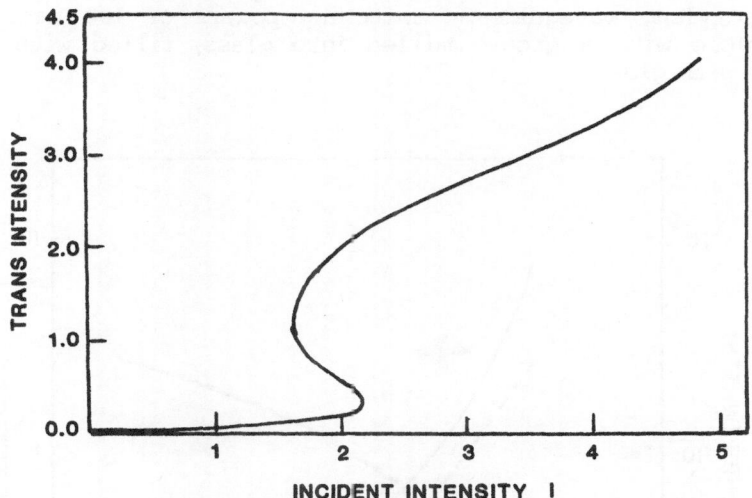

Fig 3. The transmitted versus incident power for distributed feedback bistability.

NUMERICAL ESTIMATES InSb

As an example, we chose InSb[7], which exhibits large nonlinearities at $\lambda = 5.5$ μm. We assume $L = 1$ mm, 40Å peak-to-peak surface corrugation, $n_2 = 10^{-8}$ m²/W, $n = 4.0$, and sapphire is the substrate. The results are shown for E^Y modes in Fig. 4 for both buried and ridge channel waveguides with $H/W = 0.4$. The smaller switching power for the ridge versus channel cases reflects the better field confinement obtained with a ridge waveguide.

The variation in switching power with the ratio of channel height to channel width is shown in Fig. 5. Note that in terms of channel width, E^Y modes are optimum for thin wide waveguides and the E^X modes are best for tall narrow waveguides. When compared for equal channel areas, the two modes give comparable values.

The key result is that switching powers of 10 to 100 nW are possible and that feedback parameters of unity can be achieved with surface gratings of less than 100Å.

NUMERICAL RESULTS PTS, CS$_2$

The switching power is shown for a ridge PTS[8] and a buried CS$_2$ waveguide in Fig. 6. Both of these materials are characterized by picosecond switching times. Of special interest is the low switching power (1 watt) potentially available with ridge polydiacetylene waveguides. Switching powers of 100 watts should be possible with a groove milled into glass, filled with CS$_2$, and covered with glass.

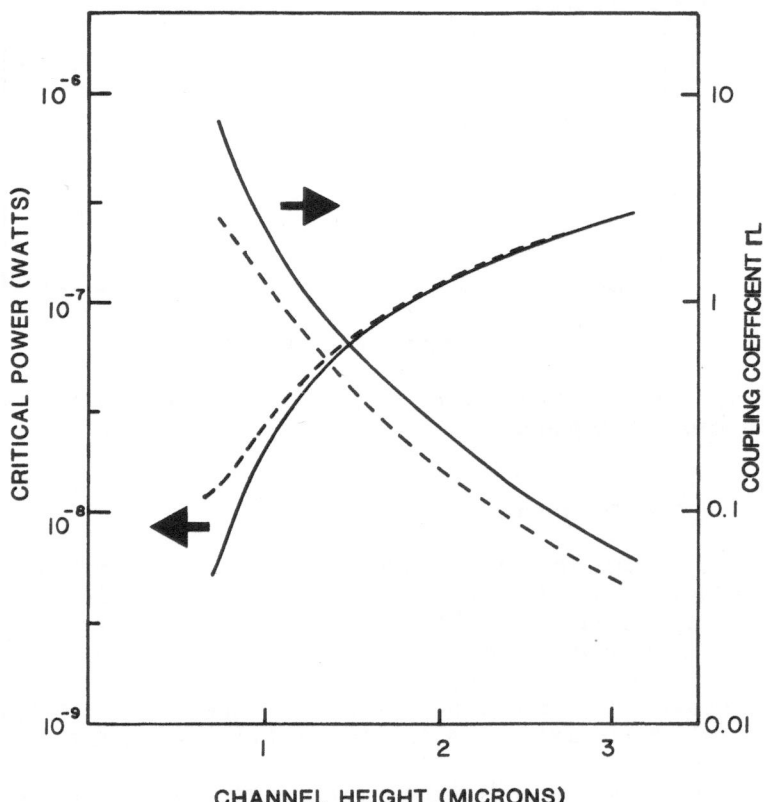

Fig. 4. The critical power $a_c(0)^2$ and the feedback parameter ΓL versus the channel height for H/W = 0.4. The solid line is for the ridge and the dashed line for the buried waveguide.

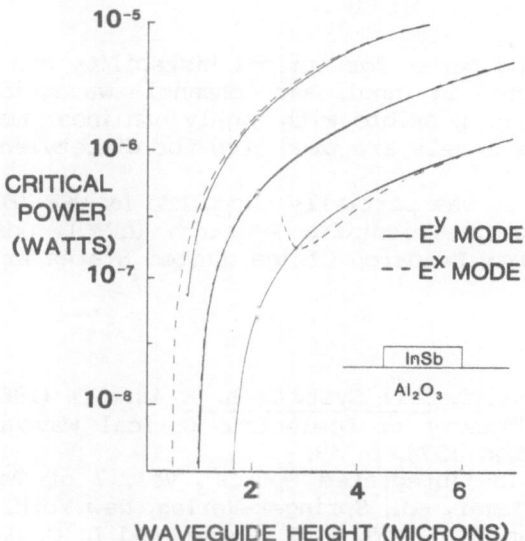

Fig. 5. Critical power versus waveguide height for both E^y and E^x modes for three different values of H/W.

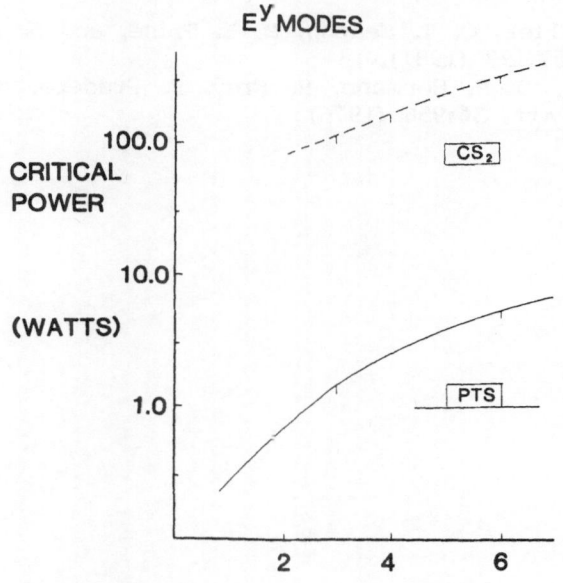

Fig. 6. The critical power for CS_2 (buried) and PTS (ridge) waveguides versus waveguide height for H/W = 0.25 and L = 1 mm.

CONCLUSIONS

The switching power for optical bistability can be minimized by guiding the light in nonlinear channel waveguides. Nanowatt switching should be possible with highly nonlinear materials such as InSb. Watt power levels are predicted for diacetylenes such as PTS.

This research was partially supported by NSF (ECS-8117483), the Air Force Office of Scientific Research (AFSC), United States Air Force, and the Army Research Office, United States Army.

REFERENCES

1. E. A. J. Marcatili, Bell Syst. Tech. J. **48**:2071 (1969)
2. D. Marcuse, "Theory of Dielectric Optical Waveguides," Academic Press, New York, 1974, p. 49.
3. H. Kogelnik, in "Integrated Optics", Vol. 7 of Topics in Applied Physics, T. Tamir, ed., Springer-Verlag, New York, 1975, p. 66.
4. G. I. Stegeman, D. Sarid, J. J. Burke, and D. G. Hall, J. Opt. Soc. Am. **71**:1497 (1981).
5. G. I. Stegeman, IEEE J. Quant. Electron. **QE-18**:1610 (1982).
6. H. G. Winful, J. H. Marburger, and E. Garmire, Appl. Phys. Lett. **35**:379 (1979).
7. D. A. B. Miller, C. T. Seaton, M. E. Prise, and S. D. Smith, Phys. Rev. Lett. **47**:197 (1981).
8. C. Sauteret, J. P. Hermann, R. Frey, F. Pradere, and J. Ducuing, Phys. Rev. Lett. **36**:956 (1976).

EFFECTS OF THE RADIAL VARIATION OF THE ELECTRIC FIELD ON SOME

INSTABILITIES IN OPTICAL BISTABILITY AND LASERS

L.A. Lugiato and M. Milani

Dipartimento di Fisica dell'Università

Via Celoria, 16 - 20133 Milano (Italy)

1. INTRODUCTION

The overwhelming majority of theories for laser and optical bistability (OB) have been worked out within the plane wave approximation. However, recently, several authors have focussed their attention on giving a description of these systems which fully includes the transverse variation of the electromagnetic field. The reason for doing that is twofold. First of all, in OB the field injected into the cavity has typically a Gaussian radial profile, and therefore it is impossible to obtain a reasonable quantitative agreement between theory and experiment without including this feature in the theory. Second, in the case of <u>instabilities</u> it has been shown that the radial variation of the field induces not only quantitative but even qualitative changes in the behaviour of the system. E.g., in the case of dispersive OB in a ring cavity with plane mirrors, the route to chaos changes from period doubling to quasiperiodic [1]. Other striking examples will be given in this paper.

A group of papers[2-4] considers the case of (ring or Fabry-Perot) cavity with <u>spherical</u> mirrors and an incident field which corresponds to the fundamental TEM_{00} mode of the cavity. They assume that the internal field at steady state keeps the same TEM_{00} also in presence of the atomic sample; this assumption will be discussed in the next section.

Using the mean field limit, they obtain analytic formulae for the steady state behaviour of the system. Another group of papers[1,5-8] studies the case of (ring or Fabry-Perot) cavity with <u>plane</u> mirrors but Gaussian incident field. Their analysis, which does not introduce

the mean field approximation nor any assumption on the radial
shape of the field internal to the cavity, is purely numerical.

It must be stressed that the two situations of plane and
spherical mirrors are basically different, because in the first
case the cavity does not support transverse modes. In this paper,
we shall consider the case of ring cavity with spherical mirrors
and shall analyze the effects of the radial field variation on
some instabilities in lasers and OB, that arise in the plane wave
theory. The first instability that we shall consider is the one
discovered by Haken and Risken, Schmid and Weidlich[10], who showed
that under proper conditions the stationary state of a homogeneously broadened, single mode ring laser can become unstable. This
instability requires a "bad cavity" situation, i.e. the cavity
damping constant K must be larger than the sum of the transverse
and longitudinal atomic decay rates $\gamma_{\|}$ and $\gamma_{\perp}$. Later, Haken[11] dem-
onstrated that such an instability coincides with the well known
Lorenz instability[12], which leads to chaotic behaviour[14] On the
other hand, Risken and Nummedal[13] and Graham and Haken[14] analyzed
the homogeneously broadened ring laser taking into account <u>all</u> the
longitudinal cavity modes. They showed that under appropriate con-
ditions in the "good cavity" case $K < \gamma_{\|} + \gamma_{\perp}$ some cavity modes,
different from the one resonant with the atomic system, can become
unstable. In this case, one has the formation of a pulse that
travels into the cavity.

In 1978, Bonifacio and Lugiato[15] discovered that in absorptive
OB, when the bistability parameter is large enough, one can have a
self-pulsing behaviour of a type similar to that found in the laser
in refs. 13, 14. This result was later generalized to the case of
mixed absorptive + dispersive OB in ref. 16. The relation between
the Bonifacio-Lugiato instability and the Ikeda instability[17] is
discussed in ref. 18. As we shall see, the transverse variation
of the electric field bears dramatic consequences on all these
instabilities.

2. TRANSVERSE MODE EQUATIONS AND ONE-TRANSVERSE MODE MODEL

Let us consider a coherent beam of frequency ω_o that is injected into a unidirectional ring cavity of length $\mathcal{L}$ (see figure). A
cylindrical sample of length L and section πd^2 contains a large
number N of two-level atoms. The atomic system is assumed homogeneously broadened, with transition frequency ω_a and longitudinal and
transverse decay rates $\gamma_{\|}$ and $\gamma_{\perp}$. We assume that the incident field
has cylindrical symmetry, and indicate by r and z the radial and
longitudinal coordinate, respectively.

Furthermore, we assume that the Fresnel number $F = \pi W_o^2/\lambda L$,
where λ is the beam waist, is much larger than unity. Hence the

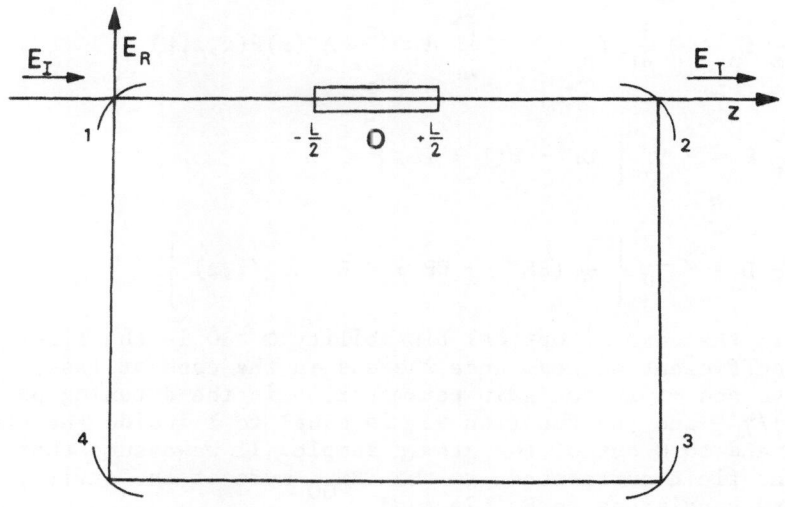

Unidirectional ring cavity with spherical mirrors. E_I, E_T, E_R are the incident, transmitted and reflected fields respectively. Mirrors 3 and 4 have 100% reflectivity, whereas mirrors 1 and 2 have transmissivity coefficient 1.

radius of the beam is uniform along the active region. This condition is met in the experiments reported in refs. 19, 20. For a more general treatment without this assumption see ref. 21.

Hence, the slowly varying envelope of the electric field in the sample is expanded in terms of transverse modes as it follows:

$$E(r,z,t) = \sum_{p'=0}^{\infty} f_{p'}(z,t) A_{p'}(r)$$

$$A_p(r) = \exp(-r^2/W_o^2) L_p(2r^2/W_o^2) \tag{1}$$

where L_p is the Laguerre polynomial of order p. As it is shown in ref. 21, the amplitude f_p, the atomic polarization P and the population difference D obey the Maxwell-Bloch equations

$$\frac{\partial}{\partial z} f_p + \frac{1}{c} \frac{\partial}{\partial t} f_p = - \alpha \int_0^\infty dr \frac{4r}{W_o^2} A_p(r)P(r,z,t) \tag{2a}$$

$$\frac{\partial}{\partial t} P = - \gamma_\perp \left\{ DE - P(1 + i\Delta) \right\} \tag{2b}$$

$$\frac{\partial}{\partial t} D = - \gamma_\parallel \left\{ \frac{1}{2} (PE^* + EP^*) + D - \chi_v(r,z) \right\} \tag{2c}$$

where in the case of optical bistability $\alpha > 0$ is the linear absorption coefficient on resonance whereas in the case of laser α is negative and $-\alpha$ is the gain parameter. Δ is the detuning parameter $(\omega_a - \omega_o)/\gamma_\perp$ and the function χ_v is equal to 1 inside the atomic sample and to 0 out of the atomic sample. If we assume that the incident field corresponds to the TEM_{00} mode of the cavity, the boundary conditions in Eq. 2a read

$$f_p(- \frac{L}{2}, t) = Ty\delta_{p,o} + (1 - T) \exp(- i\delta_p) \cdot$$

$$\cdot f_p\left(\frac{L}{2}, t - \frac{\mathscr{L} - L}{c}\right) \tag{3}$$

where y is the normalized incident field and $(k_o = \omega_o/c)$

$$\delta_p = 2\pi m - k_o \mathscr{L} - (2p + 1) \bar\delta \tag{4}$$

i.e. $\frac{c}{\mathscr{L}} \left[2\pi m - 2(2p + 1) \bar\delta \right]$ are the mode frequencies; the integer m is chosen in such a way that $|\delta_o| < \pi$.

Now, let us assume that only the fundamental TEM_{00} mode has a non-negligible amplitude (one transverse mode approximation). Hence Eqs. (2) reduce to

$$\frac{\partial}{\partial z} f_o + \frac{1}{c} \frac{\partial}{\partial t} f_o = - \alpha \int_0^\infty d\bar r \, 4\bar r \, \exp(- \bar r^2)P(\bar r,z,t) \tag{5a}$$

$$\frac{\partial P}{\partial t} = -\gamma_\perp \left\{ D f_o \exp(- \bar r^2) - P(1 + i\Delta) \right\} \tag{5b}$$

$$\frac{\partial D}{\partial t} = -\gamma_\parallel \left\{ \frac{1}{2} (Pf_o^* + P^* f_o)\exp(- \bar r^2) + D - \chi_v(\bar r,z) \right\} \tag{5c}$$

where $\bar r = r/W_o$.

Note that (5) include all the longitudinal modes of the cavity;

the one mode assumption concerns only the transverse modes.

Furthermore, if (1) we perform the mean field approximation

$$\alpha L \ll 1, \quad T \ll 1, \quad \delta_o \ll 1$$

with

$$C \equiv \frac{\alpha L}{2T} \quad \text{arbitrary}$$

$$\Theta \equiv \frac{\delta_o}{T} \quad \text{arbitrary,}$$

(6)

and (2) we assume that the time evolutions of the system occurs on a time scale much longer than the cavity roundtrip time $\mathcal{L}/c$ (note that this assumption fails whenever some off-resonance mode becomes unstable), Eqs. (5) reduce to the mean field model

$$\frac{d}{dt} f_o = - k \left\{ f_o \ (1 + i \Theta) - y + 2C \int_0^{d/W_o} dr \ 4\overline{r} \ \exp(- \overline{r}^2) P(\overline{r}, t) \right\}$$

(7a)

$$\frac{\partial}{\partial t} P = \gamma_\perp \left\{ D f_o \exp(- \overline{r}^2) - P \ (1 + i\Delta) \right\}$$

(7b)

$$\frac{\partial}{\partial t} D = \gamma_{\parallel} \left\{ \frac{1}{2} \ (P f_o^* + f_o P^*) \ \exp(- \overline{r}^2) + D - 1 \right\}$$

(7c)

where f_o depends only on t and P,D depend on $\overline{r}$ and t and k is the cavity damping constant $cT/\mathcal{L}$. We recall that d is the radius of the atomic sample. Eqs. (7) are derived from eqs. (5) and (3) by following the same procedure that is shown in the plane wave case in ref. 22. In the limit (6) the operation is restricted to a single longitudinal mode (resonant mode).

The stationary solution of eqs. (7) is easily obtained and reads[21]

$$Y = X \left\{ \left[1 + \frac{2C}{X} g(x) \right]^2 + \left[\Theta - \frac{2C\Delta}{X} g(x) \right]^2 \right\}$$

(8)

where

$$Y = y^2, \quad X = |f_o|^2$$

(9a)

$$g(x) = \ln \frac{1 + \Delta^2 + X}{1 + \Delta^2 + X \exp\left[- 2(d/W_o)^2\right]}$$

(9b)

In order to discuss the one transverse mode approximation, let us now come back to eqs. (2) and follow a procedure that is in reverse order with respect to the previous one. Precisely, let us first operate the mean field limit $\alpha L \ll 1$, $T \ll 1$ and assume that the time evolution of the system occurs on a time scale much longer than $\mathscr{L}/c$. As we shall see, only the amplitudes such that $\delta \ll 1$ survive in this limit. By defining $\theta_p \equiv \delta_p/T$ hence $\theta_o = \theta$, we obtain from (2) in the usual way

$$\frac{d}{dt} f_p = -k \left\{ f_p(1 + i\theta_p) - y\delta_{p,o} + 2C\int_0^\infty d\overline{r}\ 4\overline{r}\ A_p(\overline{r})P(\overline{r},t) \right\}$$

(10a)

$$\frac{\partial}{\partial t} P = \gamma_\perp \left\{ D\ \Sigma_p f_p A_p(\overline{r}) - P \right\}$$

(10b)

$$\frac{\partial}{\partial t} D = -\gamma_{||} \left\{ \frac{1}{2}\ \Sigma_p (Pf_p^* + f_p P^*)A_p(\overline{r}) + D - \mathscr{X}(\overline{r}) \right\}$$

(10c)

From eqs. (10) we obtain that the condition, which ensures that only the fundamental mode has a non-negligible amplitude, is that $\theta_o \lesssim 1$ whereas $\theta_p \gg 1$ for $p > 0$. In fact, the width of the transmission band of each mode is k. Since $\theta_p = (\omega_p - \omega_o)/k$, where ω_p is the frequency of the p-th mode, the modes with $\theta_p \gg 1$ do not transmit any sizeable amount of light, because they are too much out of resonance with respect to the incident field. A qualitatively similar argument, but not supported by an analysis of equations like (10), was given in ref. 2. We note also that the modes such that δ_p is not much smaller than unity do not contribute "a fortiori", because for $T \ll 1$ one has necessarily that $\theta_p = \delta_p/T \gg 1$.

We observe finally that if in eq. (10a) we restrict ourselves to one transverse mode $p = 0$, we recover eq. (7a) as expected, because they arise from the mean field limit and the one-transverse mode approximation in whatever order they are performed.

3. RESULTS AND DISCUSSIONS

In ref. 31 and 23 we performed the linear stability analysis of the stationary solutions of eqs. (5), which include the one transverse mode approximation but not the mean field approximation. In fact, even if we considered specifically the case $\alpha L \ll 1$, $T \ll 1$, we did not assume that the time evolution occurs on a time scale much longer than the cavity roundtrip time. This in order to allow also for the possibility of self-pulsing behaviour arising from instabilities of the off-resonance longitudinal modes.

For $\alpha L \ll 1$, $T \ll 1$ the steady state equation of eqs. (5) is
however still given by (8) and (9), i.e. coincides with that of eq.
(7) and the steady state field is practically uniform in the longitudinal direction along the atomic sample. Our analysis, which is performed in the resonant case $\Delta = \Theta = 0$ shows that the stability depends
crucially on the ratio d/W_o between the radius of the atomic sample
and the beam radius. For $d/W_o \ll 1$, we obtain practically the same
results of the plane wave theory, as it is obvious because in this
case the atoms interact only with the central part of the Gaussian
beam profile. On the other hand, when $d/W_o \to \infty$ so that the atoms
interact also with the wings of the Gaussian, we find that all the
instabilities predicted by the plane wave theory of refs. 9-11, 13-
15 vanish. Hence, we can conclude that in order to observe these
instabilities it is necessary to be essentially in plane wave conditions, i.e. either to realize a situation in which $d/W_o \ll 1$ or,
more realistically to use plane mirrors. In the case of optical
bistability, to fulfill the plane wave conditions, it is necessary
that the incident field radius is suitably enlarged by means of
lenses, so that only the central part of the beam is injected into
the cavity. This procedure is followed in a recent experiment by
Arecchi and collaborators[24].

There is also a third situation in OB, namely that of plane
mirrors with a Gaussian input field. In ref. 1, by numerical solutions of the Maxwell-Bloch equations it is shown that the Ikeda
instability in dispersive OB survives, even if the picture changes
dramatically. For the case of self-pulsing with $\alpha L \ll 1$, $T \ll 1$, we
have no result to illustrate what happens.

Our analysis is restricted to the situation of homogeneous
broadening. In the case of inhomogeneously broadened lasers, the
plane wave theory predicts that the instability which continues
directly the one discovered in refs. 9, 10 becomes more accessible
because it occurs the nearer to laser threshold, the larger is the
ratio between the inhomogeneous and the homogeneous linewidth[25-28].
This instability has been experimentally observed in inhomogeneously
broadened lasers[29-30]. We are presently investigating the dynamics of
this system using equations which assume a Gaussian profile for the
field internal to the cavity. While in the laser case the obtained
results seem to exclude that such instabilities can be experimentally
observed, with homogeneous broadening in the usual situation $d \gg W_o$,
in the case of optical bistability the hypothesis that the field
internal to the filled cavity corresponds to a TEM_{00} mode is ques-
tionable at least when $F \gg 1$.

Other problems in which we are studying the influence of the
transverse profile of the electric field are:
1) The self-pulsing and chaotic behaviour in the mean-field model
   for dispersive OB[31]. This system is particularly interesting from
   the viewpoint of constructing an all-optical converter of cw

coherent light into pulsed.

2) The regular and chaotic pulsations in the laser with injected signal[32].

3) The passive Q-switching in the laser with saturable absorber[31].

REFERENCES

1) J.V. Moloney, F.A. Hopf and H.M. Gibbs, Phys. Rev. A 25, 3442 (1982)

2) R.J. Ballagh, J. Cooper, M.W. Hamilton, W.J. Sandle and D.M. Warrington, Opt. Commun. 37, 143 (1981)

3) E. Arimondo, A. Gozzini, F. Lovitch and E. Pistelli in Optical Bistability, Proc. Int. Conf. Asheville, North Carolina 1980, edited by C.M. Bowden, M. Ciftan and H.R. Robl, Plenum Press, New York 1981

4) P.D. Drummond, IEEE J. Quant. Electron. QE-17, 301 (1981)

5) N.N. Rosanov and V.E. Semenov, Opt. Commun. 38, 435 (1981)

6) W.J. Firth and E.M. Wright, Opt. Commun. 40, 223 (1982)

7) J.V. Moloney, M.R. Belic and H.M. Gibbs, Opt. Commun. 41, 379 (1982)

8) J.V. Moloney and H.M. Gibbs, Phys. Rev. Lett. 48, 1607 (1982)

9) H. Haken, Z. Physik 190, 327 (1966)

10) H. Risken, C. Schmid and M. Weidlich, Z. Physik 194, 337 (1966)

11) H. Haken, Phys. Lett. 53 A, 77 (1975)

12) E.N. Lorenz, J. Atmos. Sci 20, 130 (1963)

13) H. Risken and Nummedal, J. Appl. Phys. 39, 4662 (1968)

14) R. Graham and H. Haken, Z. Physik 213, 420 (1968)

15) R. Bonifacio and L.A. Lugiato, Lett. Nuovo Cimento 21, 510 (1978)

16) L.A. Lugiato, Opt. Commun. 33, 108 (1981)

17) K. Ikeda, Opt. Commun. 30, 257 (1979)

18) L.A. Lugiato, M.L. Asquini and L.M. Narducci, Opt. Commun. 41, 450 (1982)

19) W.J. Sandle and A. Gallagher, Phys. Rev. A 24, 2017 (1981)

20) D.E. Grant and J.H. Kimble, Opt. Lett. 7, 353 (1982)

21) L.A. Lugiato and M. Milani, Z.f.Physik B 50, 171 (1983)

22) R. Bonifacio and L.A. Lugiato, in Dissipative Systems in Quantum Optics: Resonance Fluorescence, Optical Bistability, Superfluorescence, edited by R. Bonifacio, Springer-Verlag, Berlin

23) L.A. Lugiato and M. Milani, Opt. Commun. 46, 57 (1983)

24) F.T. Arecchi, G. Giusfredi, E. Petriella and P. Salieri Appl. Phys. B 29, 79 (1982)

25) M.L. Minden and L.W. Casperson, IEEE Journ. of Quantum Electronics QE-18, 1952 (1982)

26) S.T. Hendow and M. Sargent III, Opt. Commun. 40, 385 (1982)

27) P. Mandel, Opt. Commun. 44, 400 (1983)

28) L.A. Lugiato, L.M. Narducci, D. Bandy and N. Abraham, Opt. Commun., in press

29) L.W. Casperson, IEEE Journal Quantum Electron. QE-14, 756 (1978)

30) J. Beatley and N.B. Abraham, Opt. Commun. 41, 52 (1982)

31) L.A. Lugiato, L.M. Narducci, D. Bandy and C. Pennise, Opt. Commun. $\underline{43}$, 281 (1982)
32) L.A. Lugiato, L.M. Narducci, D. Bandy and C.A. Pennise, submitted for publication
33) H.J. Powell and G.J. Wolga, IEEE Journ. Quantum Electron. QE-$\underline{7}$, 219 (1971)

TRANSVERSE SOLITARY WAVES IN A DISPERSIVE RING BISTABLE CAVITY CONTAINING A SATURABLE NONLINEARITY

J.V. Moloney

Optical Sciences Center
University of Arizona
Tucson, AZ 85721

A.C. Newell and D.W. McLaughlin

Department of Mathematics
University of Arizona
Tucson, AZ 85721

ABSTRACT

Stationary and nonstationary transverse solitary waves can gradually evolve when the incident Gaussian beam amplitude exceeds threshold for switching to the high transmission branch of a bistable ring cavity. The number of transverse solitary waves scales as the square root of the effective Fresnel number and as the incident laser amplitude. Predictions from an independent theory give quantitative agreement with the numerical computations.

I. INTRODUCTION

At high Fresnel number, the switching of an incident Gaussian pulse to the high transmission branch of a ring bistable cavity exhibits strong radial dependence. The portion of the beam profile satisfying $I(x) > I_\uparrow$ ($I(x)$ is the intensity in the transverse x-coordinate and $I_\uparrow$ is the plane wave predicted switch-on intensity) switches rapidly to the high transmission branch in about 20 cavity roundtrips. The central portion of the beam on recovering from the overshoot switch causes a singular type edge to appear at the outer extremities of the on-spot. The subsequent dynamics is dependent upon whether the system is operating under self-focusing or defocusing conditions.[1] Under self-defocusing conditions, for example, the on-spot, because of weak diffraction, begins to expand slowly radially outward and stops at the plane wave switchdown intensity. The end result is a switched on beam with a sharp cutoff at a finite radius; the radius is dependent upon the extent to which the peak intensity I_p exceeds the plane wave switch-on intensity $I_\uparrow$. The situation is very different under self-focusing conditions as discussed below.

In the good cavity limit, the wave equation governing propagation through the nonlinear medium may be written

$$\frac{\partial E(x,y,z)}{\partial z} = -\frac{\alpha_0}{2} \left[\frac{1 + i\,\Delta}{1 + \Delta^2 + |E|^2} - \frac{i\,\ell n2}{4\pi\alpha_0 LF}\,\nabla_t^2 \right] E(x,y,z) \qquad (1)$$

and further, the transverse E field satisfies the following boundary conditions,

$$E(x,y,o,t) = \sqrt{T}\,E_{in}\,(x,y,o,t) + Re^{ik\mathscr{L}}\,E(x,y,L,t-\ell/c). \qquad (2)$$

The first term on the right hand side of Eq. (1) is the saturable term which gives rise to self-focusing or self-defocusing while the last term is responsible for diffraction [$\nabla_t^2 = w_0^2(\partial^2/\partial x^2)$; we assume a single transverse dimension]; $\Delta = (\omega - \omega_{ab})/\gamma_\perp$ is the normalized laser atom detuning, α_0 the linear absorption coefficient per unit length, $F = n_0 w^2_{1/2}/\lambda L$ is the Fresnel number (n_0 the background refractive index, $w_{1/2}$ is the beam waist at half-maximum, λ the wavelength and L the nonlinear medium length; $\mathscr{L} = L + \ell$ is the total cavity length.

At high Fresnel number in conventional propagation problems it is customary to drop the Laplacian term and adopt the so called uniform plane wave approximation. Optical bistable switching is unique however in that the singular edge discussed above causes the Laplacian to be important locally and when operating under self-focusing conditions a train of solitary waves begins to evolve from both outer edges towards the center of the beam. The train of solitary waves can be stationary or can undergo complicated periodic oscillations in the transverse dimension depending on the total area of the switched-on beam.

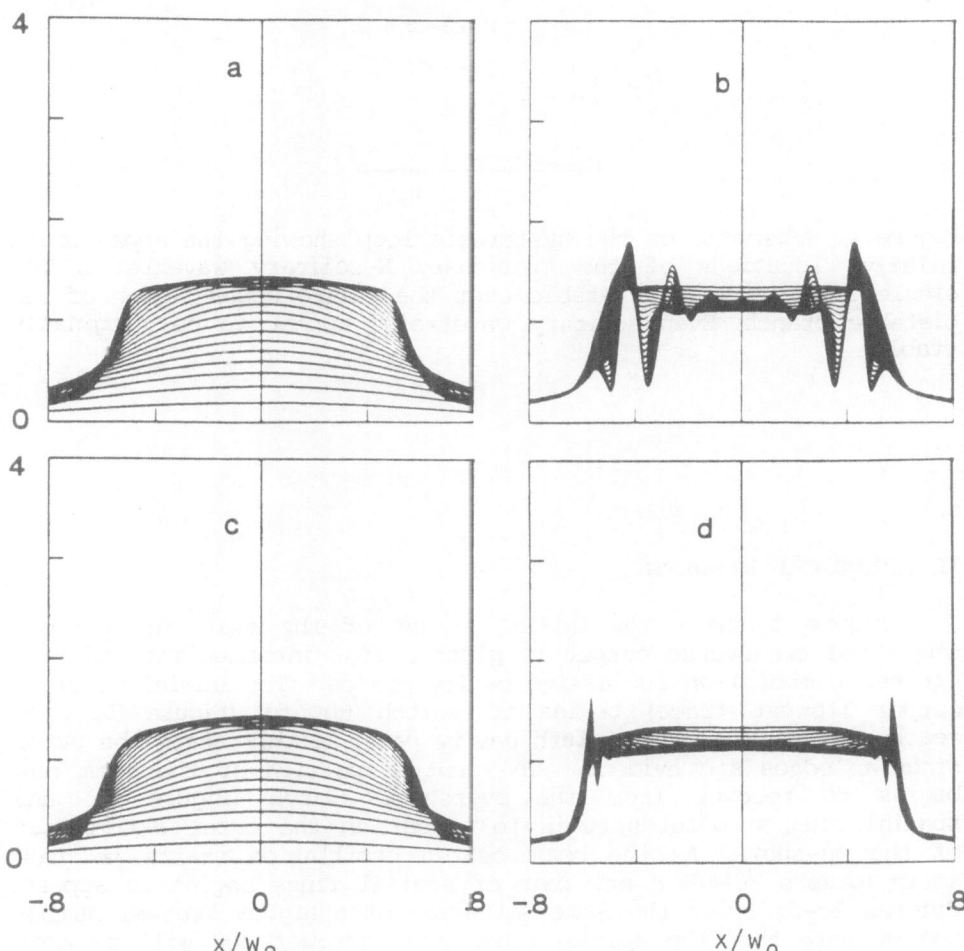

Figure 1. Transient buildup of the intracavity transverse profile on successive passes for Fresnel number F=100 and F=1000 (p=2, k𝒮=.4, a(0)=.19. The intracavity field reaches a maximum intensity on the 16th pass (Figure 1a, 1c) and begins to recover from the overshoot switch (Figure 1b, 1d). On recovery solitary waves begin to develop on the outer edges of the on-spot; for F=100 additional pairs of solitary waves have already appeared as the beam center goes into its second oscillation.

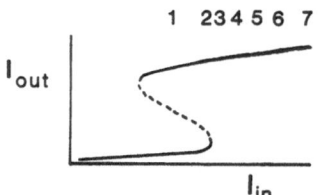

Figure 2. Schematic of the hysteresis loop showing the approximate relative locations of the stationary N-solitary wavetrains. The single solitary wave is stable over the entire upper region of the bistable branch. Even solitary wavetrains appear to be marginally stable.

II. NUMERICAL RESULTS

Figure 1 shows the initial stages of the switching process where the transverse output is plotted (for intermediate and high Fresnel numbers) on successive cavity passes. The initial Gaussian output (lowest trace) begins to switch upward (Figure 1a) and reaches a maximum on the 16th cavity pass; at this stage the outer singular edges are evident. The central portion of the beam now begins to recover from the overshoot switch (Figure 1b) and spatial ring structures begin to appear on the outer extremities of the on-spot. As the beam center oscillation starts to grow again (passes 30-40) a new pair of spatial rings begins to appear. Figures 1c-d depict the same situation at a higher Fresnel number and we note that the spatial rings are narrower and will be more numerous. Figure 2 is a schematic showing the relative location of stable N-solitary wavetrains on the high transmission branch. Between the Nth and $(N+1)^{st}$ stationary solitary wavetrains we observe complicated periodic oscillations with periods varying from tens to hundreds of cavity passes.

III. THEORY

We now briefly sketch the central points of an independent theory which establishes that the spatial rings observed numerically, are, in fact, solitary waves.   More significantly, however, the theory establishes that the stationary solitary wavetrains are fixed points of an infinite dimensional map.[2]

For convenience, we transform equation (1) to nondimensional form through the following parameter rescalings,

$$p = \frac{\alpha_0 L}{\Delta} \ , \ f = \frac{4\pi F}{\ell n 2} \ p, \quad \zeta = \frac{\alpha_0}{\Delta} z \ , \ y = \sqrt{f} \ x,$$

$$a(y) = \sqrt{\frac{T}{2\Delta^2}} \ E_{in}(x) \ G(y,\zeta) = \frac{1}{\sqrt{2\Delta}} \ E(y/\sqrt{f} , L, \zeta/p). \tag{3}$$

Assuming $\Delta \gg 1$ we drop the absorption term and Eq.(1) becomes

$$2i \ \frac{\partial}{\partial \zeta} \ G_n + \frac{\partial^2}{\partial y^2} \ G_n - \frac{G_n}{1+2|G_n|^2} = 0 \tag{4}$$

and the boundary conditions transform to

$$G_n (y,o) = a(y) + Re^{ik\mathscr{L}} \ G_{n-1}(y,p), \ n > 1 \quad G_0 = 0. \tag{5}$$

The label n refers to the discrete time index of the infinite dimensional map (5).   Eq.(5) acts as the initial data for Eq.(4) on each cavity pass.

Had we assumed a Kerr nonlinearity Eq.(4) would be identical to the nonlinear Schrodinger equation, an exactly integrable system.   We could then decompose $G_n(y,o)$ into its soliton and continuous spectrum basis and using (5) find the soliton and continuous spectrum parameters of the field envelope on the nth pass in terms of those on the $(n-1)^{st}$ pass.   Eq.(4) with the saturable nonlinearity is not integrable but nevertheless solitary wave solutions can be sought.

Before proceeding further we first physically motivate the theoretical results which follow. Assume that after (n-1) cavity passes, a single solitary wave exits from the nonlinear medium (for example the single solitary wave on the upper branch as indicated in Figure 2). This solitary wave will have the following form

$$G_{n-1}(y,\zeta) = P(\lambda y, \lambda) \ \exp(i(\lambda^2 - 1) \ \zeta/2 + i\gamma) \tag{6}$$

where the shape $P(\theta,\lambda)$ satisfies $(\theta = \lambda y)$

$$P_{\theta\theta} - P + \frac{1}{\lambda^2} \frac{2P^3}{1 + 2P^2} = 0. \tag{7}$$

[For a Kerr nonlinearity $P = \lambda$ sech θ]. We now ask what happens to the solitary wave on the next circuit around the cavity. We assume that there is no free space diffraction (i.e. a filled cavity). The solitary wave suffers a loss on reflection at both mirrors and is added to a broad Gaussian beam of much smaller amplitude (the intracavity field is much larger than the input). For the following theory to work the change in the solitary wave parameters $\delta\lambda_{n-1}$, $\delta\gamma_{n-1}$ on each pass cannot be too great. At the start of the nonlinear medium on the nth pass we have the following input (see Eq.(5))

$$a(\theta/\lambda_{n-1}) + RP(\theta, \lambda_{n-1})e^{i(\lambda^2_{n-1}-1)p/2 + i\gamma_{n-1}} \tag{8}$$

The initial propagation through the nonlinear medium will involve reshaping into a solitary wave which then propagates further without reshaping. At the end of the nonlinear medium we have completed the nth pass and we assume an output

$$P(\theta, \lambda_n)e^{i(\lambda^2_n-1)p/2 + i\gamma_n} \tag{9}$$

Projecting this solitary wave out of Eq.(4) with initial data given by Eq.(8) yields a two dimensional map in the solitary wave parameters i.e. $(\lambda_{n-1}, \gamma_{n-1}) \rightarrow (\lambda_n, \gamma_n)$. Details of the derivation of this map will be published elsewhere.[3] Here we simply write down the fixed point equation for the map (set γ_n, γ_{n-1} to γ and λ_n, λ_{n-1} to λ and solve).

$$\sin\gamma < \frac{1}{\lambda} \theta P_\theta + P_\lambda, a(\theta/\lambda)> = R\sin\left[k\mathcal{L} + \frac{P}{2} (\lambda^2-1)\right] < \frac{1}{\lambda} \theta P_\theta + P_\lambda, P>$$

$$\cos\gamma<P, a(\theta/\lambda)> = \{1 - R\cos\left[k\mathcal{L} + \frac{P}{2} (\lambda^2-1)\right]\} <P, P> \tag{10}$$

where $<P, Q> = \int_{-\infty}^{\infty} P(\theta)Q(\theta)d\theta$.

Equation (10) which needs to be solved iteratively provides the solitary wave shape to be compared with the numerical result of the first section. We find quantitative agreement with the single solitary wave (fit is ~ 1%) and a discrepancy of ~ 5% in fit when we compare with the central peak of the 7 solitary wavetrain shown in Figure 3a. This discrepancy is not surprising in the latter case as the central peak interacts with its neighbors, a fact not presently accounted for in the theory.

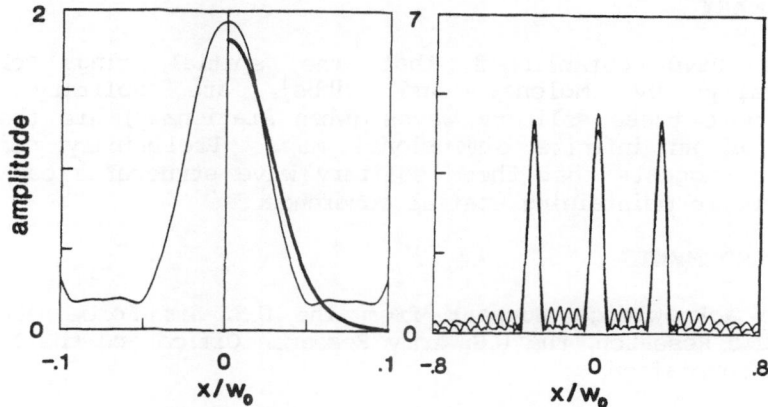

Figure 3. (a) Comparison of the solitary wave of the fixed point equation (heavy line) (eqn(10)) with the central peak of the 7-solitary wavetrain (stationary after 200 passes). The central peak interacts with its neighbours whose tails are just visible at the edges of the figure; this interaction accounts for the ~5% discrepancy in fit[2]. (b) A period 2 oscillation involving a 3-solitary wavetrain (p=8,F=25,k$\mathscr{L}$ =.4,a(0)=.1). A single wavetrain, for example, consists of the highest central peak together with its two lowest adjacent neighbors.

Can these structures period double or follow some other route to chaos?

This is an intriguing question to which we can at present provide only a partial answer. We have observed numerically a bifurcation from a stationary 3-solitary wave solution to a stable 2-cycle (see Figure 3b). There is also evidence for a further bifurcation to a 4-cycle but this appears to be very slowly modulated. The whole question of bifurcation of spatial structures has been largely ignored in the literature[4] primarily because the partial differential equations describing the underlying physical phenomena are too difficult to solve. It is well known in fluid turbulence for example that spatially coherent temporally chaotic structures exist.

IV. SUMMARY

We have established that the spatial rings observed numerically by Moloney and Gibbs[1], are solitary waves. Furthermore these solitary waves (when stationary) are the fixed points of an infinite dimensional map. Preliminary numerical evidence suggests that these solitary wave structures can period double while maintaining spatial coherence.

ACKNOWLEDGMENTS

We acknowledge support from the U.S. Air Force Office of Scientific Research, the U.S. Army Research Office and the National Science Foundation.

REFERENCES

1.   J.V. Moloney and H.M. Gibbs, Phys. Rev. Lett. 48 1607 (1982).
2.   D.W. McLaughlin, J.V. Moloney and A.C. Newell, Phys. Rev. Lett. 51 75 (1983).
3.   D.W. McLaughlin, J.V. Moloney and A.C. Newell, to be published.
4.   Exceptions are the recent articles on the driven Sine Gordon equation. A.R. Bishop, K. Fesser, P.S. Lomdahl, W.C. Kerr, M.B. Williams and S.E. Trullinger, Phys. Rev. Lett. 50, 1095 (1983); J.C. Eilbeck, P.S. Lomdahl and A.C. Newell, Phys. Lett. 87A, 1, (1981).

SELF-DEFOCUSING AND OPTICAL CROSSTALK

IN A BISTABLE OPTICAL ETALON

K. Tai, H. M. Gibbs, J. V. Moloney,
D. A. Weinberger, S. S. Tarng, and J. L. Jewell

Optical Sciences Center
University of Arizona
Tucson, Arizona 85721

A. C. Gossard and W. Wiegmann

Bell Laboratories
Murray Hill, NJ 07974

INTRODUCTION

Two transverse phenomena associated with a bistable optical etalon are discussed. The first is self-defocusing[1] of the laser beam due to the nonlinear dispersion of the medium inside the bistable Fabry-Perot etalon. Good agreement is found between GaAs data and one-dimensional numerical simulations. The second is crosstalk[2] between nearby bistable regions in the same etalon due to diffraction coupling, which is studied numerically in both one and two transverse dimensions.

SELF-DEFOCUSING

In a GaAs-GaAlAs superlattice room-temperature optical bistability[3] experiment, the intensity profile of the output beam is observed as shown in Fig. 1. A microscope objective used as an imaging lens is placed between the etalon and an aperture. Drastic changes in the profile are observed as the imaging objective is translated along the propagation direction of the laser beam. Since the image plane (i.e., aperture) stays fixed, this corresponds to selecting different object planes. Moving the objective closer to the etalon probes the virtual object planes, as suggested by the dashed beams at the bottom of Fig. 1. The experimental profile is

415

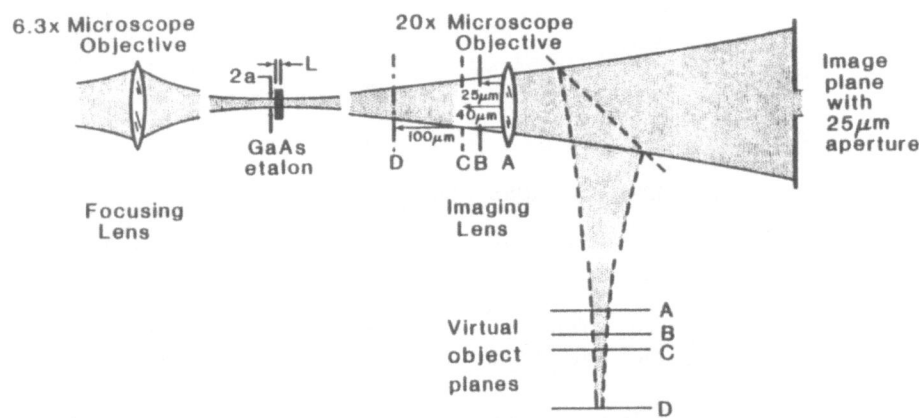

Fig. 1. Schematic setup of transverse study in bistability. L = 4.90 μm, 2a = 8 μm (FWHM), Fresnel No. F = $n_0 a^2/\lambda L$ = 12, n_0 = 3.5, λ = 0.88 μm.

mapped out by oscillating a mirror back and forth at a rate slow compared with the 40-kHz rate of the input pulse. This sweeps the output across an aperture to obtain the transverse beam profiles as in Fig. 2 for various object planes along the beam propagation direction.

Enhanced overshoot (Fig. 3b), downward-slope (Fig. 3b), and enhanced upward-slope (Fig. 3c) in the upper branch of the hysteresis loop are also seen if the power in the center of the output beam is detected instead of the total beam power. From Fig. 2, a virtual focus occurs about 100 μm before the front surface of the etalon for an input greater than the switch-on intensity (Fig. 2e), and about 75 μm for an input slightly greater than the switch-off intensity (Fig. 2d'), and the profiles are modified from Gaussian-like at various planes. A virtual focus occurs because the nonlinear dispersion in GaAs-GaAlAs defocuses the beam (the laser wavelength is on the long wavelength side of the exciton resonance). Due to the competition between the defocusing and diffraction (when the virtual focus is spatially small in diameter such that the diffraction is essential and cannot be neglected), the virtual focus occurs farther before the etalon for larger intensity than for smaller intensity input. Thus the virtual focus can be moved

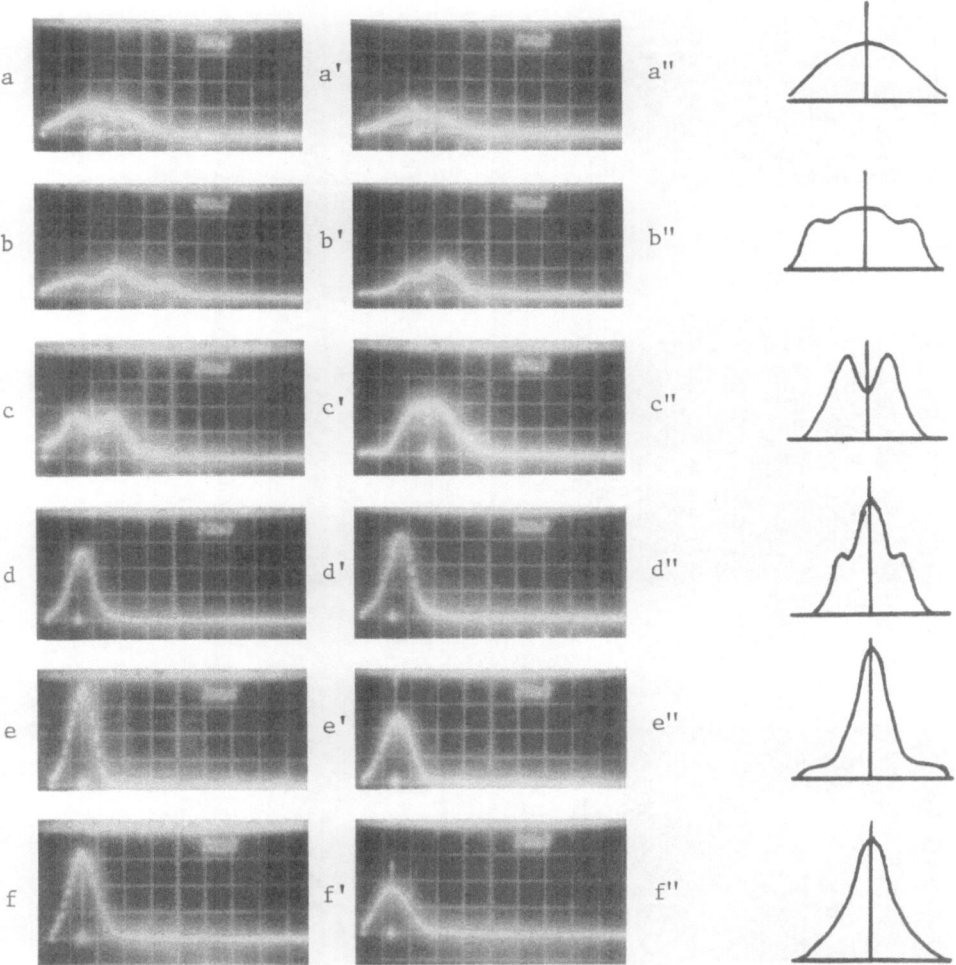

Fig. 2. Transverse intensity profiles observed in GaAs superlattice (a–f) and (a'–f'), and numerically simulated (a"–f"). (a–f) and (a"–f") are for the input greater than the switch-on intensity. (a'–f') are for the input slightly greater than the switch-off intensity. (a,a',a") are for the etalon imaged on the aperture. (b,b',b") to (f,f',f") are for virtual object planes 25, 50, ..., 125 μm before the etalon.

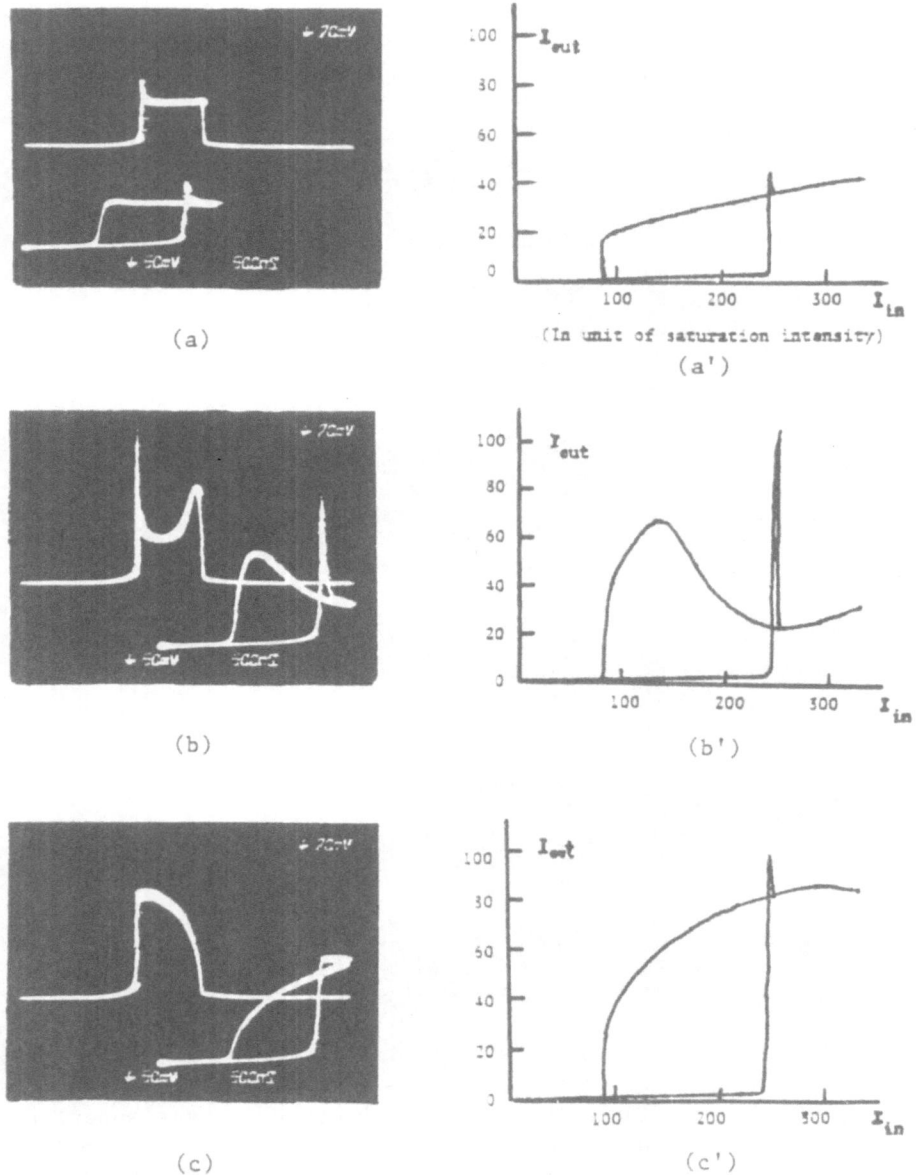

Fig. 3. Hysteresis loops (only the center of the beam is detected) observed in GaAs superlattice (bottom traces in a to c) and numerically simulated (a' to c'). (a and a') are for the etalon imaged on the aperture. (b and b') and (c and c') are for virtual object planes 50 and 100 μm before the etalon, respectively. Top traces in a to c are real-time output for a triangular input.

through a given plane by the triangle time dependence of the input used to map out the hysteresis loop. At a certain plane about 50 to 75 µm before the etalon (Fig. 2c, c', d, d') the beam is virtually focused for a smaller intensity input while it is not for a larger input. By monitoring the output beam only at the center, one can get a downward-slope (Fig. 3b) in the upper branch of the hysteresis loop as the imaging objective images the virtual object plane into the aperture. The same arguments apply for enhanced upward-slope in Fig. 3c to image the virtual object plane about 100 µm before the etalon.

The enhanced overshoot has a similar explanation. As the field switches[4] from the lower state to the upper state, not only the magnitude of the intracavity field increases, but also the exit-surface phase (with respect to the input field) changes dramatically in order to interfere constructively inside the etalon; i.e., the round-trip phase change will become close to an integer product of 2π. During the switch-on, the virtual focus moves from the etalon to the steady-state position. The dynamics of the wavefront changes cause the movement of the virtual focus and an enhanced overshoot when imaging a plane about 50 to 75 µm before the etalon.

Computer simulations have been performed and compared with the experiment. We take the case of a Gaussian input and a two-level-atom medium within a detuned ring cavity of total length 2L. Assume the good-cavity limit (T_1, $T_2 \ll T_R$ cavity round-trip time), which allows one to eliminate the medium response adiabatically. Then the Maxwell-Bloch equations can be solved for this quasi-equilibrium, reducing the equation to

$$\frac{\partial E}{\partial (z/2L)} = -\frac{\alpha_0}{2}(2L)\left[\frac{1+i\,\Delta}{1+\Delta^2+|E|^2} - \frac{i\,\ln 2\,\nabla_t^2}{4\pi\alpha_0 LF}\right]E \qquad (1)$$

where α_0 is the peak on-resonance absorption coefficient of the two-level transition, Δ is the detuning of the laser from the resonance in the unit of the resonance linewidth, and ∇_t^2 is the transverse Laplacian operator. E is the complex slowly varying total field quantity normalized to the saturation field amplitude. $F = n_0 a^2/\lambda L$ is the Fresnel number, n_0 is the background index of refraction, and 2a is the full width at half maximum (FWHM). The first term on the right-hand side of Eq. (1) describes the effect of the nonlinearity of the medium. The second term describes the diffraction effect. The fast Fourier transform (FFT) technique[5] is used to solve this equation containing a transverse Laplacian. The transmitted field profile is first calculated at the exit side of the ring cavity. Then the free-space propagation in the backward direction of this output field is computed corresponding to the movement of the imaging objective. The same equation is used to free-space propagate the field except now we set α_0 equal to zero.

The results are shown in Figs. 2 and 3 in parallel with the experimental data. Good agreement is found, although simulations are taken by using only one transverse dimension and under a good-cavity limit in order to simplify the computation, which is not the case for the experiment. The cavity round-trip time of the 5-μm-thick etalon is much less than the medium response time. However, the width of the laser pulses used in the experiment is much longer than the other two time constants (i.e., medium response time and cavity round-trip time). The transients can be neglected for this long pulse situation. Steady-state solutions for good- and bad-cavity limits will not differ significantly.

CROSSTALK

The parallel operation of many light beams causing optical bistability in the same etalon is crucial to the application of optical bistability to optical data processing. We consider now the crosstalk between the adjacent bistable regions by means of diffraction coupling and investigate how close the separation can be to ensure independent operation. To simplify the problem, we take the case of two one-dimensional Gaussian (spatially) beams incident side by side on a detuned ring cavity. The two input Gaussian beams are separated by x_0 from center to center:

$$E_{I_1}(x) = E_{I_1}(0) \exp[(-x-x_0/2)^2/w_0^2] \, e^{i\phi_1},$$

$$E_{I_2}(x) = E_{I_2}(0) \exp[(-x+x_0/2)^2/w_0^2] \, e^{i\phi_2};$$

$\Delta\phi = \phi_1 - \phi_2$ is the phase difference between the two beams. The situation is similar to that in the previous section except that now there are two beams. The FFT technique is still used to solve the problem. In the simulation we chose the Fresnel number to equal 0.55, since the Fresnel number of order unity is optimum for practical application, i.e., whole-beam switching, high transmission, and fast response.

Figure 4 gives the results of simulations for the case of two input beams with the same holding intensity ($E_I^2(0) = 200$) above the switch-off (100) and below the switch-on (230) intensities and with relative phase $\Delta\phi = 0$. Both devices are in the lower branch initially. The right one is switched to the upper branch by increasing the input intensity to 1.5 times the holding intensity until equilibrium is reached and then returning to the holding value. For $x_0 = 2.0$ (Fig. 4a) and $x_0 = 2.12$ FWHM (Fig. 4b), the diffraction coupling from the right beam makes its left neighbor switch up too, but for $x_0 = 2.35$ FWHM (Fig. 4c), the left beam stays in the lower branch. For further separation, $x_0 = 2.94$ FWHM (Fig.

4d), both intensity profiles are almost the same as for the single-beam case even in the presence of diffraction coupling. For the case of $\Delta\phi = \pi/2$ and π, a larger separation is needed to ensure independent operation. The reason is that there exists a larger phase difference[*] between the lower and upper states of optical bistability in addition to the difference in the intensity. This phase difference makes the $\Delta\phi = \pi/2$ and π cases need larger separation to prevent the constructive interference through the overlapping of the tails by means of the diffraction effect than $\Delta\phi = 0$ case, which is shown in Fig. 4.

Simulations for a <u>full two-transverse-dimension case</u>, in which one device is surrounded by four additional devices, show qualitatively similar results, i.e., for $\Delta\phi = 0$, a small separation can make devices crosstalk by means of diffraction coupling and a large separation can make devices switch independently. A separation of a few beam widths is sufficient for independent operation. The results give promise for performing parallel operation, in spite of transverse diffractive coupling. Transverse effects that are neglected in the calculation, such as exciton and free-carrier diffusion (GaAs bistability) and thermal diffusion (dye bistability[6]), could make larger separations necessary for independent operation.

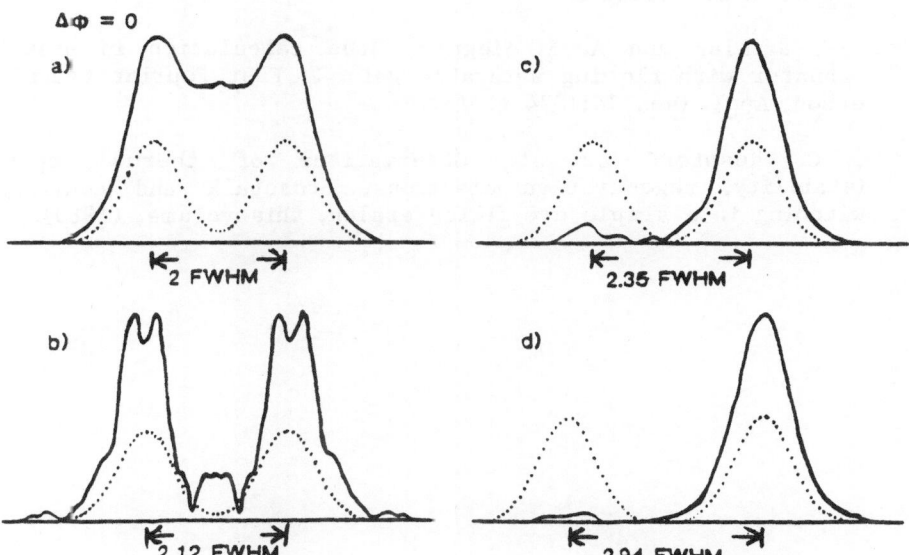

Fig. 4. Effect of diffraction coupling for the case of $\Delta\phi = 0$, and $x_0 = 2$, 2.12, 2.35, and 2.94, FWHM, respectively. In (a) and (b) the inputs (dotted lines) are too close together; the outputs are not independent. While in (c) and (d), the inputs are well separated, and the outputs are independent of each other. From Ref. 2.

ACKNOWLEDGMENT

    We acknowledge support by NSF (ECS-8020303) and the AFOSR and ARO (F49620-80-C-0022) for the Arizona portion of this research.

REFERENCES

1. H. M. Gibbs et al., Room-temperature optical bistability and self-defocusing in semiconductor etalons. International Quantum Electronics Conference '82, postdeadline paper, 1982.

2. K. Tai, J. V. Moloney, and H. M. Gibbs, Optical crosstalk between nearby optical bistable devices on the same etalon., Opt. Lett. 7:429 (1982).

3. H. M. Gibbs, S. S. Tarng, J. L. Jewell, D. A. Weinberger, K. Tai, A. C. Gossard, A. Passner, and W. Wiegmann, Room temperature excitonic optical bistability in GaAs-GaAlAs superlattice etalon, Appl. Phys. Lett. 41:221 (1982).

4. K. Tai, H. M. Gibbs, and J. V. Moloney, Intracavity phase switching and phase-plane dynamics of a bistable optical device, Opt. Commun. 43:297 (1982).

5. E. A. Sziklas and A. E. Siegman, Mode calculation in unstable resonator with flowing saturable gain. 2: Fast Fourier transform method, Appl. Opt. 14:1874 (1975).

6. M. C. Rushford et al., Observations of thermal optical bistability, regenerative pulsations, crosstalk and controlled switching in a simple dye filled etalon, this volume, (1983).

OPTICAL BISTABILITY BASED ON SELF-FOCUSING

P. W. Smith and D. J. Eilenberger

Bell Laboratories
Holmdel, N.J.   07733

In this paper we describe an experimental study of self-focusing bistable devices using a liquid suspension of dielectric particles as the nonlinear medium. We observe regions of multiple optical hysteresis in good agreement with theoretical predictions.

Two years ago, the first proposal and demonstration of self-focusing optical bistability was published.[1] In this paper it was shown that a self-focused beam in a nonlinear medium would exhibit hysteresis if the beam were reflected back on itself. This effect is illustrated in Fig. 1. A low-power incident beam diffracts so that only a small fraction of the incident power arrives at the detector. This corresponds to the low-power transmission characteristics of the P_{out} vs P_{in} plot. For an input power greater than the critical power for the onset of self-focusing, however, most of the incident light will arrive at the detector. (We ignore for the moment catastrophic self-focusing and assume that saturation of the nonlinear effectively produces a "self-trapped" or waveguided beam. We will return to this point later). As the retro reflector will return this light along the self-guided path, the power at the entrance face is now greater than that required to initiate self-focusing, and the input power can be reduced while still maintaining the system in the self-trapping high-transmission mode. Thus, optical hysteresis and bistability can be achieved.

Self-focusing optical bistability was first demonstrated using a dye laser beam and Na vapor as the nonlinear medium.[1] The optical hysteresis reported in Ref. 1 is in good qualitative agreement with a recently-published analysis of this system.[2] This analysis, based on a model which takes account of the satura-

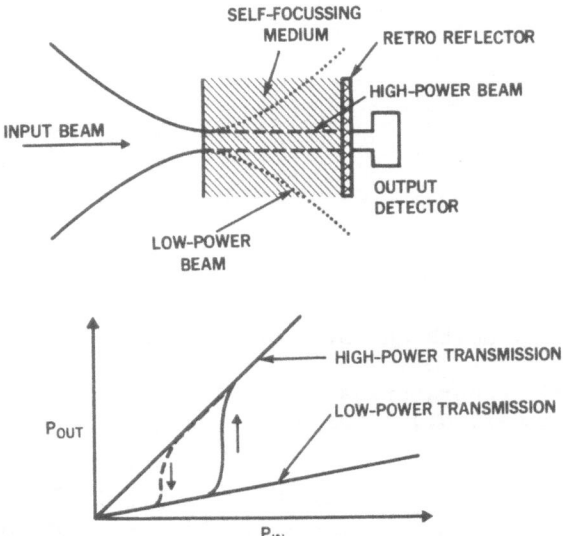

Figure 1. Bistable self-focusing configuration and idealized response characteristic

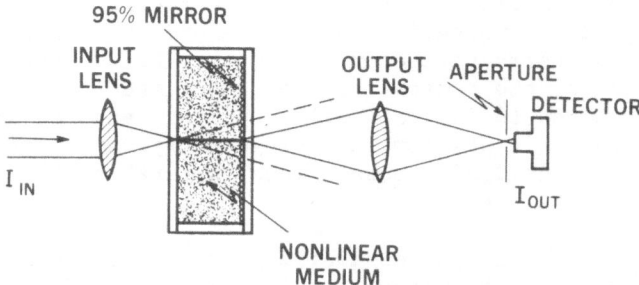

Figure 2. Experimental set-up

tion of the nonlinearity as the intensity is increased, predicts that the spot size of the self-focused beam will undergo oscillations as the self-focused beam propagates through the nonlinear medium. In this paper we report the experimental observation of optical multistability due to these spot size oscillations.

Experimental observations were made on a nonlinear system using a liquid suspension of dielectric particles as the nonlinear medium.[3] The combination of large nonlinear coefficient (calculable from first principles) and slow response time makes this medium attractive for CW and transient studies of complex nonlinear effects.[4] The experimental set-up is shown in Fig. 2. The nonlinear medium was contained in a cell with an antireflection-coated front face, and with the interior surface of the exit window coated for 95 percent reflection. The light source was the 4800 Å output of an Ar^+ laser. The cell length was 1/2 mm and the input beam was focused to a 1/2 intensity diameter of 3.5 μm.

Figure 3 shows the experimental output as input characteristic for conditions corresponding to a value of the saturation parameter, β, of 0.21. This data was obtained with a suspension of quartz particles in the organic liquid o-dichlorobenzene. For comparison, Fig. 4 shows the theoretically-calculated characteristic for the case β = 0.21. An effective reflectivity of R = 0.9 was used and the value of the effective aperture size[2] $\left(a^2/w_0^2 = 0.9\right)$ was chosen to give the best fit to the experimental curve. The intensity scan was not slow enough to obtain abrupt switching, but slower scan rates produced fluctuations due to thermally-produced turbulence in the nonlinear liquid.

In Fig. 5 we show the experimental output vs input characteristic for conditions corresponding to β = 0.11. This data was obtained with an aqueous suspension of quartz particles. Two hysteresis loops are clearly observed. For comparison, Fig. 6 shows the theoretically-calculated characteristic for the case β = 0.10. We have again used R = 0.9 and have selected the effective aperture size $\left(a^2/w_0^2 = 0.25\right)$ to give the best fit to the data. It is clear that there is good qualitative agreement between theory and experiment.

Other types of behavior were also observed. Figure 7 shows a case sometimes obtained with the aqueous suspension where the upper branch (return) of the hysteresis curve depended on the peak input intensity. When the peak input intensity was just sufficient for the initial switching to occur and was subsequently decreased, the characteristic shown in Fig. 7(a) was obtained. If the input was initially increased further to obtain a second switching, a different return path was obtained as shown in Fig. 7(b). This behavior is not understandable on the basis of

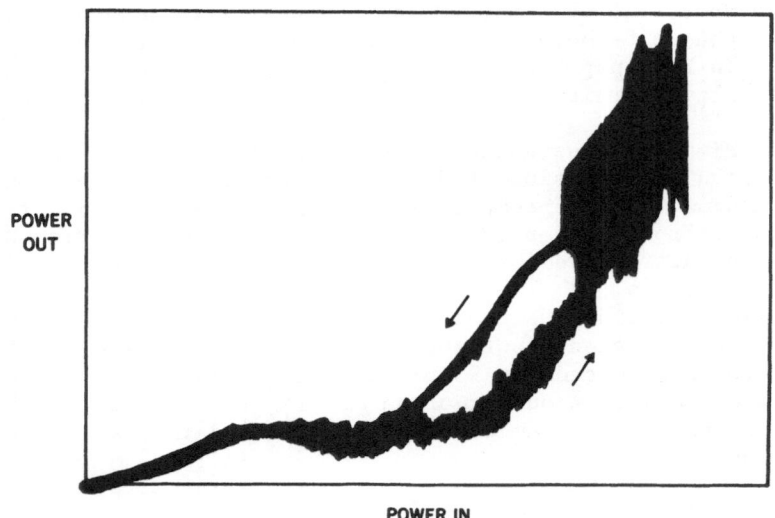

Figure 3. Experimental output vs input: o-dichlorobenzene
suspension, β = 0.21

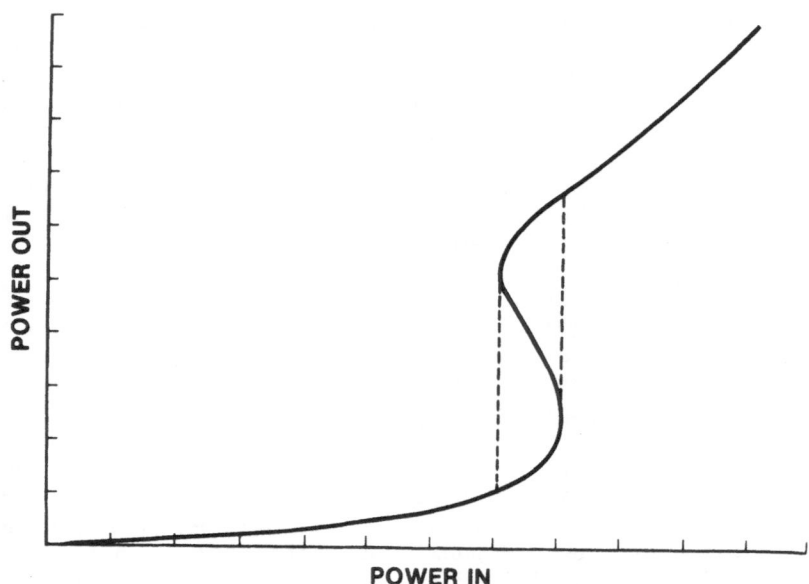

Figure 4. Theoretical output vs input: β = 0.21, R = 0.9,
a^2/w_o^2 = 0.9

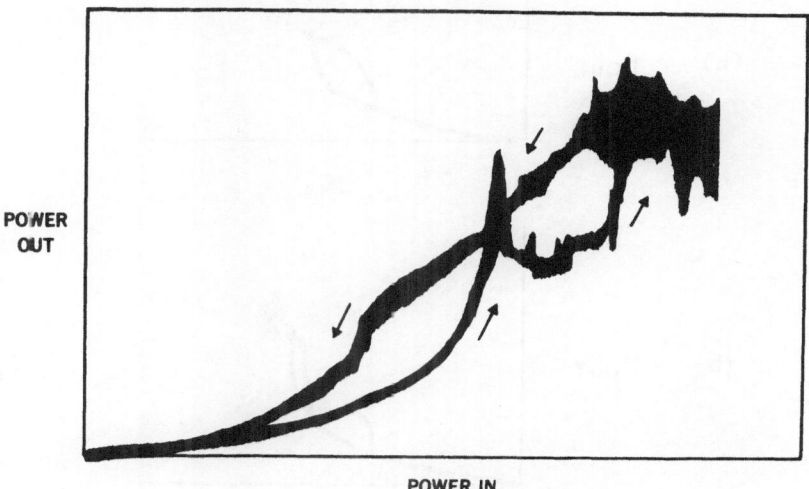

Figure 5. Experimental output vs input: aqueous suspension, $\beta = 0.11$.

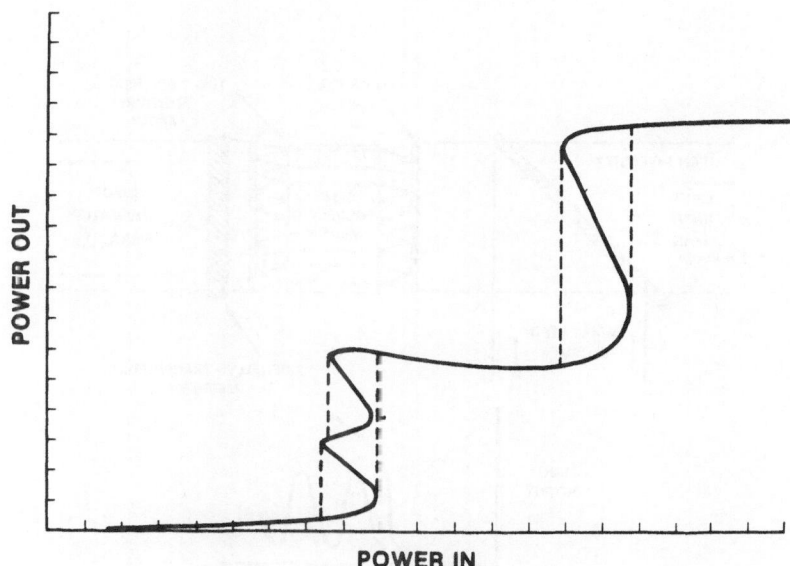

Figure 6. Theoretical output vs input: $\beta = 0.10$, R = 0.9, $a^2/w_o^2 = 0.25$.

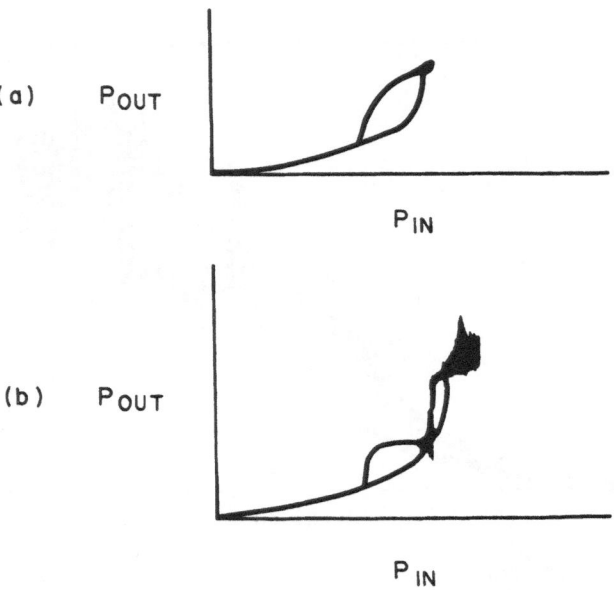

Figure 7. Experimental output vs input for aqueous suspension: return path depends on maximum input power.

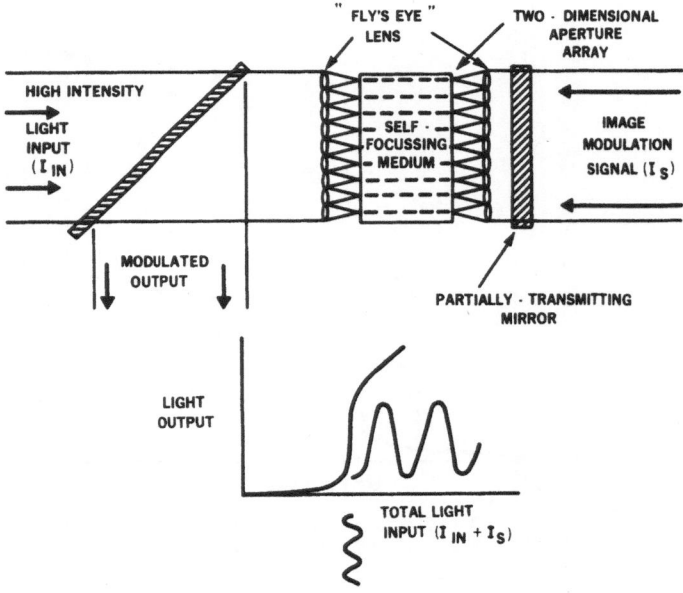

Figure 8. Proposal for an image amplifier based on self-focusing bistability (after Ref. 5).

our simple model. The oscillations and instabilities observed on portions of the characteristic curves in Figs. 3, 5 and 7 are also new effects not explained on the basis of our model, and in view of the current interest in instabilities of these systems, they deserve further study.

These studies with model systems using dielectric suspensions as the nonlinear medium are useful for developing an understanding of their complex nonlinear behavior. Devices employing a fast-responding nonlinearity should prove attractive for subpicosecond switching elements. As an example we show in Fig. 8 the use of an array of self-focusing bistable devices as an optical image amplifier.

REFERENCES

1. J. E. Bjorkholm, P. W. Smith, W. J. Tomlinson, and A. E. Kaplan, "Optical Bistability Based on Self-focusing", Opt. Lett. $\underline{6}$, 345, (1981).

2. J. E. Bjorkholm, P. W. Smith, and W. J. Tomlinson, "Optical Bistability Based on Self-focusing: An Approximate Analysis", IEEE J. Quantum Electron QE-18, 2016, (1982).

3. See, for example, A. Ashkin, J. M. Dziedzic and P. W. Smith, "Continuous-wave Self-focusing and Self-trapping of Light in Artifical Kerr Media", Opt. Lett. $\underline{7}$, 276, (1982).

4. For a description of the nonlinear characteristics of a liquid suspension of dielectric particle see P. W. Smith and W. J. Tomlinson, "Nonlinear Optical Interfaces" in this volume.

5. P. W. Smith, "On the Physical Limits of Digital Optical Switching and Logic Elements", Bell System Tech. J. $\underline{61}$, 1975 (1982).

THEORY OF OPTICAL BISTABILITY IN

COLLINEAR DEGENERATE FOUR-WAVE MIXING

Richard Lytel

Lockheed Palo Alto Research Laboratory
3251 Hanover Street
Palo Alto, CA 94304

INTRODUCTION

The objective of this paper is to investigate optical multi-stability in collinear degenerate four-wave mixing. The theory of four-wave mixing in a lossless, isotropic Kerr medium is described and solved numerically. Multiple solutions of the two-point boundary value problem are discovered when the input fields exceed a critical intensity determined by the electric susceptibility and the interaction length of the Kerr medium.

The model described below is an extension of previous work by Marburger and Lam[1], and Winful and Marburger[2]. The field equations governing the spatial evolution of the pumps, signal, and probe envelopes include several phase-matched contributions omitted in earlier studies[1,2]. The extra terms are unique to the collinear geometry and are quadratic in the pump fields. Their effect is to cause distortion of the phase conjugate of the probe[3], provide additional coupling between the relative phases of the forward and backward traveling waves and eliminate one of the conserved currents from the earlier theory[1].

The equations of motion for the electric fields in the medium are derived directly from the Maxwell equations. The Lagrangian formalism of reference 1 is extended to identify two conserved currents and construct a conserved Hamiltonian. This approach leads to a closed set of equations for the probe flux, signal flux and relative forward and backward phases. This system of equations is not equivalent to the naive theory[1,2] and cannot be solved analytically. This system is solved numerically using an initial value method to identify multiple solutions of the two-point boundary value problem. Stability analysis and evidence of hysteresis will be presented separately in a future publication.

THEORY

The geometry of collinear degenerate four-wave mixing is illustrated in figure 1. The input fields $\vec{E}_1$ (0) and $\vec{E}_2$ (L) are incident from the left and right, respectively, and have arbitrary polarization. The pump waves are taken as the $\hat{x}$ components of $\vec{E}_1$ and $\vec{E}_2$, measured at the boundaries. The probe and signal are the corresponding $\hat{y}$ components. The interaction in the medium may be thought of as two-wave mixing[4] since the two vector fields $\vec{E}_1$ and $\vec{E}_2$ may rotate and change magnitude in the medium. However, the evolution of each component is easily treated as arising from a four-wave process.

The total electric field in the medium at time t is

$$\vec{\mathcal{E}}(z,t) = \vec{E}e^{-i\omega t} + c.c. \tag{1}$$

where $\vec{E} = \vec{E}_1 \ e^{ikz} + \vec{E}_2 \ e^{-ikz} + c.c.$ is the spatial part of $\vec{\mathcal{E}}$ and $k = 2\pi/\lambda$. This field must satisfy the wave equation

$$(\nabla^2 - \mu_o \varepsilon \frac{\partial^2}{\partial t^2}) \vec{\mathcal{E}} = \mu_o \frac{\partial^2 \vec{\mathcal{P}}}{\partial t^2} \tag{2}$$

where $\vec{\mathcal{P}} = \varepsilon \vec{P} \ e^{-i\omega t} + c.c.$ and $\vec{P}$ is given by

$$\vec{P} = \chi_1 \vec{E} \cdot \vec{E}^* \vec{E} + 1/2 \ \chi_2 \vec{E} \cdot \vec{E} \ \vec{E}^* \tag{3}$$

In equation (2), μ_o is the permeability and ε is the permittivity of the medium.

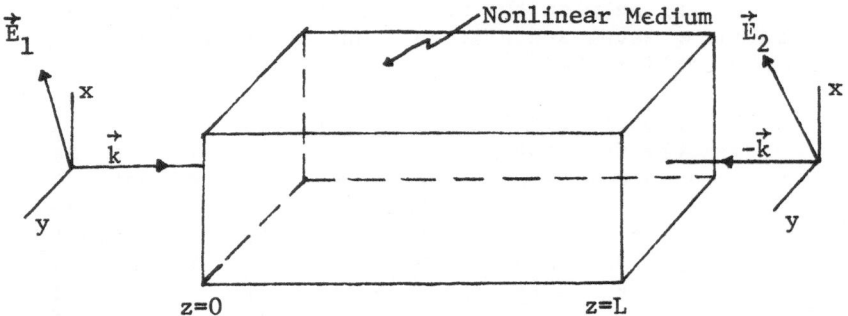

Fig. 1. Geometry for collinear degenerate four-wave mixing.

Combine equations (1), (2), and (3), and perform a short time average. The result is an equation for $\vec{E}$:

$$(\nabla^2 + \mu_o \varepsilon \omega^2) \vec{E} = -\mu_o \varepsilon \omega^2 \ [\chi_1 \vec{E} \cdot \vec{E}^* \vec{E} + 1/2 \ \chi_2 \vec{E} \cdot \vec{E} \ \vec{E}^*] \tag{4}$$

The spatial evolution equations for the envelopes $\vec{E}_1$ and $\vec{E}_2$ are found by substituting for $\vec{E}$ in equation (4), averaging over a few wavelengths and dropping the slowly varying second derivatives. The result is

$$-i \ \frac{d\vec{E}_1}{dz} = \alpha \ \{[\,|\vec{E}_1|^2 + |\vec{E}_2|^2] \ \vec{E}_1 + \vec{E}_1 \cdot \vec{E}_2^* \ \vec{E}_2 \,\}$$

$$+ \ \beta \ \{\vec{E}_1 \cdot \vec{E}_1 \ \vec{E}_1^* + 2 \ \vec{E}_1 \cdot \vec{E}_2 \ \vec{E}_2^* \,\} \tag{5a}$$

$$+i \ \frac{d\vec{E}_2}{dz} = \alpha \ \{[\,|\vec{E}_1|^2 + |\vec{E}_2|^2] \ \vec{E}_2 + \vec{E}_2 \cdot \vec{E}_1^* \ \vec{E}_1 \}$$

$$+ \ \beta \ \{\vec{E}_2 \cdot \vec{E}_2 \ \vec{E}_2^* + 2 \ \vec{E}_1 \cdot \vec{E}_2 \ \vec{E}_1^* \} \tag{5b}$$

In equations (5a) and (5b), $\alpha = \mu_o \varepsilon \omega^2 \chi_1 / 2k$ and $\beta = \mu_o \varepsilon \omega^2 \chi_2 / 4k$. These equations have been previously derived[4] for $\alpha = 2\beta$. For liquids, $\alpha < \beta$ and (5) describe the correct evolution of the fields.

It is useful at this point to introduce explicitly the forward and backward pumps, probe and signal fields E_n, with $n = F,B,P,S$:

$$\vec{E}_1 = E_F \ \hat{x} + E_P \ \hat{y}$$

$$\vec{E}_2 = E_B \ \hat{x} + E_S \ \hat{y} \tag{6}$$

Equations (5a) and (5b) become

$$i \ \frac{dE_S}{dz} = (\alpha+\beta) \ [\,|E_S|^2 + 2|E_P|^2] \ E_S + \alpha[\,|E_F|^2 + |E_B|^2] \ E_S$$

$$+ \ 2\beta E_F E_B E_P^* \ + \ \alpha E_F^* E_B E_P \ + \ \beta E_B^2 \ E_S^* \tag{7a}$$

$$-i \ \frac{dE_P}{dz} = (\alpha+\beta) \ [\,|E_P|^2 + 2|E_S|^2] \ E_P + \alpha[\,|E_F|^2 + |E_B|^2]E_P$$

$$+ \ 2\beta E_F E_B E_S^* \ + \ \alpha E_F E_B^* \ E_S + \beta E_F^2 \ E_P^* \tag{7b}$$

and similar equations for the pumps, with $S \leftrightarrow B$, $P \leftrightarrow F$.

The last two terms of (7) were omitted in an earlier treatment [1,2] and have a profound effect on the physics of the problem. This may be demonstrated by constructing a Lagrangian which generates (7) as Euler-Lagrange equations. The Lagrangian is

$$\mathcal{L} = \frac{i}{2} \sum_n \epsilon_n (E_n^* E_n' - c.c.) + (\alpha+\beta) [1/2(\sum_n |E_n|^2)^2 +$$

$$|E_F E_B|^2 + |E_P E_S|^2] + \beta[2 (E_P E_S E_F^* E_B^* + c.c.) -$$

$$(|E_F|^2 + |E_B|^2) \cdot (|E_P|^2 + |E_S|^2)] + \alpha[E_F^* E_B E_P E_S^* + c.c.]$$

$$+ \beta/2 [E_B^2 E_S^{*2} + E_F^2 E_P^{*2} + c.c.]$$

(8)

where ϵ_n = 1,-1, 1,-1 for n = F, B, P, S, in the notation of reference 1. The Lagrangian is invariant under only two global U(1)

transformations $E_F \rightarrow E_F e^{i\psi}$, $E_P \rightarrow E_P e^{i\psi}$ and $E_B \rightarrow E_B e^{i\eta}$, $E_S \rightarrow E_S e^{i\eta}$.

The corresponding Noether currents are J_1 = F + P and J_2 = B + S, where $n \equiv |E_n|^2$. The conservation of J_1 and J_2 could have been deduced directly from equations (5a) and (5b). They are just the time-averaged fluxes of the traveling waves in the medium. The last two terms in (8) imply that the current J_3 = P + S is not conserved [1].

The Lagrangian $\mathcal{L}$ is explicitly independent of z so the Hamiltonian $\mathcal{H}$ is conserved. It is given by the negative of $\mathcal{L}$, omitting the derivative terms. The Hamiltonian is useful for checking numerical calculations but cannot be used in conjunction with (7) to yield a single ordinary differential equation for the signal flux S, as was previously attempted [1]. This is a consequence of the fact that there are only two conserved currents.

These are six independent variables coupled by (7). It proves beneficial to write (7) a different way. Introduce the phases ϕ_n through $E_n = |E_n|e^{i\phi_n}$. The six variables are the four phases ϕ_n and any two of the four fluxes F, B, P, and S. The invariance of the Lagrangian implies that the relative phases $\Theta \equiv \phi_S - \phi_B$ and $\Phi \equiv \phi_P - \phi_F$ are locked together. Therefore, the system (7) may be reduced to a closed system of four equations for Θ, Φ, S and P. The result is

$$-S' = 2\sqrt{FBPS} [\sin(\Theta+\Phi) + \frac{\alpha}{2\beta} \sin (\Theta-\Phi)] + BS \sin2\Theta \qquad (9a)$$

$$P' = 2\sqrt{FBPS}\left[\sin(\Theta+\Phi) - \frac{\alpha}{2\beta}\sin(\Theta-\Phi)\right] + FP\ \sin2\Phi \qquad (9b)$$

$$\Theta' = (1+\frac{\alpha}{2\beta})(F-P) + (B-S)\sin^2\Theta - (B-S)\sqrt{\frac{FP}{BS}}[\cos(\Theta+\Phi)$$
$$+ \frac{\alpha}{2\beta}\cos(\Theta-\Phi)] \qquad (9c)$$

$$-\Phi' = (1+\frac{\alpha}{2\beta})(B-S) + (F-P)\sin^2\Phi - (F-P)\sqrt{\frac{BS}{FP}}[\cos(\Theta+\Phi)$$
$$+ \frac{\alpha}{2\beta}\cos(\Theta-\Phi)] \qquad (9d)$$

where the prime denotes $d/d\xi$, $\xi \equiv z/L$, $F = J_1-P$, $B = J_2-S$, and the fluxes are now measured relative to a critical intensity $I_c=(2\beta L)^{-1}$. Similar equations may be written for the sum phases $\Phi_S+\Phi_B$ and $\Phi_P+\Phi_F$ but they are decoupled from (9).

The system (9), the boundary conditions $P(0) = P_0$, $S(L) = S_L$, $\Phi(0) =\Phi_0$ and $\Theta(L) =\Theta_L$, and the fluxes J_1 and J_2 constitute a two point boundary value problem. Solutions of this problem are described in the next section. Multiple solutions, for a fixed set of boundary conditions, are possible when the fluxes exceed the critical intensity I_c.

MULTIPLE SOLUTIONS

Topics of interest concerning collinear degenerate four-wave mixing include small signal behavior and phase conjugation, self-oscillation and multistability. The small signal, linearized version of equations (7a) and (7b) is tractable[3] with the result that the signal exiting the medium at z = 0 is the conjugate of the probe plus some component of the probe itself. Multistability is not possible in this regime since the nonlinear feedback vanishes under linearization. The existence of multiple solutions must be discovered by solving the exact theory, equations (9).

Two-point nonlinear boundary value problems may be solved numerically using a shooting and matching technique[5]. In this way, one solution of (9) may be found[4], especially when the input fluxes are below the threshold intensity I_c for multistability. Above threshold, many solutions are possible and numerical techniques for two-point problems are difficult to implement.

An alternate technique to shooting and matching is to treat the system (9) as an initial value problem. The initial value technique may be invoked to establish the existance of multiple solutions. The input fluxes J_1 and J_2, initial probe P_0 and relative forward phase Φ_0 are held at fixed values, as in the two-point case. The signal flux S_0 and relative backward phase θ_0 are specified at $z = 0$. The initial value problem is then solved numerically. The nonlinear initial value problem has a unique solution under very general conditions[5]. Certain regions of the (S_0, θ_0) plane are mapped into regions of the (S_L, θ_L) plane. The mapping is onto and continuous. Consequently, multiple solutions of the two-point boundary value problem appear as overlap of mapped regions in the (S_L, θ_L) plane. This method yields, in principle, all multiple solutions for a given critical intensity I_c.

Figures 2, 3 and 4 illustrate three particular maps. The examples assume $\alpha = 2\beta$. In each case, the lines $0 \leq S_0 \leq J_2$, θ_0 = fixed, were mapped onto their corresponding solution sets. The initial probe flux was $P_0 = 0.4$, while the total fluxes were $J_1 = J_2 = 1$. The forward beam was assumed to be linearly polarized ($\phi_0 = 0$). The figures display the relative backward phase θ_L, in radians, plotted against the signal flux S_L for $I_c = 1.25$ (figure 2), 1.0 (fig. 3) and 0.9 (fig. 4). The actual values of θ_0 are listed under each figure. The curves in the (S_L, θ_L) plane start at the bottom ($S_0 = 0$) and increase toward the top as S_0 increases. The markers divide the domain of S_0 into thirds. The map of figure 2 exhibits unique solutions to the two-point problem for input fluxes below the critical value I_c. The curves neither cross themselves or each other.

The map of figure 3 exhibits multiple solutions at critical flux for the corresponding two-point problem. Each curve crosses itself (two solutions) at least once, while the curve with $\theta_0 = 0$ exhibits three solutions near $\theta_L = 0$. Additionally, the two curves $\theta_0 = -0.05$ and $\theta_0 = -0.025$ cross each other. In general, the (S_L, θ_L) plane is dense with solutions which cross at least twice. Thus, there exist multiple solutions to the two-point boundary value problem with multiple values of S_0 and either one or more values of θ_0. Figure 4 illustrates the case $J_1 = J_2 > I_c$, in which the maps become even more pathelogical. It suggests (but does not prove) that the solutions become chaotic for $J_1, J_2 \gg I_c$.

CONCLUSION

The theory of collinear degenerate four-wave mixing in lossless isotropic Kerr media was developed for arbitrary values of the susceptibilities χ_1 and χ_2. This treatment is the first to include all phase-matched terms in the equations of motion for the slowly-varying field envelopes. Numerical techniques were developed to prove the existence of multiple solutions for input fluxes above the critical intensity I_c. The examples were restricted to the case $\chi_1 = \chi_2$ but the result should be generally true. The results suggest, but do not prove, that hysteresis and multistability should be exhibited by an optical device based on collinear four-wave mixing. These topics and the phenomenon of self-oscillation will be discussed at length in a separate paper.

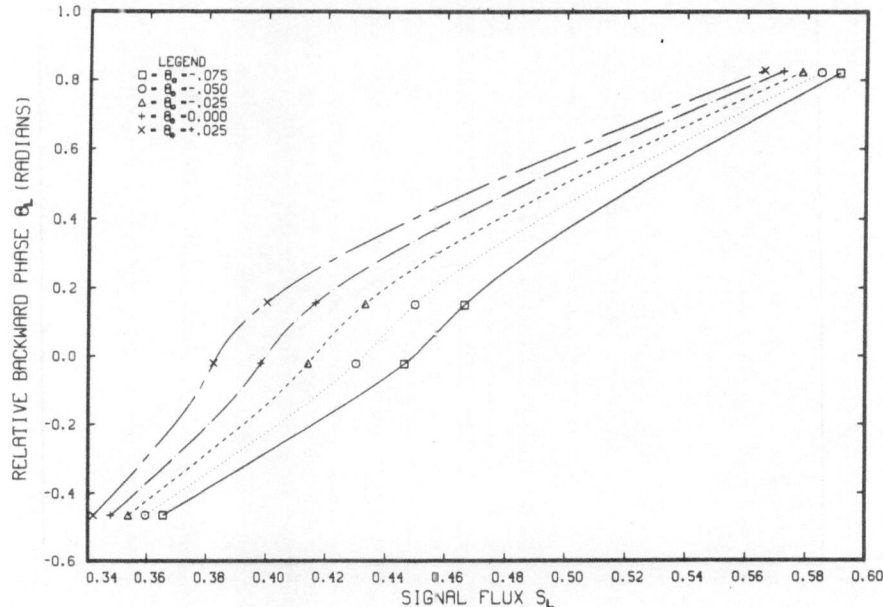

Fig. 2. Range of solution map below critical intensity.

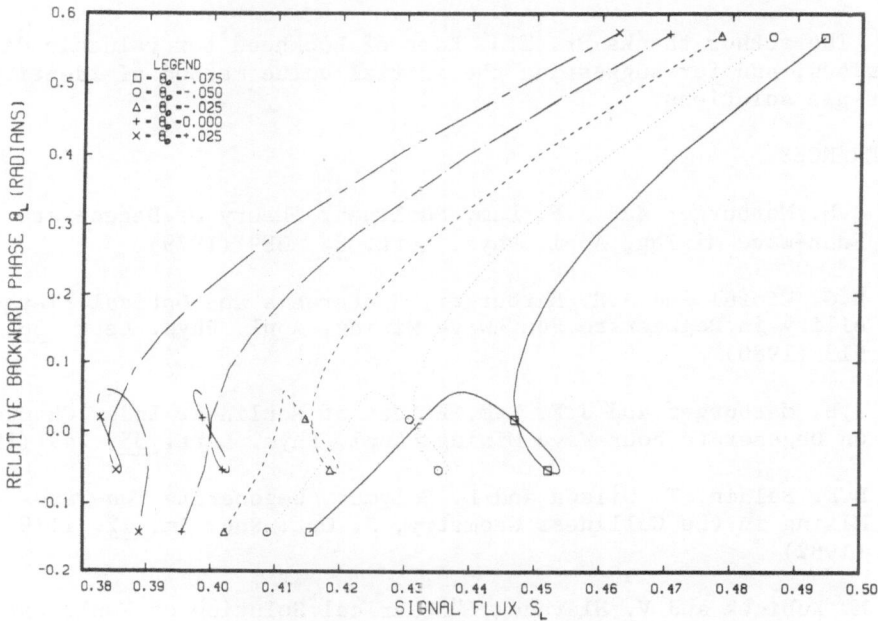

Fig. 3. Range of solution map at critical intensity.

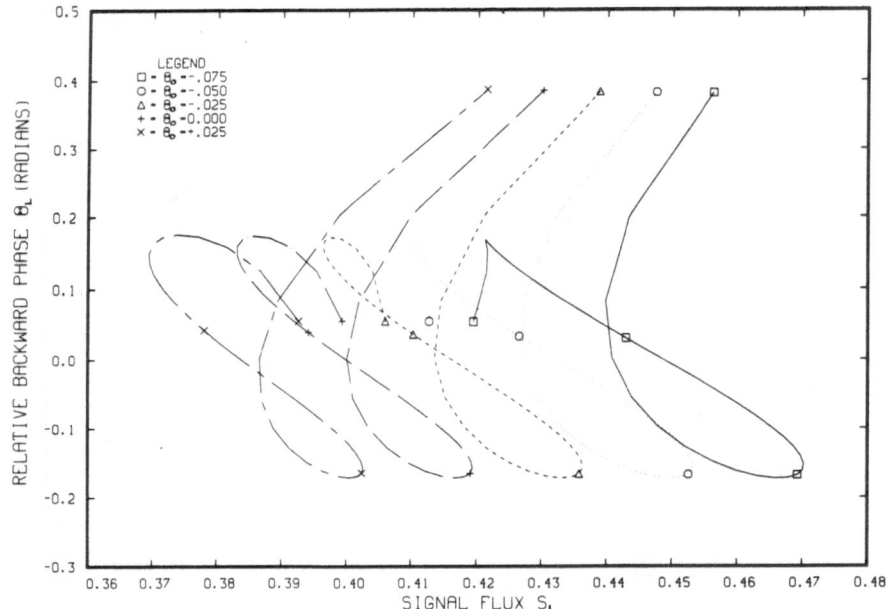

Fig. 4.   Range of solution map above critical intensity.

ACKNOWLEDGEMENT

The author thanks Dr. T.J. Karr of Lockheed for valuable discussions, and for suggesting the initial value method of identifying
multiple solutions.

REFERENCES

1. J.H. Marburger and J.F. Lam, Nonlinear Theory of Degenerate
    Four-Wave Mixing, Appl. Phys. Lett. 34, 389 (1979)

2. H.G. Winful and J.H. Marburger, Hysteresis and Optical Bista-
    bility in Degenerate Four-Wave Mixing, Appl. Phys. Lett. 36,
    613 (1980)

3. J.H. Marburger and J.F. Lam, Effect of Nonlinear Index Changes
    on Degenerate Four-Wave Mixing, Appl. Phys. Lett. 35, 249 (1979)

4. D.K. Saldin, T. Wilson and L. Solymar, Degenerate Two-Wave
    Mixing in the Collinear Geometry, J. Opt. Soc. Am. 72, 1179
    (1982)

5. M. Kubicek and V. Hlavacek, "Numerical Solution of Nonlinear
    Boundary Value Problems with Applications," Prentice Hall,
    New Jersey (1983)

SATURABLE INTERFEROMETER THEORY

OF OPTICAL BISTABILITY

N. M. Lawandy and R. Willner

Division of Engineering
Brown University
Providence, RI 02912

INTRODUCTION

Recently, a large volume of theoretical work has emerged concerning the theory of optical bistability (OB). This effort has resulted due to the intriguing nature of the stability questions which arise in OB as well as the experimental realization of optical bistable device (OBD). The demonstration of a variety of OBD configurations has added mementum to the theme of using such devices as memory and logic elements in integrated optics.

There are a variety of OB situations possible as a function of the physical mechanisms which couple the injected field to the induced polarization and boundary conditions imposed by the cavity. The simplest case of purely absorption bistability has attracted the most theoretical attention. The first treatment of this problem was given by Bonifacio and Lugiato (1). The steady state solutions for the set of coupled Maxwell-Bloch equations in the mean-field limit results in a cubic equation for the incident intensity as a function of the transmitted intensity. This yields a hysteresis curve which clearly defines the high and low transmission branches as well as the jump points for up switching and down switching. Furthermore, the conditions for a single valued cubic function can be used to determine a condition for a hysteresis loop. In addition to this type of treatment, others have considered the effects due to propagation and transverse field dependences (2-10). The questions concerning the dynamical evolution in OB have become the subject of much work recently. Various treatments have established regimes of self pulsing, period doubling and chaos (11-16). Of particular theoretical and practical interest are the switching up and switching down times in bistable devices.

Recent work on purely absorptive OB has established a nascent hysteresis under the influences of slowly varying external fields. Mandel and Erneux have theoretically examined the upswitch and downswitch times using a mean-field model of OB in the limit of large C (17,18). Without restricting the relative magnitude of the cavity and atomic relaxation times, they were able to show that as C→∞, the up switch time is proportional to the atomic relaxation time and the down switch time is proportional to the cavity decay time.

In this paper we will utilize the method of summing transmitted waves, routinely used to analyze interferometer structures, to develop an OB equation in the good cavity limit. The method yields a different state equation than the usual mean-field model. However, for small C values and or large input intensities a Taylor expansion of our state equation yields the mean-field pure absorptive OB equation. Our equation also gives a different condition for hysteresis. Due to the simplicity of this approach which utilizes the sum of a geometric series, we are able to interogate the system after a given number of round trips via the analytic expression for the sum at that point. This allows us to obtain an inverted function of time and to determine the time evolution of the system. Using these expressions, we are able to give switching times for both the up and down processes.

THEORY

The theory we present is based on a self consistent sum of partial waves emitted by a resonant structure. In these derivations we consider only the zero detuning case and therefore assume that there are only hysteresis effects due to χ_2 saturation. We will solve the case of a Fabry-Perot resonator with plane mirrors having power transmission coefficient T. The round trip length of the cavity is taken to be 2ℓ and the cavity is filled by a medium having an unsaturated absorption coefficient α_0.

The medium is assumed to be an ideal homogeneously broadened two level system with a saturation intensity I_s given by:

$$I_s = \frac{\gamma_\perp \gamma_\parallel \hbar^2 c}{2\mu^2} \tag{1}$$

where c is the speed of light in the medium. In terms of I_s, the absorption coefficient is given by

$$\alpha = \frac{\alpha_0}{1+I/I_s} \tag{2}$$

where I is the instantaneous spatially averaged intensity inside the Fabry-Perot.

Using the usual sum of partial waves method utilized for analyzing interferometers, we can express the transmitted electric field from the Fabry-Perot as:

$$E_T = TE_I e^{ik\ell} e^{-\alpha\ell} \sum_{m=0}^{M} [(1-T)\exp\{2\ell(ik-\alpha)\}]^m \tag{3}$$

This expression can be made self consistent by realizing that I or E^2 may be related to E_T in the limit of spatial averaging. The result is that

$$\frac{E^2}{A} = E_T^2 \tag{4}$$

The constant A assumes a value of 1 or 2 for the two cases of a ring laser and Fabry-Perot respectively. This results in an expression for the transmitted field after M round trips given by:

$$X = TYe^{ik\ell}\exp\left[\frac{-\rho_0}{1+X^2}\right] \sum_{m=0}^{M} (1-T)^m\exp\{2m(ik - \frac{\rho_0}{1+X^2})\} \tag{5}$$

$\rho_0 = \alpha_0\ell$ and Y and X are the dimensionless incident and transmitted electric fields. In the steady-state limit when $M \to \infty$, the sum may be evaluated using the geometric series limit and the self consistent equation which results is given by:

$$TY = X[e^\rho - e^{-\rho}] + XTe^{-\rho} \tag{6}$$

where $\rho = \dfrac{\rho_0}{1+X^2}$. This equation may be recast in the form:

$$TY = X[2\sinh(\rho) + Te^{-\rho}] \tag{7}$$

This equation gives the steady state hysteris result which we use for our calculations concerning OB. This result, it must be pointed out, hinges on the assumption that the effects of propagation given by the saturated absorption results may be ignored. In effect, this means that throughout each cavity traversal the initial and final intensities obey the equation:

$$\frac{dI}{dz} = \frac{-\alpha_o I}{1 + I_i/I_s} \tag{8}$$

and that $I_f(\ell) = I_i \exp \dfrac{-\alpha_o}{1 + I_i/I_s}$

This condition is no more stringent than the spatial averaging employed in MFT calculations. The results, however, are only valid in the adiabatic limit when $\gamma_\parallel , \gamma_\perp \gg K$.

STATIONARY STATE OB

The transmitted intensity X^2 as a function of the input intensity Y^2 may be found from the equation:

$$T^2 Y^2 = X^2 [e^\rho - Re^{-\rho}]^2 \tag{9}$$

where $R = 1 - T$. This equation yields hysteresis loops for certain values of ρ_o and T as does the MFT. The hysteresis curves for various ρ_o values and a fixed value of $R = 0.90$ are shown in figure 1. These curves correspond to the Fabry-Perot case ($A = 2$). The curves show a well defined hysteresis for $\rho_o > 0.4$. Below this value there is only a region of large differential gain.

The solution for OB which we give in equation (7) may be approximated by expanding the exponentials to give:

$$Y = X \left[1 + \frac{2\rho_o (\frac{1}{T} - \frac{1}{2})}{1 + X^2} \right] \tag{10}$$

This is exactly the form of the equation which results from the Mean Field Theory (MFT). Therefore, in this regime, the Hurwitz criterion requires that for bistability:

$$\frac{\rho_o}{T} - \frac{\rho_o}{2} > 4 \tag{11}$$

as opposed to the MFT result:

$$\frac{\rho_o}{T} > 4 \tag{12}$$

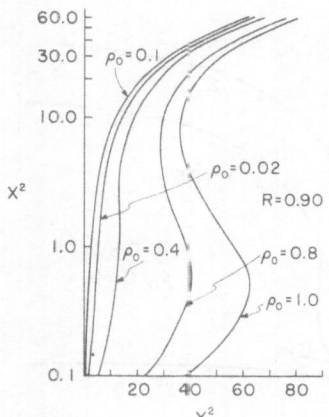

Fig. 1. Hysteresis curves for a Fabry-Perot structure.

The result in equation (11) is only a valid condition when ρ_0 is small and is not general.

DYNAMIC OB

The problem of time dependent behavior in OBD has been of great interest (11-18). In particular, the questions of switching times have received much attention. The time evolution in OB has been studied numerically and from the standpoint of asymptotic growth rates resulting from stability analysis. Attractor behavior has been shown analytically in MFT analysis by Benza and Lugiato and others (2,11). Numerical studies showing attractor behavior have also been undertaken by Bonifacio and Meystre (13,14). The model we use can yield an analytic expression relating the transmitted intensity to the time. Unfortunately the results are inverted and the time is given in terms of the intensity.

An expression which relates the output field X after a given number of round trips M to the input field Y can be found from equation (5). This is accomplished via the expression for the value of a finite geometric sum. This results in:

$$MT_c = \frac{\ln\left[\frac{1}{R}(\exp(2\rho) - (\exp(3\rho) - R\exp(\rho))\frac{X}{TY})\right]}{\ln(1-T) - 2\rho} \tag{13}$$

T_c is the cavity round trip time and $\rho = \dfrac{\rho_o}{1+AX^2}$. Equation (13) gives an expression for the time in cavity round trips it takes to develop a transmitted field X when a field Y is instantaneously

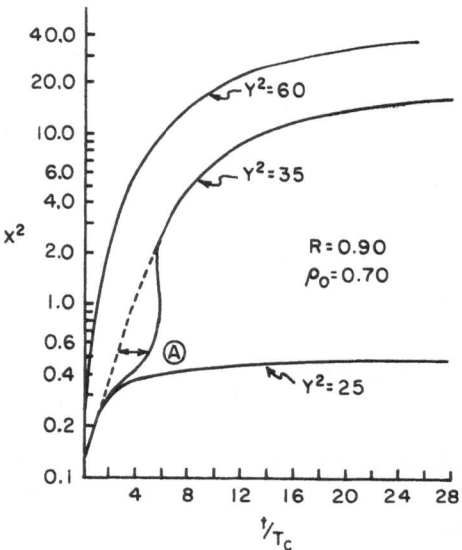

Fig. 2.   Dynamic up-switching behavior for two input intensities
          exceeding the switch-up value. The $Y^2 = 35$ curve shows
          the attractor behavior of the unstable switching point on
          the time evolution of the system.

present at the first mirror of the structure. The results for two
values of the input intensity which exceed the switch up values are
shown in figure 2. The upper curve ($Y^2 = 60$) indicates that approximately 25-30 round trips are required and that the time evolution
is a direct trajectory towards the upper branch value ($38 = X^2$ in
this case). The lower solid curve shows the results of equation (13)
when the input value just barely exceeds the up switch value of
$Y^2 = 30$. The point labelled $\underline{A}$ represents the largest deviation from
the dashed curve. This point corresponds to the value of X^2 on the
hysteresis loop which comes closest to the line $Y^2 = 35$ in the
X^2-Y^2 plane. Thus it appears that this formulation is also capable
of generating the attraction for solutions which lie on only the
upper branch if the input intensity is near the switch up value. The
time evolving system is drawn in towards the lower branch is it
approaches its steady state value in the upper branch.

Equation (13) can be used to study the effect of the controllable physical parameters ρ_0 and T on the up switch and down switching times for the optically bistable system. We define the up switch
time as the time it takes for the transmitted intensity to reach 99%
of the steady state value corresponding to an input intensity which
yields a single solution in the upper branch. The times are calculated under the assumption that the input intensity is 10% larger

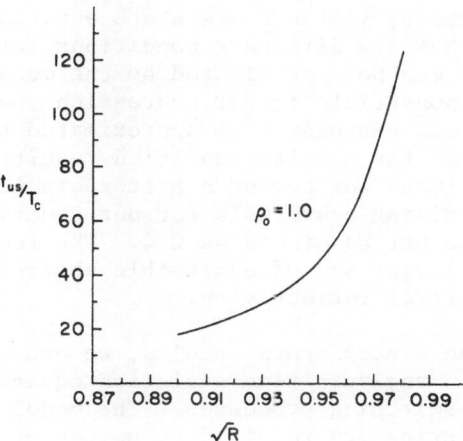

Fig. 3. Up-switch time as a function of the mirror reflectivity.

than the unstable upswitch position on the hysteresis curve. The variation of t_{us} with the mirror reflectivity R is shown in figure 3.

The problem of determining down switching times using our general method requires a reformulation of the initial problem. We now require a partial wave sum which results when an internal field already exists in the structure and is allowed to leak out under the influence of both the cavity and the saturable medium. The result of the straight forward calculation is given below:

$$t = \frac{T_c}{2} \left[\frac{\ln(X/X_o)}{\ln(1-T) - 2\rho} \right] \tag{14}$$

equation (14) relates the value of X relative to the original value X_0 (on upper branch) at different times and after Y is made instantaneously equal to zero. When the medium filling the cavity has negligible effects equation (14) becomes the usual cavity exponential decay curve. As might be expected, a larger absorption results in a shorter down switching time.

CONCLUSION

This paper has presented a new method for deriving a variety of steady state and dynamic results for the case of purely absorptive optical bistability. The method is based on a self consistent partial wave sum, similar to that used in analyzing interferometer devices. The limitations on this theory are those present in mean-field calculations, namely that propagation effects are neglected. In addition, all time evolution results only apply in the adiabatic limit or <u>good cavity</u> case.

The approach we employ yields a new state equation for the variables X and Y and has its different conditions for hysteresis. The equation we derive can be approximated by the mean field model cubic equation when exponentials in our expression are approximated to first order. When our expression is approximated by a cubic equation, application of the Hurwitz condition results in essentially the same condition (C>4) for having a hysteresis loop. The general condition for obtaining hysteresis for our equation is derived, but unfortunately it is not as simple as C>4. The results indicate our equation admits a larger set of admissible absorption-length products for a given mirror reflectivity.

In addition to the steady state results, we are able to use this approach to obtain analytic time evolution equations for the up-switching and down-switching processes. The model predicts that several tens of round trips are required to switch up to the upper branch, and that only a few roundtrips are required to drop down.

Finally, we have seen from the model results that the time evolution can have some unexpected features. In particular, the equations we derive indicate that if the intensity which causes the up-switch process to the upper branch is close to the knee of the hysteresis curve, the instability point can attract this solution as it evolves towards its steady state value in the upper branch.

REFERENCES

1. R. Bonifacio and L.A. Lugiato, Optics Comm. 19,p. 172,(1976).
2. R. Bonifacio and L.A. Lugiato, Lett.Nuovo Cimento 21,505 (1978).
3. R. Bonifacio and L.A. Lugiato, Lett. Nuovo Cimento 21,510 (1978).
4. P. D. Drummond, IEEE JQE 17, 301, (1981).
5. J. Fleck, Jr., Apl. Phys. Lett. 13, p. 365, (1968).
6. R. Bonifacio and L.A. Lugiato Phys. Rev. A18, p.1129, (1978).
7. J. A. Hermann, Optica Acta 27, p. 159, (1980).
8. S. L. McCall and H. M. Gibbs, Optics Comm.33, p. 335, (1980).
9. S. S. Hassan, S.P. Tewari and E. Abraham, Optics Comm. 40 (6), p. 461, (1982).
10. J.V. Moloney and H.M. Gibbs, Phys. Rev.Lett. 48 (23), P.1607, (1982).
11. V. Benza and L.A. Lugiato Lett. Nuovo Cimento, 26,405, (1979).
12. P. D. Drummond, Optics Comm. 40, 224, (1982).
13. R. Bonifacio and Meystre, Optics Comm., 27, 147 (1978).
14. R. Bonifacio and Meystre, Optics Comm. 29, 131 (1979).
15. M. Gronchi, V. Benza, L.A. Lugiato, P. Meystre and M. Sargent III, Phy Rev. A., 24(3), p. 1419, (1981).
16. L.M. Narducci, D.K. Bandy,C.A. Pennise and L.A.Lugiato, Optics Comm. 48 (3), p. 207, (1983).
17. P. Mandel and T. Erneux, Optics Comm. 44(1), p. 55, (1982).
18. P. Mandel and T. Erneux, Optics Comm., 42 (5), p. 362, (1982).

ABSORPTIVE OPTICAL BISTABILITY IN A BAD CAVITY

Paul Mandel
Université Libre de Bruxelles, Campus Plaine,
CP 231, Bruxelles 1050, Belgium

T.Erneux
Department of Applied Mathematics, Northwestern
University, Evanston, Illinois 60201, USA.

Optical bistability is a fascinating subject
from many points of view. Experimentally the construction
and the study of all-optical and hybrid bistable systems
open a new avenue for fundamental and applied physics.
In this respect the most striking applications of optical
bistable devices are (i) their use to test predictions of
the various properties of "deterministic chaos" such as
e.g. scaling laws, 1/f noise, roads to chaos (see the
many contributions on these subjects in this volume) and
(ii) their potential use as purely optical logic gates
and therefore all-optical circuits (see in particular
D.S.Smith's contribution in this volume). From the theoretical point of view the equations describing optical
bistability are challenging because they seem rather
simple but nevertheless they describe a huge variety of
very different behaviors which can, in principle, be
tested experimentally. This relation between theoretical
prediction and experiment is often limited because analytic studies of the coupled Maxwell-Bloch equations usually
rely on adiabatic elimination schemes. Such schemes are
based on strong inequalities between the various decay
rates of the system. In particular two classes of adiabatic schemes have been widely used in quantum optics for
the simplicity of the results they lead to. In the good
cavity limit, one assumes that the cavity decay rate
(i.e. inverse relaxation time) is much smaller than the

447

atomic decay rates. In the bad cavity limit, one assumes
that at least one atomic decay rate is much smaller than
the cavity decay rate. Adiabatic elimination was applied
in optical bistability by Benza and Lugiato /1/ (good
cavity limit in absorptive optical bistability), by
Drummond /2/ (bad cavity limit in dispersive and absorptive optical bistability) and by Ikeda /3/ (good cavity
limit in very dispersive optical bistability).

The problem with these two approximations is
that they imply the existence of inequalities involving
physical constants which are either difficult or impossible to adjust to fit the theoretical requirements. On the
other hand the power of these two limits is that they
provide a small parameter in terms of which systematic
perturbation expansions can be derived. Therefore it is
desirable to have a technique which allows us to treat
the situation in which the cavity decay rate is of the
same order of magnitude as the atomic decay rates. Two
limits can be studied with this restriction on the decay
rates: either the bistability parameter is very large and
its inverse becomes a convenient small parameter /4/ or
the bistability parameter is very near the critical value
for the onset of bistability and the small parameter is
the deviation from criticality /5/. In this contribution
we shall analyze the case of fully developed hysteresis
(large bistability parameter) in absorptive optical
bistability with the restriction of comparable atomic
and cavity decay rates.

For a single running mode absorptive optical
bistable system, the mean field approximation of the
Maxwell-Bloch equations can be written in reduced
variables as

$$dx/dt = \dot{x} = -x + y - 2Cs \qquad (1)$$
$$k_\perp \dot{s} = -s + xd \qquad (2)$$
$$k_\parallel \dot{d} = -d + 1 - xs \qquad (3)$$

where x, s and d are the field amplitude, the atomic
polarization and the atomic population difference, respectively. The constants $k_\perp$ and $k_\parallel$ are the ratios of the
cavity decay rate to the transverse and longitudinal
atomic decay rates. The control parameters are the bistability parameter C and the input field amplitude y.

When the input field is stationary (dy/dt=o)
the equations (1) to (3) have the stationary solutions

$$d = (1 + x^2)^{-1} \qquad (4)$$

$$s=x(1+x^2)^{-1} \tag{5}$$
$$y=x+2Cx(1+x^2)^{-1} \tag{6}$$

The last equation is a cubic in x which has a single real
positive root if $C<4$ (monostability) and three real
positive roots if $C>4$ (bistability). Furthermore it is
easy to verify that the branches with $dx/dy>0$ are stable
whereas the branch with $dx/dy<0$ is unstable (see fig.1).

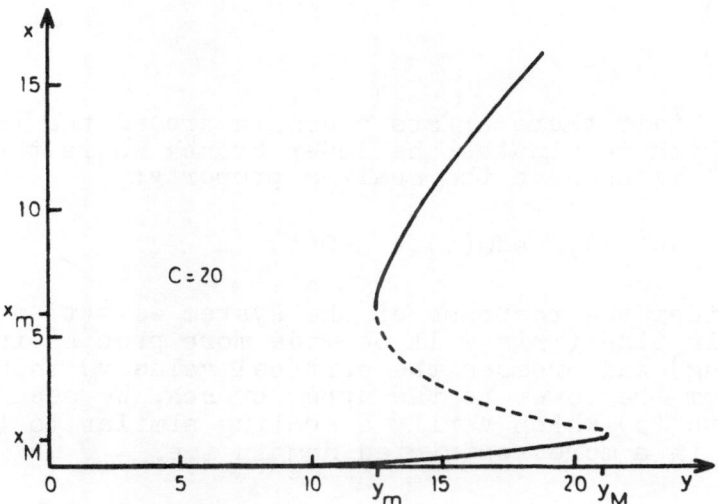

Fig.1. Stationary output field (x) versus input field (y).

At the two turning points y_m and y_M the linear stability
analysis of the stationary solutions (4) to (6) leads to
a vanishing eigenvalue /6/. This signals the breakdown
of the linearized stability analysis which must be replaced by a nonlinear stability analysis /4/; furthermore
it indicates the occurence of a critical slowing down in
the vicinity of y_m and y_M . This result is independent
of the adiabatic elimination scheme.
        From the experimental point of view the above
analysis is not always very pertinent because quite often
the input field varies in time. If this time variation
is slow compared to the relaxation time of a perturbation

from the stationary state, we expect the system to follow
closely the stationary state. However near the two turning
points, where critical slowing down appears, the relaxation
time of the bistable system diverges and no matter how
slowly the input field varies, it will act on the system
as a quickly varying field. Hence it is in the vicinity
of the two turning points that we expect dynamical
deviations from the stationary behavior. As an example
we shall focus on the turning point located at $y=y_M$ and
$x=x_M$ and we consider the limit:

$$C \rightarrow \infty$$
$$k_{\perp} = 0(1) \tag{7}$$
$$k_{\parallel} = 0(1)$$

In this limit there exists a domain around the turning
point which terminates the lower branch where the stationary solutions have the scaling property:

$$y=0(C), \quad x=0(1), \quad s=0(1), \quad d=0(1) \tag{8}$$

We consider the response of the system as $y(t)$ varies
slowly in time (this will be made more precise in the
following) and crosses the critical value y_M inducing a
jump from the lower to the upper branch. We seek solutions
of (1) to (3) which verify a scaling similar to (8) in the
same or in a more restricted domain i.e.

$$y(t)=CY(t)$$
$$Y(t), \quad x(t), \quad s(t), \quad d(t) = 0(1).$$

Correspondingly the Maxwell-Bloch equations become

$$\dot{x}=-x+\epsilon^{-1}(Y-2s) \tag{9}$$
$$k_{\perp}\dot{s}=-s+xd \tag{10}$$
$$k_{\parallel}\dot{d}=-d+1-sx \tag{11}$$

where $\epsilon = C^{-1}$. When $\dot{Y}=0$ the stationary solutions of (9) to
(11) are

$$x_{1,2}=(1\mp(1-Y^2)^{1/2})/Y +0(\epsilon) \tag{12}$$
$$d_{1,2}=Y/2x_{1,2} +0(\epsilon) \tag{13}$$

$$s_{1,2}=Y/2 \quad +0(\epsilon)\tag{14}$$

with $Y \lesssim 1+0(\epsilon)$. The linear stability analysis of these solutions yields the three eigenvalues

$$\lambda_a=-(1-x_{1,2}^2)/k_{\shortparallel} +0(\epsilon^{1/2})\tag{15}$$

$$\lambda_{b,c}=\pm i\epsilon^{-1/2}(2d_{1,2}/k_{\perp})^{1/2}-0.5(1+k_{\perp}^{-1}+x_{1,2}^2/k_{\shortparallel})+0(\epsilon^{1/2})\tag{16}$$

The eigenvector $z=col(x,s,d)$ associated with λ_a has the structure

$$z_a=col(1,0(\epsilon^{1/2}),0(1))\tag{17}$$

whereas the eigenvector associated with λ_b has the structure

$$z_b=col(1,0(\epsilon^{1/2}),0(\epsilon^{1/2}))\tag{18}$$

Hence in the limit $\epsilon \rightarrow 0$, x will contain a weighted sum of a decaying exponential and damped oscillations whereas d will only display a decaying exponential.
        From the expressions (15) and (16) we note the property

$$Re\lambda=0(1) \qquad Im\lambda=0(\epsilon^{-1/2})\tag{19}$$

This clearly indicates the existence of two widely separated time-scales and suggests that we use a multiple time-scale perturbation analysis. Let us now return to the general case $dY/dt\neq0$ and eliminate s(t) from eqs.(10) and (11) with the use of eq.(9):

$$s=Y/2 -\epsilon(x+\dot{x})/2$$

$$\epsilon k_{\perp}\ddot{x} +\epsilon \dot{x}(1+k_{\perp}) +x(2d+\epsilon)-Y-k_{\perp}\dot{Y}=0\tag{20}$$

$$k_{\shortparallel}\dot{d}+d-1+xY/2 -\epsilon(x^2+x\dot{x})/2 =0\tag{21}$$

On the basis of the previous analysis we seek solutions of (20) and (21) which are functions of the two times t and $T=\epsilon^{1/2}\omega(t)$. We further assume that the external field has only the slow time-dependence $Y=Y(t)$, $dY/dT=0$. Hence the perturbation expansion becomes

$$x(t)=x(t,T,\epsilon^{1/2})=x_0(t,T)+\epsilon^{1/2}x_1(t,T)+\dots$$

$$d(t)=d(t,T,\epsilon^{1/2})=d_0(t,T)+\epsilon^{1/2}d_1(t,T)+\ldots$$

$$\omega(t)=\omega(t,\epsilon^{1/2})=\omega_0(t)+\epsilon^{1/2}\omega_1(t)+\ldots$$

Introducing these power series into (20) and (21) and using the solvability condition /7/ we obtain

$$k_{\shortparallel}\dot{d}_0(t)=1-d_0(t)-(\ Y^2(t)+k_{\perp}Y(t)\dot{Y}(t)\ )/4d_0(t) \tag{22}$$

$$x_0(t,T)=(\ Y(t)+k_{\perp}\dot{Y}(t)\ )/2d_0(t)\ +2a(t)\cos T \tag{23}$$

$$\text{where}\ T=\epsilon^{-1/2}\int^{t}\left\{\left[2d_0(t'')/k_{\perp}\right]^{1/2}+0(\epsilon^{1/2})\right\}dt'' \tag{24}$$

$$\text{and}\ \dot{a}=-\frac{a}{4d_0}\left[\frac{Y}{2d_0 k_{\shortparallel}}(Y+k_{\perp}\dot{Y})+\dot{d}_0+2d_0(1+k_{\perp}^{-1})\right] \tag{25}$$

Eventually the problem is reduced to the analysis of a single nonlinear differential equation (22) which describes the dynamicsof the system around (x_m,y_M). It is noteworthy that the fundamental variable is the population difference $d_0(t)$. The field amplitude $x_0(t,T)$ contains two terms: the first term describes the adiabatic entrainment by the time evolution of the atomic population; this contribution could have been derived, together with (22), from a direct and naive one-time analysis of (9) to (11). But the second term in (23) requires the more elaborate two-time analysis to be generated. It describes a correction to the adiabatic evolution whose peculiarity is to oscillate at a high frequency of order $\epsilon^{-1/2}$ but with an amplitude of order 1; the two contributions to $x_0(t,T)$ are therefore comparable and none of them may be neglected. In essence this information was already contained in the expressions (17) and (18) and this justifies, a posteriori, our study of the eigenvectors and eigenvalues of the linearized equations derived from (9) to (11). Note that the structure of the oscillatory term in (23) is of the form $a(t)\cos T$ where $a(t)$ is the damping function. Hence the oscillation frequency is of order $\epsilon^{-1/2}$ but, to leading order in ϵ, it is damped over a time which is of order 1. Furthermore the frequency is not a strict constant since, according to (24), it is a function of the slow time-dependent fundamental variable $d_0(t)$.

One may (correctly) guess that the technique we have just described can be applied to the other domains defined on the stationary solution x=x(y). In this way the whole curve can be analyzed, domain by domain. As a rule it turns out that on the lower branch the fundamental variable which verifies a closed equation is the population difference whereas the output field is the sum of an adiabatic contribution and a fast oscillating term with $O(1)$ amplitude. On the contrary, in the domains belonging to the upper branch, it is the output field which becomes the fundamental variable and the population difference is the sum of an adiabatic contribution and a fast oscillatory term (whose frequency is related to the time-dependent Rabi frequency) with $O(1)$ amplitude. As a consequence a jump corresponding to an up-switching or a down-switching is a very complex process in which the role of fundamental variable and the attribution of oscillatory terms will be exchanged between the output field $x(t)$ and the population difference $d(t)$.

References

1. V. Benza and L. A. Lugiato, Lett. Nuovo Cimento 26(1979)405.

2. P. D. Drummond, Optics Comm. 40(1982)224.

3. K. Ikeda, Optics Comm. 30(1979)257.

4. T. Erneux and P. Mandel, Phys. Rev. A (in press).

5. P. Mandel and T. Erneux, Optics Comm. 44(1982)55.

6. R. Bonifacio and L. A. Lugiato, Phys. Rev. A18(1978)1129.

7. A. H. Nayfeh, "Introduction to perturbation techniques", J. Wiley, New York (1981).

MIRRORED AND MIRRORLESS OPTICAL BISTABILITY: EXACT

C-NUMBER THEORY OF ATOMS FORMING A FABRY-PEROT CAVITY

R.K. Bullough, S.S. Hassan[†], G.P. Hildred and R.R. Puri

Department of Mathematics
UMIST, P.O. Box 88
Manchester M60 1QD, U.K.

[†]Ain Shams University
 Cairo, Egypt

INTRODUCTION

We report a fundamental theoretical study of the optical bistability (OB) of a set of 2-level atoms forming by their own geometry
a Fabry-Perot cavity. Together with numerical work currently in
hand the theory reported here should form a good basis for a detailed
and quantitative comparison with experimental observations. Apparently
suitable experiments are currently being performed[1].

The theory is a c-number one, but otherwise it assumes no more
than the following: the collection of 2-level atoms occupies a slab-
like region $0 \leq z \leq L$; they are driven by a c.w. laser field
$E_o e^{-i(\omega t - k_o z)}$ + cc. where $ck_o = \omega$ is the dispersion relation of free
space; outgoing boundary conditions <u>at infinity</u> are imposed on the
radiation from the atoms. Thus there is no appeal to phenomenological
Maxwell boundary conditions at the surfaces $z = 0$, L themselves. We
show nevertheless that the slab has a natural non-linear refractive
index $m(\omega)$ depending on the incident intensity $|E_o|^2$: the problem
of spatial dependence within the slab is handled specifically, and
no 'mean field' assumption is made. Except that transverse effects
and inhomogeneous broadening are not yet included the theory seems
to be comprehensive enough for the quantitative comparison with
experiments we intend.

From the technical point of view the theory assumes a slowly
varying envelope approximation (in space alone) together with r.w.a.;
there is supposed a smooth distribution of atoms, number density n,

inside the slab, but many-body corrections are indicated and introduce the Lorentz field. We develop and use a natural generalisation of the 'optical extinction theorem' to handle the surface effects[2],[3]: this means these can be handled self-consistently in terms of the non-linear refractive index $m(\omega)$; but then we show that, providing we express all these surface effects in terms of $m(\omega)$ the usual reflexion and transmission coefficients arise and the usual Fabry-Perot cavity action appears. The non-linearity is then solely in $m(\omega)$: it is a complex number, and we can expect that the theory contains the conventional theories of both absorptive and dispersive OB within it. We show this is the case. Then apart from the comprehensiveness of the analysis the particular new features are the handling of the spatial dependences, the effect of these on the non-linear refractive index $m(\omega)$, the internal consistency of the treatment of the surfaces, and the fact that cavity action arises solely from the slab-like geometry.

The last means the theory is "mirrorless" but generates its own mirrors. But it is also mirrorless in the deeper sense that in the non-linear theory the Lorentz field shifts the atomic resonance in an intensity dependent way. The theory here compares with Ref.[4] though the additional roots produced this way depend on the usual sources of OB. The Lorentz field is linear in n and therefore significant to solids but not vapours. Additional local field effects correct this by a complex number[6] so the radiation damping important to absorptive OB is also modified. These several effects are described in the following.

2. THE EQUATIONS OF MOTION

These are the coupled Bloch-Maxwell equations without in the first instance any slowly varying envelope approximation. In a frame rotating at angular frequency ω the Bloch equations for the atoms are

$$\dot{r}_+(\underset{\sim}{x},t) + \gamma_0(1+i\delta)r_+(\underset{\sim}{x},t) = 2i\{E^*(\underset{\sim}{x},t) - \Omega e^{-ik_0 z}\}r_3(\underset{\sim}{x},t) \qquad (2.1a)$$

$$\dot{r}_3(\underset{\sim}{x},t) + 2\gamma_0(\tfrac{1}{2}+r_3(\underset{\sim}{x},t)) = -i\{E^*(\underset{\sim}{x},t) - \Omega e^{-ik_0 z}\}r_-(\underset{\sim}{x},t)+cc. \qquad (2.1b)$$

with $\dot{r}_- = (\dot{r}_+(\underset{\sim}{x},t))^*$: * denotes complex conjugate. The detuning $\delta \equiv (\omega-\omega_0)\gamma_0^{-1}$ is scaled against one half of the A-coefficient $\gamma_0 \equiv \frac{2}{3}p^2\hbar^{-1}\omega_0^3 c^{-3}$; ω_0 is the atomic resonance; $\Omega \equiv p\hbar^{-1}E_0$ is the laser Rabi frequency; p is the magnitude of the dipole matrix element $p\hat{\underset{\sim}{u}}$.

In the laboratory frame the dipole density is $np\hat{\underset{\sim}{u}}(r_+e^{i\omega t}+r_-e^{-i\omega t})$ $\equiv n\underset{\sim}{P}(\underset{\sim}{x}, t)$. In this frame Maxwell's equation is

$$\text{curl curl } \underset{\sim}{E}(\underset{\sim}{x},t) + c^{-2}\partial^2\underset{\sim}{E}(\underset{\sim}{x},t)/\partial t^2 = -4\pi n\partial^2\underset{\sim}{P}(\underset{\sim}{x},t)/\partial t^2 \qquad (2.2)$$

As long as we are concerned with time dependences $e^{\pm i\omega t}$ alone we can work with Green's functions $\underline{F}(\underline{x},\underline{x}';\omega) \equiv (\nabla\nabla + k_o^2\underline{U}) e^{ik_o r} r^{-1}$ and $\underline{F}^*$: $r = |\underline{x}-\underline{x}'|$ and $\underline{U}$ is the unit tensor. Then, expressed as a Rabi frequency, the total field driving the Bloch equations (2.1) has negative frequency part $E^*(\underline{x},\omega) - \Omega e^{-ik_o z}$ with

$$E^*(\underline{x},\omega) \equiv np^2\hbar^{-1} \int_{V-v} \underline{F}^* (\underline{x}, \underline{x}'; \omega): \hat{\underline{u}}\hat{\underline{u}}\, r_+(\underline{x}', \omega)\, d\underline{x}' \qquad (2.3)$$

For simplicity all fields lie in the direction $\hat{\underline{u}}$ of the atomic dipole matrix elements. The $e^{\pm i\omega t}$ dependence means that the $r_+(\underline{x},t)$ do not depend on t and can be relabelled $r_+(\underline{x},\omega)$; E^* in (2.1) is $E^*(\underline{x},\omega)$ and (2.1) is reduced to the corresponding steady state relation.

The integral (2.3) is adequately defined only by defining behaviours near $r = 0$: it is therefore convenient to extract a small sphere v centred at $r = 0$ from the total region V (the slab $0 \le z \le L$) of integration. It is now possible to move the operator in $\underline{F}^*$ outside the integral to get[5]

$$4\tfrac{\pi}{3} np^2\hbar^{-1} r_+(\underline{x},\omega) + np^2\hbar^{-1}\hat{\underline{u}}\hat{\underline{u}}: (\nabla\nabla + k_o^2\underline{U}) \int_V e^{ik_o r} r^{-1} r_+(\underline{x}',\omega)\, d\underline{x}'$$

$$\qquad (2.4)$$

The first term is exactly the Lorentz field term. It has arisen here from the small sphere and appears to be arbitrary. However, more complete analysis[5,6] shows that inter-atomic correlation introduces at least 2-body correlation functions $g_2(\underline{x},\underline{x}')$ and the local field correction $np^2\hbar^{-1} \int_{V-v}\hat{\underline{u}}\hat{\underline{u}}: \underline{F}^*(\underline{x},\underline{x}'\omega) r_+(\underline{x}',\omega) [g_2(\underline{x},\underline{x}') - 1] d\underline{x}'$. The rotational and translational invariance of the integrand then justifies the choice of the small sphere for v[5,6]. But this local field correction is then roughly of the form $s\, r_+(\underline{x},\omega)$ where s is a complex number. Moreover $\text{Im}(s) \sim \tfrac{1}{2}\gamma_o n \int (g_2(\underline{x},\underline{x}') -1) d\underline{x}'$[5,6] and this modifies the radiation damping $\tfrac{1}{2}\gamma_o$. In short the generalised Lorentz field in (2.4) is $np^2\hbar^{-1} r_+(\underline{x},\omega) (4\pi/3 - s)$ where s is complex. The remaining integral in (2.4) then breaks into two parts: surface integrals which via the extinction theorem determine the surface transmission properties and the expressions like $4\pi np^2\hbar^{-1} r_+(\underline{x},\omega)/(m^2(\omega) - 1)$ which appear in precise form in (3.3) below. In introducing the refractive index $m(\omega)$ here we are already ahead of ourselves however so we say more about this now.

In linear theory $r_3(\underline{x},t) = -\tfrac{1}{2}(1 + 0(|E_o|^2)$ and the intensity dependent term is dropped since the intensities are weak. In this case one introduces $m(\omega)$ by looking for solutions of (2.1) with (2.3)

$$pr_+(\underline{x},\omega) = P_o(\omega)(e^{-imk_o z} + \Lambda(\omega)e^{+imk_o z}) \qquad (2.5)$$

One then finds[3,7]

$$m^2(\omega) - 1 = 4\pi np^2\hbar^{-1}/(\omega_o - \omega - i\gamma_o - np^2\hbar^{-1}(4\pi/3 + s)) \qquad (2.6)$$

$$\Lambda(\omega) \quad = \quad (m(\omega)-1)(m(\omega)+1)^{-1}\exp(-2im(\omega)k_oL) \tag{2.7a}$$

$$P_o(\omega) \quad = \quad (4\pi n)^{-1}(m^2(\omega)-1)\,E_o\,\{2(1+m)^{-1}\exp(-i(m-1)k_oL)\,\times$$

$$\times\,[1-(m-1)^2(m+1)^{-2}\exp-2imk_oL]\}. \tag{2.7b}$$

Technically we have made use of the extinction theorem[2,3,5,6] which, in effect, asserts that the incident free field, which has wave number k_o, and which would traverse the slab $0 \le z \le L$ in the absence of the atoms, becomes extinguished when the atoms are present and is replaced by the system of two dipole waves (2.5). Since $m \ne 1$ all waves $e^{\pm imk_oz}$ and $e^{\pm ik_oz}$ are independent of each other. The condition of consistency for waves $e^{+ik_oz}_{\pm imk_oz}$ yields (2.7a) and that for e^{-ik_oz} yields (2.7b): that for e^{+imk_oz} yields (2.6) which is the Lorentz-Lorenz relation[5,6] for 2-level atoms in r.w.a. Thus the solution (2.5) is determined and $m(\omega)$ has been fixed by an eigen condition.

The action of the extinction theorem in linear theory is therefore to obtain the usual reflexion and transmission coefficients, and to relate the transmitted field $E_{out}(\omega)e^{-ik_oz}$, through (2.7b), to the input $E_o e^{-ik_oz}$ by

$$E_{out}(\omega) = E_o\{2(m+1)^{-1}\}\{2m(m+1)^{-1}\}\exp-i(m-1)k_oL\,\times$$

$$\times\,[1-(m-1)^2(m+1)^{-2}\exp-2imk_oL]\quad. \tag{2.8}$$

Fabry-Perot (FP) cavity action is contained in the last term. These expressions depend only on $m(\omega)$ fixed, by the action of the theorem, through (2.6). Note that the Lorentz term in (2.6) shifts the resonance ω_o. Then because the local field term s is complex this both shifts the resonance and the radiation damping $\tfrac{1}{2}\gamma_o$.

The next §3 reports a natural non-linear generalisation of this linear theory.

3. THE NONLINEAR THEORY

This is complicated by the fact that r_3 is no longer $-\tfrac{1}{2}$ throughout the slab and depends explicitly on z. This means that (2.6) must be replaced by

$$pr_+(\underset{\sim}{x},\omega) = P_f(z,\omega)e^{-imk_oz} + \Lambda(\omega)P_b(z,\omega)e^{+imk_oz} \tag{3.1}$$

where m and Λ are constants to be determined.

We have managed to extend the application of the extinction theorem to the non-linear theory even though (3.1) now depends on spatially varying envelopes $P_{f,b}(z,\omega)$. Some details are in Ref.3. The key results are that equations (2.7) and (2.8) all survive intact (with one standing wave correction - see §6). However the eigen condition for $m(\omega)$ is replaced by

$$\gamma_o(1+i\delta)\, P_{f,b}(z,\omega) = \frac{2i4\pi np^2}{\hbar(m^2-1)}\, r_3(z,\omega)\left[\frac{m^2+2}{3}\, P_{f,b}(z,\omega) \mp \right.$$

$$\left. \mp \frac{2im}{k_o(m^2-1)}P'_{f,b}(z,\omega)\right] \qquad (3.2)$$

Both m and $P_{f,b}(z,\omega)$ are now to be determined from (3.2), for it is an eigen condition only if $P'_{f,b} = 0$.

The way to proceed is to treat $m(\omega)$ as a disposable parameter and choose it so that FP cavity action survives. For this it is sufficient that $P_f(0) = P_b(0) = P_f(L) = P_b(L) (= P_o(\omega)$ and this is achieved by

$$(m^2-1)(m^2+2)^{-1} = (-2R_3)4\pi np^2\hbar^{-1}/3(\omega_o-\omega-i\gamma_o) \qquad (3.3)$$

where

$$R_3^{-1} \equiv L^{-1}\int_o^L [r_3(z,\omega)]^{-1}\, dz. \qquad (3.4)$$

Conversely if dispersive OB is to be interpreted in terms of a non-linear refractive index $m(\omega)$ it is given by (3.3) and R_3 must be chosen as in (3.4). Evidently R_3 is an average of $r_3(z,\omega)$; but it is not the mean field average.

For $r_3(z,\omega)$ itself we find

$$r_3(z,\omega) = -\tfrac{1}{2} - \tfrac{1}{2}(r_3(z,\omega))^{-1}[e^{-i(m-m^*)k_o z}|P_f(z,\omega)|^2$$

$$+ e^{+i(m-m^*)k_o z}|P_b(z,\omega)||\Lambda|^2] \qquad (3.5)$$

One can check that it is the absorptive bistability (the damping) which induces the z-dependence in (3.5). The problem is solved numerically by a self-consistent choice of R_3: R_3 is chosen, and determines $m(\omega)$; then $r_3(z,\omega)$, used to express $P_{f,b}(z,\omega)$ via (3.2), is calculated from (3.5), and R_3 is calculated from (3.4) and checked against its initial choice. We report on numerical results obtained this way elsewhere. Here we report how conventional OB lies within the present theory (§4) and on some effects of the Lorentz field (§5).

4. SMALL CO-OPERATION NUMBER THEORY

The result (2.8), which survives in the non-linear theory connects output intensity $|E_{out}|^2$ with input intensity $|E_o|^2$. For simplicity (only) we shall assume $m^2-1 \approx 2(m-1)$, a result true only if n is small. If $2(m-1)k_oL \equiv A - iB$ one easily finds

$$|1 - (m-1)^2(m+1)^{-2} \exp - 2imk_oL|^2$$

$$= (1 - |R|^2 e^{-B})^2 + 4|R|^2 \sin^2 \tfrac{1}{2}(\theta+A) \qquad (4.1)$$

where $R \equiv (m-1)(m+1)^{-1} = |R|e^{i\lambda}$ and $\theta = 2(\lambda+\nu\pi-k_oL)$ with ν an integer. Evidently $\Lambda = R \exp -2imk_oL$ by (2.7a). We keep R in (4.1) and in what follows: it has been found consistently from the 2-level atom theory as such; but it could become applicable to real many-level atoms by adding in the effects of all other levels linearly, and these probably dominate the non-linear feature.

Here we focus attention on the second term: evidently A depends on R_3 and hence, via (3.4), (3.5), (2.7b) and (2.8) on $|E_o|^2$. Consequently this second term is the source of dispersive OB. We introduce the co-operation number $C = 2\tau_R^{-1}\gamma_o^{-1}(1+\delta^2)^{-1}$ where $\tau_R \equiv \hbar(4\pi np^2k_oL)^{-1}$ is the co-operative super-fluorescence time[7]: C can be small either because n is small or k_oL is small or both. We suppose n small so that the Lorentz field can be neglected. Then, for small enough C, one finds from (3.5)

$$R_3 \approx -\tfrac{1}{2}(1+\delta^2) / (1+\delta^2+\gamma_o^{-2}I_o) \qquad (4.2)$$

where $I_o \equiv p^2\hbar^2|E_{out}|^2(1+|R|^2)|(1+m)/2m|^2$ and is essentially the intensity inside the cavity. For small n one finds $A = \delta(4\pi np^2\hbar^{-1})R_3/(1+\delta^2)$. Then for $4\sin^2\{\tfrac{1}{2}(\theta+A)\} \approx (\theta+A)^2$ and $B \approx 0$, the use of (4.2) and insertion of (4.1) in (2.8) yields

$$p^2\hbar^{-2}|E_o|^2(1+|R|^2) = |m|^2I_o\left[1 + \left\{\frac{|R|^2}{1-|R|^2}\left(\theta - \frac{\delta(2\pi np^2\hbar^{-1})}{1+\delta^2+I_o\gamma_o^{-2}}\right)\right\}\right] \qquad (4.3)$$

We use $(1-|R|^2)^2 \approx |2m/(1+m)|^2$. The result (4.3) is essentially the usual expression for dispersive OB. It becomes multiply branched if sine terms as such are retained from (4.1). If $B \neq 0$ (4.3) is corrected by additional absorptive bistability.

5. EFFECT OF THE LORENTZ FIELD

If n is not small enough (3.4) means that

$$m^2(\omega) - 1 = 4\pi np^2\hbar^{-1}(-2R_3)/(\omega_o-\omega-i\gamma_o+np^2\hbar^{-1}(-2R_3)(4\pi/3+s)) \qquad (5.1)$$

The resonance ω_o is therefore shifted by a term linear in R_3: because s is complex γ_o is also changed similarly. According to Bowden et al.[4] the interesting behaviour arises for $k_oL \ll 1$: C is then small and a generalisation of (4.2) applies. For, from (5.1) $\delta \to \Delta = \delta - (n^2p^2\hbar^{-1}\gamma_o^{-1})(-2R_3)(4\pi/3 + \text{Re}(s))$ and $\gamma_o \to \Gamma = \gamma_o + (np^2\hbar^{-1})(-2R_3)\text{Im}(s)$. We focus attention on the shifted detuning Δ and ignore s. Since C

is small the generalisation of (4.2) proves to be this expression
with $\delta \to \Delta$. This however means that

$$I_o \gamma_o^{-2} = -R_3^{-1}(\tfrac{1}{2}+R_3)(1+\Delta^2) \qquad (5.2)$$

which is a cubic in R_3, divided by R_3.

Because C is small $(1 - |R^2|e^{-B})^2 \simeq (1 - |R|^2)^2 \{1 + B/(1 - |R|^2)\}^2$
and $B = (4\pi n p^2 R_3)p^2 k_o L(\hbar\gamma_o)^{-2}(1 + \Delta^2)^{-1}$. Since $k_o L \ll 1$ we neglect the
dispersive contribution. This way we reach

$$p^2\hbar^{-2}|E_o|^2|2(1+m)^{-1}|^2 = I_o\{1 + B/(1 - |R|^2\}^2 \qquad (5.3)$$

in which the curly brackets is $(1 + \Delta^2 + DR_3)/(1+\Delta^2)^2$ where D absorbs
the various constants. If we substitute I_o from (5.2) into (5.3)
the right side is a cubic in R_3 divided by R_3 multiplied into a
quartic divided by the quartic $(1 + \Delta^2)^2$. Evidently one factor
$(1+\Delta^2)$ cancels and the right side is a quintic divided by a cubic.
Thus there can be up to five distinct roots for R_3 given the value
of $|E_o|^2$. Evidently there are only three such roots when the Lorentz
field is ignored. To this extent the two extra roots corroborate
the thesis of Bowden et al.[4]. But we have not yet done the numerical
work to establish the conditions under which these roots arise and
where they will be in terms of the parameters of the system.

The main point of this §5 is to indicate the comprehensive
character of the unapproximated theory. Note that the extra roots
associate with $B \neq 0$ and so with absorptive OB even if the roots are
effective in a very different range of parameters. Evidently dispersive OB could also provide extra roots via the Lorentz field.

6. STANDING WAVE EFFECTS

Unfortunately the analysis of the steady state is in general
incomplete because $r_3(z,\omega)$ actually depends on an intensity in the
slab proportional to $|P_f(z,\omega)e^{-imk_o z} + P_b(z,\omega)\Lambda(\omega)e^{imk_o z}|^2$ and cross
terms $e^{\pm i(m+m^*)k_o z}$ arise. The ansatz (3.1) must then be extended
to include all odd spatial harmonics of $e^{\pm imk_o z}$. The use of the
extinction theorem is still possible[3]. But now the surface effects
derived are more complicated (involving e.g. $m(\omega)$, $3m(\omega)$, etc.) and
coupled equations of motion from the expansions of $P_{f,b}(z,\omega)$ and
$r_3(z,\omega)$ also involving $m(\omega)$ arise. These are described briefly in
Ref. 3 and their consequences must be explored elsewhere. One
result[2,3] is that (2.8) gains the extra factor $[r_3(0,\omega)/r_3(L,\omega)]^2$
on the right side but the effects of this must also be reported
elsewhere.

REFERENCES

1. H. J. Kimble, A. T. Rosenberger and P. D. Drummond, "Optical
        Bistability with Two-Level Atoms" in: "Coherence and Quantum
        Optics 5", L. Mandel and E. Wolf, eds. Plenum Press, New
        York (1983).
2. R. K. Bullough, S. S. Hassan and S. P. Tewari, "Refractive Index
        Theory of Optical Bistability", in: "Quantum Electronics and
        Electro-Optics," pp. 229-232, P. L. Knight ed., John Wiley
        & Sons Ltd., London (1983).
3. R. K. Bullough and S. S. Hassan, "The optical extinction theorem
        in the non-linear theory of optical multistability", proc.
        SPIE 369:257-263 (1982).
4. C. M. Bowden, F. A. Hopf and W. H. Louisell, "Mirrorless Intrinsic
        Optical Bistability due to the Local Field Correction in the
        Maxwell-Bloch Formulation", Paper THA5 in these Proceedings.
5. L. Rosenfeld, "Theory of Electrons", Dover Publications Inc.,
        New York (1965).
6. R. K. Bullough, Phil. Trans. Roy. Soc. A 254:397 (1962);
        J. Phys. A 1:409 (1968); 2:477 (1969); 3:708, 726, 751 (1970).
7. For example R. K. Bullough, R. Saunders and C. Feuillade,
        "Theory of Far Infra Red Superfluorescence", in: "Coherence
        and Quantum Optics IV," L. Mandel and E. Wolf, eds., Plenum
        Press, New York (1978) and the references there.

THEORY OF OPTICAL BISTABILITY IN NONLINEAR MEDIA

FOR SLOWLY VARYING INCIDENT INTENSITY

Y. B. Band

Allied Corporation
7 Powderhorn Drive
Mt. Bethel, New Jersey 07060

Many treatments of bistability in nonlinear media describe the phenomena in the regime when the incident intensity varies slowly compared with the material response time by writing down the transmission of a Fabry-Perot interferometer, τ, as a function of the intracavity phase shift in a roundtrip pass, δ : $\tau = (1+F \sin^2 \delta/2)^{-1}$ where $F = 4r/(1-r)^2$ and r is the reflectivity from the surface of the material[1,2,3]. When the index of refraction is a function of the intensity inside the cavity (I_{in}), the phase shift is intensity-dependent, and it appears that one can substitute the expression for the intensity-dependent phase shift into the expression for the transmission thereby obtaining an expression for the transmission as a function of the intensity inside the cavity. When this equation is solved in conjunction with an expression for the transmission as a function of the ratio of I_{in} to the incident intensity (I_0), one obtains an expression for $\tau(I_0)$ which may be multivalued. The flaw in this argument is that the expression for the transmission as a function of phase shift is derived assuming linear wave propagation where the superposition principle applies[2]. But the propagation equation is nonlinear, and the superposition principle is not valid!

For thin slabs (kL $\ll$ 1), the euristic Fabry-Perot approach is inadequate, even for very low incident intensities. This can be shown by the following argument. By solving the boundary value problem for a linear dielectric medium one obtains

$$E(Q) = E_0 T [\cos (nkQ) - in^{-1} \sin (nkQ)] , \qquad (1)$$

where

$$T = \frac{2e^{-ikL}}{2 \cos (nkL) + i(n+1/n) \sin (nkL)} \qquad . \qquad (2)$$

Letting $\tau = |T|^2$ and taking n as a function of the average intensity inside the medium, I_{in}, $n(I_{in})$, we find

$$\tau = \frac{1}{1 + \frac{(n^2(I_{in})-1)^2}{4n^2(I_{in})} \sin^2 (n(I_{in})kL)} \qquad . \qquad (3)$$

From Eq. (1), the intensity as a function of position in the medium is given by

$$I(Q) = I_0 \tau n [\cos^2 (nkQ) + n^{-2} \sin^2 (nkQ)]. \qquad (4)$$

The average value of intensity inside the medium, $I_{in} = \langle I(Q) \rangle$, depends on how the average is performed. Averaging over large Q (say, because the light bounces around in the medium many times), yields $I_{in} = I_0 \tau n [1+n^{-2}]/2$ whereas the averaging over the length L, where $kL \ll 1$, yields $I_{in} = I_0 \tau n$. These two relations for I_{in} vs I_0 will yield different values of reflectivity vs I_0. The standard euristic procedure[2,3] (applied when $kL \gg 1$) is to neglect the I_{in} dependence in the prefactor of $\sin^2 (n(I_{in})kL)$ in Eq. (3) and to use the relation $I_{in} = I_0 \tau n [1+n^{-2}]/2$ within sine. It should be mentioned that for thick slabs the euristic Fabry Perot approach is also inadequate as can be shown by using the slowly-varying envelope approximation (SVEA)[4]. An additional coherence term changes the argument of sine in Eq. (3) by a factor of 1.5 (see below). The dashed line in Fig. 1 shows the reflectivity as calculated by the standard approach for a thin slab of nonlinear material InSb of thickness, L, equal to 0.18 µm wherein the nonlinear susceptibility is given by the model proposed by Weaire et. al.[5] $\chi(I_{in}) = n^2 - 1 + 2n n_2 I_{in}$ with n = 4, $n_2 = -6 \times 10^{-9}$ m^2/W for 1886 cm^{-1} (5.3 µm) radiation. The dotted line is obtained by neglecting the I_{in} dependence of the prefactor in Eq. (3) and using $I_{in} = I_0 \tau n$. The dashed-dotted curve results from using Eq. (3) with $I_{in} = I_0 \tau n [1+n^{-2}]/2$.

In Reference 6, the exact propagation equations are solved for the reflection and transmission coefficients (R,T) for a plane wave with field strength, $|E_0|^2 = y$, incident normally on a slab of material of length L with an intensity-dependent refractive index. Using the same notation as in Section II of Reference 1, the nonlinear wave equation inside the slab ($Q \epsilon (0,L)$) takes the form

$$(d^2/dQ^2 + k^2)E(Q) = V(Q),|E(Q)|^2)E(Q) , \qquad (5)$$

and the boundary conditions can be written as[6]

$$E(Q) = \sqrt{y}[e^{-ikQ} + R(L,y)e^{+ikQ}], \qquad Q \geq L \qquad (6)$$

$$E(Q) = \sqrt{y}\, T(L,y)e^{-ikQ}, \qquad\qquad Q \geq 0 \qquad (7)$$

where (in MKS units) k is the wavevector in vacuum, $k^2 = \epsilon_0\mu_0\omega^2$, and V is given in terms of the susceptibility, X, by

$$V(Q, |E(Q)|^2) = -k^2 X(Q, |E(Q)|^2) . \qquad (8)$$

The propagation equations for the reflection and transmission coefficients as a function of varying slab thickness, x , can be derived using the method of invariant imbedding. They take the form

$$\tilde{R}_x(x,y) = (2ik-iV/k) \tilde{R}(x,y)-iV/(2k)(1+\tilde{R}^2(x,y))$$

$$+ Vy/k\, \tilde{R}_y(x,y)\, \text{Im}\, \tilde{R}(x,y), \qquad (9)$$

$$T_x(x,y) = -iV/(2k)\, T(x,y)\, (1+\tilde{R}(x,y)) + Vy/k\, T_y(x,y)$$

$$X\, \text{Im}\, \tilde{R}(x,y) \quad , \qquad (10)$$

where

$$\tilde{R}(x,y) \equiv e^{2ikx}\, R(x,y) \qquad (11)$$

and

$$V \equiv V(x, |E(x,y)|^2) = V(x,y\, |e^{-ikx} + \tilde{R}(x,y)e^{ikx}|^2)$$

$$= V(x,y\, |1+\tilde{R}(x,y)|^2) \quad . \qquad (12)$$

The initial conditions for $\tilde{R}$ and T are

$$\tilde{R}(0,y) = 0 \qquad (13)$$

$$T(0,y) = 1 \qquad (14)$$

In Figure 1 the solid curves show the reflectivity vs. incident intensity calculated using this approach. The partial differential equation solver of Ref. 7 was used. Two paths are shown: one with maximum incident intensity equal to 15.38 mW/cm^2 before reduction of the intensity back down to zero, and another with maximum intensity equal to 20 mW/cm^2. Note that optical bistability is predicted for both paths, but that the bistability is different for the two paths. The euristic Fabry-Perot approaches do not yield optical bistability in this regime of kL >> 1.

In the SVEA we let $E(Q) = A_+(Q)e^{ikQ} + A_-(Q)e^{-ikQ}$, where K = nk. The following first-order differential equation is obtained

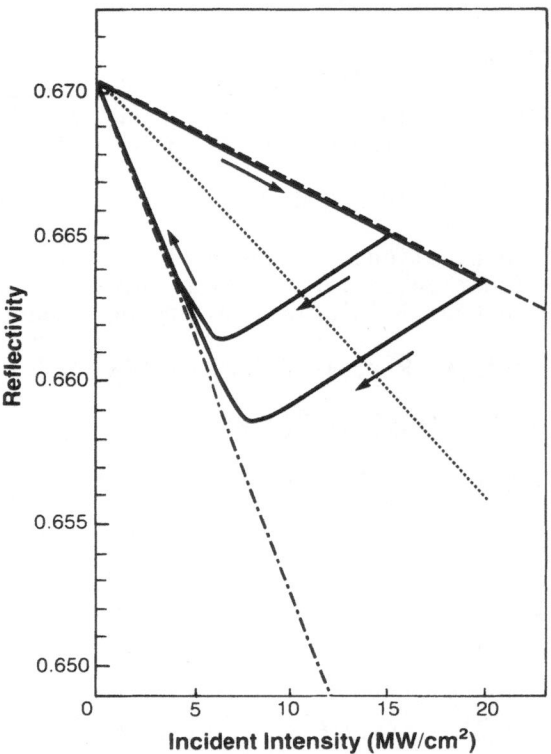

Fig. 1.   Reflectivity vs. incident in a thin slab in InSb of thick-
          ness 0.18 μm (for which kL << 1).   The dashed, dotted and
          dashed-dotted curves are various euristic Fabry-Perot
          transmission model calculations.   The solid curves are
          numerical solutions of the nonlinear boundary value prob-
          lem for maximum intensities of 15.38 and 20 mW/cm^2, re-
          spectively.

$$A_+^{'} e^{iKQ} - A_-^{'} e^{-iKQ} = \frac{-k^2}{2iK} \chi_{n\ell}^{'} \left(|A_+ e^{iKQ} + A_- e^{-iKQ}|^2 \right)$$

$$X \left(A_+ e^{iKQ} + A_- e^{-iKQ} \right) \quad , \tag{15}$$

where $\chi_{n\ell} \left(|E(Q)|^2 \right)$ is the nonlinear susceptibility.   Matching at
the boundary interface yields

$$A_+(0) = \sqrt{y}\, T(1-k/K)/2 \quad , \tag{16}$$

$$A_-(0) = \sqrt{y}\, T(1-k/K)/2 \quad , \tag{17}$$

$$A_+(L) = \sqrt{y}\, [e^{ikL}(1-k/K) + e^{ikL} R(1+k/K)]/2e^{ikL} , \tag{18}$$

$$A_-(L) = \sqrt{y}\, [e^{-ikL}(1+k/K) + e^{ikL} R(1-k/K)]/2e^{-ikL} . \tag{19}$$

For kL < 1 (as is the case here) it is not valid to drop in the propagation equation the rapidly varying terms on the right hand side of Eq. (15), and furthermore, the second derivative terms of A_+, A_-, should be retained if a significant effect is present in a fraction of a wavelength. Therefore, the SVEA is appropriate only for kL >> 1. Only if an _analytic_ solution can be obtained for $A_+(Q), A_-(Q)$ is it possible to easily solve the two point boundary value problem for T, R and the other two integration constants in the analytic expressions for $A_+(Q)$, $A_-(Q)$ from the four Eqs. (16) - (19). Otherwise iterative shooting methods for the solution of (15) - (19) must be used. For kL >> 1, and the present, simple form of susceptibility, an analytic solution for $A_+(Q)$ and $A_-(Q)$ can be found and a transcendental equation for R and T is obtained from (19) - (22):

$$A_+(Q) = A_+(0) \exp\,[ik^2 \xi(|A_+(0)|^2 + 2|A_-(0)|^2)\, Q/(2K)], \tag{20}$$

$$A_-(Q) = A_-(0) \exp\,[-ik^2 \xi(2|A_+(0)|^2 + |A_-(0)|^2)\, Q/(2K)], \tag{21}$$

where $\xi = nn_2 \sqrt{\epsilon_0/\mu_0}$. Substituting into Eqs. (16) - (19) the following equation for R results,

$$(1+\tilde{R}/p) \exp\,[-i3k^2 \xi yL(1- |\tilde{R}|^2)\, [(1-k/K)^2 + (1+k/K)^2]/(8K)]$$

$$-e^{2iKL}\,(1+\tilde{R}\,p) = 0 , \tag{22}$$

where $p = (1-k/K)/(1+k/K)$. This equation gives results identical to those found by Marburger and Felber[4a], but one set of branches of solution for $|R|^2$ obtained by Marburger and Felber is not a solution of Eq. (22) for R as these branches are in fact false solutions to the problem[4b]. In Eq. (22) an additional coherence term not present in the euristic Fabry-Perot approach arises from scattering of the forward (backward) propagating wave from the grating formed by the forward and backward waves to produce a wave which propagates in the backward (forward) direction.

Figure 2 shows the results of calculations using the standard Fabry-Perot transmission model for L = 182.2 μm (KL equals a large multiple of π so the reflectivity at zero intensity equals zero). Figure 3 shows the reflectivity calculated using Eq. (22). Thus,

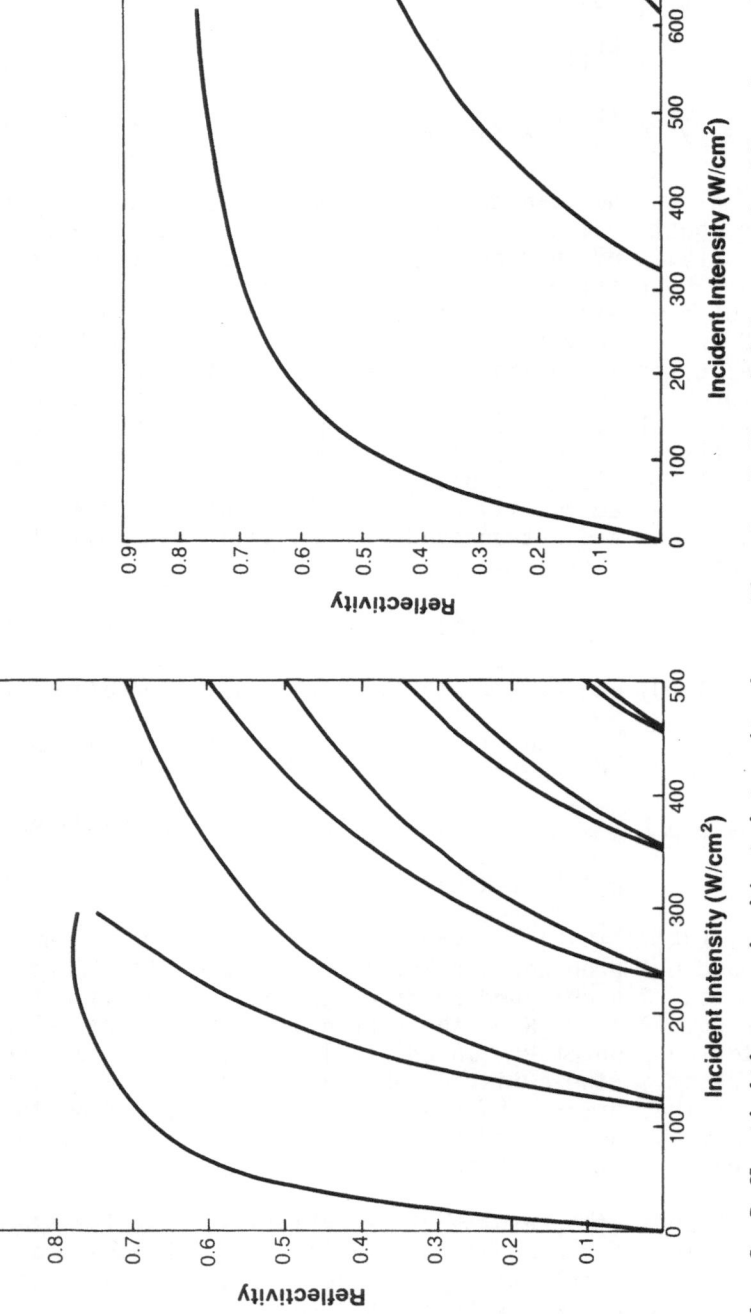

Fig. 3. Reflectivities vs. incident intensity in a slab of InSb of thickness 182.2 μm calculated using the slowly-varying envelope approximation

Fig. 2 Reflectivities vs. incident intensity in a slab of InSb of thickness 182.2 μm calculated using the standard Fabry-Perot euristic approach.

even for kL >> 1, there is a different scale factor for the incident intensity dependence of the reflectivity predicted by the euristic Fabry-Perot approach then by the SVEA. A more complete study of the optical bistability obtained by the SVEA will be presented elsewhere[4b].

REFERENCES

1. M. Born and E. Wolf, Principles of Optics, 6th ed. (Pergamon, N.Y., 1980), pp 323-329.
2. F.S. Felber and J.H. Marburger, Appl. Phys. Lett. 28, 731 (1976).
3. D.A.B. Miller, S.D. Smith and A. Johnson, Appl. Phys. Lett. 35, 658 (1979).
4. (a) J.H. Marburger and F.S. Felber, Phys, Rev. A17, 355 (1978). (b) Y.B. Band, to be published.
5. D. Weaire, B.S. Wherrett, D.A.B. Miller, and S.D. Smith, Opt. Lett. 4, 331 (1979). See also, D.A.B. Miller, S.D. Smith, and C.T. Seaton, in Optical Bistability, ibid, pp 115-126.
6. Y.B. Band, J. Appl. Phys. 53, 7240 (1982).
7. R.F. Sincovec and N.K. Madsen, "Software for Nonlinear Partial Differential Equationa," ACM Transactions on Mathematical Software, September 1975, p. 232 ff.; G. Sewell, "IMSL Software for Differential Equations in One Space Variable," IMSL Technical Report Series No. 8202, January 1982. "Coherence and Quantum OpticsV," L. Mandel and E. Wolf, eds., Plenum Press, New York (1983).

STRICTLY QUANTUM OPTICAL BISTABILITY

R.R. Puri, G.P. Hildred, S.S. Hassan[†] and R.K. Bullough

Department of Mathematics
UMIST, P.O. Box 88
Manchester M60 1QD, U.K.

[†]Department of Applied Mathematics
Ain Shams University
Cairo, Egypt

INTRODUCTION

Theories of dispersive optical bistability in cavities of
finite extent indicate that strictly quantum (quantal) aspects need
play no role (cf. eg.[1]). Correspondingly the Dicke model driven
by a single mode coherent state field displays many wholly quantal
features but does not show optical bistability[2,3,4]. This model,
due to R.H. Dicke[5], consists of N 2-level atoms on the same site
and there is no cavity: in the coherent single mode field of amplitude Ω there is a phase transition in a thermodynamic limit in which
$N \to \infty$ and $\Omega\gamma_o^{-1} \to \infty$ in such a way that the order parameter $\Theta \propto \Omega N^{-1}\gamma_o^{-1}$
remains finite[2,3]: (γ_o is the A coefficient): as Θ increases the
system moves from an atomic coherent state and zero quantum fluctuations in the collective inversion per atom to a state in which such
quantum fluctuations are finite and have value 1/12 per atom: at the
same time the intensity-intensity correlation function $g^{(2)}(0)$ for
resonance fluorescence moves from 1 to 1.2, that is from coherence
to <u>partial</u> coherence.

The question then arises as to whether this manifestly quantal
system can display optical bistability in any circumstances. We
show in this paper that by coupling the model into a cavity that it
can. Indeed we show that both the cavity and the quantal character
of the model are essential to the optical bistability, and that
either alone is not sufficient for it. This otherwise paradoxical
result (paradoxical when compared with behaviour in the extended

cavity) seems to be closely associated with conservation of total spin $\vec{S}^2$: $\vec{S}^2$ is a constant of the motion for the Dicke model but is not conserved in the corresponding system of N 2-level atoms occupying an extended region of space. Moreover if $\vec{S}^2$ = constant is deliberately broken by a de-correlation of the Dicke model itself absorptive optical bistability becomes a consequence[4,6].

These results all suggest that experimental studies of a coherently driven Dicke-like atomic system, both coupled to a cavity and not so coupled, would be particularly rewarding. We have shown elsewhere how to include the effects of a cavity on the Dicke model in a broad band chaotic field[7]. The results agree astonishingly well with those recently observed on high Rydberg Na atoms in single mode black-body fields[8]. The single mode lies inside a cavity and its wavelengths ~ 2mm. so that the point-like Dicke model becomes in this respect a realistic physical model. On the other hand the model is not the limit of any real extended atomic system and indeed $\vec{S}^2$ is no longer a constant once the atoms occupy a region of finite extent however small. Thus its experimental investigation inside a cavity and otherwise would be particularly interesting as a test of the model as well as of the consequences of the model.

Our results for the Dicke model in a single mode coherent state field without a cavity are already available[2,3,4]. The point of the present paper is therefore to determine the effects of the cavity, to show that quantum optical bistability is a consequence of it, and to stimulate experiments which might observe it.

2. THE EQUATIONS OF MOTION

We work first of all with the total density operator ρ for a system of N 2-level atoms on the same site coupled by dipole interaction to a single cavity mode on resonance with the atoms. The cavity mode is driven by an external c-number field E(t): the coupling of a coherent external field to the field inside a cavity is conceptually simple but technically involved[9] and we take a simpler viewpoint here. Nevertheless we believe this simplified analysis contains the essential feature which allows us to conclude that, given the validity of the model, quantum optical bistability can be observed inside a coherently driven cavity.

We take the equation of motion for ρ to be

$$\frac{d\rho}{dt} = -i[(H_o + H_1 + H_2), \rho(t)] + \Lambda_a \rho(t) + \Lambda_f \rho(t) \quad ; \tag{1}$$

$H_o = \omega_s S_z + \omega_s a^\dagger a$; $H_1 = -igaS_+ + iga^\dagger S_-$; $H_2 = i\kappa(a^\dagger E(t) - aE^*(t))$ in which $S_\pm$, S_z are the usual collective Dicke operators and satisfy $[S_+, S_z] = \mp S_\pm$, $[S_+, S_-] = 2S_z$[2,3,4,5]; g is a real coupling constant, and there is a preferred mode on exact resonance with the atomic resonance at $\omega = \omega$. The situation compares with the incoherently driven case [7], but the new feature is the coherent driving term H_2.

The two remaining terms are

$$\Lambda_a \rho = \gamma_o n_{cav} [2S_-\rho S_+ - S_+ S_-\rho - \rho S_+ S_-] \tag{2}$$

and

$$\Lambda_f \rho = \kappa [2a\rho a^\dagger - a^\dagger a\rho - \rho a^\dagger a] \qquad . \tag{3}$$

In these γ_o is again the A-coefficient, n_{cav} plays the role of Purcell's cavity factor [7,10] and, κ is a cavity damping constant. The term $\Lambda_a \rho$ in $\gamma_o n_{cav}$ describes collective spontaneous emission and its general form accords with that derived in the incoherent case [7] Conceptually it differs profoundly from that obtained in the theory of super-radiance [11] with which it might be compared — the important point being that here it concerns a <u>point</u> atomic system. The remaining term $\Lambda_f \rho$ describes the damping of the cavity mode as before[7,11].

It is easy to show from (1) that, with $\eta_{cav} = 1$, an arbitrary atomic operator Q and the field mode operator a satisfy respectively the equations of motion

$$\frac{d\langle Q\rangle}{dt} = +i\omega_s \langle[Q, S_z]\rangle + g\langle[Q, aS_+]\rangle - g\langle[Q, S_- a^\dagger]\rangle$$
$$+\gamma_o \langle[2S_+ QS_- - QS_+ S_- - S_+ S_- Q]\rangle \tag{4}$$

$$\frac{d\langle a\rangle}{dt} = +i\omega_s \langle a\rangle - \kappa(\langle a\rangle - E) + g\langle S_-\rangle \qquad . \tag{5}$$

So far we are unable to solve this system exactly. But the work on the coherently driven Dicke model [2,3,12] indicates that for large enough N ($N \to \infty$) we can de-correlate $\langle aA\rangle = \langle a\rangle\langle A\rangle$ if A is an arbitrary atomic operator. De-correlating equations (4) this way and moving to a frame rotating at ω_s, we find that (with $\alpha \equiv \langle a\rangle$)

$$\frac{d\langle Q\rangle}{dt} = g\alpha\langle[Q, S_+]\rangle - g\alpha^*\langle[Q, S_-]\rangle + \gamma_o\langle[2S_+ QS_- - QS_+ S_- - S_+ S_- Q]\rangle \tag{6}$$

$$\frac{d\alpha}{dt} = -\kappa(\alpha - E) + g\langle S_-\rangle \qquad . \tag{7}$$

These are the equations of motion of the theory.

3. THE STEADY STATE SOLUTION

For a theory of optical bistability we need the steady state solution of equations (6) and (7). Because (6) coincides with the corresponding equation for the Dicke model without a cavity we can use the exact results for the coherently driven model we have obtained already [2,3,4]. Accordingly for arbitrary N we know that a steady solution of (6) for $Q \equiv S_-$ can be obtained in the form

$$N^{-1}<S_-> = \tfrac{1}{2}z[1 - (N+1)D^{-1}] \quad ; \quad D = \sum_{m=0}^{N}\left(\frac{Nx}{2}\right)^{-m}\frac{(N+m+1)!(m!)^2}{(N-m)!(2m+1)!} \quad (8)$$

where $z = -2\alpha g/\gamma_o N$ and $x = |z| = 2|\alpha|g/\gamma_o N$. On the other hand the steady solution of (7) yields

$$N^{-1}<S_-> = + (\kappa\gamma_o/2g^2)(z - 2Eg/\gamma_o N) \quad (9)$$

so we introduce $y \equiv 2|E|g/\gamma_o N$ which is proportional to the incident field amplitude $|E|$. From (9) and (8) we then find

$$y = x\{1 + C[1 - (N+1)D^{-1}]\} \quad (10)$$

where the co-operation number $C \equiv g^2\kappa^{-1}\gamma_o^{-1}$. This shows bistable behaviour even for $N=1$, but for this value of N our de-correlation of equation (4) to reach (6) must be in doubt. For large N, (10) becomes

$$y = (1 + C)x + O(N^{-1}) , \quad x \lesssim 1$$

$$y = x[1 + C(1 - \sqrt{x^2 - 1}/x^2 \sin^{-1}(1/x)] + O(N^{-1}) , \quad x \gtrsim 1 \quad (11)$$

and for $N \to \infty$ two distinct regions $x < 1$, $x > 1$ become well defined, while there is a cusp at $x = 1$ itself.

Plots of x against y for different C for $N = 20$ and $N \to \infty$ are shown in Figs. 1(a) and (b) respectively and show conventional bistable

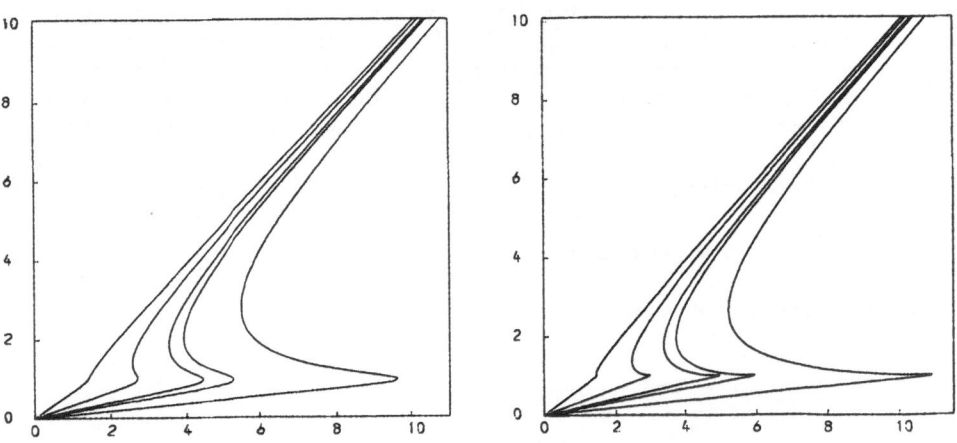

Fig. 1. (a) x versus y for $N = 20$ and $C = 0.5, 2, 4, 5, 10$; C increases from left to right. (b) the same as (a) but for $N \to \infty$.

behaviour for $C \gtrsim 0.1$. Of course this is $C \ll 4$, and the curves are not the usual cubic catastrophe curves. Still the portions of negative slope are unstable and hysteresis must occur in practice.

Despite this the bistability is strictly quantal in character: we can use the solution of (6) obtained previously[3] to compute the intensity-intensity correlations $g^{(n)}(0) \equiv \langle S_+^n S_-^n \rangle / \langle S_+ S_- \rangle^n$ as a function of x. And from (10) we can find $g^{(n)}(0)$ as a function of y. For small y, x is small and (from [3])

$$\langle S_+ S_- \rangle = |y|^2 [(1+C)^2 + 2|y|^2]^{-1} \quad , \quad \langle S_+^2 S_-^2 \rangle = 0 , \qquad (12)$$

for $N = 1$ (and small y — given that the theory of bistability is valid for $N = 1$); but for $N \geq 2$

$$\lim_{y \to 0} g^{(n)}(0) = 1 \quad , \quad n = 1, 2, \cdots, N$$

while

$$\lim_{y \to \infty} g^{(2)}(0) = 1.2 \quad (\text{for } N \to \infty) \qquad . \qquad (13)$$

Thus for $N \geq 2$ the state on the lower branch of the optical bistability curve tends to a coherent state as $N \to \infty$. On the other hand the upper branch, as well as the unstable branch, is partially coherent but certainly not incoherent (i.e. $g^2(0)2$). The Fig. 2 shows $g^{(2)}(0)$ as a function of y for $N = 4, 20, 50$ and fixed $C = 10$: evidently there is anti-bunching for $N \lesssim 4$ near to the phase transition. In Fig. 3 we plot the fluctuations $\sigma_{zz} = \{\langle S_z^2 \rangle - \langle S_z \rangle^2\} N^{-2}$ per

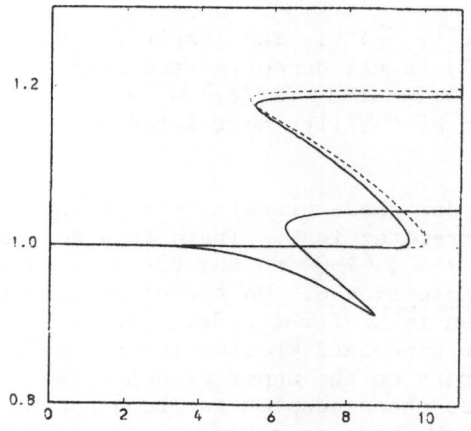

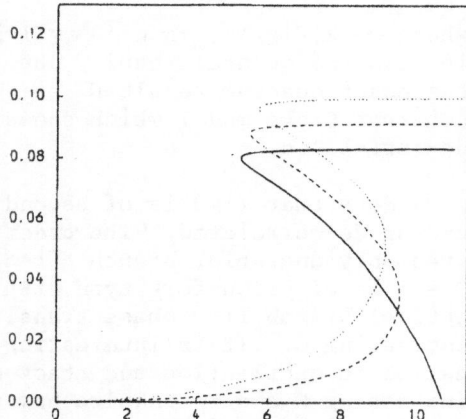

Fig. 2. $g^{(2)}(0)$ versus y for $C = 10$ and $N = 4, 20, 50$; $N = 4$ shows the antibunching; $N = 50$ is dotted.

Fig. 3. σ_{zz} versus y for $C = 10$ and $N = 10$ (dotted line), 20 (dashed line), ∞ (full line).

atom as a function of y: for $N \to \infty$, $\sigma_{zz} = 0$ on the whole of the lower branch and this is consistent with the coherent state on this branch. On the upper branch σ_{zz} rises steadily to values $\geq 1/12$ even for $N \to \infty$.

For the resonance fluorescence spectrum on the lower branch we so far have to follow the work on the Dicke model in Ref.[12]. We have not yet solved the model for transients, but for $N \gg 1$ semi-classical theory is rigorously applicable for $x < 1$, and the spectrum, which is then a δ-function, was found this way[12]. In the present case adiabatic elimination of $<a> = \alpha$ in (4) yields the same equation in semi-classical approximation, so the system will have the same spectrum in semi-classical approximation (the equations are not identical more generally). Evidently the same methods do not apply to the upper branch where $g^{(2)}(0) \to 1.2$ and the quantum fluctuations σ_{zz} persist. We have calculated the spectrum analytically for $N = 1$[13] (when it has the well known 3-peaked form) and numerically for $N = 2$ where it gains two weak additional side-bands. And we also know from semi-classical analysis[4],[14] that for N atoms the spectrum gains $2(N-1)$ weak side-bands. Beyond this we have no results.

It is important to stress again that although for $N \to \infty$ the system is in a coherent state on the lower branch and behaves semi-classically there, the system remains strictly quantal: for if we try to solve equations (6) in semi-classical approximation by de-correlating the atomic operators, we find the solution of the resulting equations in the steady state is

$$N^{-1}<S_z> = -\tfrac{1}{2}\sqrt{1 - \Theta^2} \quad , \Theta < 1; = 0 \quad , \quad \Theta \geq 1 \quad ; \tag{14}$$

$$N^{-1}<S_y> = \tfrac{1}{2}\Theta \quad , \quad \Theta < 1 \quad ; \quad = 1/2\Theta \quad , \quad \Theta \geq 1 \quad ; \tag{15}$$

where $\Theta = 2|E|g/N(\gamma_0 + g^2\kappa^{-1}) = y(1 + g^2\kappa^{-1}\gamma_0^{-1})^{-1}$. The result for $<S_z>$ is also the quantal result, but (15) is the de-correlated form of the exact quantum result $N^{-1}<S_y> = \Theta[1 - (N+1)(N\Theta/2)^{2N}D^{-1}]$ for the coherent Dicke model which shows no bistability correlated or de-correlated.

Note that (14) is of second order phase transition type correlated or de-correlated. The order parameter is Θ. There is a second, preumably unstable, branch $N^{-1}<S_z> = +\tfrac{1}{2}\sqrt{1 - \Theta^2}$ so the bifurcation at $\Theta = 1$ is of pitch-fork type with decreasing Θ. On the other hand the optical bistability phase transition is of first order type with increasing Θ. It is interesting to speculate whether further bifurcation to oscillation and chaos occurs on the upper branch with increasing Θ and to enquire how this chaos couples to the quantum fluctuations on this upper branch. We hope to report on this aspect elsewhere.

4. CONCLUSION

Our conclusion from this strictly quantal model coupled to a cavity must be that quantum fluctuations and cavity feed back are together sufficient for optical bistability but that either separately is not. But our model is strictly collective: it contrasts with recent work on optical bistability[15] where spontaneous decay is the sum of individual atom decays: thus, typically, $C = Ng^2/\kappa\gamma_o$ in Refs.[15] but is $C = Ng^2/\kappa N\gamma$ in our work — which means that $\gamma_o \to N\gamma$ as is usual for collective action. This is closely associated with the fact that $\hat{S}^2$ is a constant of the motion, a feature broken only by atomic de-correlation. On the other hand in extended systems (see e.g.[9]) semiclassical theory, in which $\hat{S}^2$ is not a constant, gives a good description of both absorptive and dispersive optical bi- and multi-stability. The precise significance of quantum fluctuations to optical bistability in general is therefore to be determined: a relevant paper is[16].

Although we note in §1 that the Dicke model is not the limit of any system of two-level atoms occupying a finite region of space, the success of the present model in describing the recent observations[8] for N Rydberg atoms inside a cavity containing a single mode chaotic field suggests to us that the same model will describe equally well the behaviour of N Rydberg atoms inside a cavity driven by a single mode coherent field. We therefore expect to see strictly quantum optical bistability in such an experiment.

REFERENCES

1. A. Dorsel, H. Walther, P. Meystre and E. Vignes, "Radiation Pressure induced optical bistability", Paper FC4-1 in these Proceedings.
2. R. R. Puri and S. V. Lawande, Phys. Lett. 72A:200 (1979) Physica 101A:599 (1980); S. V. Lawande, R. R. Puri and S.S. Hassan J. Phys. B 14:4171 (1981); S. S. Hassan, G. P. Hildred, R. R. Puri and R. K. Bullough, "The driven Dicke models" in: "Coherence and Quantum Optics 5", L. Mandel and E. Wolf, eds., Plenum Press, New York (1983).
3. S. S. Hassan, R. K. Bullough, R. R. Puri and S. V. Lawande, Physica 103A:213 (1980).
4. S. S. Hassan and R. K. Bullough in: "Optical Bistability", C. M. Bowden, M. Ciftan and M. R. Robl, eds., Plenum Press, New York (1981).
5. R. H. Dicke, Phys. Rev. 93:99 (1954).
6. H. J. Carmichael and D. F. Walls, J. Phys. B 10:L685 (1977); S. S. Hassan and D. F. Walls, J. Phys. A 11:L87 (1978).
7. R. R. Puri, G. P. Hildred, S. S. Hassan and R.K. Bullough, "Exact Quantum Theory for N Rydberg Atoms in a Cavity" in:

"Coherence and Quantum Optics V", L. Mandel and E. Wolf, eds.,
Plenum Press, New York (1983).

8. J. M. Raimond, P. Goy, M. Gross, C. Fabre and S. Haroche, Phys.
    Rev. Lett. 49:117 (1982); 49:1924 (1982); M. Gross and
    S. Haroche, Physics Reports 93:301 (1982).

9. R. K. Bullough, S. S. Hassan, G.P. Hildred and R. R. Puri,
    "Mirrored and Mirrorless optical bistability: exact c-number
    theory of N atoms in a Fabry-Perot cavity", paper in these
    proceedings (1983).

10. E. M. Purcell, Phys. Rev. 69:681 (1941).

11. R. Bonifacio, P. Schwendimann, and F. Haake, Phys. Rev. A4:302,
    854 (1971).

12. S. S. Hassan, G. P. Hildred, R. R. Puri and S. V. Lawande,
    J. Phys. B 15: 1029 (1982).

13. S. S. Hassan and R. K. Bullough, J. Phys. B 8:L147 (1975).

14. P. D. Drummond and S. S. Hassan, Phys. Rev. A22:662 (1980).

15. R. Bonifacio and L. Lugiato, Optics Comm. 19:172 (1976); G.S.
    Agarwal, L. M. Narducci, R. Gilmore and D. M. Feng, Phys.
    Rev. A18:620 (1978). Also the references in "Optical Bistability",
    Ref. [4].

16. P. D. Drummond, "Critical Quantum Fluctuations" in: "Optical
    Bistability", Paper WD 3-1 in these proceedings.

PANEL DISCUSSION: TOPICAL MEETING ON OPTICAL BISTABILITY

University of Rochester
Rochester, N.Y.
June 17, 1983

PRESIDER:

Professor Elsa Garmire
University of Southern California
Center for Laser Studies
Los Angeles, California 90089

TRANSCRIBED BY:

Doreen Weinberger and Jack Jewell
Optical Sciences Center
University of Arizona
Tucson, Arizona 85721

PART ONE: PHYSICS

1. In general, what have we learned in the last two years?
   What do we not yet understand?

2. Optical bistability: implications for quantum optics
   - what can we {learn from} other fields
     {teach to}

3. Atomic and molecular physics/spectroscopy
   - what can O.B. {learn from} this field
     {teach to}

4. Solid state physics
   - what can O.B. {learn from} this field
     {teach to}

GARMIRE:

Let's begin by discussing the physics of optical bistability. I'd like to solicit comments on the above questions, or any other questions you may have.

HOPF:

One of the disappointing things that came out of Asheville was the conclusion on the part of some people that there was nothing further to learn about the physics of optical bistability; that optical bistability was only an engineering matter and should be exclusively pursued as a matter of materials research. I don't want to put down materials research; I think that materials research has been very exciting. But I do think that the people that attended this conference have absolutely refuted the idea that there is nothing to learn in the physics of optical bistability. In particular, the entire field of chaos and Ikeda instabilities (which, to my knowledge, was not even mentioned at the last bistability conference) has blown completely apart. As a spin-off I would cite what I call the "synthetic mode," a very novel concept of a mode in a cavity that is strongly nonlinear. To my knowledge, that is a novel concept in quantum optics: the idea that one can solve the boundary conditions of a cavity, using the fact that we know how this strong nonlinear change has affected the boundary conditions. Then we have what I call the Max Planck gambit (whether done by Russians or Swiss), involving entirely new kinds of applications for optical bistability in ring cavities, with this enhancement perhaps helping us to look for gravitational waves or even the metric in space. Kaplan discussed new ideas of spectroscopy related to this. My only recent endeavor into the mirrorless bistability was strongly motivated by the idea that we may be able to do "doable" experiments with "doable" theory to examine issues that have been mostly a matter of theorists arguing with each other. For example, there are questions like: what do you mean by retardation? When can you neglect quantum correlations? This has been a matter of a great deal of heat, but very little light. It begins to look like we're going to be able to do experiments now, and this is a direct outgrowth of our progress in optical bistability. So I offer these three examples (I'm sure I left others out) which are enough since the Asheville conference to convince me that there is no question that physics continues to play an important role in optical bistability, and the whole field is important to physics, not just engineering. We have to pursue both aspects of the problem, and we will learn from each other in the process.

SANDLE:

Certainly in the fields with which we've been most closely associated, the importance of atomic physics effects in optical bistability is becoming quite clear: with such effects there are possibilities for many modes, and there are new types of behavior which have not yet been fully explored. For example, consider polarization switching. This was predicted for a three-state model and it now appears in a four-state model. Is polarization switching a type of behavior that is generally observable? Also, types of self-oscillations should be able to be categorized in some way. We need to have a better understanding of the regions of applicability for particular kinds of behavior. That's a whole field which remains to be developed. I think there's a possibility of very fruitful collaboration between quantum optics and the more conventional laser spectroscopy/laser physics in this joint area.

GARMIRE:

I was interested in the question: is there something we can teach spectroscopists? Is there something about doing bistable experiments that helps you understand more about the basic atomic physics?

SANDLE:

Certainly there is in principle, because you can come very close to having an atom interact with a single mode. The cavity picks out a single mode of the radiation field so that you have, in principle, a cleaner system than in free space, where you have an enormous number of modes. In practice, as well, I think bistability experiments can bring out new points in atomic physics. This is because the emphasis is different; we're approaching problems in spectroscopy or laser physics from a slightly different viewpoint. Whether the statistics of an atom in a cavity is or is not different from that in a free space radiation field -- I think that remains to be sorted out.

McCALL:

I just realized that Haroche, who is not here, solved this interesting experimental problem of atoms in a single mode cavity. One could observe optical bistability simply by shining in radiation at a microwave frequency.

NARDUCCI:

I'd just like to point out something that struck me today: when we met in Asheville, there was a certain amount of theoretical discussion on instabilities and self-pulsing, although barely any mention of chaos. This morning I was pleased and impressed by a paper by the Hannover group reporting the observation of critical slowing down in magnetically-induced self-pulsing. What really struck me was not only the observation of self-pulsing in a laboratory set-up, but also the demonstration of a flexible way to modulate and demodulate an optical wave. It seems to me that in spectroscopy, these modulated beams could become quite a useful tool.

STENHOLM:
    In trying to discuss the connection between optical
bistability and quantum optics there's one thing I believe is very
interesting: this was brought up by Meystre's most beautiful and
novel paper on the free mirror system. It really has very little
quantum mechanics; if you think about it, you only use dispersion,
and of course that inevitably comes from the fact that you have
discrete levels. (To some extent you can say that the levels in a
solid come from that.) After you consider dispersion, it's all
classical. I would like to know if experimentalists know of any
quantum effects that one can actually see in these systems, or if
any theoretician has good justification why we should use quantum
theory here. Theoreticians claim a monopoly on understanding
quantum mechanics, so we are very eager to find out why we need
quantum mechanics here, because to me it all seems horribly
classical.

McCALL:
    The conditions of Haroche's experiment were very interesting.
He had relatively few atoms interacting with a single mode in a
cavity in a way such that the interaction was the same for each
atom. This is the standard Jaynes/Cumming problem. (Some people
call it the Dicke problem.) In this case the atoms are initially
inverted. You could also tune the cavity so the atoms are in the
ground state and shine into the cavity microwaves of an
appropriate frequency and have a bistable system.

GARMIRE:
    Do you agree with that?

STENHOLM:
    Yes, but that remains to be seen.

KAPLAN:
    I don't think that we need any dispersion, even classical, to
talk about optical bistability as something that belongs to the
world of two-level systems. There are many bistable systems which
don't use any quantum levels, for example, a particular resonance,
a nonlinear Kerr material, or a nonlinear interface: extremely
simple systems, classical only. But on the other hand, I think the
simplest system is one that uses just a single physical element to
demonstrate bistability: I refer to my paper on instabilities in
the cyclotron resonance by a single electron in free space.

STENHOLM:
    But that's a classical effect.

KAPLAN:
    Well, the most interesting thing is that you can observe
quantum effects; it's the closest thing to a quantum system. And in

my understanding of the quantum limit, there is no such thing as optical bistability.

GARMIRE:

There is some question as to whether or not there is bistability in quantum systems.

HOPF:

That's a problem that has bothered me a lot. It's clear that there is no such thing as bistability in quantum mechanics; wave functions are unique objects. We believe that the world is quantum mechanical, and it therefore follows that there is no such thing as bistability; and yet there is! The reason is one of perspective. Bistability is a transient effect. If I put any device in the upper state and wait for an infinity of time, then it will have "glitched" an infinite number of times, and I will have established some sort of probability distribution. In quantum mechanics I will be in the ground state. But that doesn't bother me, because I want to work within the time of one glitch. There's no conflict between these; it's a conflict in application. This, I think, poses a challenge to the quantum people, who are in the best position to try to understand some of these quantum problems in terms of tunneling phenomena that can describe the dynamical decay of a system to its quantum state. I'd like to have a lot more discussion on this at the next bistability conference.

MOLONEY:

I'd like to make a comment on what optical bistability can teach people in other fields. Fluid dynamicists for hundreds of years have been trying to understand the transition to turbulence. As their experiments are becoming more and more refined they're seeing these period doubling sequences, but more commonly they're seeing sequences which involve periodic and quasi-periodic motions. In many cases their experiments take days or even weeks, and they're often not sure whether they're observing a transient behavior or an asymptotic state. Optics, and when I say optics I mean Gaussian beam optics (that's the real situation), I think can teach an awful lot in this area. There is evidence that one can actually see these transitions in a cavity. The big advantage is that you can do the theory quantitatively (it's difficult but you can do it), and you can do the experiments. In fluids the Navier-Stokes equations, even in the linear regime, are extremely difficult to solve, and are probably impossible to solve in the fully nonlinear regime. The Lorentz equations are a severe truncation of a Fourier mode decomposition of the original Navier-Stokes equations; but if you increase the number of Fourier modes to get more equations you see very different dynamics. In the nonlinear wave equation we have a real quantitative system that we can actually experiment on, and also do theory on, and I think teach people in fluids in particular a lot.

MATTAR:

    An analogy exists between fluid calculations and laser calculations in that there is a one-to-one relationship between the density of the fluid and the intensity of the laser, and the velocity of the fluid and the field gradient of the laser. Obviously whatever we have learned in the measurements of optical bistability of interference of several waves can be connected back to the physics. But we should also use what the hydrodynamics people have done in properly calculating the interference of two types of waves, or similar waves, whether the wave results from gas dynamics, or whether the wave is an electromagnetic wave. We should be able to learn from them and carry on the physics.

GARMIRE:

    In the next section we're going to talk about optical engineering. Would you say that one of the engineering applications of bistable devices is an analog computer for fluid dynamicists?

HOPF:

    They could use it.

HARRISON:

    I'd like to ask a question, with great humility: why are there experiments to look at chaos and turbulence in bistability inside gain medium systems (i.e. lasers)? It seems to me that these systems are extremely complex, making the problem of analysis very, very difficult. Why bother to deliberate on this problem when there are very simple systems in which we understand the physics of these processes (if that's what we're trying to do)? Could I get some comments on that?

MEYSTRE:

    We're not completely sure yet, but we probably have the simplest system for looking at that, too. When we allow both mirrors to move, we think we get chaos. So we're going to investigate that.

GARMIRE:

    Is there someone who has been doing experiments on chaos in lasers who would like to answer the question?

NARDUCCI:

    Neal Abraham is the person who should answer this question, but unfortunately, he is not here. He has been doing experiments directed towards understanding the origin of chaos in ordinary lasers and in lasers with injected signals. I believe that he would agree with the following assessment: even under carefully controlled conditions, there are many apparently different ways by which chaos emerges in optical systems; they have to be sorted out, due to the fact that our theoretical understanding is very limited. I am not sure that I share Ruelle's optimism with regard to the solution of the weak turbulence problem, mostly because I have a hard time deciding where weak turbulence ends and strong turbulence begins. As usual, careful experiments can shed a lot of light on the path that the theory should take.

HOPF:
    I think in human terms the answer's a lot simpler than that. We have to remember that the concept of chaos in optics as a real thing in the real world is something like two years old, and as a widespread experimental phenomenon is only about one and a half years old. It's early days for us. Four years ago, chaos in optics was unknown and largely disbelieved. So the fact that it's being seen in many different systems is giving people confidence that it's really there. The one thing that I liked about the Bryn Mawr experiment is that it was another voice in the wilderness at the time. But I think you're absolutely right that in the long run, if we're going to learn something about the nature of optical turbulence by doing very good experiments, it will have to be done in passive systems because noise is a killer in any kind of good quantitative work. I think people in active systems are going to be in trouble due to noise from the discharge. You don't believe so, Matt?

DERSTINE:
    No. There are mathematicians that are doing this right now: developing good ways for stable systems with noise in which to separate the noise from the chaos.

HOPF:
    Let me amplify my statement. It is not purely a matter of noise -- if it were just one external Gaussian signal messing things up, that would be fine. The problem is when you put noise together with drift, for example, when the plasma changes your gain medium, or something like that. Mathematicians don't seem to have much confidence that in the near term they are going to come up with ways of dealing with systems in which the gain isn't stable to one part in 10^8. They can live with quantum noise, but they can't live with the drifts, and I'm not sure they can live with noise on the scale of a laser. Lasers are very noisy things.

GARMIRE:
    It seems to me that we are, however, making progress on lasers, in stabilizing lasers. I think it's interesting to point out that bistability is probably helping us understand laser performance better and lasers are behaving better now than they were 10 years ago. Some of these experiments can be understood a lot better today than they could have in the past.

E. ABRAHAM:
On the other hand, historically the first connection between turbulence in optics and hydrodynamics was made by Haken in 1975. But the condition for observing chaos in a single-mode laser was too stringent, since it demanded that the loss in the cavity be greater than the sum of $\gamma_{\parallel} + \gamma_{\perp}$. Now there is new hope that this can be done, because with the observations by Heroche, for high Rydberg atoms you can get $\gamma_{\parallel}, \gamma_{\perp}$ very low, and you can get a reasonable cavity for the observation of chaos according to the predictions of Haken.

STENHOLM:
I want to comment on why we want to do this type of work in laser cavities. I think some of us who have been in laser physics for a long time remember around 1970, we were working hard to understand mode locking: if you take a 2-meter He-Ne laser you have about 20 modes, and it's very easy, actually, to lock them. But it is a horrible theoretical problem to actually handle this, and it's essentially a problem that we cannot solve with present-day computers. So I've been interested in this problem for a long time, trying to understand what happens when these modes choose to cooperate. I find it very exciting that even when they choose not to cooperate, they display these very simple, theoretical concepts like period doubling, etc. These actually can occur in a laser. I don't know if we're going to learn anything useful, but it's certainly fascinating that in this field which has been with us so long and is still essentially an unsolved problem, we can see such totally unexpected behavior. Because if you take 20 modes with a semiclassical Lamb theory and put it on a computer, the computer goes crazy. I can tell you!

McCALL:
I once did some interesting work in laser mode coupling. With three running modes there are regions where, experimentally, the output is chaotic. I know that from experience.

S.D. SMITH:
I have a few comments: two points made by solid state physics. (1) What can bistability teach to this field? There are some technical things already: the ability to make an AND gate, even a delayed AND gate, is the most effective method for measuring carrier lifetimes. Carrier lifetimes are very fundamental: they're one of the least known parameters of semiconductors. They are vital to understanding dynamic, nonlinear effects, so the two are connected in this way. (2) The other thing bistability is doing to solid state physics is highlighting a completely new area which the semiconductor community hasn't looked at very actively yet, namely the response of the susceptibility on timescales from picoseconds upwards. We can answer questions about what happens on picosecond timescales with a lot more sense than we had before

by utilizing these devices. Finally, I'm very surprised that at this meeting there's been done so little microscopic solid state physics related to applications of bistability. I would expect a lot more work coming in that area.

JEWELL:

I'd like to give a word of caution about deriving fundamental quantities such as carrier lifetimes from a device. You can scan a Fabry-Perot peak across the laser wavelength much faster than a carrier lifetime. I think you have to be careful about these things.

MILLER:

I agree with what Des Smith said. I think optical bistability has stimulated new interest in solid state physics in an area that was being neglected. I'd like to make some general comments about the theory of nonlinear optical phenomena in solid state physics. Semiconductor systems, which in our view at the moment are the most interesting in terms of being practical devices, are too complicated to model by nice "first principles" models. The only way to tackle the physics of the solid state is to put in most of the parameters as being empirical, and pursue the understanding of only one effect that you can't understand. There is a great deal of knowledge from other branches of solid state physics of many things that go on in semiconductors. A lot of these things we still don't understand, but life is too complicated to try to solve all of them at once. For example, we don't understand what happens at the band edge, even in linear absorption; that's a semi-classical effect, and we're all working at the band edge. We don't really understand all the details of recombination, and there's a great deal of effort in semiconductor lasers trying to understand that problem. We don't understand in a lot of detail what's going on in the shapes of some of the exciton lines and the like. My belief, then, is that we should not try to model what is a very complicated situation by trying to solve the whole problem at once from first principles. Maybe this is a general criticism: that we should try to discern the general effects in whatever particular system we study. A lot of things we see in bistability are general effects: chaos is a general effect; obviously the switching of a Fabry-Perot in dispersive optical bistability is a general effect. I think it's a mistake for us to put too much emphasis on two-level systems, which admittedly are nice systems where we think we can understand everything. The justification for working with the two-level system is that because it is physically simple it will enable us to see the general effects without confusing us with the particular details. But when we do work with semiconductors, all we can hope to do is look for the general effects and try to get some phenomenological idea of what is going on in the system. General conclusions from other physical systems are very useful in understanding the semiconductor situation, and conversely, the observation of particular phenomena in semiconductor systems (such as whole-beam switching) proves their generality because of the extreme complexity of the semiconductor system. I hope that might stimulate somebody to tell me I'm wrong.

HARRISON:
But do we really understand two-level systems? Until we can understand these, how can we hope to perceive anything? If we haven't got systematic experimental evidence in support of the simplest of theories describing two-level systems, we can't hope to proceed. It would seem to me that experimentally you should investigate such systems, then add an additional complication to the system, an additional parameter, see what happens, and proceed so forth. You don't start with a state of "turbulence" in your experiment to find a solution.

MILLER:
But we do!

HARRISON:
You do. I don't.

KIMBLE:
A brief comment: it seems to me that what is often lost in this debate between experiment and theory is an issue on which many of the experimentalists in the room may disagree with David Miller, who seems to be concerned only with the nonlinear susceptibility χ. There's a lot more to bistability and chaos than the idea that if you know χ, you know it all. What a lot of people here are trying to understand is not what's going to happen next year with any given material, but rather what is the inherent range of dynamical behavior that bistable systems can exhibit. For this one really needs to explore the underlying microscopic processes in very simple systems. It is not reasonable to ask a theorist to model such processes in anything but systems for which there is at least an understanding at the deterministic level. So I'd like to put in a plug for exploring the whole of the physics of optical bistability (and there's lots of physics at a level below χ: self-pulsing instabilities, stability to quantum noise, relationships to non-equilibrium statistical mechanics, etc.). There's a lot of theoretical literature but very little experimental work in this area, and some day these issues will have to be faced, long after we understand what χ is for a particular material.

MILLER: (Note added in proof)
I must reply to this comment. First of all let me say that I applaud the experimental work on two-level systems, of which Dr. Kimble's is an outstanding example. My opinion is certainly not that work on two-level systems (and other related ideal systems)

is pointless, but merely that it is at its best when it illuminates the general phenomena in the areas of, for example, microscopic phenomena, noise, and nonequilibrium statistical mechanics to which Dr. Kimble alludes. But I do profoundly disagree with the reductionist approach to physics that one must understand in ever increasing detail the microscopic physics in order to make progress in understanding the macroscopic world. Chaos is a recent shining example of the opposite view of physics. Its beauty lies in that it describes phenomena at a macroscopic level independently of the microscopic detail. Thus, for example, electronic oscillators, fluids, and yes, two-level systems can all show the same classes of chaotic behavior, dependent on macroscopic, phenomenological parameters. The assertion that all I want is χ is something that is obviously not true, but I will let that pass with liberal application of dialectic license. (I also wish incidentally that I could get someone to calculate χ for me!) What worries me more, however, is the implicit assumption that there will be a field of optical bistability in 15 years time to which the more detailed microscopic processes are relevant in anything other than an academic sense. There is a danger of trying to run before we can walk. Let us not lose sight of the fact that there is no optically bistable system of any kind operating in a practical application anywhere. A major reason for that has been the historical lack of a material with sufficiently nonlinear χ to make a practical device. Unless that problem is solved, and solved soon, there is a danger that the whole field will lose much of the momentum it has gained over the past few years as the probability of practical relevance of albeit excellent academic research fades.

GARMIRE:
        I just wanted to say that personally I think your experiment is absolutely magnificent, Jeff. We really need those very pure, clean experiments. But I'm also very aware of the needs of Bell Labs and other such places, and obviously there's room for both, and we need both.

MILLER: (Note added in proof)
        If you're aware of the needs of Bell Labs as far as bistability is concerned, please let me know, Elsa!

BATES:
        I think that there's obviously a very rich interaction between the field of optical bistability and these other areas mentioned. The experimental results and the development of theory in optical bistability will probably stimulate other efforts in atomic and molecular theory, as well as in solid state physics. One thing that I noticed in looking at the various experimental papers is that people by and large were satisfied if they could somehow achieve a hysteresis curve. They were then satisfied that they had

demonstrated optical bistability in a system. Of course the field is rather young, and at this early stage to get a hysteresis curve in some system is certainly worthwhile. But I think there is a lot of work that needs to be done to study the transient nature of the phenomenon: switching from one state to another. Do the observed transients really fit what theory predicts? Does theory predict something?

GARMIRE:
I totally agree with you. There was a paper by de Shazer about 1970 to look for bistability in dyes. He concluded he could see any pulse shape he wanted to depending on the particular levels of the atoms or the molecules.

LUGIATO:
I wish to express my opinion as a theoretician about this connection between theory and experiment in bistability. There is a big effort now in pursuing the understanding of the basic mechanism of bistable systems that hopefully will lead our community to reach the final goal of having practical devices that will replace electronic devices. Of course to do that there must include a theoretical effort in this direction, a different kind of theory that is more interested in answering the fundamental questions. I fully understand and sympathize with the people who are aiming towards developing practical devices. But on the other hand I don't think that this is the only theory that we can develop. In fact, I think there is still a great imbalance between the status of the theory and the status of the corresponding experiments. I think that most of the theory has been worked out, and I don't expect any more big surprises. It's true that since Asheville there has been a dramatic interest in chaotic behavior, but already this field has been quite extensively exploited. We shall certainly have progress in investigating these types of problems: we should understand better the connection between the various instabilities and so on. But I think that to continue developing the theory we now need input from experiments, because for example, I remember that when Sandle did his experiment in sodium, it became clear that the plane wave theory was not suitable to take into account the experimental data. So it immediately stimulated interest in transverse effects, which are now being actively investigated. And so I think now we really need something from experimentalists in order to go on with the theory; otherwise it's difficult.

S. D. SMITH
I don't believe we understand the temporal development of switching, or the spatial development of switching. In contrast to Lugiato's point, there are some experiments which theory has not yet explained. Almost every switching experiment hasn't been done well enough in time or space to characterize it, but it's certainly

not been explained quantitatively by theory, either. I suggest we haven't yet done this well, and we need to.

PART TWO: ENGINEERING

1. How will optical bistability be useful in
    - optical computing
    - optical signal processing
    - optical beam shaping (space, time)

2. Regimes of speed, power and size
    - what is currently feasible
    - etalons vs. integrated optics
    - projections for future

3. What work needs to be done in next two years in engineering development (theory and experiment)?

GARMIRE:
      Let's move on now and discuss the engineering aspects of optical bistability with regard to the above questions.

MEYSTRE:
      I was a bit disappointed at this meeting to see that bistability is finally doing what everybody else is doing: namely following fashions. The big fashion now is to make solid state devices small and quick.

GARMIRE:
      You want them big and slow?

MEYSTRE:
      I think there is more to life than computers. I think what we need is a few more crazy people like Alex and other people around here to look for completely different applications of this beautiful nonlinear theory.

McCALL:
      I'd like to comment on Huang's recipe for parallel processing. I'm a little disappointed because it involves parallel processing on a scale of perhaps 1000 bits. That sounds like a lot more than 32 bits or 64 bits. But next comes 256, 512, or 1024 bits. So I appeal to the computer software people for algorithms that will process Megabits in parallel! Then maybe we'll have a real motivation to work with small, efficient photon devices.

P. SMITH:

A couple of comments. As you know, I've done some thinking about question #1, and at least on the basis of the type of physical processes that I've looked at, there seems to be two main areas that one can identify where optical switching elements might have significant application. One of them people have talked about a lot: the fact that with fast nonlinearities you can do switching in picosecond or subpicosecond times. It's going to be very hard for any other switching technologies to get down to this time scale, so certainly in the area of very fast switching we can do things optically that we cannot do with other technologies. There is, however, another major area, and I think that it's perhaps more pertinent in view of the number of papers and discussions I have heard at this conference; that is the area of parallel processing. Optics seems to be rather uniquely special in that we can envisage devices which can process in parallel many, many bits of information at the same time. In order to utilize this capability and maybe use this as an impetus for important discussion, I'll state a crazy calculation that David Miller and I did recently: if you take a 1-cm^2 element which would have, perhaps, mirrors coated on it so that it would be a bistable Fabry-Perot, you can in principle focus down to a spot size on the order of 1 μm, and in that spatially resolved element make a bistable device. We're not too sure exactly how far away we can put the next element (the Arizona group has done some nice work in looking at that), but we know it's on the order of a few microns. You can then calculate how many resolvable spots you have in this 1-cm^2 element. Then you can ask, how fast can you switch? Well, with multiple quantum wells, it's about 20 nsec. And so with things already demonstrated in the laboratory you can compute a number which is the throughput, that is the number of bits/sec, that you could process with this device, in terms of the number of resolvable spots multiplied by the number of operations per second. And if you look at that rather large number, that would correspond to simultaneous telephone conversations by every person in the world! This is a little bit silly perhaps, but I say this to emphasize the fact that there really is a tremendous potential for parallel processing of information. We really don't know how to utilize this potential yet. The talk that Alan Huang gave us was very nice, and that's beginning to tell us ways that we can start to use this kind of capability. I think the kind of studies that I saw on one of the poster papers, where you start to see how close you can put these spots together so that they don't interact, is important in understanding these types of devices. The kind of question I feel is important for people to think about is this: how can we use this capability, how can we optically address individual elements in some such an array as this? How can we use our ability to optically address this in order to do something useful? I don't like to call it a computer, but at least make some optical signal processing device that will do some useful functions.

HUANG:

I came across the same question that Sam McCall brought up: these computers would be very unusual. Most computers you can only change 16 or maybe 32 bits at any given time, and here's a computer that can change its entire memory every cycle period. I started to wonder: how am I ever going to use this? So I thought about it for a while, and first I naively tried to simulate a regular computer with it. If you look at the state space, only a fraction of a percent of the memory ever changes. I thought: this is terrible! I have all this parallelism and I'm not using it. So I spent some time and came up with some very strange absolutely parallel algorithms. But when I started talking about these things, people said, "Why would you ever do things like that? There's no hardware to ever do it!" So I went back and was very pleased when I came up with what I call a synthetic way of using all this parallelism: taking your ordinary circuit and redrawing it in a form that you can put into a pipeline, and then overlaying this on the optics. I agree that it does not use the full parallelism of optics, but it's something that you can give to an engineer and say: "See, this is something useful you can do. You don't have to fight two battles at the same time of proving the utility, as well as proving the technology."

S.D. SMITH:

I'd like to agree with Peter Smith that the first thing that we'd really like to look at is the two-dimensional question. One thing that is being looked at in the United Kingdom is a real-time, optically addressable spatial filter, used with two lenses to do Fourier transform optics in real-time in some way. There is also some military interest in using lenses in this sort of way to preprocess signals. The point is, if you do it optically you don't lose phase information. Once you convert it to an electronic signal, you lose phase information. In addition I see another very useful and yet very simple outgrowth of our bistability work which is just becoming apparent: despite our knowledge of quantum electronics and laser physics, lasers still aren't very stable devices. But with the fast processing devices we are talking about today which could operate at all sorts of ranges of intensities, it seems to be possible to stabilize the intensity of laser pulses and improve the quality of intensity-dependent, that is nonlinear, experiments. That's one point. The second point concerns noise gain. Once again, lasers are noisy devices, but if these devices have a power limit they may also have a noise limit. This might be one way in which we could improve our experiments by requiring better equipment which these devices could bring.

GARMIRE:

I agree with you -- there was a nice paper several years ago, I think from China, about stabilizing a laser using a liquid crystal modulator. They took their standard 3% noise down to considerably below 1% with a very simple type of modulator.

P. SMITH:

I think there's another question which might be interesting for some of the theorists to think about. A number of us who worry about question #2 (regimes of speed, power, size, etc.) have looked into extrapolating current devices down to waveguide geometries and small sizes and so on. But even if we do that using existing materials and existing experimental measurements for the parameters, we get to numbers which are low enough so that it looks as though the fundamental limit on the power is going to be determined by our perception that we need a certain number of photons, large compared to unity, to do our switching. This relates to the glitching that Sam McCall mentioned and that Fred Hopf has talked about: namely that you've got to use a lot of photons to define a state if you want to define it very well, that is with a low error rate. Now the question I'd like to ask the theorists is: is this really true? Is there some way that we could define a stable state of a bistable device or an optical switch without using a number of photons large compared to unity?

BATES:

I'd like to make a comment relevant to the possibility of these optical digital computers. I was quite impressed with the new ideas that Alan Huang mentioned this morning with regard to parallel processing in contrast to sequential or serial processing. Normally those of us who are now involved at all with computers think serially. Programs are written serially, so the software developers are more or less locked into these ideas of serial concepts in their algorithms. It seems to me that for some time to come they're going to still be doing serial processing, but there are certain key areas in the field right now that critically need parallel processing. One of the areas I can think of is speech synthesis and recognition. It's quite conceivable that what we might envision as a useful optical computer right now is not the whole computer, but just perhaps a subsystem that does the areas that critically need parallel processing. Most speech synthesis and speech recognition systems are totally bottlenecked in the current sequential processing mode. So if we have a subsystem which could do the speech synthesis or recognition using optical computer techniques, even though it may not be quite as fast as one might eventually envision, it could really add something.

PARK:

Regarding the statement the previous speakers made about the utilization of the parallel processing ability of optical bistability, I'd like to mention another 10-20 years downstream application. Nowadays, many people talk about so-called "supercomputers." There are two ways in which to consider supercomputers; one is the scientific computer which needs extremely fast signal processing ability. Picosecond switching potential of the bistable optical device appeals to this type of application. The other is the direct optical image processor tuned to artificial intelligence and robotics applications. In these applications we are dealing with the optical image as a whole, that is, in an analog form. I think the parallel processing (or real time image processing ability) of bistable optical devices has a unique advantage over other technologies in these applications. Perhaps we may be able to see optical image preprocessors made of bistable optical devices sooner than the optical computer.

McCALL:
    One quick statement on question #2: a nice place to be would be nsec switching times, fJ switching energies; we're working now with a few nsec and very close to a pJ.

MILLER:
    I wanted to agree with that comment on the parallel processing of images. Electronics basically cannot handle images, and that's become increasingly important in processing for robotic vision and the like. If you want to compress a video signal to send down a communications channel, to try to resolve the difference between the last frame and the next one is basically almost impossible even with high-speed electronic circuitry. It's a vast area, it's one the electronics people won't talk about because they know they can't solve it, and it's one which is ideally suited for parallel processing. One other point about parallel processing that might interest theorists: it was very impressive to see a paper in the poster session on self-organization in parallel feedback systems. I think that it's a fascinating area of physical theory, one which really should not be overlooked in the context of chaos and spatial organization as well.

BOWDEN:
    It seems the field has matured a great deal since the last meeting three years ago. In particular, we've discovered the field of chaos which has mushroomed since then. The question now is, what can we see projected for the next three years? From the statements given in the talks and so forth at this meeting, it seems that exciting things can be looked forward to in the area of transverse effects. This has been stated in several papers in terms of understanding self-pulsing and switching, both temporal and spatial. This seems to be one area we ought to be thinking about in connection with the fundamental understanding of what's going on, as Jeff has well-stated, and in the applications thereof. In the area of solid state physics we have also seen this in terms of theory which has come to bear on solid-state experiments. Also in this meeting we've seen several very new optical bistability

experiments reported in CdS and CuCl. Surely there is need for further development and understanding of the theory of solid state optically bistable devices, as well as new experiments in this area. The only thing that doesn't seem to have been touched on here is the effects of propagation, as well as transverse effects, in solid state materials.

CONTRIBUTORS

Eitan Abraham
Department of Physics, Heriot-Watt University, Edinburgh, U.K.

Govind P. Agrawal
Bell Laboratories, Murray Hill, NJ 07974

G. S. Agrawal
School of Physics, University of Hyderabad, Hyderabad 500 134, India

I. A. Al-Saidi
Department of Physics, Heriot-Watt University, Edinburgh, U.K.

F. T. Arecchi
Istituto Nazionale di Ottica - Firenze and Universita di Firenze - Italy

E. Arimondo
Istituto di Fisica Sperimentale, Universita di Napoli, Italy and Gruppo Nazionale di Struttura dilla Materia del C.N.R., Pisa, Italy

D. Armbruster
Institute for Information Sciences, University of Tübingen, Tübingen, F.R. Germany

R. J. Ballagh
Physics Department, University of Otago, Dunedin, New Zealand

Y. B. Band
Allied Corporation, 7 Powderhorn Drive, Mt. Bethel, NJ 07060

I. Bar-Joseph
Department of Electronics, Weizmann Institute of Science, Rehovot, Israel

Jean-Yves Bigot
Laboratoire de Spectroscopie et d'Optique du Corps Solide, Université Louis Pasteur 5, rue de l'Université, 67000 Strasbourg (France)

Bruno Bosacchi
Western Electric, Engineering Research Center, Princeton, NJ 08540

C. M. Bowden
Research Directorate, US Army Missile Laboratory, US Army Missile Command, Redstone Arsenal, Alabama 35898

E. Brun
Physik-Institut, Universität Zürich, CH-8001 Zürich

R. K. Bullough
Department of Mathematics, UMIST, P.O. Box 88, Manchester M60 1QD, U.K.

H. J. Carmichael
Department of Physics, University of Arkansas, Fayetteville, AR 72701

D. S. Chemla
Bell Laboratories, Rm 4B-417, Crawfords Corner Road, Holmdel, NJ 07733

Y. Cho
Fachbereich Physik, Universität Essen, D-4300 Essen - W. Germany   (Permanent
address: Institute of Scientific and Industrial Research, Osaka University, Uama-
dakami, Suita, Osaka 565, Japan)

G. Cooperman
GTE Laboratories Incorporated, 40 Sylvan Road, Waltham, MA 02254

E. J. D. Cummins
Department of Physics, Heriot-Watt University, Edinburgh, U.K.

M. Dagenais
GTE Laboratories Incorporated, 40 Sylvan Road, Waltham, MA 02254

R. Daley
Royal Signal and Radar Establishment, Malvern, Worcestershire, U.K.

B. Derighetti
Physik-Institute, Universität Zürich, CH-8001 Zürich

M. W. Derstine
Optical Sciences Center, University of Arizona, Tucson, AZ 85721

B. M. Dinelli
Istituto Ricerca Onde Elettromagnetiche (C.N.R.), Firenze, Italy

A. Dorsel
Sektion Physik, Universität München am Coulombwall 1, D-8046 Garching, F.R. Germany

Peter D. Drummond
Department of Physics and Astronomy, University of Rochester, Rochester, NY 14627

T. G. Dziura
University of Rochester, Institute of Optics, Rochester, NY 14627

D. J. Eilenberger
Bell Laboratories, Holmdel, NJ 07733

C. A. Emshary
Department of Physics, Heriot-Watt University, Edinburgh, U.K.

T. Erneux
Department of Applied Mathematics, Northwestern University, Evanston, IL 60201

William J. Firth
Department of Physics, Heriot-Watt University, Edinburgh, U.K.

J. A. C. Gallas
Max-Planck-Institut für Quantenoptik, D-8046 Garching, F.R. Germany

C. W. Gardiner
University of Waikato, Hamilton, New Zealand

Elsa Garmire
Center for Laser Studies, University of Southern California, Los Angeles, CA 90089

H. M. Gibbs
Optical Sciences Center, University of Arizona, Tucson, AZ 85721

A. C. Gossard
Bell Laboratories, Murray Hill, NJ 07974

R. Graham
Fachbereich Physik, Universität Essen, D-4300 Essen - F.R. Germany

D. G. Hall
University of Rochester, Institute of Optics, Rochester, NY 14627

M. W. Hamilton
Physics Department, University of Otago, Dunedin, New Zealand   (Permanent
address: Joint Institute for Laboratory Astrophysics, University of Colorado and
National Bureau of Standards, Boulder, CO 80309)

S. Hammel
Program in Applied Mathematics, University of Arizona, Tucson, AZ 85721

Ch. Harder
Applied Physics 128-95, California Institute of Technology, Pasadena, CA 91125

R. G. Harrison
Department of Physics, Heriot-Watt University, Edinburgh, U.K.

S. S. Hassan
Department of Applied Mathematics, Ain Shams University, Cairo, Egypt

H. Haug
Institut für Theoretische Physik der Universität Frankfurt, Robert-Mayer Str. 8, D-
6000 Frankfurt-Main, F.R. Germany

G. Häusler
Physikalisches Institut der Universität Erlangen, Erwin-Rommel-Str. 1, D-8520 Erlangen

G. P. Hildred
Department of Mathematics, UMIST, P.O. Box 88, Manchester M60 1QD, U.K.

Bernd Hönerlage
Laboratoire de Spectroscopie et d'Optique du Corps Solide, Université Louis Pasteur 5, rue de l'Université, 67000 Strasbourg (France)

F. A. Hopf
Optical Sciences Center, University of Arizona, Tucson, AZ 85721

K. Ikeda
Department of Physics, Kyoto University, Kyoto 606, Japan

Jack L. Jewell
Optical Sciences Center, University of Arizona, Tucson, AZ 85721

Warren W. Johnson
Department of Physics and Astronomy, University of Rochester, Rochester, NY 14627

C. R. T. Jones
Program in Applied Mathematics, University of Arizona, Tucson, AZ 85721

K. Y. Lau
Applied Physics 128-95, California Institute of Technology, Pasadena, CA 91125

N. M. Lawandy
Division of Engineering, Brown University, Providence, RI 02912

C. Liao
Optical Sciences Center and Arizona Research Labs, University of Arizona, Tucson, AZ 85721

Richard Lytel
Lockheed Palo Alto Research Laboratory, 3251 Hannover Street, Palo Alto, CA 94304

D. L. Kaplan
Optical Sciences Center, University of Arizona, Tucson, AZ 85721

H. J. Kimble
Department of Physics, University of Texas, Austin, Texas 78712

S. W. Koch
Institut für Theoretische Physik der Universität Frankfurt, Robert-Mayer Str. 8, D-6000 Frankfurt-Main, F.R. Germany

M. Kus
Institute of Theoretical Physics, Warsaw University, Warsaw 00-681, Poland

W. Lange
Institut für Quantenoptik, Universität Hannover, Welfengarten 1, D-300 Hannover 1, F.R. Germany

Roland Levy
Laboratoire de Spectroscopie et d'Optique du Corps Solide, Université Louis Pasteur 5, rue de l'Université, 67000 Strasbourg, France

Chun-fei Li
Department of Physics, Harbin Institute of Technology, Harbin, People's Republic of China (Permanent address: Harbin Institute of Technology, Harbin, People's Republic of China)

L. A. Lugiato
Dipartimento di Fisica Dell'Universita, Via Celoria, 16-20133 Milano, Italy

Ai-qun Ma
Department of Physics, Harbin Institute of Technology, Harbin, People's Republic of China

Paul Mandel
Université Libre de Bruxelles, Campus Plaine, CP 231, Bruxelles 1050, Belgium

S. L. McCall
Bell Laboratories, Murray Hill, NJ 07974

J. D. McCullen
Physics Department, University of Arizona, Tucson, AZ 85721

D. W. McLaughlin
Department of Mathematics, University of Arizona, Tucson, AZ 85721

D. Meier
Physik-Institut, Universität Zürich, CH-8001 Zürich, Switzerland

E. Menchi
Istituto di Fisica dell'Universita di Pisa, Italy

Robert B. Meyer
Department of Physics, Brandeis University, Waltham, MA 02254

Pierre Meystre
Max-Planck-Institut für Quantenoptik, D-8046 Garching, F.R. Germany

M. Milani
Dipartimento di Fisica Dell'Universita, Via Celoria, 16-20133 Milano, Italy

A. Miller
Royal Signal and Radar Establishment, Malvern, Worcestershire, U.K.

D. A. B. Miller
Bell Laboratories, Rm 4B-417, Crawfords Corner Road, Holmdel, NJ 07733

F. Mitschke
Institut für Quantenoptik, Universität Hannover, Welfengarten 1, D-300 Hannover 1,
F.R. Germany

J. Mlynek
Institut für Quantenoptik, Universität Hannover, Welfengarten 1, D-300 Hannover 1,
F.R. Germany

J. V. Moloney
Optical Sciences Center, University of Arizona, Tucson, AZ 85721

M. A. Muriel
E.T.S. Ingenieros de Telecomunicacion U.P.M., Ciudad Universitaria, Madrid-3, Spain

Barry Muhlfelder
Department of Physics and Astronomy, University of Rochester, Rochester, NY 14627

Lorenzo M. Narducci
Drexel University, Department of Physics, Philadilphia, PA 19104

A. C. Newell
Department of Mathematics, University of Arizona, Tucson, AZ 85721

Hiap Liew Ong
Department of Physics, Brandeis University, Waltham, MA 02254

S. Ovadia
Optical Sciences Center, University of Arizona, Tucson, AZ 85721

G. Parry
Royal Signal and Radar Establishment, Malvern, Worcestershire, U.K.

A. Passner
Bell Laboratories, Murray Hill, NJ 07974

N. Peyghambarian
Optical Sciences Center, University of Arizona, Tucson, AZ 85721

C. D. Poole
Center for Laser Studies, University of Southern California, Los Angeles, CA 90089

R. R. Puri
Department of Mathematics, UMIST, P.O. Box 88, Manchester M60 1QD, U.K.

A. T. Rosenberger
Department of Physics, University of Texas, Austin, Texas 78712

M. C. Rushford
Optical Sciences Center, University of Arizona, Tucson, AZ 85721

W. J. Sandle
Physics Department, University of Otago, Dunedin, New Zealand

D. Sarid
Optical Sciences Center, University of Arizona, Tucson, AZ 85721

C. M. Savage
Physics Department, University of Waikato, Hamilton, New Zealand

Axel Schenzle
Universität Essen, 4300 Essen, F.R. Germany

H. E. Schmidt
Institut für Theoretische Physik der Universität Frankfurt, Robert-Mayer Str. 8, D-6000 Frankfurt-Main, F.R. Germany

M. L. Steyn-Ross
Physics Department, York University, Toronto, Ontario M3J 1P3

Y. Silberberg
Department of Electronics, Weizmann Institute of Science, Rehovot, Israel

Surendra Singh
Department of Physics, University of Arkansas, Fayetteville, AR 72701

P. W. Smith
Bell Laboratories, Rm 4B-417, Crawfords Corner Road, Holmdel, NJ 07733

S. D. Smith
Physics Department, Heriot-Watt University, Riccarton, Edinburgh EH14 4AS, U.K.

George I. Stegeman
Optical Sciences Center and Arizona Research Labs, University of Arizona, Tucson, AZ 85721

N. Streibl
Physikalisches Institut der Universität Erlangen, Erwin-Rommel-Str. 1, D-8520 Erlangen

C. C. Sung
Department of Physics, University of Alabama in Huntsville, Huntsville, Alabama 35899

K. Tai
Optical Sciences Center, University of Arizona, Tucson, AZ 85721

S. S. Tarng
Optical Sciences Center, University of Arizona, Tucson, AZ 85721

Thomas Thel
Universität Essen, 4300 Essen, F.R. Germany   (Permanent address: University of Budapest)

J. Tomlinson
Bell Laboratories, Holmdel, NJ 07733

F. A. P. Tooley
Physics Department, Heriot-Watt University, Riccarton, Edinburgh EH14 4AS, U.K.

T. Venkatesan
Bell Laboratories, Murray Hill, NJ 07974

E. Vignes
Max-Planck-Institut für Quantenoptik, D-8046 Garching, F.R. Germany

D. F. Walls
Physics Department, University of Waikato, Hamilton, New Zealand

H. Walther
Max-Planck-Institut für Quantenoptik, D-8046 Garching, West-Germany and Sektion Physik, Universität München am Coulombwall 1, D-8046 Garching, West-Germany

Doreen A. Weinberger
Optical Sciences Center, University of Arizona, Tucson, AZ 85721

W. Wiegmann
Bell Laboratories, Murray Hill, NJ 07974

Charles R. Willis
Physics Department, Boston University, Boston, MA 02215

R. Willner
Division of Engineering, Brown University, Providence, RI 02912

Herbert G. Winful
GTE Laboratories Incorporated, 40 Sylvan Road, Waltham, MA 02254

K. Wodkiewicz
Department of Physics and Astronomy, University of Rochester, Rochester, NY 14627 (Permanent address: Institute of Theoretical Physics, Warsaw University, Warsaw 00-681, Poland)

E. M. Wright
Department of Physics, Heriot-Watt University, Edinburgh, U.K.

A. Yariv
Applied Physics 128-95, California Institute of Technology, Pasadena, CA 91125

Zhen Fu Zhu
Center for Laser Studies, University of Southern California, Los Angeles, CA 90089
(Permanent address: Beijing Institute of Environmental Features, People's Republic
of China)

INDEX